Carbohydrases from *Trichoderma reesei* and Other Microorganisms

Structures, Biochemistry, Genetics and Applications

Carbohydrates from *Trichoderma Reesei* and Other Microorganisms

Structures, Biochemistry, Genetics and Applications

Edited by

Marc Claeyssens

Wim Nerinckx

Kathleen Piens

Department of Biochemistry, Microbiology and Physiology, University of Ghent, Ghent, Belgium

The proceedings of the Tricel '97 meeting held in Ghent, Belgium on 28–31 August 1997

Special Publication No. 219

ISBN 0-85404-713-1

A catalogue record for this book is available from the British Library

Published by The Royal Society of Chemistry,
Thomas Graham House, Science Park, Milton Road,
Cambridge CB4 4WF, UK

For further information see our web site at www.rsc.org

Printed by Bookcraft (Bath) Ltd.

Preface

US troops in the Pacific called it "the green fungus among us" and strain QM6a was isolated from a deteriorated shelter in Bougainville Island at the end of World War II. It was identified as *Trichoderma viride* but eventually recognised as a new species belonging to the *Trichoderma longibrachiatum* aggregate. However, being morphologically distinct, it was finally named *Trichoderma reesei* in honour of Elwyn Reese who, together with Mary Mandels, pioneered research in the US Army lab in Natick, and developed the first cellulase enhanced mutants of this microorganism in the late sixties. Other hyper-producing mutants were further isolated, *e.g.* at Rutgers (US) and VTT (Finland) in the 70's and 80's.

The advantages of *Trichoderma reesei* as a source of cellulase were recognised for its complete system of enzymes capable to hydrolyse crystalline cellulose and for its high yield of extracellular protein. The late Elwyn Reese proposed his famous and still debated C_1-C_X theory and endo- and exo-acting cellulases were soon identified.

Since WW II, research aimed at the development of hypercellulolytic and hyperhemicellulolytic systems or complexes from various microorganisms, active at extreme temperatures or pH. Indeed, as a result of the energy crises, several programs were set up in the 70's which led to new industrial applications of these enzymes, now produced at ton scales.

Austrian colleagues organised the first TRICEL meeting (Vienna, 1989), when important new findings about the biochemistry and structure of several cellulolytic enzymes and more specifically those from *T. reesei* were published. The typical modular architecture of cellulases from *T. reesei* and *Cellulomonas fimi* was recognised almost simultaneously and important developments in the molecular biology studies of these enzymes at both sides of the Atlantic led to a cascade of new findings. The cellulolytic complexes of anaerobic bacteria were recognised and work on enzyme systems from mesophilic and thermophilic fungi and bacteria was initiated.

By the time of the second TRICEL meeting (Majvik, 1993) the research field had extended world-wide and interest in this class of biomass degrading enzymes kept growing. Important structural knowledge became available after the first X-ray and NMR studies were published and successful classification of this enzymes in families had been described. Results also on mechanistic aspects of these enzymes had led to a better understanding of their specificity and catalytic particularities.

Continuing the tradition of gathering the "cellulase people" at regular intervals, we took the initiative to organise TRICEL 97 in Ghent, considering also the vast amount of structural information which accumulated in recent years. We were particularly fortunate in bringing together a representative crowd of biochemists, molecular biologists, X-ray crystallographers and industrial representatives.

We hereby dedicate this book to two pioneers in cellulase research, Dr. E.T. Reese (†1993) and Prof. Dr. J.P. Aubert (†1997).

Marc Claeyssens
Ghent, 24 December 1997

The front cover shows a model of intact Cellobiohydrolase I from *Trichoderma reesei* (*i.e.* catalytic domain, linker domain and cellulose-binding domain) bound to a cellulose surface. An accessible chain end is sequestered by the enzyme's 50 Å long cellulose-binding tunnel, and cellobiose units split off processively. The model is based on the crystal structure of CBHI in complex with cellohexaose (Divne *et al.*, 1998), the solution structure of the cellulose-binding domain (Kraulis *et al.*, 1989) and a modelled linker region (C. Divne, unpublished results). Modelling of the intact enzyme was done manually with the program O (Jones *et al.*, 1991).

This picture was created by Christina Divne (Uppsala University, Sweden) with the programs MOLSCRIPT v1.4 (Kraulis, 1991) and the Raster3D rendering program v2.2beta (Bacon & Anderson, 1988; Merritt & Murphy, 1994).

References

Divne C., Ståhlberg J., Teeri T.T. and Jones T.A. (1998) *J. Mol. Biol.* **275**, 309.
Kraulis P.J. *et al.* (1989) *Biochemistry* **28**, 7241.
Jones T.A., Zou J.-Y., Cowan S.W. and Kjeldgaard M. (1991) *Acta Cryst.* **A47**, 110.
Kraulis P.J. (1991) *J. Appl. Cryst.* **24**, 946.
Bacon D.J. and Anderson W.F. (1988) *J. Molec. Graphics* **6**, 219.
Merritt E.A. and Murphy M.E.P. (1994) *Acta Cryst.* **D50**, 869.

Contents

Structure/Function of Carbohydrases

Substrates and Industrial Applications

Gene Regulation and Expression

Protein-Linked Glycosyl structures in Lower Eucaryotes

In Memoriam

Elwyn Thomas Reese, 16 January 1912 - 17 December 1993

Elwyn Reese was active in cellulase research for 50 years. He was 81 when he died December 17th, 1993, following a stroke a year earlier. Elwyn spent his career studying microbial polysaccharases, and is especially recognized for developing the concept of synergism in cellulolysis via the C_1-Cx theory.

Elwyn was born in Scranton, PA, and was proud of the fact that his father was a coal miner born in Wales. He first worked as a high school science teacher (1934-39), and for a brief time with the Knaust Brothers Mushroom Company, Coxsackie, NY. He then joined the U.S. Army Quartermaster Tropical Research Laboratory, Philadelphia* 1944-1945. During this time-frame he was able to develop his Ph.D. studies under Drs. Overholts and Sinden, and the degree was awarded by the Pennsylvania State University in 1946 (Reese, 1946). He briefly worked for the J.T. Baker Company (1946-1948) on the production of penicillin, an era when the fermentation vessels were milk bottles! He returned to the Quartermaster Laboratories (1948) and devoted his career to the fundamental study of cellulolysis and of microbial degradation. The Quartermaster Research Laboratory evolved in World War II as a result of microbial degradation of Army materiel especially in the South East Asian tropical zone. The original program was to determine the nature of the degradative microbes and of their deteriorative activities, and through fundamental study to develop control measures preferably not based on the use of fungicides. This was well before the publication of Rachel Carson's *Silent Spring*. Elwyn's research held true to the original tenets of basic research.

Elwyn epitomized the classic microbiologist. He was keen to consider diverse microbes, be they bacteria, fungi or actinomycetes. He aided in the development of the Quartermaster (Natick) Culture Collection* (Reese *et al.*, 1950a), and had several hundred cultures made available to him during the bi-annual maintenance transfer of the Collection. He kept many of them at hand for immediate experimentation, this being reflected in his studies of such diverse fungi as *Aspergillus terreus*; *Chrysosporium pruinosum (syn. Phanerochaete); Humicola grisea*; *Myrothecium verrucaria*; *Penicillium iriensis*; *Pestalotiopsis westerdijkii*; *Scopulariopsis brevicaulis* and *Thermoascus aurantiacus*, besides a range of actinomycetes and bacteria. Indeed, it was through the differences in these microbial cellulases that the diversity of their components became apparent, and thus the central idea of the requirement of their synergistic interaction to yield extensive degradation of cellulose developed: the C_1-Cx theory (Reese et al., 1950b). Biodiversity has since become one of today's scientific buzzwords.

In Elwyn's physiological studies, the importance of trace elements in cellulase production became clear, and also the role of a basal level of enzyme synthesis with regard to induction with insoluble substrates was illustrated by his group's key finding of sophorose as a transglycosylation product formed in the preparation of glucose via acid hydrolysis of starch. As a result, when commercial glucose was purified of this disaccharide, it no longer acted as a cellulase inducer. In the early 1960's, he broached the idea of a cellobiohydrolase through consideration of ustilagic acid as a substrate. The variety of Elwyn's research whets one's palate: from the importance of the anomeric status of the products of cellulolysis, to cellulase inhibitors such as gluconolactone and nojirimycin, besides such practical applications as the use of enzymes in the purification of natural products. And absolutely central to cellulolysis was the development of meaningful assays with recalcitrant crystalline cellulose as the substrate, besides assessment of individual components - for instance, exo-glucanase in crude cellulase mixtures. Although Elwyn's name is synonymous with cellulase, his research interests were eclectic as illustrated in enzymology alone by his study of benzoyl esterase, chitosanase, fructosanase, laminaranase, levansucrase, mannanase, mycodextranase, purine-β-ribosidase, sucrase, thioglucosidase and xylanase. The Natick group extended the application of other enzymes - for instance, the use of *Trichoderma* protease as a meat tenderizer. The electrophoretic purification of many of these enzymes was carried out using the crude starch block system, the latter method having been developed at the Pioneering Research Laboratory. In these days of "relevance", clearly Elwyn's insightful, basic studies had broad practical payoff. They spearheaded the concept of energy conservation through recycling of waste cellulosics, a notion important today *vis-à-vis* the Greenhouse Effect. Further practical payoff was Elwyn's notable lead in the selection of hypercellulolytic mutants including the high yielding *Trichoderma reesei* QM9414 and MCG77, besides the very successful large scale fermentative preparation of both enzymes and of sugars from waste cellulosics, accomplishments which clearly steered industrial companies into this area. All these basic investigations of *Trichoderma* later gave considerable practical understanding in such programs as the green mold scourge of the mushroom industry, and the application of *Trichoderma* as a biocontrol agent.

Elwyn was an especially hard working bench scientist and his co-workers responded very positively to the model that he set. He taught and inspired. He had that innate intuition to focus on a singular vital fact buried in a mass of data. Not surprisingly, he attracted a steady stream of visiting scientists, post-doctoral fellows and students to his laboratory, from national and international spheres. He travelled and lectured widely. Throughout this world-wide web he was known for his generosity: distribution of cultures and donations of purified enzymes and their products, and always open discussion of unpublished results. Though he officially retired in 1972, he remained active in the laboratory for the next 20 years. Elwyn was a private individual and few knew of his investment skills, his mycological bibliophilic expertise, or his keen sense of nature and his love of hiking with his family and friends.

Elwyn Reese was widely recognized for his scientific contributions including: the American Chemical Society Marvin Johnson Award (1982) and Biotechnology Award (1990), and the Secretary of the Army Fellowship (1964, 1965) with travel to Great Britain, India, Israel and Japan. He was particularly gratified to have been invited to the Vatican (Reese, 1969). The original Pioneering Research Laboratory mycologist, M.

Lawrence White, clearly noted that the *Trichoderma* sp. QM6a did not fit the description of any known *Trichoderma* species. In 1977, Emory Simmons recognized it as a distinct species and, in naming it *Trichoderma reesei*, honored Elwyn T. Reese as " ... student, innovator and mentor....".

* In 1954, this laboratory later moved to Natick, MA., and through reorganization received several name changes including the better known Pioneering Research Laboratory (PRL). The original name is retained in the QM Culture Collection (14,000 species), now conserved at the USDA Northern Regional Laboratory, Peoria, IL.

E.T. Reese, 1946, 'Decomposition of cellulose by microorganisms at temperatures above 40°C', Ph.D. thesis, Pennsylvania State University, University Park, PA.

E.T. Reese, 1969, 'Microbial decomposition of soil polysaccharides in organic matter and soil fertility', *Pontificiae Academiae Scientiarum Scripta Varia* (Rome, Italy), **32**, 540.

E.T. Reese, H.S. Levinson, M. Downing and M.L. White, 1950a, 'Quartermaster Culture Collection', *Farlowia*, **4**, 45.

E.T. Reese, R.G.H. Siu and H.S. Levinson, 1950b, 'The biological degradation of soluble cellulose derivatives and its relationship to the mechanism of cellulose hydrolysis', *J. Bact.*, 59, 485.

E. Simmons, 1977, 'Classification of some cellulase-producing *Trichoderma* species', Second International Mycological Congress, Abstracts, p. 618, University of South Florida, Tampa, FL.

Douglas Eveleigh
Mary Mandels

In Memoriam

Jean-Paul Aubert, 7 November 1924 - 9 October 1997

"Science is like bingo: 100 % of the winners had bought a ticket". By paraphrasing a well-known advertising slogan, Jean-Paul Aubert reiterated his conviction that creative science requires to take chances. Circumstances proved him right on several occasions.

Upon setting out to study cellulases, he chose to do it with *Clostridium thermocellum*, which was known to be an excellent cellulase producer, and whose thermophilic properties were a desirable feature for industrial applications. Little did it matter that almost nothing was known on the biology of *C. thermocellum*, that there were no tools for genetic manipulations (there are still none), and that the organism was strictly anaerobic, which made its cultivation rather tedious. Neither was he deterred by the early observation that the cellulolytic enzymes formed what seemed at the time to be hopeless aggregates. The saga of the cellulosome was to show that *C. thermocellum* wasn't such a bad choice, after all.

A little later, he realised that the project wouldn't go very far without recombinant DNA technology. At that time, no cellulase gene had been cloned, and the endeavour was far more ambitious than it has become since. The easy activity tests based on carboxymethylcellulose and Congo Red or methylumbelliferyl-β-cellobioside did not exist; more importantly, there were serious doubts whether genes from a Gram-positive, thermophilic and strictly anaerobic bacterium such as *C. thermocellum* would be expressed in an organism as distantly related as *Escherichia coli*. Well, the Earth isn't flat, Columbus didn't fall off from the edge of the world, and the *C. thermocellum* endoglucanase genes were spontaneously expressed in *E. coli*, which gave us the opportunity to study them before the topic became totally burnt out.

While subcloning the endoglucanase gene *celD*, we found by chance an overproducing clone, which allowed the facile production of more than 10 mg of highly pure enzyme in less than a week. Some wouldn't have asked for more, being satisfied with an excellent opportunity to study the biochemistry of the enzyme. But not Jean-Paul Aubert. "If your enzyme is that easy to purify, you should try to crystallise it and get its three-dimensional structure". Thus, CelD became the first cellulase to be crystallised, and we started a most fruitful collaboration with the Unit of Structural Immunology at the Pasteur Institute.

It soon became clear that the scaffolding protein CipA, or S_L/S1 as it was known at the time, was an essential component of the cellulosome. We knew that progress had been made early on at MIT in the cloning and sequencing of the gene, and there wasn't much incentive to start a race while running three years behind. The objection was brushed aside: "You need that gene. You don't know if and when it will be made available to you. If I were you, I would go ahead and clone it". As a matter of fact, we only got the 3' end of the gene, and MIT rightly came first with the sequence of the whole gene. But we didn't come out empty-handed either. Jean-Paul had suggested that we screen our clones using ^{125}I-labelled CelD, whose dockerin domain would bind to polypeptides harbouring cohesin domains. As a result, we picked a second gene, *olpA*, whose product turned out to be a member of a new family of polypeptides probably involved in anchoring individual cellulases and cellulosomes to the surface of *C. thermocellum*.

Jean-Paul Aubert was probably best known to participants of the TRICEL meeting for his contribution to the study of cellulases; however the latter represented but one facet of the many interests he had in his career. Space wouldn't permit to recount in detail a scientific life of fifty years, and the author of these lines only knows first hand about the last part of it. Suffice it to mention that he made lasting contributions on topics as diverse as calcium metabolism in the rat, the triggering of sporulation in *Bacillus*, and endo- and exocellular proteases. From 1980 until his retirement in 1994, his research unit focused on several projects connected with biotechnology: next to cellulose degradation, other groups in his lab were studying nitrogen fixation, methanogens, lactic acid bacteria, and methylotrophs.

Loath as he was to shirk anything he could do for the common good, it is no wonder that his professional life was rather busy. Among the numerous responsibilities he took on during his career, his tireless commitment to teaching and to the dissemination of knowledge was particularly noteworthy. Professor of Microbiology at the University of Paris 7, he also headed the Department of Teaching at the Pasteur Institute. He also organised several meetings, which he did with a keen judgement to set up the programme and a painstaking care for details. Thus, the 1987 FEMS Symposium on Biochemistry and Genetics of Cellulose Degradation afforded the first synthesis of new basic science topics in the field, which would experience an explosive development during the next decade.

A native of the North of France, Jean-Paul Aubert didn't indulge in the exuberant display of his emotions. He was deeply honest and knew that telling the truth is the first token of consideration. Thus, he may have appeared somewhat intimidating at first. Yet, to many he became a warm friend, whose unfailing, tactful support could always be counted upon. In short, he was a gentleman.

Pierre Béguin

Biochemistry of Glycanases

CELLULOSE-CELLULASE INTERACTIONS OF NATIVE AND ENGINEERED CELLOBIOHYDROLASES FROM *TRICHODERMA REESEI*

Tuula T. Teeri*, Anu Koivula, Markus Linder, Gerd Wohlfahrt, Laura Ruohonen, Janne Lehtiö, Tapani Reinikainen, Malee Srisodsuk, Karen Kleman-Leyer[1], T. Kent Kirk[1] and T. A. Jones[2]

VTT Biotechnology and Food Research, P.O. Box 1500, FIN-02044 VTT, Finland
*Current address: Royal Institute of Technology, Department of Biochemistry and Biotechnology, S-10044 Stockholm, Sweden
[1] Department of Bacteriology, University of Wisconsin, Madison, WI 53706, U.S.A.
[2] Department of Molecular Biology, BMC, P.O. Box 590, S-75124 Uppsala, Sweden.

1 INTRODUCTION

The cellobiohydrolases, CBHI and CBHII, are the most abundant cellulases produced by the filamentous fungus *Trichoderma reesei*, and clearly the key enzymes in crystalline cellulose degradation.[1,2] Traditionally, cellobiohydrolases have been described as exoglucanases releasing cellobiose from the non-reducing ends of the cellulose glucan chains.[1] Recent evidence indicates that CBHI and CBHII have opposite chain end preferences, CBHI acting at the reducing end and CBHII at the non-reducing end of the chain.[3,4] Both enzymes seem to be processive enzymes catalysing several bond cleavages before the dissociation of the enzyme substrate complex.[5-8]

As typical cellulases, both CBHI and CBHII consist of distinct catalytic and cellulose-binding domains.[9-11] So far no structures have been obtained for intact fungal cellulases, probably due to the properties of the extended, glycosylated linker peptide joining the two functional domains. However, structures of the isolated catalytic domains of both the *T. reesei* cellobiohydrolases, and an endoglucanase EGI as well as the cellulose-binding domain of CBHI have been determined facilitating detailed structure-function studies.[5-7,12-14] The two-domain structure and tight binding of cellulases have been shown to be important determinants of their ability to attack crystalline cellulose.[15,16] Thus, the secret of crystalline cellulose degradation can be at least partially unravelled by striving to understand the domain interplay of the cellobiohydrolases.

2 CATALYTIC DOMAIN FUNCTION OF CBHI AND CBHII

2.1 Catalytic domain structures

High resolution crystal structures of the catalytic domains of *T. reesei* CBHI and CBHII show that, in spite of completely different folding motifs, both enzymes have extended active site tunnels for substrate binding and catalysis.[5-7] In CBHII, a 20 Å long tunnel is formed by the side chains of two well ordered loops, each stabilised by a disulphide bridge.[5] Complex structures of CBHII with several different ligands have revealed four clearly defined glucosyl binding sites (-2, -1, +1, +2) within its active site tunnel.[5,17] In the subsites -2, +1 and +2 tryptophan side chains W135, W367 and W269, respectively, make significant contributions to the formation of the sugar binding sites (Figure 1). In site -2, in particular, this results in a narrower tunnel cross section. In addition, the ligand structures have revealed a 20° twist in the chain between subsites +1 and +2. Besides the tryptophan residues, the active site tunnel contains many residues which form hydrogen bonds to the substrate.[5] Most of these interactions involve charged residues, which are conserved within the CBHII family (glycosyl hydrolase family 6).

The active site tunnel of CBHI is about 50Å long, formed by four loops partly stabilised by disulphide bridges.[6,7,12] Similar to CBHII, the active site tunnel of CBHI is flattened to accommodate up to 10 glucose units (subsites -7 to +3) of a single glucan chain, and has a narrow point in the centre which forces the chain to twist about 130°.[6,7] Binding of the glucose units occurs via hydrogen bonding and, in four subsites, stacking onto the tryptophan residues W376, W367, W38 and W40.

2.2 Catalytic reaction mechanism of CBHII

Cellulases catalyse the hydrolysis of the glycosidic bonds by general acid catalysis. CBHII has been shown to be an inverting glycosidase.[18,19] It is postulated to proceed through a single-displacement reaction involving a general acid to donate a proton, and a base to assist the nucleophilic attack of water.[20] Two aspartic acids, D221 and D175 and a tyrosine, Y169 have been identified at the centre of the active site tunnel of CBHII, very near the scissile bond between the subsites -1 and +1 (Figure 1).[5] Severe loss of activity of CBHII has been observed upon mutagenesis of D221 which is the likely proton donor in the reaction.[21] Enzymes with a mutated D175 residue have low but significant residual activity, especially on longer substrates. It has been suggested that this residue has a dual role in ensuring the protonation of D221 and perhaps stabilising the charged reaction intermediates.[21] A candidate for the general base has been more difficult to appoint, and kinetic studies with cellobiose fluorides have cast doubt on a classical single-displacement mechanism for CBHII.[22] It is thus possible, although speculative, that D221 and D175 are the only catalytic amino acid residues needed in the reaction catalysed by CBHII.

The subsite -1 of CBHII is structurally different from the other binding sites. It lacks the sugar-binding tryptophan and has a protrusion which may permit alternative sugar conformations.[23] A conserved tyrosine, Y169, is situated at hydrogen bonding distance from both D175 and the hydroxymethyl group of the glucose ring at this subsite (Figure 1). Removal of a single hydroxyl group in the mutant Y169F resulted in improved binding but decreased the catalytic rate on soluble cello-oligosaccharides.[23] Determination of a complex structure of the CBHII Y169F mutant with a modified tetrasaccharide shows a clearly distorted sugar conformation at subsite -1. We have thus proposed that the Y169 in the wild-type CBHII facilitates catalysis by stabilising the distorted ring conformation in the transition state through hydrogen bond formation.[23]

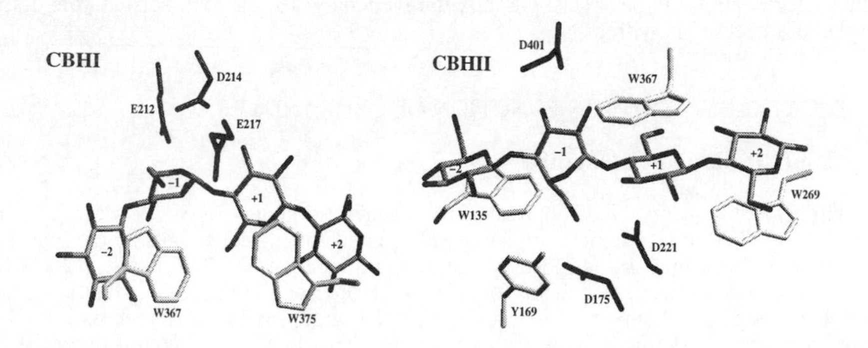

Figure 1 *Arrangement of the proposed catalytic residues in the active sites of* T. reesei *CBHI and CBHII (see text for details). The structural models are based on the X-ray structures of CBHI*[7] *and CBHII (Zou and Jones* et al., *manuscript in preparation) complexed with different ligands.*

2.3 Catalytic reaction mechanism of CBHI

T. reesei CBHI is a retaining enzyme thought to use a double-displacement mechanism for catalysis.[24,25] The active site of *T. reesei* CBHI contains three carboxylates, E212, D214 and E217 that are positioned around the scissile bond between subsites -1 and +1 (Figure 1).[6] The spatial arrangement of the carboxylates observed in complex structures of CBHI hinted that E217 acts as the proton donor and E212 as the nucleophile in the double-displacement reaction. According to this scheme, the role of D214 would be to position and/or control the protonation state of E212.[6] Three different mutants (E212Q, D214N and E217Q) were constructed and their behaviour analysed on 2-chloro-4-nitrophenyl β-lactoside (CNP-Lac) and bacterial cellulose (BMCC).[12] Mutation of the putative nucleophile, E212, and the proposed proton donor, E217, abolished the activity of CBHI on BMCC while clear residual activity was detected with the mutant D214N. On CNP-Lac all the three mutants had a decreased k_{cat} but an unchanged K_M. The drop in the k_{cat} was largest with D212Q suggesting that deglycosylation is the rate limiting step for this substrate. The activity of the mutant D214N was least affected which is consistent with its proposed supporting role in the reaction catalysed by CBHI.[12] Similar to CBHII, molecular modelling based on several independent structures of CBHI with different oligosaccharides suggests the presence of a distorted sugar conformation at the subsite -1 (Figure 1).[7]

3 EXOACTIVITY ON CRYSTALLINE CELLULOSE?

The exo-mode of action of CBHI and CBHII have long been subject to debate since both have been seen to exhibit endoglucanase-like behaviour on some substrates. Endoglucanase activity is usually assayed following the cleavage of internal glycosidic bonds in soluble substituted celluloses such as CMC or HEC, or *e.g.* the loss of staining by Congo red of barley ß-glucan. Enzymes homologous to *T. reesei* CBHI and CBHII have been observed to cleave a soluble, modified tetrasaccharide in a manner indicating endo-activity.[26] However, for pure preparations of CBHI, no endoglucanase activity is detected on CMC[27] nor on barley ß-glucan in plate assays.[28] For CBHII, activity on barley ß-glucan has been clearly demonstrated on plate assays and by detailed analyses of degradation patterns.[28,29] Nevertheless, pure CBHII produced in a *T. reesei* host devoid of the major endoglucanase genes shows no activity on HEC (our unpublished results). The discovery of the unique, tunnel shaped topology of their active sites allowed for a structural interpretation of the exo-activities of CBHI and CBHII.[5-7] Examination of the structures revealed that the loops forming each tunnel are stabilised by disulphide bridges. Moreover, no significant differences have so far been observed in the loop conformations of free and ligand-bound forms of CBHI and CBHII. It has therefore been proposed that a single glucan chain enters the tunnel from one end, is then threaded through the entire tunnel, followed by bond cleavage at the far end of the tunnel.[5-7]

Endoglucanases homologous to CBHI and CBHII apparently lack the long active site loops and thus have more open active sites allowing them to cleave bonds in the middle of the cellulose chains.[5-7,14] However, it is becoming apparent that the division into endo- and exoglucanases may not be absolute, and that some cellobiohydrolases exhibit increasing degrees of endoglucanase activity due to fewer loops or loops that are shorter or able to undergo conformational changes to open their active sites for endoactivity.[2,30,31] In order to achieve a better understanding of the endo- and exoglucanase activities of the *T. reesei* cellulases, we have used gel permeation chromatography to determine the degree of polymerisation (DP) of the insoluble reaction products released from crystalline celluloses.[32,33] Table 1 summarises our data obtained on bacterial microcrystalline cellulose (BMCC) after a one day incubation with individual *T. reesei* EGI, CBHI and CBHII, and Figure 2 shows the data on dewaxed cotton linters incubated for two days with EGI alone or in a synergistic mixture with CBHI.

Table 1 *Degradation of bacterial microcrystalline cellulose by* T. reesei *EGI, CBHI and CBHII. In each experiment, 1 nmol of enzyme was incubated with 1 mg of cellulose in 50mM sodium acetate pH 5.0 at 39°C for 24 h.* DP_w, *weight average degree of polymerisation;* DP_n, *number average degree of polymerisation (data obtained from references 32 and 33).*

Enzyme	*Weight loss %*	DP_w	DP_n	*Polydispersity* DP_w/DP_n
Control	0	480 - 497	96 - 113	4.4 - 5.0
EGI	17	231	78	3.0
CBHI	65	351	120	2.9
CBHII	39	384	125	3.1

The data collected in Table 1 show that while CBHI and CBHII are much more efficient in releasing soluble sugars from BMCC, EGI is able to decrease its degree of polymerisation much more rapidly than either of the cellobiohydrolases. On the other hand, both cellobiohydrolases alone were practically inactive against dewaxed cotton cellulose while EGI is able to dramatically decrease its DP with 15 % weight loss in 24 h (Figure 2).[32,33] However, when the cellobiohydrolases were incubated together with EGI, both were clearly able to promote the release of soluble products from cotton.[32,33] These and our earlier data provide strong support for the primary action at the chain ends by both cellobiohydrolases on these crystalline substrates but do not exclude the possibility that the active site loops sometimes 'breathe' to allow for the low endoglucanase activities occasionally detected on soluble substrates. However, such a loop opening seems to be a rare event and does not seem to play a significant role in the degradation of crystalline cellulose.

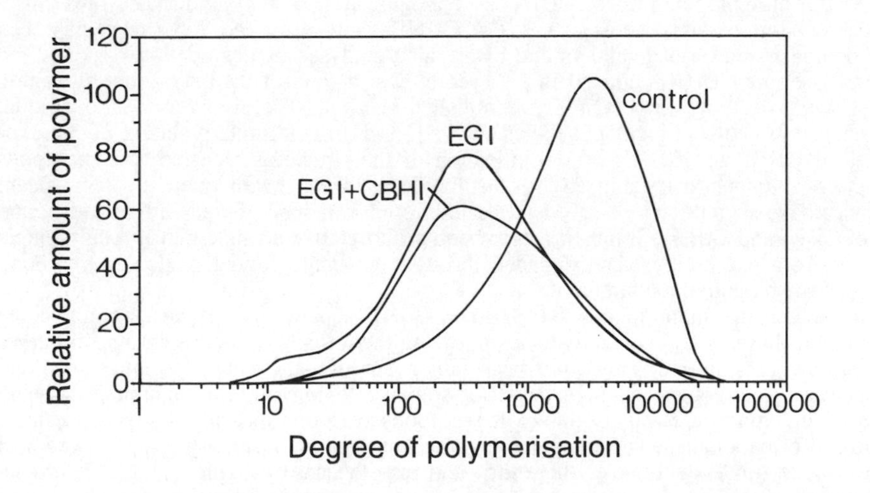

Figure 2 *Changes in the molecular size distribution of cotton cellulose as a result of hydrolysis for 24 h by* T. reesei *EGI or EGI+CBHI. 2 nmol of each enzyme was incubated alone or in a 1:1 mixture of EGI and CBHI with 10 mg of dewaxed cotton in 50 mM sodium acetate buffer, pH 5.0 at 39°C (data from references 32 and 33).*

4 ROLE OF THE TRYPTOPHAN RESIDUES IN THE CBHII 'TUNNEL'

Hydrophobic stacking interactions between the sugar rings and aromatic amino acid residues are considered important for substrate binding by both the catalytic and substrate-binding domains of various carbohydrases.[16,34,35] In the active site of CBHII, such stacking interactions are evident for W135 at subsite -2, W367 at subsite +1, and W269 at subsite +2. Modelling studies with longer oligosaccharides suggest hydrophobic stacking interactions between the sixth glucosyl unit with yet another tryptophan residue, W272, forming a putative binding site +4 at the very entrance of the tunnel.[17,23] This, however, requires another twist of 110° in the cellulose chain between the fourth and sixth glucosyl units, and no obvious sugar-protein interactions can be seen at the putative subsite +3. We have used site-directed mutagenesis to explore the roles of two tryptophan residues, W272 at the entrance and W135 at the exit of the active site tunnel CBHII.

Proton NMR studies of the wild-type CBHII have demonstrated that cellobiose but no glucose is released from the non-reducing end of the chain implying that the subsite -2 must be occupied for hydrolysis to occur.[21,36] Our current data show that replacement of the indole ring of W135 to the phenolic ring in the mutant W135F reduces both the binding and catalytic activity of CBHII (Table 2, and our unpublished data). If the aromatic ring structure is completely abolished, as in the mutant W135L, an even greater loss of binding and activity is observed, confirming that tight binding of the terminal glucose at site -2 is indeed an important determinant of the catalytic efficiency of CBHII.

Table 2 *Catalytic constants for the CBHII wild-type and selected mutants on cello-oligosaccharides. Hydrolysis experiments were performed in 10 mM sodium acetate buffer, pH 5.0 at 27°C. Samples were taken at different time points and analysed by HPLC as described earlier.[21,36] Kinetic constants were calculated by a non-linear regression analysis (Enzfitter) or by analysing whole progress-curves as described.[36] The experimental error is estimated to be 10 %. wt = wild-type; nd = not determined.*

Protein	k_{cat} (min^{-1})			*Reference*
	Glc_4	Glc_5	Glc_6	
CBHII wt	220	60	840	8, 21
W135F	75	nd	nd	21
W135L	3	nd	nd	21
W272A	240	480	≈1500	37

The catalytic rates measured for the wild-type CBHII on oligosaccharides ranging from four to six glucose units show that the rate of cellopentaose hydrolysis is lower than that of cellotetraose but increases again as cellohexaose is used as substrate (Table 2).[8] This led us to suggest that there might be more than four subsites in the active site tunnel, and that some of the oligosaccharides show non-productive binding modes to the CBHII catalytic domain.[8] Based on an earlier molecular modelling study[17], W272 at the proposed subsite +4 seemed the likely candidate responsible for such non-productive interactions. To verify it's role, W272 was mutated to an alanine (W272A).[37] In response to this change, the activity of the mutant was increased on small soluble oligosaccharides, and the characteristic drop of activity observed with wild-type CBHII on cellopentaose vanished (Table 2). Apparently, removal of the aromatic side chain of W272 relieved some of the proposed non-productive binding modes for the oligosaccharides. However, removal of this tryptophan selectively impaired the enzyme function on highly ordered crystalline cellulose suggesting that W272 is involved in the rate-limiting step of CBHII on crystalline cellulose (data not shown).[37] Since W272 is located on a loop readily exposed to the enzyme's surface, and at the open entrance of the active site tunnel of CBHII, we propose that it has a role in catching and guiding a single glucan chain end into this tunnel (Figure 3).

Owing to the apparent importance of the W135, the chain must then be translocated through the entire tunnel for catalysis (see above). A similar mechanism can be predicted for CBHI, based on studies of complex structures with a series of overlapping oligosaccharides spanning its entire active site tunnel.[7]

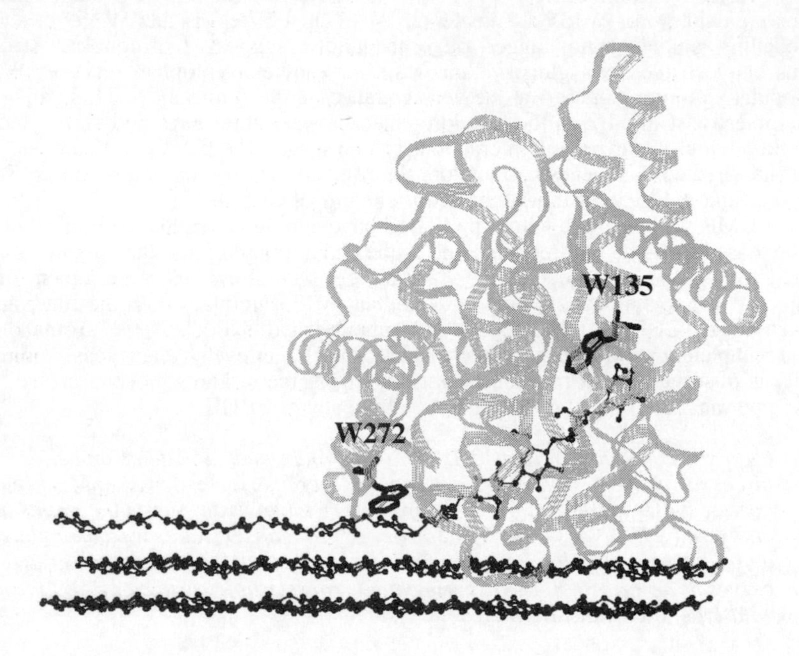

Figure 3 *A hypothetical illustration of the CBHII catalytic domain acting on crystalline cellulose surface. Our results indicate that the side-chain of W135 is important for efficient catalysis and suggest that the side-chain of W272 is important for the crystal breaking capacity of CBHII.*

5 THE ROLE(S) OF THE CBD?

Many enzymes active against insoluble, polymeric carbohydrates have domain structures consisting of distinct substrate-binding domains.[15,16] In the case of CBHI and CBHII, removal or mutagenesis of their cellulose-binding domains (CBDs) reduces their activities on crystalline cellulose but not on soluble substrates.[9,11,38,39] All *Trichoderma* and other fungal CBDs so far studied are small compact domains with a flat binding surface that depends on three aromatic amino acids for relatively tight binding to cellulose.[13,38-41] It is commonly assumed that the main function of the CBD is to tether the catalytic domain onto the solid substrate to maintain a high local enzyme concentration in the vicinity of the microbial cell or hyphae.[15,42] It has also been suggested that a CBD may assist the catalytic function of cellulases by promoting the release of single glucan chains from the crystalline cellulose surface.[10,39,43] However, clear-cut experimental evidence for such an activity has been difficult to obtain. In most cases, physical linkage of the two functional domains is required for full positive effect of a CBD on crystalline cellulose hydrolysis.[16] A synergistic effect between separated catalytic and cellulose-binding domains on cellulose degradation has been reported only in one case, and even then it was observed on one particular substrate only.[44]

5.1 Binding of a double CBD

Intact cellobiohydrolases seem to bind to crystalline cellulose practically irreversibly while the isolated catalytic domains can be easily eluted.[16] Therefore, it has been suggested that the binding of their CBD is very tight, if not in itself irreversible.[45] However, when considering the function of an intact, processively acting cellobiohydrolase, irreversible binding and subsequent immobilisation of the enzyme on the cellulose surface do not seem likely. We have investigated the role of the CBD by constructing a double CBD with two different fungal CBDs joined by a relatively long linker.[46] This two-domain arrangement is analogous to and serves as a simplistic model of an intact cellulase. Determination of the binding isotherms on crystalline cellulose revealed that the affinity of the double CBD was much better than the added affinities of the corresponding single CBDs, and an analysis of the binding parameters suggested co-operativity between the domains.[46] In analogy to data obtained with intact cellulases, these results imply that the apparent irreversibility of binding of intact CBHI may also result from synergistic binding of their two functional domains. The model also suggests a dynamic interplay between the two domains, both undergoing continuous but mutually independent cycles of binding and desorption.[16,46]

5.2 The exchange rates of the individual CBDs

The rate of desorption and rebinding at equilibrium is a key parameter for understanding the dynamics of the domain interplay in the cellobiohydrolases. Considering the extreme cases first, one could imagine that a very low exchange rate binding of the CBD would lead to various non-productive enzyme-substrate complexes. Either the enzyme is bound to an area with no chain ends to start from, or the processive action of an initially productively bound cellobiohydrolase is prevented by the tightly bound CBD. On the other hand, an enzyme with extremely rapidly exchanging or no CBD might not have sufficient time to initiate hydrolysis before desorption from the substrate surface. One would thus expect an efficient cellobiohydrolase to have a CBD with a moderate exchange rate, allowing enough time for the catalytic domain to capture a loose chain end and to initiate hydrolysis but also permitting its subsequent processive movement along the chain.

We have investigated these aspects of the CBD-cellulose interaction using the CBHI CBD labelled with tritium (CBD-^{3}H) to permit accurate and sensitive measurement of its binding.[47] In a typical experiment, the CBD-^{3}H was first allowed to reach equilibrium with cellulose. Thereafter, the same molar amount of non-labelled CBD was added at a concentration matching that of the free CBD-^{3}H remaining in the supernatant. Because the bound and free protein are constantly exchanging, the amount of non-labelled CBD decreased and the amount of ^{3}H-labelled CBD increased in the supernatant. By following the increase of the radioactivity in the supernatant, the rate of exchange between the bound and free CBDs could be determined accurately. Although the value obtained in this way is not an off-rate in the traditional sense, it is directly related to the half-life of the bound state. Comparison of the values obtained (Table 3) with the published catalytic rates of CBHI indicates that binding of the CBD should not be rate-limiting for the progression of its catalytic domain on crystalline cellulose.[47]

However, not all CBDs have the same moderate affinity, and *e.g.* EGI CBD has a significantly higher affinity than CBHI CBD on crystalline cellulose.[41] Somewhat surprisingly, a recombinant CBHI with the EGI CBD had close to wild-type hydrolytic activity on crystalline cellulose in spite of its significantly improved binding.[48] Therefore, apart from the CBD affinity other mechanisms may also contribute to the mobility of the cellobiohydrolase catalytic domains on the crystal surfaces.

Table 3 *Rates of CBHI CBD exchange on cellulose at different temperatures. The values were obtained by competition between tritium labelled CBD and non-labelled CBD.*

Temperature (°C)	Half-time of the CBD on the surface (s)	Exchange rate of the CBD (s^{-1})
4	140	0.005
20	60	0.012
30	24	0.029

6 A COMPREHENSIVE MODEL OF CELLOBIOHYDROLASE ACTION

Figure 4 summarises our current working model for the action of the *T. reesei* cellobiohydrolases. In this model, we have attempted to involve all of the information so far accumulated both on the natural substrate and on various native and genetically engineered variations of CBHI and CBHII.

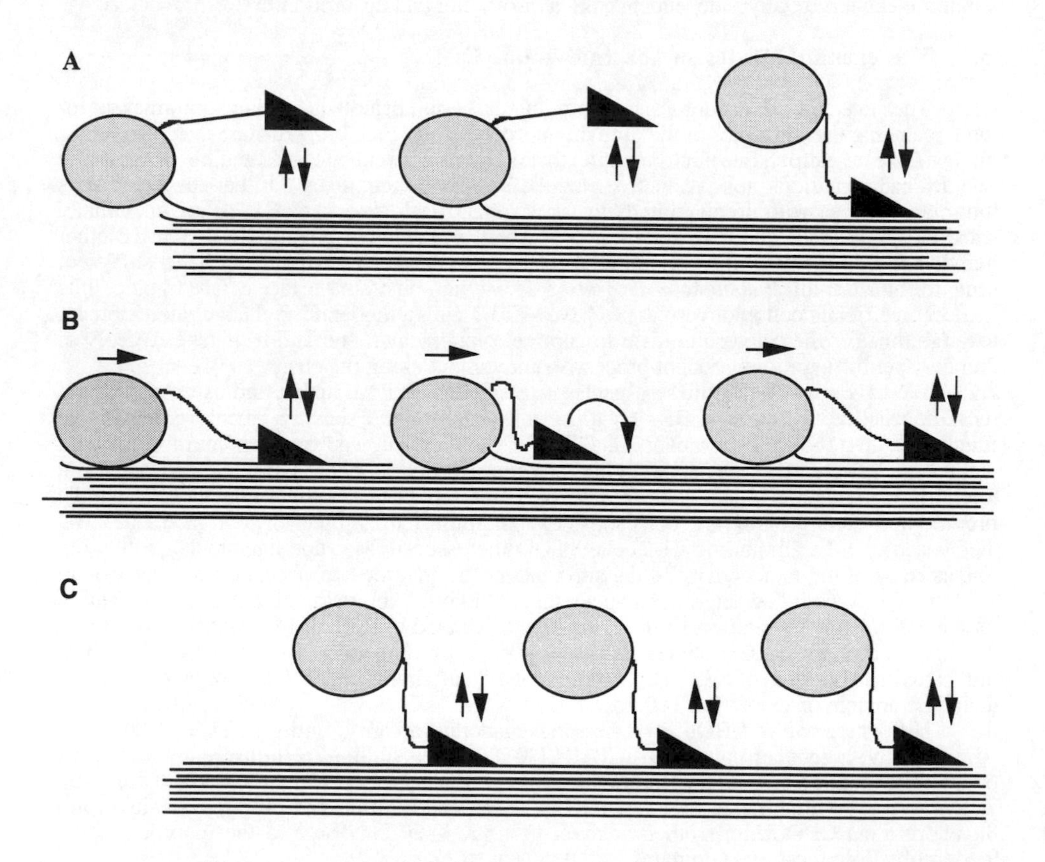

Figure 4 *A schematic model of the different stages of crystalline cellulose degradation by a cellobiohydrolase. See text for details.*

The cellulose in plant cell walls is a heterogeneous substrate presenting both highly crystalline and less ordered, amorphous regions.[49] In addition to the obvious chain ends in the crystal termini, chain ends hidden within the crystals have been assumed.[2,50] The two functional domains of *T. reesei* cellobiohydrolases are both capable of binding to the substrate, although probably in different regions.[16,38,39,42,51] Early in the process, the enzymes approach the substrate and bind to the surface through only one or both of the domains. As illustrated in Figure 4 A, depending on the site of interaction, the initial binding event is followed by productive or non-productive interaction of the catalytic domain. There is some evidence that productive binding of the catalytic domain may occur on chain ends presented by easily accessible, disordered regions of the substrate.[39,42,51] Structural, biochemical and mutagenesis studies of CBHII support the hypothesis that crystalline cellulose degradation begins by introduction of a single glucan chain end at the entrance of the active site tunnel (see above). The chain is then apparently translocated to fill the entire tunnel with subsequent release of cellobiose from the opposite end of the tunnel. The mechanism of the chain translocation within the tunnels is still poorly understood, but may involve the large twists observed along the glucan chain axes in several different complex structures of both CBHI and CBHII[7] (Zou and Jones *et al.*, manuscript in preparation). It seems thus that it might be the mechanism of initiation together with the unusual active site architecture, rather than explicit chemical chain end recognition that condemn CBHI and CBHII to action at the ends of the glucan chains of crystalline cellulose.

The presence of a CBD facilitates the degradation of highly crystalline substrates, but its exact mechanism of action is still largely unknown. In the productive mode (Figure 4 B), the enzyme is tightly bound to the substrate due to synergistic binding interactions of two distinct domains which are nevertheless physically linked to each other.[42,46] In spite of the apparently very high affinity of the intact enzyme, the catalytic domain is able to carry out processive hydrolysis which - at least in the case of CBHI - is facilitated by a relatively fast exchange rate of its CBD.[47] However, the presence of a higher affinity CBD, introduced by genetic engineering,[48] did not impair the activity of CBHI on crystalline substrates. Whether this reflects a more active mechanism for the desorption of the CBD, or perhaps sliding of the CBD without an actual detachment from the surface, remains to be seen after further investigations.

When CBHI is acting alone, non-productive binding through the CBD seems evident towards the end of the hydrolysis and/or in high enzyme coverage[39,48,51] (Figure 4 C). As demonstrated by the data obtained on cotton cellulose (see section 3), in the absence of endoglucanases, the substrate surface is apparently gradually depleted of chain ends required for the initial productive binding of the catalytic domain.[32,33] With a fully functional CBD, binding can nevertheless occur all along the substrate surfaces. Although detrimental for the activity of an isolated enzyme, such non-productive binding can be easily tolerated in the natural synergistic enzyme mixtures in which the endoglucanases can provide new chain ends for the bound cellobiohydrolases.

References

1. T.M. Wood and V. Garcia-Campayo, *Biodegradation*, 1990, **1**, 147.
2. T.T. Teeri, *TIBTECH*, 1997, **15**, 160.
3. M. Vrsanská and P. Biely, *Carbohydrate res.*, 1992, **227**, 19.
4. B. Barr, Y.-L. Hsieh, B. Ganem and D. Wilson, *Biochemistry*, 1996, **35**, 586.
5. J. Rouvinen, T. Bergfors, T.T. Teeri, J. Knowles and T.A. Jones, *Science*, 1990, **249**, 380.
6. C. Divne, *et al.*, *Science*, 1994, **265**, 524.
7. C. Divne, J. Ståhlberg, T.T. Teeri and T.A. Jones, *J. Mol. Biol.*, 1997, *in press.*
8. V. Harjunpää *et al.*, *Eur. J. Biochem.*, 1996, **234**, 278.
9. H. van Tilbeurgh, P. Tomme, M. Claeyssens, R. Bhikhabhai and G. Pettersson, *FEBS Lett.*, 1986, **204**, 223.
10. T.T. Teeri, P. Lehtovaara, S. Kauppinen, I. Salovuori and J. Knowles, *Gene*, 1987, **51**, 42.

11. P. Tomme *et al.*, 1988, *Eur. J. Biochem.*, **170**, 575.
12. J. Ståhlberg *et al., J. Mol. Biol.,* 1996, **264**, 337.
13. P. Kraulis *et al.*, *Biochemistry*, 1989, **28**, 7241.
14. G. Kleywegt et al, *J. Mol. Biol.,* 1997, *in press.*
15. P. Tomme, R.A.J.. Warren and N. Gilkes, *Adv. Microb. Physiol.*, 1995, **37**, 1.
16. M. Linder and T.T. Teeri, *J. Biotechnol.*, 1997, **57**, 15.
17. J. Rouvinen, Doctoral Thesis, University of Joensuu, 1990.
18. J. Knowles, P. Lehtovaara, M. Murray and M. Sinnott, *J. Chem. Soc. Chem. Comm.*, 1988, 1401.
19. M. Claeyssens, P. Tomme, C. Brewer and E.J. Hehre, *FEBS Lett.*, 1990, **263**, 89.
20. D. Koshland, *Biol. Rev.*, 1953, **28**, 416.
21. L. Ruohonen *et al.*, 'Trichoderma reesei cellulases and other hydrolases' (P. Suominen and T. Reinikainen, Eds.), Foundation for Biotechnical and Industrial Fermentation Research. Helsinki, 1993, Vol **8**, 87.
22. A. Konstantinidis, M. Marden and M. Sinnott, *Biochem. J.*, 1993, **291**, 883.
23. A. Koivula *et al,. Protein Engin.*, 1996, **9**, 691.
24. M. Sinnott, *Chem. Rev.*, 1990, **90**, 1171
25. G. Davies, M. Sinnott and S. Withers, 'Comprehensive Biological Catalysis', Academic Press 1997, pp. 119-211.
26. S. Armand, S. Drouillard, M. Schülein, M. Henrissat and H. Driguez, *J. Biol. Chem.*, 1997, **272**, 706.
27. D. Irwin, M. Spezio, L. Walker and D. Wilson, *Biotechnol. Bioeng.*, 1993, **42**, 1002.
28. M. Penttilä, L. André, P. Lehtovaara, M. Bailey, T.T. Teeri and J. Knowles, *Gene*, 1988, **45**, 253.
29. K. Henriksson *et al.*, *Carbohydrate Polym.*, 1995, **26**, 119.
30. R.A.J. Warren, *Ann. Rev. Microbiol.*, 1996, **50**, 183.
31. S. Chang and D. Wilson, *J. Biotechnol.*, 1997, **57**, 101.
32. K. Kleman-Leyer, M. Siika-aho, T.T. Teeri and T.K. Kirk, *Appl. Env. Microbiol.*, 1996, **62**, 2883.
33. M. Srisodsuk, K. Kleman-Leyer, T.K. Kirk and T.T. Teeri, *Eur. J. Biochem.*, 1997, accepted for publication.
34. N. Vyas, *Curr. Opinion Struct. Biol.*, 1994, **1**, 737.
35. E.J. Toone, *Curr. Opinion Struct. Biol.*, 1994, **4**, 719.
36. A. Teleman *et al.*, *Eur. J. Biochem.*, 1995, **231**, 250.
37. A. Koivula *et al.*, 1997, manuscript in preparation.
38. T. Reinikainen *et al.*, *Proteins*, 1992, **14**, 475.
39. T. Reinikainen, O. Teleman and T.T. Teeri, *Proteins*, 1995, **22**, 392.
40. M. Linder *et al., Protein Science*, 1995, **6**, 294.
41. M. Linder, G. Lindeberg, T. Reinikainen, T.T. Teeri and G. Pettersson, *FEBS Lett.*, 1995, **372**, 96.
42. J. Ståhlberg, G. Johansson and G. Pettersson, *Bio/technology*, 1991, **9**. 286.
43. Tormo *et al.*, *EMBO J.*, 1996, **15**, 5739.
44. N. Din et al., *Proc. Natl. Acad. Sci., USA*, 1994, **91**, 11383.
45. B. Henrissat, *Cellulose*, 1994, **1**, 169.
46. M. Linder, I. Salovuori, L. Ruohonen and T.T. Teeri, *J. Biol. Chem.*, 1996, **271**, 21268.
47. M. Linder and T.T. Teeri, *Proc. Natl. Acad. Sci. USA*, 1996, **93**, 12251.
48. M. Srisodsuk J. Lehtiö, M. Linder, E. Margolles-Clark, T. Reinikainen and T.T. Teeri, *J. Biotechnol.*, 1997, **57**,49.
49. R. Atalla, 'Trichoderma reesei cellulases and other hydrolases' (P. Suominen and T. Reinikainen, Eds.), Foundation for Biotechnical and Industrial Fermentation Research. Helsinki, 1993, Vol **8**, 1.
50. N. Gilkes, E. Kwan, D. Kilburn, R. Miller, and R.A.J. Warren, *J. Biotechnol.*, 1997, **57**, 83.
51. M. Srisodsuk, T. Reinikainen, M. Penttilä and T.T. Teeri, *J. Biol. Chem.*, 1993, **268**, 20756.

REVERSAL OF CELLULOSE HYDROLYSIS CATALYSED BY CELLOBIOHYDROLASE II OF *TRICHODERMA REESEI*

Michael L. Sinnott and Nizar S. Sweilem

Department of Chemistry (M/C111), University of Illinois at Chicago, 845 West Taylor Street, Chicago, Illinois 60607-7061 USA (both authors) and
Department of Paper Science, UMIST, POB 88, Sackville Street, Manchester M60 1QD, UK (MLS)

1 INTRODUCTION

Despite extensive structural data, including the X-ray crystal structures of both enzymes,[1,2] the cellobiohydrolases of *Trichoderma reesei* present the mechanistic enzymologist with a number of puzzles. Firstly, why does Nature go to the trouble of making two cellobiohydrolases, with opposite stereochemistries of action,[3] which appear to attack the cellulose chain from opposite ends: the retaining CBHI from the reducing end and the inverting CBHII from the non-reducing end?[4] Secondly, why are the catalytic efficiencies of both enzymes so low when acting on soluble substrates? For example, the k_{cat} value for β-cellobiosyl fluoride hydrolysed by CBHII is a mere 4.0 s^{-1},[5] whereas that for hydrolysis of a-glucosyl fluoride by the glucoamylase of *Aspergillus niger* is 730 s^{-1}.[6] These two reactions are directly comparable; in both cases an exo-acting, inverting glycohydrolase is acting on a glycosyl fluoride corresponding to the liberated saccharide unit. Given that the spontaneous hydrolysis of a-glycosyl fluorides is some 40 times slower than that of β-glycosyl fluorides,[6] the glucoamylase apparently is the more efficient catalyst by nearly 4 orders of magnitude. Admittedly, the use of larger substrates can result in k_{cat} values for cellooligosaccharide hydrolysis by CBHII approaching more usual values[7] (*e.g.* 12 $^{s-1}$ for cellohexaose), but the contrast with glucoamylase, which acts on a non-crystalline substrate, is still striking. Similar arguments can be made for CBHI, although as a retaining enzyme it has two chemical steps, formation and hydrolysis of the glycosyl-enzyme intermediate, and simple kcat values are less informative than those obtained for the single-displacement enzyme CBHII.[5]

Considerations along these lines led us to propose some years ago that the biological function of CBHI and CBHII was not simply the hydrolysis of glycosidic bonds.[5] In making this suggestion we were strongly influenced by work on the forced evolution of the *ebg* β-galactosidase of *Escherichia coli*, which indicated that a single evolutionary event could alter kinetic parameters for hydrolyses catalysed by this glycosidase by an order of magnitude.[8] It was most unlikely, therefore, that the CBHI and CBHII loci were under any selection pressure to increase efficiency in the cleavage of glycosidic bonds *per se*. Rather, we speculated that both enzymes might be work-producing enzymes which used the free energy of hydrolysis of the glycosidic bond to bring about disruption of the cellulose crystallite. It is known that both CBHI and CBHII are essential in degrading crystalline cellulose, even though amorphous cellulose can be degraded by endoglucanases.[9] The basic hypothesis is shown in Scheme 1.

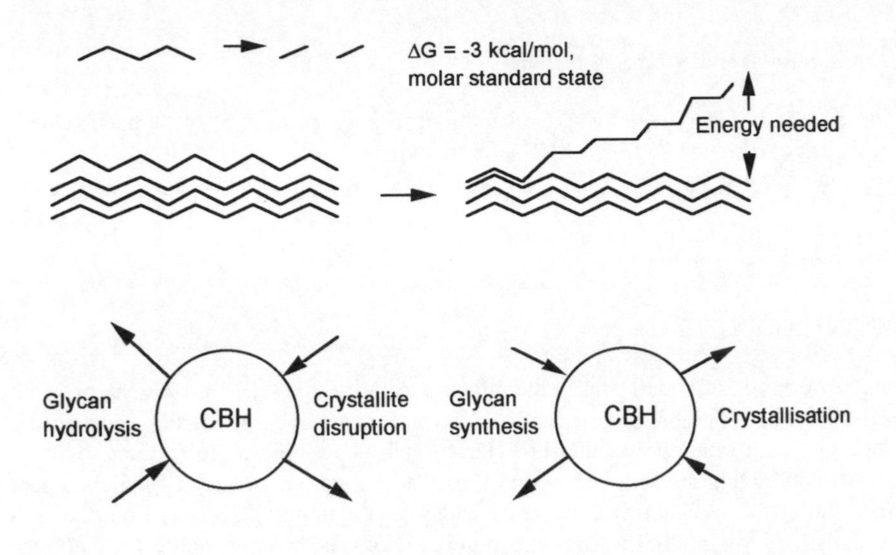

Scheme 1 *Cartoon illustrating the hypothesis that cellobiohydrolases use the free energy of hydrolysis of the glycan link to disrupt the cellulose, and its necessary corollary.*

If CBHI and CBHII indeed coupled glycoside hydrolysis to crystallite disruption, they would belong to a class of enzymes which bring about coupled vectorial processes. These processes involve the coupling of a thermodynamically downhill process (commonly the hydrolysis of ATP) to an energy-requiring activity, such as ion transport or the performance of mechanical work. The detailed molecular mechanisms whereby the processes are coupled remain as yet ill-defined, but a commonly suggested model for work processes coupled to the hydrolysis of ATP involves the existence of two conformational states of the catalytic region of the appropriate enzyme. In the first conformational state, that before the performance of the work, the equilibrium constant *on the enzyme* is close to unity, and the intrinsic binding energy of the substrate is used to bind it tightly. Since the substrate is bound more tightly than products in this state, the equilibrium constant between substrate and products on the enzyme is closer to unity than in free solution. In the second conformational state of the enzyme, after the work has been done, the equilibrium constant on the enzyme is close to that in free solution. Because Km values for substrate are very low in the "pre-work" conformational state, there is little chemical flux through this conformational state: only after the work has been done and the tight binding of substrate has been relaxed in the "post-work" conformational state are Km values large enough for there to be appreciable chemical flux through the system.[10,11]

These ideas, which, it must be emphasised, are wholly speculative in terms of the mechanism of the coupling process with CBHII, are illustrated in Scheme 2.

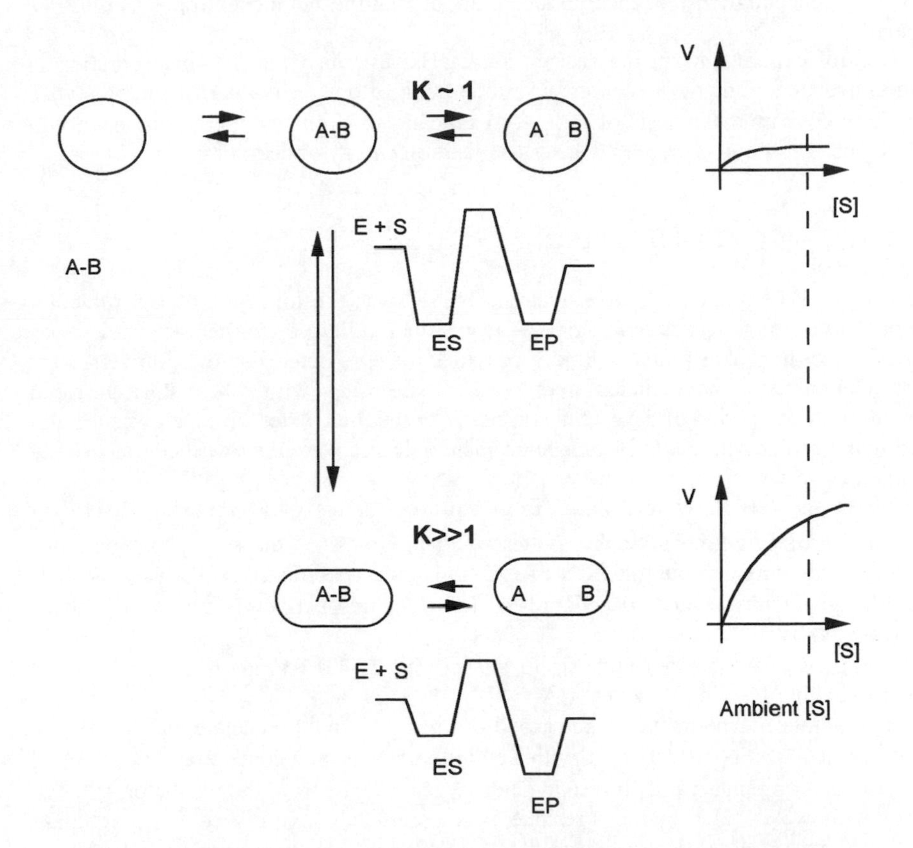

Scheme 2 *Speculative mechanism whereby a cellobiohydrolase could couple the hydrolysis of a glucan link to crystallite disruption, based on other systems where a hydrolysis mechanism is used to perform work. The tense form of the enzyme is denoted by true ellipses and the relaxed form by flattened ellipses. Work is done when the tense form changes to the relaxed form. The free energy profiles for the tense and relaxed form are drawn with the same* k_{cat}/K_m, *to show how at constant ambient substrate concentration higher flux is achieved the higher the* K_m *value.*

A key feature of this, and of any sound molecular mechanism for coupling a chemical change to a work-producing activity, is that it is reversible: it must in principle be possible to synthesise the A-B bond from A and B, even though the equilibrium between them in free solution is very unfavourable, by applying work to the system of enzyme, A and B. Synthesis of ATP from ADP and inorganic phosphate during oxidative phosphorylation, driven by a proton motive force (*i.e.* a concentration gradient) of course lies at the heart of all biochemistry.

We now report that we have successfully driven cellulose hydrolysis backwards in the presence of crystalline cellulose: in the presence of CBHII, radioactive cellobiose, at millimolar concentrations, is incorporated into crystalline cellulose from *Acetobacter xylinum.*

Control experiments in the absence of CBHII establish that the incorporation is enzyme-mediated. The most economical explanation of our results is that the otherwise contrathermodynamic synthesis of a $\beta(1\rightarrow4)$ glucan link from its sugar components in dilute aqueous solution is driven by the crystallisation energy of cellulose.

2 METHODS AND MATERIALS

Cellulose from *Acetobacter xylinum* (sample NQ-5) was the kind gift of Dr. Malcolm Brown, University of Texas at Austin. Any residual cellulose synthetase activity was removed according to a protocol kindly provided by Dr. Peter Tomme, University of British Columbia. The pellicles were washed six times with 4% NaOH at room temperature over a period of days, then washed with distilled water, and then washed and boiled under reflux with 3N HCl for 1 hour. The pellicles were then washed extensively with distilled water.

CBHI and CBHII were isolated from culture filtrates of *T. reesei* QM9414 by DEAE and affinity chromatography as described previously;[5] the several preparations used appeared homogeneous on SDS-PAGE and gave specific activities (against 3,4-dinitrophenyl lactoside and α-cellobiosyl fluoride, respectively) similar to those previously observed.

Cellobiose was labelled with ^3H to an activity of 4.0 Ci/mmole at the National Tritium Labelling Facility, Berkeley, CA.

Incorporation experiments were carried out in 0.1M sodium acetate buffer, pH 5.0, at 30°C (300µL) in Eppendorf tubes, with a cellulose suspension concentration of 0.439 g mL^{-1}. At each time interval an Eppendorf tube was removed from the incubator, filtered, washed with water (2L), and the residue was completely hydrolysed by a cellulase cocktail (Sigma). A 50µL portion of the aqueous solution was transferred to Sigma FLUOR Universal liquid scintillation cocktail for aqueous solutions and counted.

3 RESULTS AND DISCUSSION

Figure 1 displays the incorporation of radioactive cellobiose at various concentrations into cellulose by CBHII as a function of time. Despite the noise in the data, it is clear that incorporation is proceeding in a time and concentration-dependent manner. The equilibrium constant for the hydrolysis of lactose is around 180 M;[12] on the assumption that the free energy of hydrolysis of the Gal$\beta(1\rightarrow4)$Glc linkage is similar to that for the hydrolysis of a Glc$\beta(1\rightarrow4)$Glc linkage, at thermodynamic equilibrium, depending on the cellobiose concentration, 1-10 ppm of the cellobiose will be present as (soluble) cellotetraose, and correspondingly less (on the order of parts per billion) as insoluble cellohexaose. Yet the highest incorporations of cellobiose correspond to incorporation of 1 part in 10^4 of total radioactivity present into the insoluble cellulose phase. Simple reversal of glycoside hydrolysis in free solution cannot, therefore, explain our results.

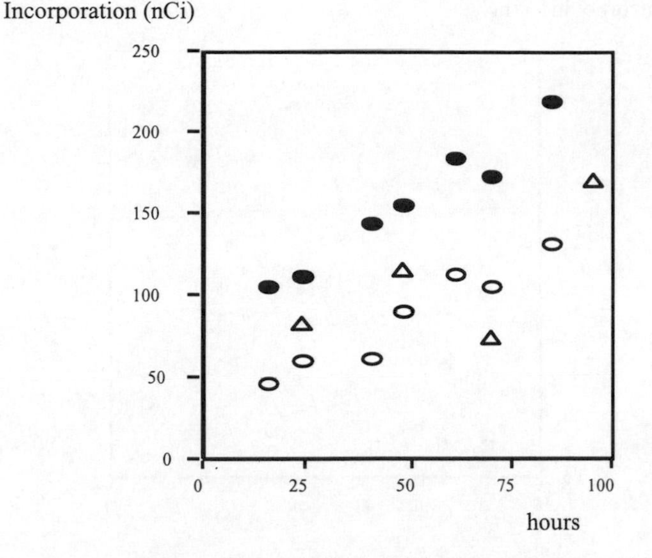

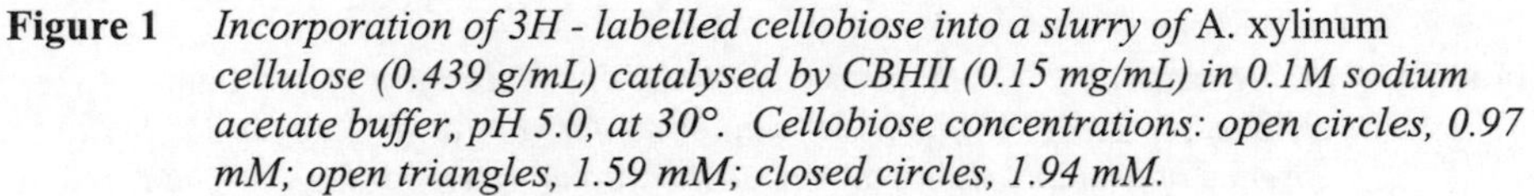

Figure 1 *Incorporation of 3H - labelled cellobiose into a slurry of* A. xylinum *cellulose (0.439 g/mL) catalysed by CBHII (0.15 mg/mL) in 0.1M sodium acetate buffer, pH 5.0, at 30°. Cellobiose concentrations: open circles, 0.97 mM; open triangles, 1.59 mM; closed circles, 1.94 mM.*

Figures 2 and 3 represent obvious control experiments. Figure 2 shows the time course of apparent incorporation in absence of any protein. Simple physical absorption onto the cellulose surface gives around 50-fold less incorporation than in the presence of CBHII. Figure 3 addresses the objection that precipitation of protein in the experiment with CBHII could entrain radioactive cellobiose and this would not be removed by washing. A protein with no cellulolytic function, bovine serum albumin, was used at a concentration 2.5 times that of CBHII.

Whilst the apparent incorporation may be very marginally increased, at the highest concentrations and times it is still 25 times less than observed with CBHII.

More subtle objections are addressed by the experiment described in Figure 4, which describes the analogous experiment performed with CBHI. It is seen that at longer times there is no incorporation of cellobiose. A possible objection to the reality of the incorporation seen with CBHII is that it represents additional physical absorption consequent upon the opening of cracks in the cellulose by the catalytic action of the enzyme. However, similar cracks should also be opened by CBHI, yet little or no additional incorporation is seen.

The apparent transient incorporation possibly observed with CBHI (although the data is ambiguous) might arise as a consequence of "snap-back".[13] The glycosyl enzyme formed by CBHI from a cellulose chain on the crystal could conceivably react by transglycosylation rather than transfer as shown in Scheme 3. If CBHI then came off the cellulose, the result would be incorporation.

However, the fleeting nature of any incorporation is in accord with the idea that CBHI acts processively on the crystallite. Nonprocessive action, with CBHI coming on and off the crystal, should result in steady increase in incorporation with time, as seen with CBHII.

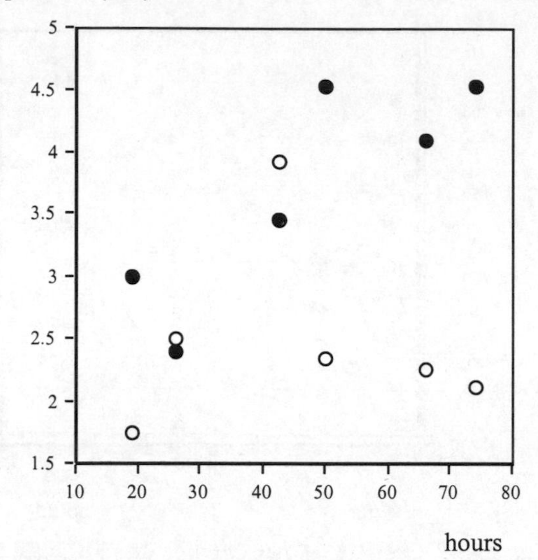

Figure 2 *Incorporation of 3H - labelled cellobiose into a slurry of* A. xylinum *cellulose (0.439 g/mL) in 0.1M sodium acetate buffer, pH 5.0, at 30° in the absence of protein. Cellobiose concentrations: open circles, 0.97 mM; closed circles, 1.94 mM.*

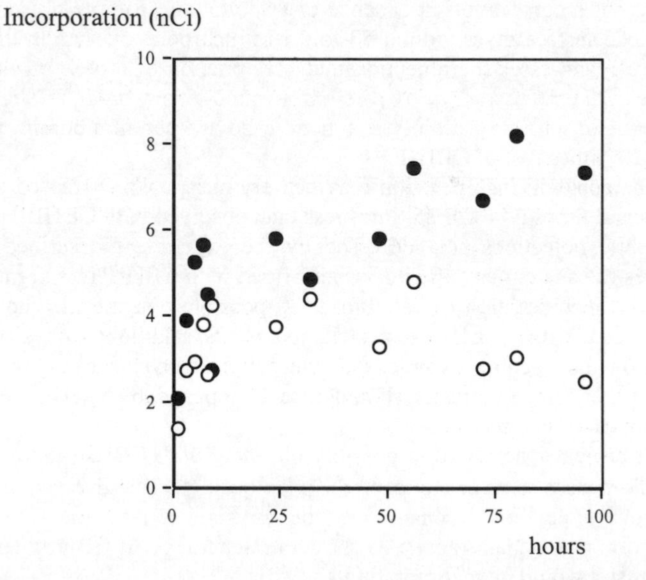

Figure 3 *Incorporation of 3H - labelled cellobiose into a slurry of* A. xylinum *cellulose (0.439 g/mL) in 0.1M sodium acetate buffer, pH 5.0, at 30° in the presence of bovine serum albumin (1 mg/mL). Cellobiose concentrations: open circles, 0.97 mM; closed circles, 1.94 mM.*

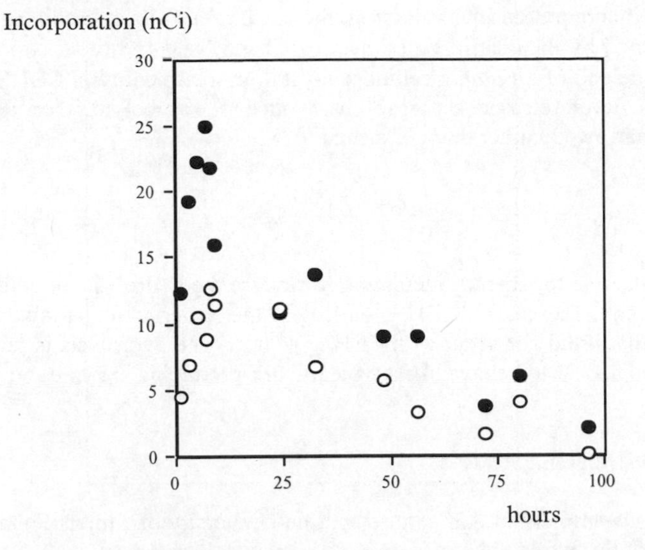

Figure 4 *Incorporation of 3H - labelled cellobiose into a slurry of* A. xylinum *cellulose (0.439 g/mL) in 0.1M sodium acetate buffer, pH 5.0, at 30° in the presence of CBHI (0.4 mg/mL). Cellobiose concentrations: open circles, 0.97 mM; closed circles, 1.94 mM.*

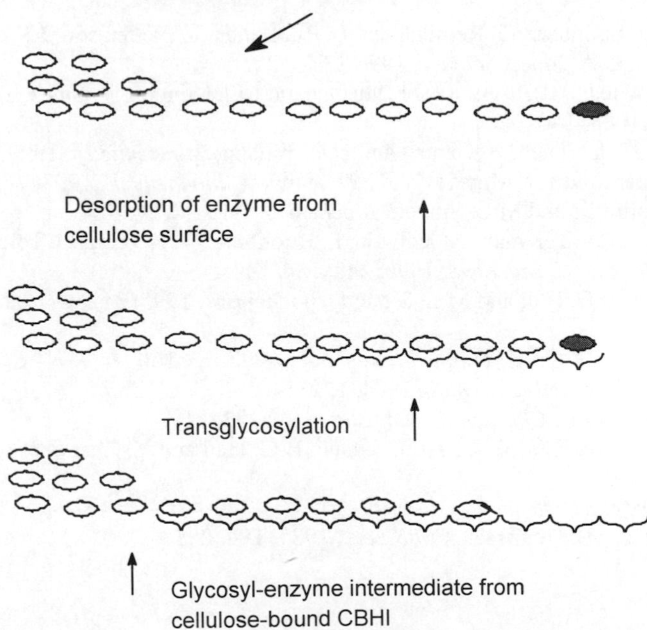

Scheme 3 *Cartoon showing how transglycosylation with the retaining CBHI could lead to incorporation of cellobiose into crystalline cellulose.*

That CBHII is nonprocessive, at least on the scale of a *Valonia* cellulose crystal, is of course established by the classic experiment of Chanzy and Henrissat, who observed "sharpening" of one end of a *Valonia* cellulose crystal on treatment with CBHII[14]. If the CBHII molecules never released a glycan chain once it was bound, then the crystals would be slowly narrowed, rather than sharpened.

4 CONCLUSION

Although the results of experiments attempting to reverse the hydrolysis of cellulose with CBHI are ambiguous, those with CBHII, barring some experimental problem of such extreme subtlety that it did not occur with CBHI, are not. The hydrolysis is reversed and CBHII does indeed appear to behave like a typical work-producing enzyme.

5 ACKNOWLEDGEMENT

We thank the US Department of Agriculture and the University of Illinois Foundation for financial support of this work.

References

1. J. Rouvinen, T. Bergfors, T. Teeri, J.K.C. Knowles and T.A. Jones, *Science*, 1990, **249**, 380.
2. C. Divne, J. Ståhlberg, T. Reinikainen, L. Ruohonen, G. Pettersson, J.K.C. Knowles, T.T. Teeri and A. Jones, *Science*, 1994, **265**, 524.
3. J.K.C. Knowles, P. Lehtovaara, M. Murray and M.L. Sinnott, *J. Chem. Soc. Chem. Commun.*, 1988, 1401.
4. B.K. Barr, Y.-L. Hsieh, B. Ganem and D.B. Wilson, *Biochemistry*, 1996, **35**, 586.
5. A.K. Konstantinidis, I. Marsden and M.L. Sinnott, *Biochem. J.*, 1993, **291**, 883.
6. A. Konstantinidis and M.L. Sinnott, *Biochem. J.*, 1991, **279**, 587.
7. V. Harjunpää, A. Teleman, A. Koivula, L. Ruohonen, T.T. Teeri, O. Teleman and T. Drakenberg, *Eur. J. Biochem.*, 1996, **240**, 584.
8. S. Krishnan, B.G. Hall and M.L. Sinnott, *Biochem. J.*, 1995, **312**, 971; and previous papers.
9. Reviewed by T.T. Teeri, *Trends. Biotechnol.,* 1997, **15,** 160.
10. W.P. Jencks, *Methods Enzymol.,* 1984, **171**, 145.
11. D.W. Urry, *Angew. Chem. Intl. Edn. Eng.*, 1993, **32**, 819.
12. A.C. Elliott, L. Sinnott, P.J. Smith, Z. Guo, B.G. Hall and Y. Zhang, *Biochem. J.*, 1992, **282**, 155.
13. A.A. Klyosov, *Biochemistry,* 1990, **29**, 10577.
14. H. Chanzy and B. Henrissat, *FEBS Lett.*, 1985, **184**, 285.

MUTATIONAL ANALYSIS OF SPECIFICITY AND CATALYSIS IN *BACILLUS* 1,3-1,4-β-GLUCANASES

Antoni Planas
Laboratori de Bioquímica,
Institut Químic de Sarrià, Universitat Ramon Llull
08017-Barcelona, Spain

1 INTRODUCTION

Endo-β-glucanases are classified according to their substrate specificity on β-glucan polysaccharides. Among them, 1,3-1,4-β-glucanases (or lichenases, EC 3.2.1.73) are highly specific enzymes able to depolymerise mixed-linked glucans containing β-1,3 and β-1,4 glycosidic bonds such as barley β-glucan and lichenan, but they do not hydrolyse cellulose or laminarin derivatives. 1,3-1,4-β-glucans are abundant in cereals, specially in barley, where they are major components of the starchy endosperm cell walls. It makes 1,3-1,4-β-glucanase an important enzyme for processing barley grain, as in brewing and animal feeding industries.

In water-soluble barley β-glucan, up to 90% of each polysaccharide is composed of cellotriose and cellotetraose units linked by single β-1,3 glycosidic bonds and randomly distributed along the chain sequence.[1] The title enzyme specifically hydrolyses β-1,4 glycosidic bonds in 3-O-substituted glucopyranose units to yield the final trisaccharide 3-O-β-cellobiosyl-D-glucopyranose (**1**) and the tetrasaccharide 3-O-β-cellotriosyl-D-glucopyranose (**2**)[2,3] (Figure 1).

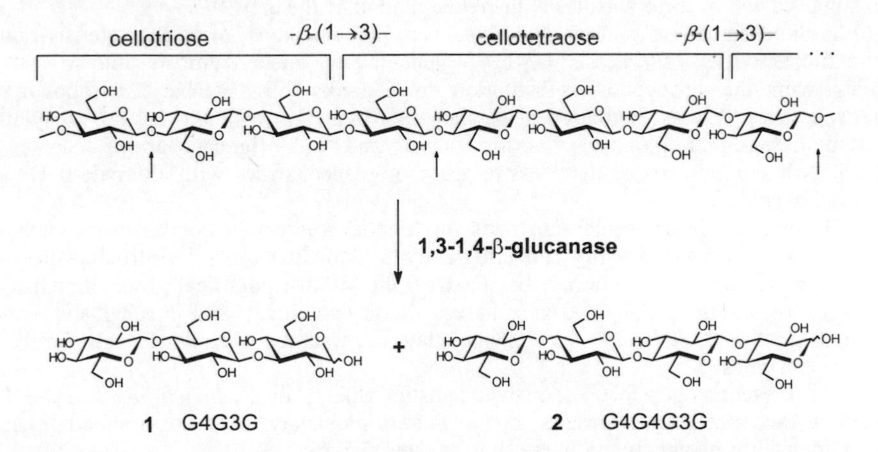

Figure 1 *1,3-1,4-β-glucanase-catalysed degradation of barley β-glucan*

The 1,3-1,4-β-glucanases cloned so far belong to two distinct families, plant and microbial enzymes, with neither sequence similarity nor related three dimensional structures. Plant 1,3-1,4-β-glucanases are essentially expressed during germination and facilitate the action of amylolytic enzymes on starch by breaking down the 1,3-1,4-β-glucan rich amylaceous cell walls.[1] The function of bacterial 1,3-1,4-β-glucanases is less well defined. The fact they usually are extracellular enzymes seems to indicate an involvement in the degradation of polysaccharides that can be used as an energy source by these microorganisms. Genes encoding bacterial 1,3-1,4-β-glucanases have been cloned and sequenced from different *Bacillus* species, *Fibrobacter succinogenes*, *Ruminococcus flavofaciens*, and *Clostridium thermocellum*.[4] Together with bacterial 1,3-β-glucanases ("laminarinases"), all bacterial 1,3-1,4-β-glucanases share a high degree of sequence similarities, and have been classified as members of family 16 glycosyl hydrolases.[5,6] They are monodomain proteins, with molecular masses in the range of 25-30 kDa, are active in a wide pH range, have a basic pI (8-9), and are quite thermostable compared to the plant isozymes.

The three dimensional structures of two wild type 1,3-1,4-β-glucanases from *Bacillus* (*B. licheniformis*[7] at 0.18 nm resolution, and *B. macerans*[8] at 0.16 nm resolution), as well as several hybrids of *B. amyloliquefaciens* and *B. macerans*[9], have been solved by X-ray crystallography. They show nearly identical jellyroll β-sandwich folds with the carbohydrate-binding cleft located on the concave face of a β-sheet formed of consecutive antiparallel β-strands.

The *B. licheniformis* enzyme has been kinetically characterised using a new family of chromogenic low molecular weight oligosaccharides[10,11] of general structure $[\beta Glcp(1\rightarrow4)]_n\beta Glcp(1\rightarrow3)\beta Glcp$-aryl (n=1 to 4, aryl= 4-methylumbelliferyl, series of substituted phenyl aglycons). These substrates have the basic core substructure -4G3G-X according to the requirements of natural polysaccharide substrates (barley β-glucan and lichenan) to be hydrolysed by the enzyme on a single glycosidic bond in a 3-*O*-substituted glucopyranose unit with release of the chromophoric aglycon. Kinetic studies have shown that the *B. licheniformis* enzyme is a retaining glycosidase[3] which has four glucopyranose-binding subsites at the non-reducing end from the scissile glycosidic bond, with subsite -III having the larger contribution to transition-state stabilisation.[12,13] Molecular modelling also suggests two or three subsites at the reducing end of the cleft.[14]

In addition to its natural hydrolase activity, we have also characterised other enzymatic activities of *Bacillus* 1,3-1,4-β-glucanases. The enzyme is able to hydrate glycals with the proper 3-*O*-substitution to 2-deoxy oligosaccharides following a stereospecific 1,2-trans hydration. It also catalyses the efficient transglycosylation reaction between a glycosyl fluoride donor and an oligosaccharide acceptor, a regiospecific reaction useful to access to gluco-oligosaccharides with alternate β-1,3 and β-1,4 linkages.[15]

The use of chromogenic substrates for kinetic studies, in combination with site-directed mutagenesis and x-ray crystallography provide the tools for structure-function analyses of the enzyme action. Because of the almost identical three dimensional structures of *Bacillus* 1,3-1,4-β-glucanases, the biochemical and mechanistic results obtained for the *B. licheniformis* enzyme may be assumed to hold for all family 16 *Bacillus* isozymes.

The present paper focuses on mechanistic aspects in *B. licheniformis* 1,3-1,4-β-glucanase, and preliminary results on the mutational analysis of enzyme-carbohydrate interactions defining specificity in the *B. macerans* enzyme.

2 PROBING THE CATALYTIC MECHANISM

2.1 Mechanism

The mechanism of a retaining glycosidase involves a double displacement reaction assisted by general acid/base catalysis (Figure 2). In the first step (*glycosylation*) the amino acid residue acting as a general acid protonates the glycosidic oxygen with concomitant C-O breaking of the scissile glycosidic bond, while the deprotonated carboxylate functioning as a nucleophile attacks the anomeric center to give a covalent glycosyl-enzyme intermediate. The second *deglycosylation* step involves the attack of a water molecule assisted by the conjugate base of the general acid to render the free sugar with overall retention of configuration. Both steps proceed *via* transition states with considerable oxocarbenium ion character.[16,17]

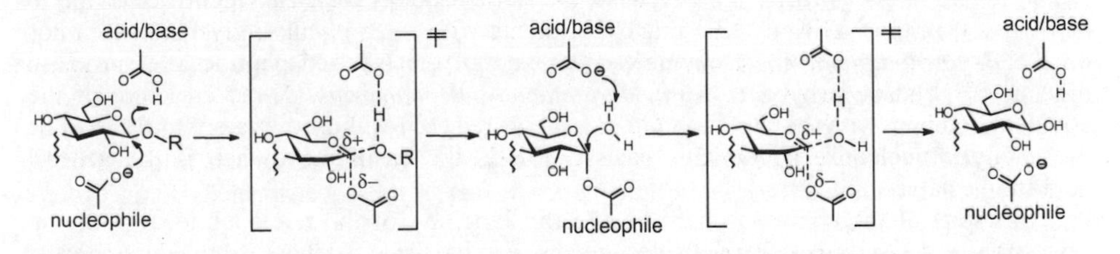

Figure 2 *Mechanism of retaining glycosidases involving a glycosyl-enzyme intermediate*

A Hammett analysis on the wild type *B. licheniformis* enzyme has been performed using a set of substituted aryl glycosides with different phenol-leaving group ability as measured by the pK_a of the free phenol released upon enzymatic hydrolysis. The steady-state kinetic parameters show no simple dependence of reactivity on aglycon acidity. The Hammett plot (log k_{cat} *vs.* pK_a of the aglycon) in Figure 3 is biphasic with an upward curvature at low pK_a values.[18] While the mechanistic interpretation of this unusual behaviour, not seen before for glycosidases, remains obscure (suggesting a change in transition-state structure depending on the aglycon), the overall shape indicates that the *glycosylation* step is at least partially rate determining even for highly activated substrates such as the 2,4-dinitrophenyl glycoside; for *deglycosylation* to become rate limiting, k_{cat} would level off for activated substrates (low pK_a of the phenol).

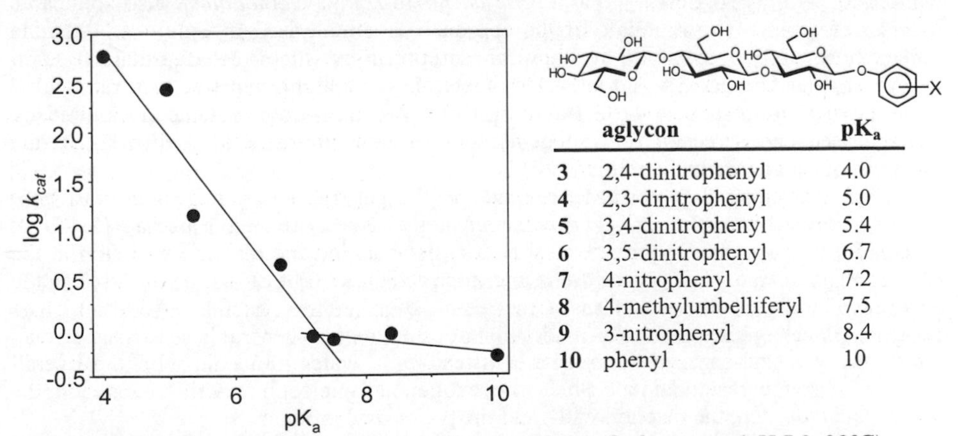

	aglycon	pK_a
3	2,4-dinitrophenyl	4.0
4	2,3-dinitrophenyl	5.0
5	3,4-dinitrophenyl	5.4
6	3,5-dinitrophenyl	6.7
7	4-nitrophenyl	7.2
8	4-methylumbelliferyl	7.5
9	3-nitrophenyl	8.4
10	phenyl	10

Figure 3 *Hammett plot for the* B. licheniformis *1,3-1,4-β-glucanase (pH 7.2, 30°C)*

The pH dependence of the kinetic parameters for the hydrolysis of the chromogenic substrates, and the pH-dependent enzyme inactivation by a water-soluble carbodiimide (EAC) are consistent with a general acid/base mechanism and suggest two essential catalytic groups with pK_a values of 5.5 and 7.0 in the free enzyme. The latter value is shifted up to 1.5 pH units upon binding of substrate in the non-covalent enzyme-substrate Michaelis complex.[13]

2.2 Identification of catalytic residues

Initially, in the absence of structural information, we approached the search of the catalytic residues by a scanning mutagenesis where all the Asp and Glu residues that are conserved among the *Bacillus* 1,3-1,4-β-glucanases were replaced by the isosteric Asn and Gln residues. It was concluded that E138 and E134 are the essential catalytic residues in the *B. licheniformis* enzyme.[19,20] Likewise, the same residues were also identified in the *B. macerans* isozyme.[8] Affinity labelling experiments with an epoxyalkyl glycoside inhibitor on the *B. amyloliquefaciens* isozyme (sequencing of the labelled peptide after protease digestion)[21], and on a hybrid *B. amyloliquefaciens-B. macerans* (X-ray structure of the covalent complex with the same inhibitor)[22] suggested the residue equivalent to E134 to be the catalytic nucleophile. On this basis, E138 is the candidate to act as the general acid/base catalyst.

In spite of the common claim of specific labelling of the nucleophile in retaining glycosidases by epoxyalkyl glycosides, recent results on a xylanase from *Trichoderma reesei* have shown that either the nucleophile or the general acid residues are chemically derivatised by this type of affinity labels depending on the length of the alkyl chain bearing the epoxide function.[23] Therefore the above assignment is not final.

Summarising, the mutational analysis identified E134 and E138 as the essential catalytic residues. They are at 5.3Å apart to each other in the x-ray structure of the free enzyme[7] and suitably disposed for catalysis. Although E134 is the most likely candidate to be the catalytic nucleophile, there is no direct functional proof to assign E138 as the general acid/base residue. In principle, the basic question is: these two glutamates are the catalytic residues, but which is the general acid/base and which is the nucleophile?

2.3 Chemical rescue of E138A and E134A by exogenous nucleophiles

Here we extend the *kinetic analysis approach* first proposed by Withers and co-workers on *exo*-glycanases[24,25] to a general *chemical rescue methodology* also applicable to *endo*-enzymes. The rationale of the approach is summarised in Figure 4. Alanine replacement of both catalytic residues are prepared by site-directed mutagenesis to produce the inactive E138A and E134A mutants. Using a highly reactive substrate with a good leaving group such as the 2,4-dinitrophenyl glycoside, both mutants are kinetically characterised and screened for the generation of new products of hydrolysis in the presence of exogenous nucleophiles.

For a retaining β-glycosidase, removal of the general acid/base residue will have little or no effect on the first *glycosylation* step for an activated substrate (2,4-DNP glycoside) that does not require general acid assistance, but the reaction will stop at the *deglycosylation* step. The glycosyl-enzyme intermediate will be slowly hydrolysed in the absence of the general base to assist the attack of a water molecule. Addition of an exogenous nucleophile (such as an azide) that does not require general base assistance may attack the glycosyl-enzyme intermediate instead of a water molecule with the overall effect of enzyme reactivation. Since the exogenous nucleophile will operate on the *deglycosylation* step, the reaction will yield the β-glycosyl product.

On the other hand, mutation of the nucleophile should render an inactive mutant where the first *glycosylation* step has been drastically slowed down because of the absence of the nucleophilic residue to form the glycosyl-enzyme intermediate. In this case, a cavity has been created in the active site on the α-face (Glu to Ala mutation) which can accommodate a small exogenous nucleophile such as an azide. The enzyme might be reactivated through a different mechanism, following a single inverting displacement to give the α-glycosyl product. Therefore, if reactivation occurs, the stereochemistry of the new glycosyl azide product formed upon chemical rescue of both alanine mutants will indicate which residue acts as the general acid/base and which operate as the nucleophile in the catalytic mechanism of the wild type enzyme.

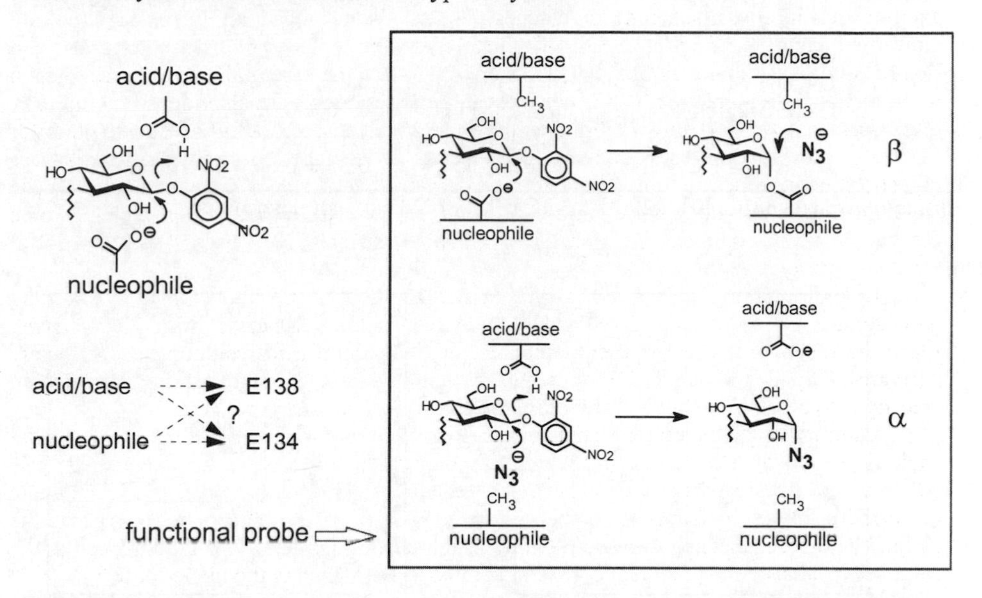

Figure 4 *Chemical rescue of mutants to assign the function of the catalytic residues*

Table 1 compiles the kinetic parameters of the wt enzyme and both alanine mutants at the proposed essential catalytic residues (E138A and E134A) using two aryl glycoside substrates: the 4-methylumbelliferyl trisaccharide G4G3G-MU with a poor leaving group (pK_a [4-methylumbelliferone] of 7.5), and the 2,4-dinitrophenyl trisaccharide G4G3G-2,4DNP with a good leaving group (pK_a [2,4-dinitrophenol] of 4.0). Both mutants are very inactive as compared to the wild type enzyme. k_{cat}/K_M for these two substrates are vastly different, that for the activated substrate being much higher than that for the other.

Azide had no activation effect on the wild type enzyme, rather it behaves as a competitive inhibitor with a K_I of 0.48 ± 0.15 M (G4G3G-2,4DNP substrate).

Addition of azide reactivates both mutants in a concentration dependent manner using the activated 2,4-dinitrophenyl glycoside substrate. For E138A (Figure 5a) k_{cat} increases with azide concentration following a saturation curve, but surprisingly, K_M decreases down to the micromolar range. At 1 M azide, k_{cat}/K_M increases 4000-fold and reaches the wild type k_{cat}/K_M value (Table 1). On the other hand, E134A is also rescued by azide but no saturation on k_{cat} is observed up to 3.3 M azide (Figure 5b). Now K_M increases 7-fold compared to the mutant with no added azide. In terms of k_{cat}/K_M, the reactivation is less efficient than for the E138A mutant (only 10-fold) still being 10^4-fold lower than the wild type value.

Table 1
Kinetic parameters for wild type and mutant B. licheniformis *1,3-1,4-β-glucanases*

enzyme	substrate	k_{cat} (s^{-1})	K_M (mM)	k_{cat}/K_M ($M^{-1}s^{-1}$)
wt	G4G3G-MU	0.67 ± 0.02	0.79 ±0.04	850 ±70
	G4G3G-2,4DNP	603 ± 8	0.20 ±0.01	(3.0 ±0.2)·10^6
E138A	G4G3G-MU	(6.7 ± 0.4)·10^{-4}	0.34 ±0.07	2.0 ± 0.5
	G4G3G-2,4DNP	0.256 ± 0.005	0.22 ± 0.02	(1.2 ± 0.1)·10^3
	" + 1M NaN_3	15.4 ± 0.4	0.003 ± 7·10^{-4}	(4.0 ± 0.8)·10^6
	" + 4 M HCOONa	no reactivation		
E134A	G4G3G-MU	(4 ± 2)·10^{-6}	n.d.	--
	G4G3G-2,4DNP	(7.1 ± 0.1)·10^{-4}	0.16 ± 0.02	4.0 ± 0.5
	" + 3.3 M NaN_3	(5.6 ± 0.2)·10^{-2}	1.1 ± 0.7	(5 ± 1)·10^1
	" + 4 M HCOONa	2.10 ± 0.15	0.35 ± 0.04	(6 ± 1)·10^3

citrate/phosphate buffer pH 7.2, 0.1 mM $CaCl_2$, 30°C. n.d., not determined.

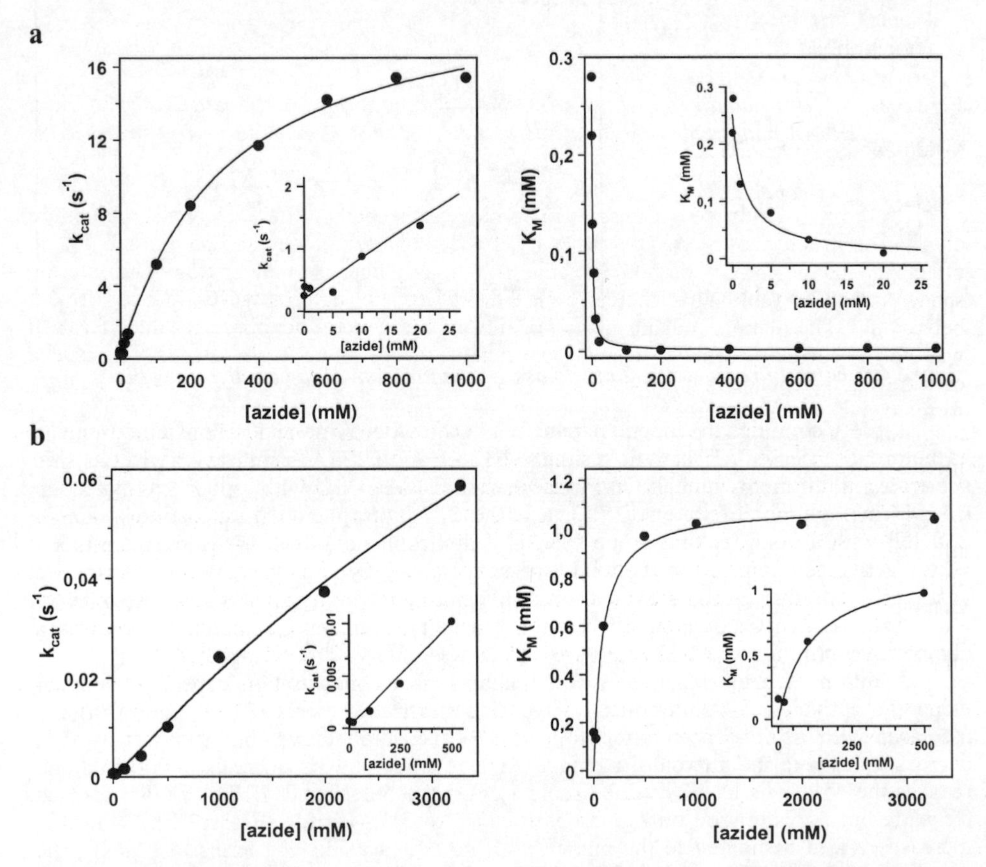

Figure 5 *Kinetics of mutant reactivation by added sodium azide: a) E138A, b) E134A. Conditions: citrate/phosphate buffer pH 7.2, 0.1 mM $CaCl_2$, 30°C, substrate: G4G3G-2,4DNP*

The enzymatic reactions were monitored by ^{1}H-NMR spectroscopy for structure determination of the final products from azide reactivation. As shown in Figure 6, the E138A mutant yields the β-glycosyl azide (δ (H1) 4.76 ppm, *J* 8.7 Hz), whereas E134A gives the α-glycosyl azide (δ (H1) 5.52 ppm, *J* 3.6 Hz). Therefore, based on the rationale of the chemical rescue methodology, the results provide a functional evidence of E138 being the general acid/base catalyst, and confirm the proposal of E134 as the catalytic nucleophile.

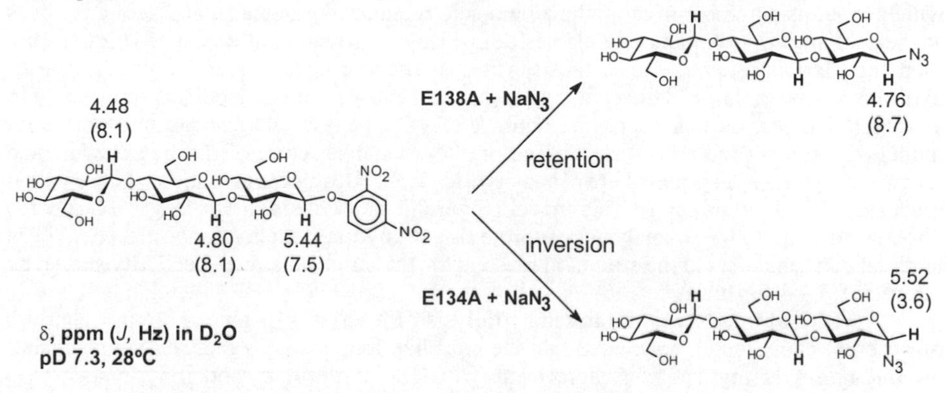

Figure 6 *^{1}H-NMR monitoring of the chemical rescue reactions of the inactive E138A and E134A mutants by sodium azide. Chemical shifts for the anomeric protons are given.*

Carboxylic acids were also assayed as exogenous nucleophiles. They resemble more closely the structure of the removed glutamate side chains. Acetate and propionate had no effect on enzyme activity on both mutants, presumably because they could not fit into the space created by removal of the carboxyl side chain. Formate, however, had a different behaviour. The alanine mutant in the position of the general acid/base residue (E138A) was not rescued by formate, whereas the mutant at the nucleophile (E134A) showed a concentration-dependent reactivation. k_{cat} did not reach saturation up to 4 M sodium formate, but the magnitude of the activity was higher than that obtained with azide as exogenous nucleophile (Table 1). K_M, on the other hand, only changed slightly, with an increase of 2-fold relative to the nucleophile-free E134A mutant.

The final product of formate reactivation was the same hydrolysis product of the normal wild type reaction, the trisaccharide **1**. Interestingly, a transient compound with a life-time of approximately 2 h was detected by ^{1}H-NMR monitoring of the E134A + formate reaction with the 2,4-dinitrophenyl glycoside substrate in D_2O. It is tentatively assigned to the transient α-glycosyl formate intermediate **11** that is slowly hydrolysed. This is the first case where a non-modified sugar gives a long-lived covalent intermediate that mimics the proposed glycosyl-enzyme intermediate of retaining glycosidases.

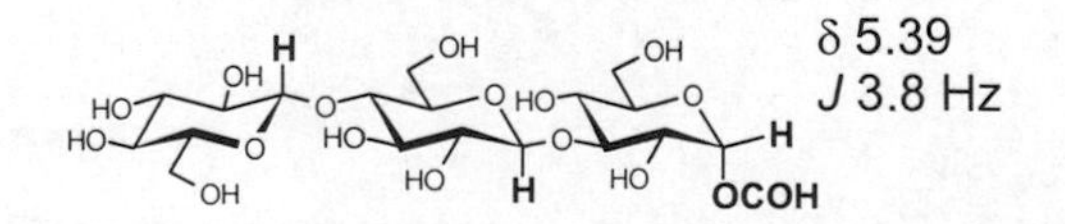

11

3 DISSECTING PROTEIN-CARBOHYDRATE INTERACTIONS

The three-dimensional structures of a number of *Bacillus* 1,3-1,4-β-glucanases (wild type enzymes from *B. licheniformis*[7] and *B. macerans*[8], several hybrids and circularly permuted proteins[9]) have been solved by X-ray crystallography. The crystal structure of an enzyme-inhibitor complex is not yet available since no good inhibitors have been reported.[26]

A deep channel runs across the concave side of the 1,3-1,4-β-glucanase jellyroll which contains a number of aromatic amino acid residues on its walls, and acidic residues at the bottom. This binding site cleft is defined by two structural elements (Figure 7): a loop that partially covers the distant subsites at the non-reducing end from the site of hydrolysis, and a large β-sheet at the concave face of the molecule composed of 6 antiparallel β-strands that shape the entire cleft. A hexasaccharide substrate has been modelled into the binding site cleft based on the crystal structure of the free enzyme and that of a covalent complex between a hybrid 1,3-1,4-β-glucanase and an epoxyalkyl glycoside suicide inhibitor.[22] This model (Figure 7, and details in Figures 9, 11 and 12) directed the mutational analysis of protein-carbohydrate interactions that has been undertaken to understand the structural reasons for the strict substrate specificity shown by *Bacillus* 1,3-1,4-β-glucanases.

Kinetics of the enzyme-catalysed hydrolysis of a series of β-glucan oligosaccharides of different chain length concluded that the cleft has four glucopyranose-binding subsites on the non-reducing end[12] in agreement with the molecular modelling, which also proposes two subsites at the reducing end from the scissile glycosidic bond.

The mutational analysis has been conducted on two *Bacillus* enzymes: loop residues on the *B. licheniformis* protein, and binding-site residues proposed by modelling on the *B. macerans* enzyme.

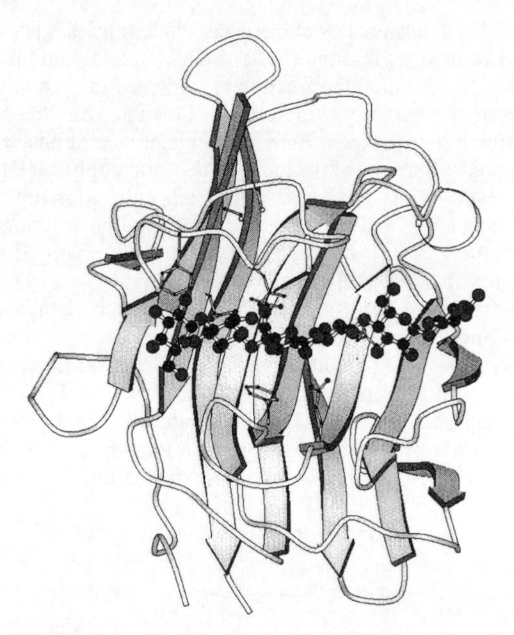

Figure 7 *Model of the E·S complex based on the X-ray structure of the free enzyme*

3.1 Mutational analysis of loop residues (*B. licheniformis*)

The carbohydrate-binding cleft is partially covered by a surface loop between residues 51 to 67, which is linked to a β-strand-(87-95) of the minor β-sheet III of the protein core by a single disulfide bond at Cys61-Cys90. An alanine scanning mutagenesis approach has been applied to analyse the role of loop residues from D51 to R64 in substrate binding and protein stability by means of equilibrium urea denaturation, enzyme thermotolerance, and kinetics.

Table 2 summarises the data for the 13 single alanine mutants.[27] The disulfide bond has a contribution of 5.3 ± 0.2 $kcal \cdot mol^{-1}$ to protein stability as the average of $\Delta\Delta G_U$ values (urea denaturation) between oxidised (S-S) and reduced (SH) enzyme forms for all wild type and mutant proteins. Most of the mutations have an effect on activity. k_{cat}/K_M values are plotted against stability data (ΔG_U(oxidised form)) in Figure 8. Inspection of this plot suggests that mutants can be classified in different groups. One group is formed by those enzymes showing a good correlation between catalytic efficiency and enzyme stability. The decrease in catalytic efficiency is mainly due to k_{cat} since K_M values are < 2-fold larger than the wild type K_M value. The mutated residues in this group have no specific role in substrate binding as proposed from the structure of the modelled enzyme-substrate complex. Therefore, the reduction in k_{cat}/K_M is interpreted as the result of local rearrangements in the protein structure induced by the mutations, which also have a proportional effect on protein stability. Second, the single mutant M58A is surprisingly more active, with a k_{cat} value 7-fold higher than that of the wild type enzyme. Third, mutants Y53A, N55A, F59A and W63A have a pronounced deleterious effect on enzyme activity, with K_M > 2-fold and k_{cat} < 5% of the wild type kinetic parameters. These mutated residues are directly involved in substrate binding or in hydrophobic packing of the loop.

Table 2
Kinetic parameters and stability data for wild type and loop mutants of B. licheniformis *1,3-1,4-β-glucanases*

enzyme	enzyme activity			equilibrium urea denaturation		
	K_M (mM)	k_{cat} (s^{-1})	k_{cat}/K_M ($mM^{-1}s^{-1}$)	ΔG_U (ox) ($kcal \cdot mol^{-1}$)	ΔG_U (red) ($kcal \cdot mol^{-1}$)	$\Delta\Delta G^{S-S}$ ($kcal \cdot mol^{-1}$)
wt	1.8	4.0	2.2	10.50	4.98	5.6
D51A	1.8	2.0	1.1	8.30	3.42	4.9
G52A	1.7	2.5	1.5	8.79	3.63	5.2
Y53A	5.7	0.2	0.04	8.76	3.32	5.4
S54A	2.8	2.5	0.9	9.19	4.00	5.2
N55A	4.1	0.14	0.034	9.48	4.30	5.2
G56A	2.2	2.7	1.2	8.94	3.63	5.3
N57A	2.0	2.8	1.4	10.10	4.78	5.3
M58A	3.7	27.8	7.5	9.59	3.89	5.7
F59A	9.5	0.083	0.009	9.44	3.94	5.5
N60A	1.8	3.1	1.7	9.28	3.90	5.4
T62A	1.9	2.8	1.5	8.78	3.35	5.4
W63A	4.7	0.21	0.05	8.34	3.00	5.3
R64A	2.0	2.8	1.4	8.88	3.52	5.4

kinetics: citrate/phosphate buffer pH 7.2, 0.1 mM $CaCl_2$, 45°C, substrate: G4G3G-MU
denaturation: citrate/phosphate buffer pH 7.2, 0-8 M urea, 37°C.
ox: oxidised disulfide form; red: reduced disulfide form; $\Delta\Delta G^{S-S} = \Delta G_U (ox) - \Delta G_U (red)$

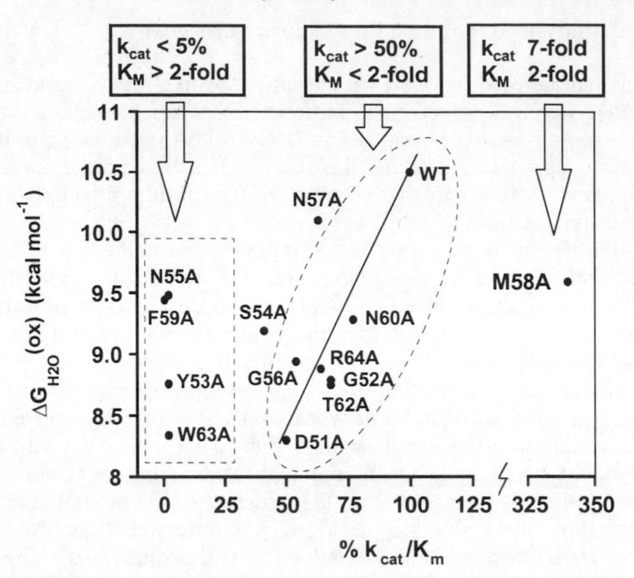

Figure 8 *Catalytic efficiency (k_{cat}/K_M) vs. enzyme stability (ΔG_U) plot for wt and mutant* B. licheniformis *1,3-1,4-β-glucanases*

Tyr53 forms a hydrogen bond with the 3-OH of the glucopyranose unit of the substrate in subsite -III, whereas the amide nitrogen of Asn55 forms a hydrogen bond with the 6-OH of the glucopyranose unit in subsite -II according to the modelled enzyme-substrate complex (see below, Figure 9). On the other hand, Phe59 has the aromatic side chain pointing toward the core of the protein, and it has a strong hydrophobic (stacking) interaction with Trp213, which belongs to a β-strand of the major β-sheet on the concave face of the molecule. This interaction might be important to position the loop and to create a hydrophobic environment in this portion of the cleft when the substrate binds. Trp63 also has the aromatic side chain interacting with the main core and very close to Phe59, contributing to the structural integrity of the loop. Residues at positions 53 and 55 have been analysed in further detail introducing other mutations and analysing them with a set of substrates of different chain length (data not shown).

3.2 Mutational analysis of the binding-site cleft (*B. macerans*)

A complete mutational analysis of all residues that may interact with the substrate is currently under investigation. The map in Figure 9 corresponds to the putative protein-carbohydrate interactions proposed by molecular modelling. In subsite -IV only a possible stacking interaction with Tyr22 is observed. Subsites -III and -II may have extensive hydrogen bonding interactions involving a number of residues: Tyr22, Asn24, Glu61, Arg63 in subsite -III, and Phe90, Tyr92, Asn180, Trp182 in subsite -II. For subsite -I the model might not be appropriate since ring distortion is expected for the bound glucopyranose unit bearing the glycosidic bond to be hydrolysed (as proposed for other glycosidases[28]). The current model places a $^4C_1$ chair conformation in this subsite for which the glycosidic bond to Glc*p*+I is strained into a conformation of about 3 kcal·mol^{-1} above the global energy minimum for a relaxed β-1,4 linkage (from conformational analysis calculations[9,29]). The observed protein-carbohydrate contacts in this model involve Ser88, Glu109 and Tyr121. Therefore results on this subsite are only preliminary awaiting a refined structural model.

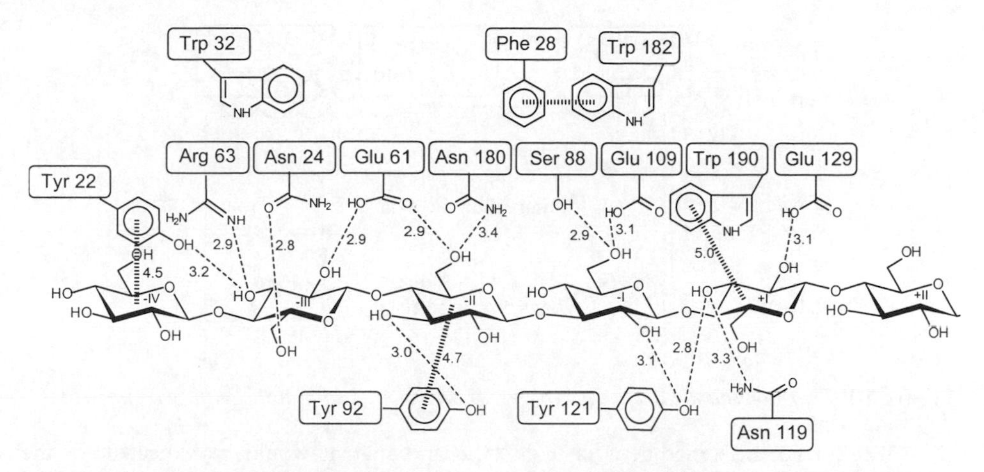

Figure 9 *Protein-carbohydrate interactions in the modelled E·S complex (*B. macerans *numbering; for* B. lich., *add +31). H-bonds and staking interactions (d in Å) are indicated*

Protein-carbohydrate interactions are probed by site-directed mutagenesis. Different mutants are designed to evaluate hydrogen bonding and hydrophobic (stacking) interactions. For instance, mutation of a tyrosine to phenylalanine will remove the hydrogen bond interaction with an OH group of the substrate, and a further mutation to alanine will remove the stacking interaction with the glucopyranose ring of the substrate.

3.2.1 Subsite mapping model: thermodynamic cycles from kinetic analysis. The contribution of individual protein-carbohydrate interactions to transition-state stabilisation is evaluated from the kinetic parameters of wild type and mutant enzymes with two homologous substrates. The model, previously developed and tested for subsite -III mutants of the *B. licheniformis* enzyme (data not shown), is illustrated in Figure 10.

- The contribution of subsite n+1 (-III in Figure 10) to transition-state stabilisation is given by the difference in substrate transition-state activation energy between two substrates, (n) and (n+1), differing in one glucopyranose unit according to:[12]

$$\Delta\Delta G^{\ddagger}_{\text{subsite n+1}} = - RT \ln [(k_{cat}/K_M)_{n+1}/(k_{cat}/K_M)_n] \quad (1)$$

Likewise, when comparing the wild type and the mutant with the same substrate, the overall effect of the mutation is given by:

$$\Delta\Delta G^{\ddagger}_{\text{mut-wt}} = - RT \ln [(k_{cat}/K_M)_{mut}/(k_{cat}/K_M)_{wt}] \quad (2)$$

- To evaluate the contribution of a residue that may interact with the substrate in subsite n+1 (-III in Figure 10), both the wild type and the mutant at this residue are kinetically characterised with two chromogenic oligosaccharides of different chain length: a trisaccharide (substrate n+1) that will occupy subsite -III, and a shorter disaccharide (substrate n) which leaves the mutated subsite empty in the productive complex. The transition-state stabilisation energies $\Delta\Delta G^{\ddagger}$ are calculated from k_{cat}/K_M values for each reaction (eq.1,2) to build the thermodynamic cycle that relates them.

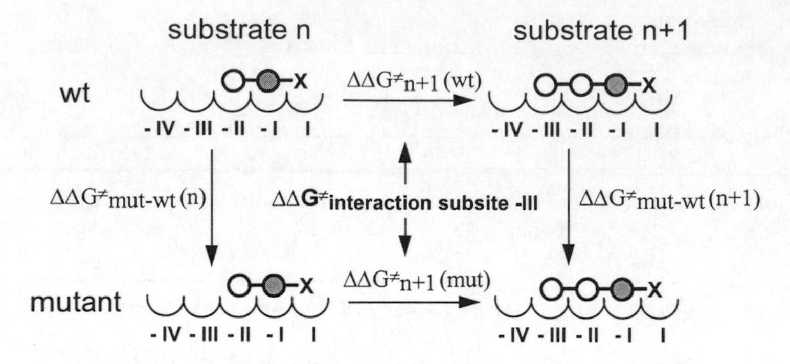

Figure 10 *Thermodynamic cycle that relates $\Delta\Delta G^{\ddagger}$ of wt and mutant with two substrates*

Ideally, no difference between wild type and mutant would be expected for the disaccharide (substrate n) since the subsite where the mutation has been introduced remains empty. So, $\Delta\Delta G^{\ddagger}_{mut\text{-}wt}(n)$ for the disaccharide is the "background" effect that accounts for small solvent and structural rearrangements introduced by the mutation. Then, the $\Delta\Delta G^{\ddagger}$ assigned to the individual interaction in subsite n+1 (-III in this example) will be the difference of $\Delta\Delta G^{\ddagger}_{mut\text{-}wt}$ between trisaccharide and disaccharide, or what is the same, the difference between the contribution of subsite -III to transition-state stabilisation in the mutant and the wild type:

$$\Delta\Delta G^{\ddagger}_{interaction} = \Delta\Delta G^{\ddagger}_{n+1}(mut) - \Delta\Delta G^{\ddagger}_{n+1}(wt) \quad (3)$$

3.2.2 Subsites -IV and -III. Figure 11 illustrates the protein-carbohydrate interactions in the modelled E·S complex at subsites -IV and -III. The mutants at residues that may interact with the substrate are designed to probe their function (hydrogen bonding and/or hydrophobic (stacking) interactions), and kinetically analysed with the set of chromogenic substrates $[G4]_nG3G$-MU (n = 0 to 3, MU = 4-methylumbelliferyl). From the steady-state k_{cat}/K_M values for each substrates, $\Delta G^{\ddagger}$ values are calculated according to the subsite mapping model presented above. The kinetic parameters of a given mutant and wild type enzymes with trisaccharide and disaccharide substrates will give the energetic contribution to transition state stabilisation of the mutated residue in subsite -III. Likewise, the contribution in subsite -IV will be obtained from tetrasaccharide and trisaccharide kinetics, whereas pentasaccharide and tetrasaccharide kinetics will give information of an eventual new subsite -V, only when analysing residues at the edge of the binding site cleft at the non-reducing end. Table 3 summarises the results expressed as $\Delta G^{\ddagger}$ values for each subsite. The interaction between the side chain of the mutated residue and the carbohydrate substrate (as proposed by molecular modelling) is also indicated, as well as the assigned energetic contribution of the interaction from the subsite mapping calculations.

Tyr22 interacts through a hydrogen bond with the 3-OH of glucopyranose in subsite -III, and has a stacking interaction with the glucopyranose ring in subsite -IV. Removal of the phenolic hydroxyl group by mutation to phenylalanine has the larger effect on subsite -III, +2.0 kcal·mol^{-1}, and a negligible effect on subsite -IV. The +2.0 kcal·mol^{-1} are then assigned to the H-bond Tyr22-OH···3-OH(Glc*p*-III). A further mutation to alanine will remove the stacking interaction with the glucopyranose unit in subsite -IV. Comparing the phenylalanine and alanine mutants, a +0.61 kcal·mol^{-1} destabilisation is obtained for subsite -IV, value assigned to the lost stacking interaction in the Ala mutant. However, a stabilisation of 0.5 kcal·mol^{-1} is observed in subsite -III, indicating that the aromatic side chain had unfavorable contacts with the substrate or that the small structural rearrangement induced by the Phe to Ala mutation allows better binding to subsite -III.

Table 3
Effects on transition-state stabilisation energy in binding mutants of B. macerans *1,3-1,4-β-glucanase*

	Mutation	$\Delta\Delta G^{\ddagger}_{SUBSITE}$ (kcal·mol^{-1}) -III	-IV	(-V)	Interaction	$\Delta G^{\ddagger}$ interaction (kcal·mol^{-1})
Tyr22	Y22F	+2.00	- 0.02	+0.00	H-bond Tyr-OH···O3(Glc*p*-III)	2.00
	Y22A					
	F22A	- 0.50	+0.61	+0.04	stacking Tyr///Glc*p*-IV	0.61
	Y22W	+0.52	- 0.97	- 0.40	improved stacking Trp///Glc*p*-IV	
Asn24	N24A	+2.20	- 0.12	+0.03	H-bond Asn-NH_2···O6(Glc*p*-III)	2.20
	N24Q	+0.86	+0.04	- 0.39		
Glu61	E61A	+3.58	- 0.38	--	H-bond Glu··O2(Glc*p*-III) & O6(Glc*p*-II)	3.58
	E61Q	+2.92	- 0.06	--		
	E61D	--	+0.05	--		
Arg63	R63A	+1.62	+0.51	--	H-bond Arg-NH···O3(Glc*p*-III)	1.62
	R63K	+1.73	+0.09	--		

citrate/phosphate buffer pH 7.2, 0.1 mM $CaCl_2$, 50ºC
Kinetic parameters (k_{cat}, K_M, and k_{cat}/K_M) are determined for each mutant with substrates $[G4]_n$G3G-MU n = 0 to 3. $\Delta\Delta G^{\ddagger}$ values are calculated using equations (1) to (3).

Introduction of a larger aromatic ring in position 22 (Y22W mutation) has interesting effects: a) the indolic nitrogen seems to maintain partially the hydrogen bond interaction with the 3-OH (Glc*p*-III), since destabilisation at subsite -III is only +0.52 kcal·mol^{-1} as compared to the effect of a tyrosine to phenylalanine mutation (+2.0 kcal·mol^{-1}); b) the stacking interaction in subsite -IV is improved (-0.97 kcal·mol^{-1} relative to the wild type enzyme), and c) the higher catalytic efficiency (in terms of k_{cat}/K_M) of the tryptophan mutant with a pentasaccharide substrate results in a stabilising effect of 0.40 kcal·mol^{-1} for a virtual subsite -V, suggesting that the larger aromatic side chain of the Trp residue is able to interact with an additional glucopyranose unit, as if a new subsite -V had been created.

A similar analysis of mutants at positions 24, 61, and 63 allow to infer the following conclusions:
- Asn24: the side chain amide function of asparagine 24 interacts through a hydrogen bond with the 6-OH of Glc*p* in subsite -III, with a contribution of 2.2 kcal·mol^{-1} to transition state stabilisation.
- Glu61: the γ-carboxylate of Glu61 forms a bidentate hydrogen bond with the 2-OH group of Glc*p*-III and the 6-OH group of Glc*p*-II. The $\Delta\Delta G^{\ddagger}$ of 3.58 kcal·mol^{-1} corresponds to the contribution of the H-bond between the 2-OH of Glc*p*-III and the charged γ-carboxylate acceptor. Mutation to Gln does not change the size of the side chain but replaces a charged residue for an uncharged residue, with a destabilising effect of +2.92 kcal·mol^{-1}.

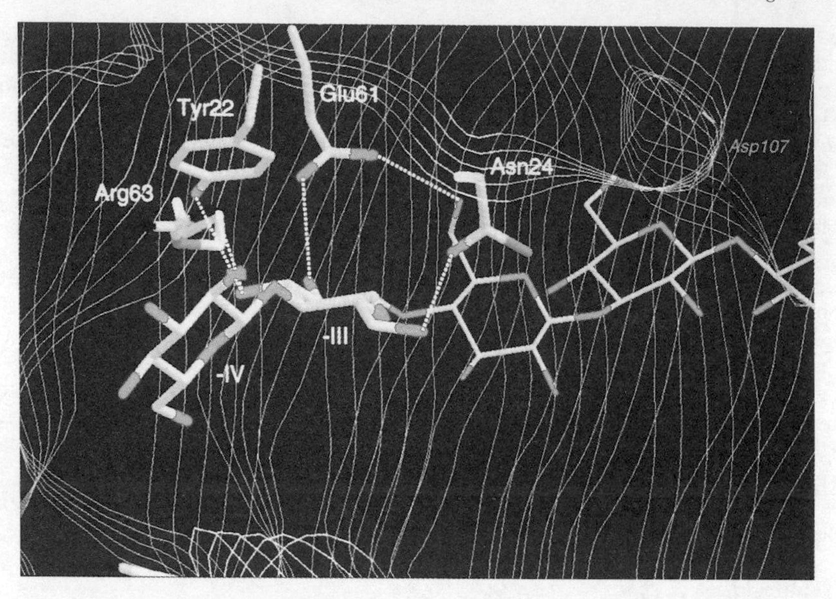

Figure 11 *View of the hydrogen bond interactions at subsites -IV and -III in the modelled enzyme-substrate complex. Amino acid numbering is for the* B. macerans *1,3-1,4-β-glucanase. (For* B. licheniformis *add + 31)*

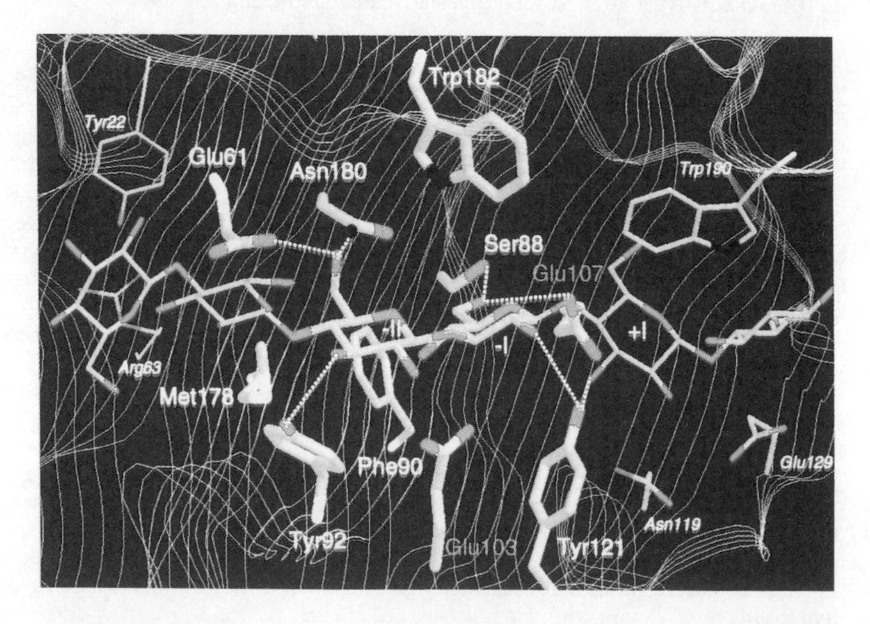

Figure 12 *View of the hydrogen bond interactions at subsites -II and -I in the modelled enzyme-substrate complex. Amino acid numbering is for the* B. macerans *1,3-1,4-β-glucanase. (For* B. licheniformis *add + 31)*

- Arg63: removal of the arginine side chain eliminates a hydrogen bond interaction with the 3-OH of Glc*p*-III (1.62 kcal·mol^{-1}), but apparently, the large cavity that has been created has also a destabilising effect on subsite -IV, which is restored when introducing a lysine side chain.

The novelty of this mutational approach is the ability to assign the effect of individual residues to individual subsites since the overall deleterious effect of the mutation is subtracted as background effect using a second substrate of shorter chain length that will not occupy the subsite where the mutation has been introduced.

3.2.3 Subsites -II and -I. The complete analysis based on the subsite mapping model used for subsites -IV and -III can not be applied to subsites -II and -I. For subsite -II it would require the determination of kinetic parameters with a disaccharide (G3G-MU) and a monosaccharide (G-MU), but the latter is not a good substrate for 1,3-1,4-β-glucanases since it is very slowly hydrolysed by the wild type enzyme. Subsite -I, on the other hand, will always be occupied by a hydrolysable substrate. Therefore, evaluation of the protein-carbohydrate interactions through kinetic analysis of mutants at residues proposed by molecular modelling is essentially qualitative.

Preliminary kinetic data are presented in Table 4, and Figure 12 illustrates the enzyme-substrate interactions in the modelled E·S complex at subsites -II and -I.

Table 4
Kinetic parameters for wild type and mutant B. macerans *1,3-1,4-β-glucanases*

mutant		substrate [a]	k_{cat} (s^{-1})	K_M (mM)	k_{cat}/K_M ($M^{-1}s^{-1}$)	interaction
wt		di-	$4.1 \cdot 10^{-2}$	15	2.68	
		tri-	1.63	1.6	$1.0 \cdot 10^{3}$	
Phe90	F90A	di-	$6.8 \cdot 10^{-5}$	13	$5.4 \cdot 10^{-3}$	positioning
		tri-	$2.2 \cdot 10^{-3}$	3.7	0.59	
Tyr92	Y92A		unstable			H bond Tyr-OH...O3(Glc*p*-II)
	Y92W	di-	$8.2 \cdot 10^{-3}$	15	0.53	and stacking Tyr///G*l*cp-II
		tri-	$2.3 \cdot 10^{-1}$	2.6	87	
Asn180	N180A	di-	---	---	$5.5 \cdot 10^{-4}$	H bond Asn-NH_2···O6(Glc*p*-II)
		tri-	$8.4 \cdot 10^{-4}$	6.6	0.13	
	N180Q	di-	---	---	$5.9 \cdot 10^{-4}$	
		tri-	$1.7 \cdot 10^{-3}$	7.1	0.21	
Trp182	W182Y	di-	$1.7 \cdot 10^{-4}$	42	$4.0 \cdot 10^{-3}$	no direct interaction
		tri-	$2.9 \cdot 10^{-2}$	10	2.8	
Ser88	S88A	di-	$7.5 \cdot 10^{-3}$	13	0.58	H bond Ser-OH···O6(Glc*p*-I)
		tri-	0.25	1.7	$1.5 \cdot 10^{2}$	
Tyr121	Y121F	di-	0.12	17	7.1	H bond Tyr-OH··O2(-II)&O4(-I)
		tri-	1.53	0.8	$1.9 \cdot 10^{3}$	no staking
	Y121A	di-	---	---	n.d.	
		tri	0.26	4.2	61	

[a] substrates: di-: G3G-MU; tri-: G4G3G-MU. n.d.: not determined
Citrate/phosphate buffer pH 7.2, 0.1 mM $CaCl_2$, 50°C

The 6-OH and 3-OH groups of the glucopyranose ring in subsite -II have tight hydrogen bonds with the side chains of Glu61, Asn180, and Tyr92. Phe90 is perpendicularly oriented to the plane of the Glc*p* ring and it may be important to position the sugar in the binding cleft and help to create an hydrophobic environment. Tyr92 also has a stacking interaction with the Glc*p*-II. These residues contribute importantly to stabilisation of the substrate transition-state as deduced from the kinetic parameters of the alanine mutants using both di- and trisaccharide substrates. k_{cat}/K_M values are decreased about 1000-fold or more relative to those of the wild type enzyme with both susbtrates.

In subsite -I the model is only tentative and preliminary since it does not take into account the expected ring distortion as argued before. Ser88 may have a hydrogen bond with the 6-OH group of Glc*p*-I, but the S88A mutant shows only a 6-fold reduction in k_{cat} and almost no change in K_M for di- and trisaccharide substrates. Tyr121 may also have a hydrogen bond with the 2-OH group of Glc*p*-I. Again, reduction in k_{cat}/K_M for the Y121A mutant is only 16-fold relative to the wild type value, but the phenylalanine for tyrosine has an odd effect, Y121F being more active than the wild type enzyme. At this stage no conclusions can be drawn from these results. Work is in progress to better define the interactions at subsites -I and -II, both refining the enzyme-substrate complex model, and performing more specific kinetic measurements.

3.2.4 Substrate specificity. After examining most of the enzyme-substrate interactions that stabilise the productive complex, some light can be given to the underlaying questions concerning the strict substrate specificity shown by *Bacillus* 1,3-1,4-β-glucanases: Why does the enzyme require a β-1,3 linkage between subsites -I and -II, and a β-1,4 linkage between subsites -II and -III? Which are the differences compared to cellulases and laminarinases?

Conformational analysis of β-glucan oligosaccharides containing β-1,3 and β-1,4 linkages[29] have shown that two glucopyranose units linked by a β-1,3 bond have the Glc*p* rings in the same orientation, *i.e.* the 6-CH_2-OH side chains pointing to the same face, whereas a β-1,4 linkage renders both Glc*p* in alternate conformations, the 6-CH_2-OH side chains at approximately 180°. Figure 13 illustrates the relative orientation that different oligosaccharides may have when binding to hypothetic subsites -I to -IV. Barley β-glucan and the pneumococal polysaccharide (a glucan with alternating β-1,3 and β-1,4 glycosidic bonds[30]) are substrates, whereas laminarin and cellulose derivatives are not. If the Glc*p* ring in subsite -I is oriented for proper catalysis (the side chain up in the scheme), the Glc*p* in subsite -II should have the same orientation (up), as corresponds to a β-1,3 linkage. For subsite -III, the required β-1,4 bond will position the Glc*p* ring in the alternate orientation (down), while subsite -IV can accommodate both. When comparing the relative ring orientations of laminarin and cellulose derivatives, which are not substrates, the major differences are in subsite -II and -III. The reported mutational analysis shows that the glucopyranose rings at subsites -III and -II are tightly bound with a number of hydrogen bond interactions with most of the hydroxyl groups of the sugar, thus positioning the rings in a specific orientation. By contrast, the single stacking interaction between a Tyr and the Gluc*p*-IV ring may allow binding of both orientations in subsite -IV.

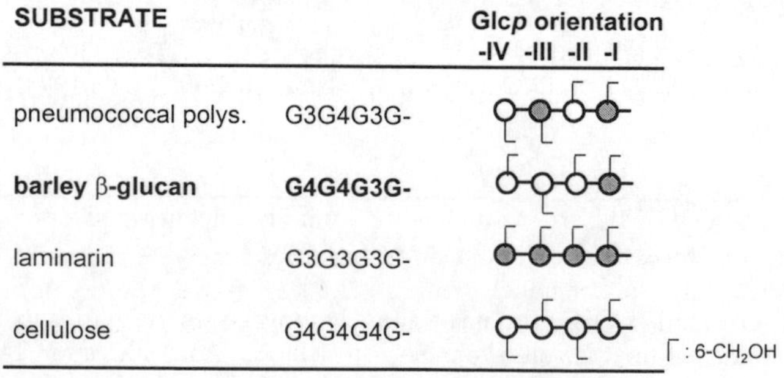

SUBSTRATE		**Glc*p* orientation** -IV -III -II -I
pneumococcal polys.	G3G4G3G-	
barley β-glucan	**G4G4G3G-**	
laminarin	G3G3G3G-	
cellulose	G4G4G4G-	

⌐ : 6-CH_2OH

Figure 13 *Relative orientation of glucopyranose rings in different polysaccharides*

4 ACKNOWLEDGMENTS

The author would like to thank the co-workers and collaborators (*) who made this work possible. *Probing the catalytic mechanism*: J.L. Viladot, O. Durany, E. de Ramón, M. Abel, J. Palasí and C. Pallarés in the Institut Químic de Sarrià, Barcelona. *Mutational analysis of loop residues in the* B. licheniformis *enzyme*: J. Pons and E. Querol* in the Universitat Autònoma de Barcelona, B. Fité and V. Serra in the Institut Químic de Sarrià. *Mutational analysis of the binding-site cleft in the* B. macerans *enzyme*: K. Piotukh and R. Borriss* from the Humboldt Universität, Berlin, and V. Serra and J.L. Viladot in the Institut Químic de Sarrià. This work was supported in part by Grants BIO94-C0912-C02-02 and BIO97-0511-C02-02 from the Comisión Interministerial de Ciencia y Tecnologia (CICYT), Spain.

5 REFERENCES

1. B.A. Stone and A.E. Clarke, 'Chemistry and Biology of (1→3)-β-glucans', La Trobe University Press, Australia, 1992.
2. F.W. Parrish, A.S. Perlin and E.T. Reese, *Can. J. Chem.*, 1960, **38**, 2094.
3. C. Malet, J. Jiménez-Barbero, M. Bernabé, C. Brosa, and A. Planas, *Biochem. J.*, 1993, **296**, 753.
4. R. Borriss, *Current Topics in Mol. Genet.*, 1994, **2**, 163.
5. B. Henrissat and A. Bairoch, *Biochem. J.*, 1993, **293**, 781.
6. G.J. Davis and H. Henrissat, *Structure*, 1995, **3**, 853.
7. M. Hahn, J. Pons, A. Planas, E. Querol and U. Heinemann, *FEBS Letters*, 1995, **374**, 221.
8. M. Hahn, O. Olsen, O. Politz, R. Borriss and U. Heinemann, *J. Biol. Chem.*, 1995, **7**, 3081.
9. U. Heinemann, J. Aÿ, O. Gaiser, J.J. Müller and M.N. Ponnuswamy, *Biol. Chem.*, 1996, **377**, 447.
10. C. Malet, J.L. Viladot, A. Ochoa, B. Gállego, C. Brosa and A. Planas, *Carbohydr. Res.*, 1995, **274**, 285.
11. C. Malet, J. Vallés, J. Bou and A. Planas, *J. Biotechnol.*, 1996, **48**, 209.
12. A. Planas and C. Malet, in 'Carbohydrate Bioengineering' (S.B. Petersen, B. Svensson, S. Pedersen, Eds.), Elsevier Science, Amsterdam, 1995, Chap. 8, p. 85.
13. C. Malet and A. Planas, *Biochemistry*, 1997, in press.
14. M. Hahn, T. Keitel and U. Heinemann, *Eur. J. Biochem.*, 1995, **232**, 849.
15. J.L. Viladot, V. Moreau, A. Planas and H. Driguez, *J. Chem. So., Perkin Trans. I*, 1997, 2383.
16. M.L. Sinnott, *Chem. Rev.*, 1990, **90**, 1171.
17. J.D. McCarter and S.G. Withers, *Curr. Opin. Struct. Biol.*, 1994, **4**, 885.
18. A. Planas, J.L.Viladot, J. Palasí, O. Millet, C. Pallarés and M. Abel, *Carbohydr. Res.*, submitted.
19. A. Planas, M. Juncosa, J. Lloberas and E. Querol, *FEBS Lett.*, 1992, **308**, 141.
20. M. Juncosa, J. Pons, T. Dot, E. Querol and A. Planas, *J. Biol. Chem.*, 1994, **269**, 14530.
21. P.B. Høj, R. Condron, J.C. Traeger, J.C. McAuliffe and B.A. Stone, *J. Biol. Chem.*, 1992, **267**, 25059.
22. T. Keitel, O. Simon, R. Borriss and U. Heinemann, *Proc. Natl. Acad. Sci. USA*, 1993, **90**, 5287.
23. R. Havukainen, A. Törrönen, T. Laitinen, T. and J. Rouvinen, *Biochemistry*, 1996, **35**, 9617.
24. A.M. MacLeod, T. Lindhorst, S.G. Withers and R.A.J. Warren, *Biochemistry*, 1994, **33**, 6371.
25. S.G. Withers, in 'Carbohydrate Bioengineering' (S.B. Petersen, B. Svensson, S. Pedersen, Eds.), Elsevier Science, Amsterdam, 1995, Chap. 9, p. 97.
26. V. Moreau, J.L. Viladot, E. Samain, A. Planas and H. Driguez, *Biorg. Med. Chem.*, 1996, **4**, 1849.

27. J. Pons, E. Querol and A. Planas, *J. Biol. Chem.*, 1997, **272**, 13006.
28. G. Sulzenbacher, H. Driguez, B. Henrissat, M. Schülein and G.J. Davis, *Biochemistry*, 1996, **35**, 15280.
29. M. Bernabé, J. Jiménez-Barbero and A. Planas, *J. Carbohyd. Chem.*, 1994, **13**, 799.
30. M.A. Anderson and B.A. Stone, *FEBS Lett.*, 1975, **52**, 202.

CELLULOSOME STRUCTURE: FOUR-PRONGED ATTACK USING BIOCHEMISTRY, MOLECULAR BIOLOGY, CRYSTALLOGRAPHY AND BIOINFORMATICS

Edward A. Bayer[1,*], Ely Morag[1], Raphael Lamed[2], Sima Yaron[3] and Yuval Shoham[3]

[1]Department of Biological Chemistry, The Weizmann Institute of Science, Rehovot; [2]Department of Molecular Microbiology and Biotechnology, Tel Aviv University, Ramat Aviv; [3]Department of Food Engineering and Biotechnology, Technion-IIT, Haifa, Israel

1 INTRODUCTION

Many cellulolytic microorganisms produce an intricate type of multi-enzyme complex — termed the *cellulosome*.[1] The organization of the cellulases into the cellulosome is believed to play a crucial role in their particularly high activity on insoluble forms of cellulose. Understanding the structural basis for the various cellulosomal components on the molecular level is thus important for understanding cellulosome assembly and function.

The cellulosome was first identified and described in the anaerobic, thermophilic, cellulolytic bacterium, *Clostridium thermocellum,* on the basis of combined biochemical, immunochemical, ultrastructural and genetic techniques.[2,3] The initial breakthrough involved the isolation of an adherence-defective bacterium, the preparation of mono-specific antibodies and the characterization of a cellulose-binding factor. This factor was shown to be a multi-protein complex and to contain various cellulolytic activities.

The complex is composed of a conglomerate of subunits, each of which comprises a set of interacting functional domains. Insight into the structural organization of the major non-enzymatic subunit and the various cellulosomal enzymatic subunits was accomplished first by their cloning and sequencing.[4-10] This approach has led to the identification of numerous functional domains, and supportive biochemical data have indicated their various functions.[11-16]

One of the major cellulosomal subunits, called *scaffoldin*, is responsible for organizing the cellulolytic subunits into the multi-enzyme complex. Scaffoldin thus plays a central role in cellulosome structure and function. This is accomplished by the interaction of two complementary classes of domains, located on the two separate interacting subunits, *i.e.*, a *cohesin* domain on scaffoldin and a *dockerin* domain on each of the enzymatic subunits. The cohesin-dockerin interaction defines the cellulosomal structure. The scaffoldin subunit of the *C. thermocellum* cellulosome also mediates the adhesion of the bacterium to cellulosic substrates, by virtue of a cellulose-binding domain (CBD). The adhesion process is a primary event which precedes the degradation process.

Due to the complexity of this system, insight into the structural organization of the cellulosome is dependent on a multidisciplinary approach. Within the past decade, the original contributions of biochemistry to this area have been advanced greatly, first by molecular biology and sequence analysis, and more recently through the benefits of three-dimensional structure determination. In this context, crystal structures have recently been described for two of the major non-hydrolytic domains of the scaffoldin subunit from the cellulosome of *C. thermocellum.* Finally, the wealth of sequencing data for the various domains provides a comparative modeling approach for predicting various sites of interaction among the cellulosomal domains.

The purpose of this chapter is not necessarily to review comprehensively the entire cellulosome area. Many such reviews have been published during the past decade, and the reader will benefit greatly by following the development and history of the field.[1,17-27]

Instead we focus on recent structural evidence and the type of information that comparative sequence analysis and homology modeling can provide.

This communication concentrates on our current views concerning the structures and functions of the non-enzymatic domains of the cellulosome. The information which has accumulated from such a combined approach promises to enrich our knowledge and understanding of cellulosome action on cellulosic substrates.

2 STRUCTURAL ORGANIZATION OF THE CELLULOSOME

The exact nature of cellulosome structure is still unknown. Nevertheless, it was even clear at an early stage that the cellulosome is a complex consisting of many different species of enzymes and a very special subunit which plays a multiple role in cellulosome structure and function. This subunit (historically termed the S1 subunit,[3] the S_L subunit,[28] CipA[9] and CelL[29]) has earned the descriptive term "scaffoldin".[24]

Even in the initial studies on cellulosome structure and function,[1-3,30] it was suggested that the various activities of scaffoldin include (1) the cellulose-binding function, (2) the organization of the component parts of the cellulosome into the complex, and (3) the anchoring of the complex into the cell surface. These predictions have withstood the test of time, and all three functions have indeed been demonstrated to be associated with specified domains of the scaffoldin subunit. Thus, the scaffoldin subunit includes a CBD for the binding of the cellulosome to the cellulosic substrate,[12] multiple copies of cohesin domains to incorporate the enzymatic subunits into the complex,[13,15] and a special type of dockerin domain for the attachment of the cellulosome to a complementary type of non-cellulosomal cohesin, which is positioned on the cell surface.[31]

A simplified view of the cellulosome architecture in *C. thermocellum* is shown schematically in Figure 1. The structure of the scaffoldin subunit and its multi-domain structure is based upon the results of recombinant DNA technology. The nature of the enzymatic subunits as shown in the figure is a much simplified model, particularly regarding their exact position and content.

A few words about cellulosome terminology: The generic terms "scaffoldin", "cohesin" and "dockerin" were instituted, in order to prevent imminent confusion in the area, due to the unbridled variety of abbreviations bestowed upon the different domains by different research groups. Thus, in addition to the simple description of a relevant band during SDS-PAGE (*i.e.*, S1, S_L, and J_1),[32] the scaffoldin subunit was dubbed by various groups as CbpA (cellulose-binding protein A in *C. cellulovorans*),[10] the scaffolding protein[33] or CipA (cellulase-integrating protein A in *C. thermocellum),*[9] and CipC in *C. cellulolyticum.*[34] The cohesin subunit has been referred to in the literature as SBD (subunit-binding domain),[21] HBD (hydrophobic repeated domain),[35] IRE (internally repeated elements),[9,36] R (repeat),[29,37] and RD (receptor domain).[14] The dockerin subunit has been called RS (reiterated sequence),[21] DS (alternatively duplicated segment[14] or duplicated ligand sequence),[37] DD (docking domain),[38] and CDR (conserved duplicated region[36] — not to be confused with the complementarity determining region of antibodies)! In addition, the term anchoring or anchorage has been used either to describe the binding of the cellulosome to the cell surface[30] or the integration of cellulosomal enzyme subunits into the complex.[29] The infatuation with abbreviations probably stems from the one successful and lasting abbreviation in the cellulase field: the CBD. In any event, the generic terms were introduced[24] simply to unify the other terms without assigning new and potentially misleading abbreviations.

The hydrolytic subunits of the cellulosome of *C. thermocellum* are relatively large, multi-domain enzymes which range in size from about 50 to 170 kDa.[1,3,39,40] All of the enzymatic subunits, of course, contain a catalytic domain. Nearly twenty enzymatic subunits, considered to be associated with the cellulosome, have thus far been sequenced and can be classified into many different hydrolytic families of glycosyl hydrolases.[26,41] Some cellulosomal enzymes (*e.g.*, CelJ)[38] are known to carry at least two different catalytic domains. At least one enzyme (CelF) is known to have a CBD.[42] Other domains of unknown function have also been detected (by sequence homology) in the enzymatic subunits.

All of the enzymatic subunits contain a single dockerin domain, which is considered to be the means by which the enzymes are incorporated into the cellulosome complex. For this purpose, the dockerin domains of the enzymatic subunits partake in a very stable type of interaction with the cohesin domains of the scaffoldin subunit.

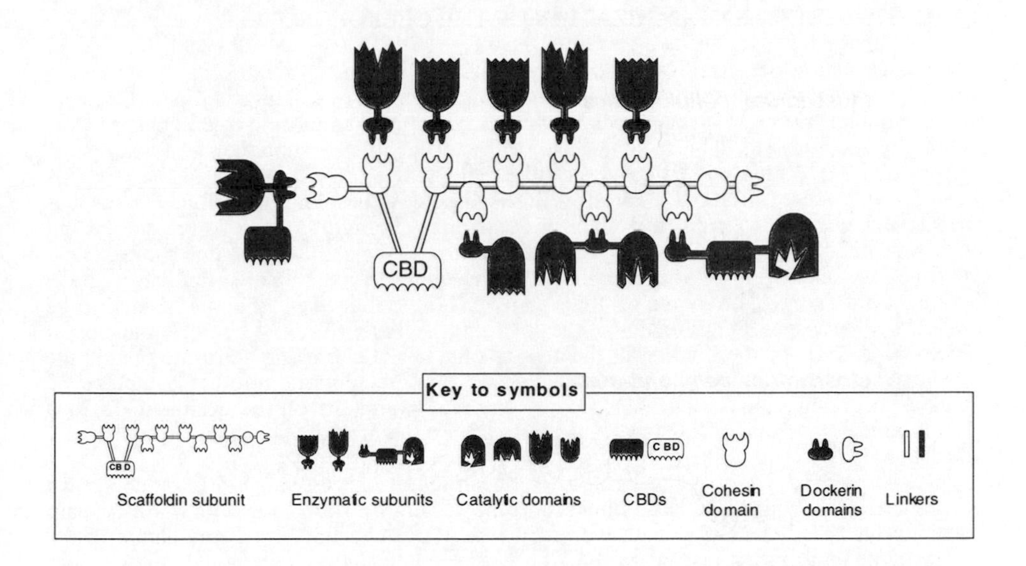

Figure 1 *Schematic model of the cellulosome from* C. thermocellum. *All of the subunits are composed of multiple domains (see Key). The non-hydrolytic scaffoldin subunit integrates the enzymatic subunits (shaded structures) into the complex via the reciprocal intersubunit cohesin-dockerin interaction. The scaffoldin subunit itself bears a dockerin domain, which is believed to mediate its implantation into the cell surface. The enzymatic nature of the hydrolytic subunit is dictated by one or more catalytic domains. The various domains of the various subunits are usually separated by distinct linker sequences. The enzymatic subunits contain other domains (not shown), the functions of which are still unclear .*

Recent evidence from several laboratories has demonstrated that there seems to be little or no specificity in the binding among the various cohesins and the various dockerins.[29,37,43,44] Within a given cellulosome, the different cohesins seem to recognize nearly all of the dockerin domains in an equivalent manner. As a consequence of this observation, we would expect a random incorporation of the enzymatic subunits into the cellulosome. However, the observation does not preclude the possibility that an unrelated mechanism would regulate the incorporation of a specific enzyme into the complex. Another possibility is that the enzyme subunits may be exchanged among the various cohesin domains during the degradation process. Further biochemical information is needed to clarify these issues.

Clostridium thermocellum

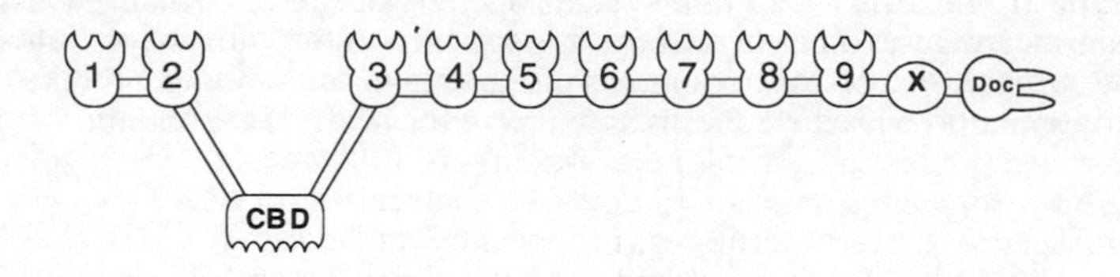

Clostridium cellulovorans

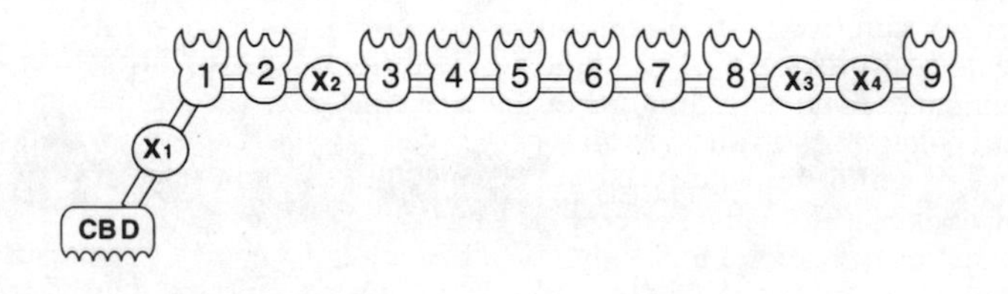

Clostridium cellulolyticum

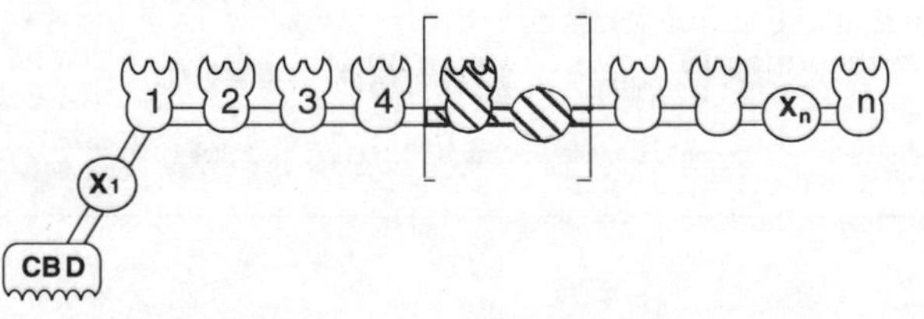

Figure 2 *Comparative organization of the domains of the scaffoldin subunits from different clostridial species. The genes for the cellulosomal scaffoldin subunits from the thermophilic* C. thermocellum *and from the mesophilic* C. cellulovorans *have been totally sequenced. Both contain a single CBD and 9 cohesin domains. The scaffoldin subunit from* C. cellulovorans *also contains 4 hydrophilic domains (X1-X4) whereas that from* C. thermocellum *exhibits only one such domain. The scaffoldin gene from another mesophilic strain,* C. cellulolyticum, *has also been sequenced, save for a ~1-kb segment (see crosshatched portion) which could accommodate one or two cohesin and/or hydrophilic domains. The general organization of the scaffoldin domains from* C. cellulovorans *and* C. cellulolyticum *are quite similar but differ from that of* C. thermocellum. *The latter also bears a single C-terminal dockerin domain, lacking in both of the two mesophilic scaffoldins.*

3 THE SCAFFOLDIN SUBUNIT

The first scaffoldin gene to have been sequenced was the *cbp*A gene in *Clostridium cellulovorans.*[10] The gene for the corresponding scaffoldin in *C. thermocellum,* termed *cip*A, was determined the following year.[9] The sequence of a third scaffoldin gene (*cip*C of *Clostridium cellulolyticum*)[34,45] is nearly complete (J.P. and Anne Belaich, personal communication).

The overall structural arrangement of the three known scaffoldin subunits is shown in Figure 2. All three are very large, multi-domain proteins. All three contain a CBD, eight or nine cohesin domains and one to four copies of a hydrophilic domain of unknown function (provisionally called domain X).[23] The scaffoldin of *C. thermocellum* contains a C-terminal dockerin domain which is absent in the other two scaffoldins. This particular dockerin serves to anchor the cellulosome of this species into the cell surface by interacting with a special type of cohesin — the type-II cohesin domain. The presence of the type-II dockerin — in this species only — raises the question as to whether the cellulosomes from the other two strains are attached to the cell surface, and if so, how?

The scaffoldins of the two mesophilic strains are clearly more similar between themselves than they are to that of *C. thermocellum*. Both *C. cellulovorans* and *C. cellulolyticum* possess a CBD at their N-terminus, followed first by a hydrophilic domain and then the cohesin domains. In contrast, the scaffoldin of *C. thermocellum* has two cohesin domains at the N-terminus, followed by an "internal" CBD and the other seven cohesins. The *C. cellulovorans* and *C. cellulolyticum* scaffoldins differ mainly in the distribution of the cohesins and hydrophilic domains. The scaffoldin subunit of *C. cellulovorans* features a hydrophilic domain between the second and third cohesin domains which is lacking in *C. cellulolyticum.* All three scaffoldins display a penultimate hydrophilic domain (a twin in *C. cellulovorans*).

As described previously,[46-48] cellulosome-like complexes have been noted in other microorganisms. One telltale sign of a cellulosome is the detection of dockerin- or cohesin-like sequences in the genome of a given microorganism, which would indicate the presence of cellulosomal enzymes or a scaffoldin subunit, respectively. An increasing number of dockerin domains have indeed been reported to grace glycolytic enzymes in several examples of bacteria and fungi.[49-51] In a similar manner, the presence of a type II-like cohesin may also be indicative of a protein which anchors a cellulosome to the cell surface via a dockerin of a scaffoldin subunit. Thus far, however, none has been reported other than those of *C. thermocellum.*

It is clear that the sequencing of such large and intricate scaffoldin genes, which encode for polypeptides of around 1800 amino acid residues and a multiplicity of repetitive domains, is no simple feat. Nevertheless, as the field of cellulosome research progresses, it is hoped that other such scaffoldin genes will be identified and sequenced (particularly in non-clostridial species), in order to determine the diversity of scaffoldin composition and architecture.

4 ANCHORING THE *C. THERMOCELLUM* CELLULOSOME ONTO THE CELL SURFACE

A 10-kb region of the *C. thermocellum* genome was sequenced downstream from the C-terminal portion of the scaffoldin gene.[52] This achievement revealed three new genes which encoded for three cell-surface proteins: OlpB, Orf2p and OlpA (originally termed AncA), in order of their appearance on the chromosome. The calculated sizes of these proteins were 178, 75 and 48.5 kDa, respectively. Later, a gene encoding a fourth protein, called SdbA (68.6 kDa),[31] was sequenced; this gene was located in a different part of the chromosome.

All four of these proteins are associated with the cell surface, by virtue of a special module (Figure 3). For this purpose, the proteins contain three consecutive SLH (S-layer homology) domains at their C-terminus, which is also present in S-layer proteins from other gram-positive bacteria.[53] Indeed, immunochemical and ultrastructural evidence have since supported the location of these proteins on the *C. thermocellum* surface.[54,55]

All four proteins bear cohesin domains. Originally, it was thought that OlpA would be a primary candidate for binding the scaffoldin subunit via its dockerin domain.[52] However, later experiments proved that OlpA binds to the dockerins of the *enzymatic* subunits from the cellulosome.[14] Thus, the single cohesin domain of OlpA appears to bind individual cellulolytic components and incorporate them separately onto the cell surface.[54]

The three other proteins, SdbA, Orf2p and OlpB, appear to bind selectively the dockerin domain of the scaffoldin subunit. The cohesins of these proteins are similar among themselves but clearly different from those of the scaffoldin subunit and OlpA. On the basis of sequence homology and dockerin specificity, we can discriminate between the two types of cohesin. Those that recognize the dockerins of the enzymatic subunits are classified as type-I cohesins, and those that recognize the scaffoldin dockerin are now referred to as type-II cohesins.[31]

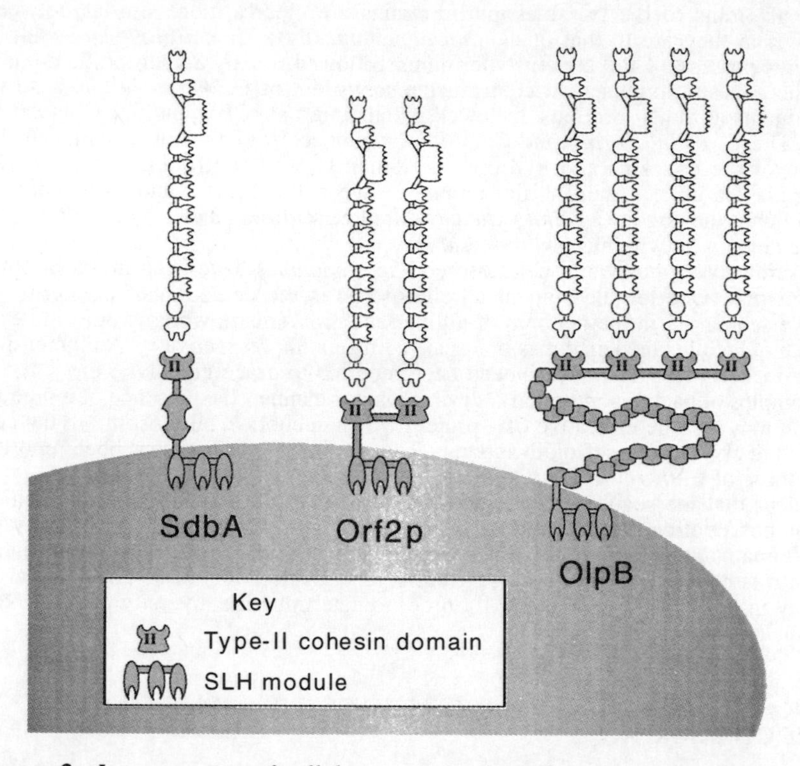

Figure 3 *Incorporation of cellulosomes into the cell surface via selective interaction between the scaffoldin subunit and type-II cohesin domains of cell-surface proteins. Three such proteins have been characterized, which bear one, two and four copies of the type-II cohesin domain. The type-II cohesin has been shown previously to be selective for the type-II dockerin of the scaffoldin subunit. Each of these proteins bears a triple-domained SLH module which is commonly associated with surface-layer proteins in gram-positive bacteria. SdbA, Orf2b and OlpB (shaded structures) would thus be able to incorporate one, two and four copies, respectively, of the scaffoldin subunit (white structures) and its complement of enzymes (not shown) onto the cell surface. This figure is based on articles of Béguin and colleagues.*[27,31,52,55]

SdbA contains a single type-II cohesin domain, Orf2p contains two such domains and OlpB contains four (Figure 3). Assuming that the capacity for binding to scaffoldin subunits (and hence cellulosomal complexes) is equal to the number of cohesin domains, the three proteins can carry the corresponding number of cellulosomal units. When the scaffoldin subunit was sequenced, it was noted that the nine enzymatic subunits (averaging roughly 100 kDa, each) plus the 210-kDa scaffoldin subunit could account for only about half of the estimated size of the cellulosome (2.1 MDa). In this respect, purified cellulosome preparations may contain multiple forms, *i.e.*, monomeric, dimeric and tetrameric complexes, by virtue of the interaction of the scaffoldin subunits with Orf2 and OlpB. This could also help explain the broad peaks which characterized the gel filtration pattern in the early studies.[1,3] Notably, two extraneous bands were consistently observed, particularly in cell-surface, rather than extracellular, preparations of the cellulosome. The molecular size of one of these bands (~70 kDa) was consistent with either SdbA or Orf2p. It would be interesting to determine whether these surface proteins co-purify with the cellulosome.

As discussed in earlier publications and reviews,[18,21,27,30,46,54-57] these proteins seem to be packaged into cell-surface organelles, called polycellulosomal protuberances, which house multiple copies of the cellulosome. In particular, we wish to refer to the detailed discussion and diagrams presented in previous reviews.[18,21]

5 THE CBD

5.1 Structure of the scaffoldin CBD and proposed binding residues

The cellulosomal scaffoldin CBD from *C. thermocellum* belongs to the Family-III CBDs, classified according to sequence alignment.[58] The CBD has been subcloned and expressed,[59] and the crystal structure of the recombinant form has recently been determined (PDB ID code 1NBC).[60]

The 155-residue scaffoldin CBD was the first crystal structure of a bacterial CBD to have been determined. A second crystal structure has recently been solved for another CBD from the same family, *i.e.*, a non-cellulosomal CBD from cellulase E4 of *Thermomonospora fusca*.[61] An NMR structure of a Family-II CBD (exoglucanase C_{ex} from *Cellulomonas fimi)* was described earlier,[62] which was our first glimpse into the secrets of the bacterial CBD structure. More recently, another CBD from Family IV (CBD_{N1} from *Cellulomonas fimi* β-1,4 glucanase C) was also determined by NMR.[63] Historically, the first NMR structure of a CBD to have been determined was the small, 36-residue Family-I CBD from a fungal cellulase.[64]

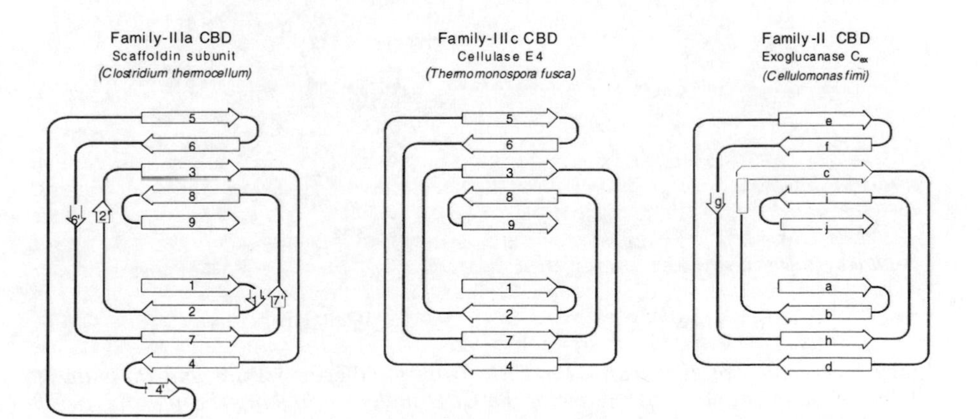

Figure 4 *Comparative topologies of the bacterial CBDs. The schematic diagrams show the β-sheet architecture and labeling of secondary structure elements. The three types of CBD exhibit essentially the same fold and jellyroll topology, forming a β sandwich. The simplest structure is the Family* ***IIIc*** *CBD of E4, in which all of the β strands are arranged in the two β sheets. The Family* ***IIIa*** *CBD includes several additional short strands which decorate the molecule but do not form part of the β sheets.*

Like the Family-II and IV CBDs, the structure of the scaffoldin CBD belongs to the all-β family of proteins.[60] The CBD consists of 9 β strands, which assume a jellyroll topology and form two antiparallel β sheets (Figure 4). In fact, the two known Family-III CBDs share the same basic topology which is essentially the same as that described earlier for the CBD from Family II. The Family-IV CBD also exhibits the same topology. Thus it is clear that the CBDs from Families II, III and IV all share the same basic jellyroll structure. It is interesting that, despite the lack of sequence identity among the three families, all of the bacterial CBDs from these three families would attain the same topology.

Conserved surface residues were mapped on the scaffoldin CBD of *C. thermocellum.* The conserved residues were mainly distributed on two opposite sides of the molecule — *i.e.*, a flat planar surface on the "bottom" of the molecule and a shallow groove on the top (Figure 5).

These two faces were subjected to a docking procedure with the known structure of cellulose. This process revealed a series of surface-exposed amino acid residues which potentially interact with the cellulose surface. The evidence suggests that the planar face of the CBD molecule interacts with 3 successive chains on the cellulose surface. Upon this face, a planar strip of aromatic residues (W118, H57, Y67 and the D56-R112 salt bridge) aligns precisely along one of the cellulose chains. The stacking interactions, formed between the planar strip residues and the glucose rings along the cellulose chain, are considered to be the major reason for the specificity and strong binding of the CBD to crystalline cellulose substrates. The interaction with the planar strip residues is bolstered by up to 5 additional hydrogen bonds (N16, N10, and Q110; N12 and S133) along two adjacent cellulose chains. The latter residues have been proposed to anchor the surface of the CBD to the surface of the substrate. Site-directed mutagenesis is currently being employed to systematically replace and analyze the contribution of each of the ten suspected binding residues in the scaffoldin CBD.

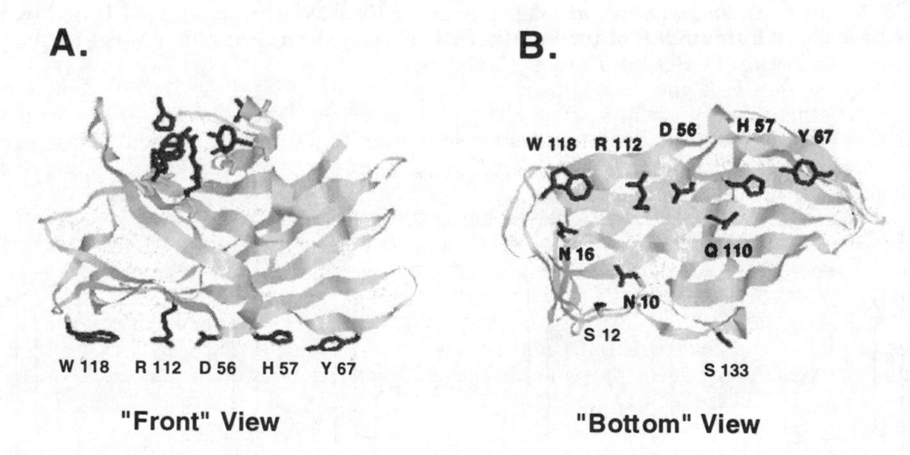

Figure 5 *Structure of the scaffoldin CBD from* C. thermocellum *and the putative cellulose-binding residues.* ***A:*** *View of the CBD molecule looking through the two β-sheets that form the sandwich, displaying the distribution of the two sets of conserved residues. On top, four tyrosine residues among others fill a shallow groove of yet unknown function. On the bottom, a planar strip of three aromatic amino acids (W118, H57 and Y67) and a salt bridge (formed between D56 and R112) are proposed to form stacking interactions with four glucose rings along a single cellulose chain.* ***B:*** *90° rotation of* **A** *about the horizontal axis showing the cellulose-binding surface as proposed to be viewed from the surface of the substrate. In addition to the five planar strip residues, five additional residues are proposed to anchor the CBD via hydrogen bonds to two adjacent chains of the crystalline cellulose substrate. Three residues (N16, N10 and Q110) are believed to interact with a central chain, and the other two (S12 and S133) purportedly form hydrogen bonds with glucose residues of a third chain.*

Similar analyses of the Family-I and Family-II CBDs suggested a comparable type of interaction with cellulose.[60,65,66] In both cases, the CBD was proposed to interact with 3 adjacent chains. A planar strip of aromatic residues was implicated, together with additional anchoring residues. Although this may represent a general theme of CBD interaction with crystalline cellulose, other types of CBD (*e.g.*, those of Family III**c** and Family IV) seem to interact very differently with the substrate.

The role of the shallow groove is not clear at the present time. It is interesting to note that the CBDs of Family-II lack this structural element, which is present in the Family-III and Family-IV CBDs. Conversely, the Family-IV CBDs lack the planar surface and the cellulose-binding residues proposed for the Family-II and Family-III CBDs. In this context, the shallow groove or cleft seems to provide the structural explanation for the preference of the Family-IV CBD for amorphous cellulose and soluble oligosaccharides, and its failure to bind strongly to crystalline cellulose.[67]

The function of the shallow groove in the Family-III CBDs is less evident. All of the CBDs of this family clearly contain all the amino acid residues consistent with the shallow groove, but these residues appear to be different than those from Family IV. Moreover, interaction with soluble oligosaccharides has not been demonstrated. Nevertheless, we cannot rule out at this time a saccharide-binding function for the shallow groove of the Family-III CBDs. On the other hand, the groove of the scaffoldin CBD apparently recognizes peptide segments (Tormo, *et al.*, unpublished results), *e.g.*, those of the linker regions, and the groove might serve to form secondary protein-protein interactions with other components of the scaffoldin subunit or with the enzymatic subunits of the cellulosome.

5.2 The Three Subfamilies of the Family-III CBDs

On the basis of the two known structures for Family-III CBDs, we can now reclassify this family into three subfamilies (Figure 6). Previous classification schemes initially included all three subfamilies in a single family,[68] after which the division into two subfamilies was adopted.[58] Further division into three subfamilies was suggested when the crystal structure of the scaffoldin CBD from *C. thermocellum* was reported.[60]

5.2.1 Subfamily IIIa. The three known scaffoldin CBDs (from *C. thermocellum*, *C. cellulovorans* and *C. cellulolyticum*) all belong to the Family-III CBDs. They are all very similar amongst themselves and are clearly different from the other members of this family. We can thus classify them in a separate subfamily — *i.e.*, Family III**a**.

5.2.2 Subfamily IIIb. The known CBDs listed in Family III**b** are all components of free cellulases and are very similar in their sequences to those of the scaffoldin CBDs in subfamily III**a**. Most of the proposed cellulose-binding residues are highly conserved in this subfamily. There are two notable exceptions: An entire 8- or 9-residue segment — the scaffoldin loop — is entirely absent from the CBDs of subfamily III**b**. This loop includes the important tyrosine residue (Y67), believed to participate in cellulose binding. In addition, and presumably linked to the latter phenomenon, H57 of Family III**a** is replaced by tryptophan in each of the Family-III**b** CBDs. These differences are sufficient to classify such CBDs in their own subfamily. For additional discussion of the first two subfamilies, see section 9.1.

In any case, members from Family III**b** would clearly be expected to bind strongly to crystalline cellulose, and this would apparently be their major function. Most of the planar strip residues are conserved, including those which form the salt bridge, as are most of the proposed anchoring residues. In our laboratory (Yaron *et al.*, unpublished results), we have subcloned and expressed a representative Family-III**b** CBD (from CelI of *C. thermocellum*).[69] The CBD indeed binds strongly to crystalline cellulose. We are currently crystallizing this CBD, in order to compare its fine structure with those of the two other subfamilies.

5.2.3 Subfamily IIIc. The CBDs in Family III**c** are clearly distinct from those of subfamilies III**a** and **b**. Many, but not all, of the purported cellulose-binding residues are absent. Thus, the equivalent of H57, W118 and most of the "anchoring" residues are lacking. On the other hand, the analogues of D56, R112, Q110 and Y67 are retained in most of the Family-III**c** CBDs, with the lone exception of the cellulosomal CelF CBD from *C. thermocellum*.[42] The apparent nonconformity of the latter CBD may be somewhat deceptive, since the immediate positions of many residues may have been exchanged (*e.g.*, note the tyrosine in the CelF CBD which replaces the Q110 equivalent). In any case, eccentric forms of a given family or subfamily deserve further attention, and the significance of such deviations may be resolved only through structure determination by comparative crystallographic or NMR studies.

FAMILY III CBDs

```
Family IIIa
CipA  C.th   NLK VEFY NSNPS -DTTN SINPQ FKVT N-TGS SAIDL SKLTL RYYY TVDGQ -KDQT FWCDH AAII GSNGS YNGIT SNVKG TFVK MSSST N
CbpA  C.cv   SMS VEFY NSNKS -AQTN SITPI IKIT N-TSD SDLNL NDVKV RYYY TSDGT -QGQT FWCDH AGAL LGN-S YVDNT SKVTA NFVK ETASP T
CipC  C.cl   GVV SVQF NNGSS PASSN SIYAR FKVT N-TSG SPINL ADLKL RYYY TQDAD -KPLT FWCDH AGYM SGS-N YIDAT SKVTG SFKA VSPAV T
Family IIIb
CelZ2 C.st   VIQ IQMF NGNTS -DKTN GIMPR YRLT N-TGT TPIRL SDVKI RYYY TIDGE -KDQN FWCDW SSVG S--------- NNITG TFVK MAEPK E
CelA  B.la   DLV VQYK DGDRN NATDN QIKPH FNIQ N-KGT SPVDL SSLTL RYYF TKDSS -AAMN GWIDW AKLG G--------- SNIQI SFGN HNGAD S
CelV  E.ca   DVV LQYR NVDNN P-SDD AIRMA VNIK N-TGS TPIKL SDLQV RYYF HDDGK -PGAN LFVDW ANVG P--------- NNIVT STGT PAAST D
CelA  C.sa   QIK VLYA NKETN -STTN TIRPW LKVV N-SGS SSIDL SRVTI RYWY TVDGE -RAQS AISDW AQIG A--------- SNVTF KFVK LSSSV S
CelI2 C.th   EVV LQYA NGNAG -ATSN SINPR FKII N-NGT KAINL SDVKI RYYY TKEGG -ASQN FWCDW SSAG N--------- SNVTG NFFN LSSPK E
Family IIIc
CelI1 C.th   EIF VEAG VNASG NN--- FIEIK AIVN NKSGW PARVC ENLSF RYFI NIEEI VNAGK SASDL QVSS ----S YNQGA KLSDV KHYK DN---  -
CelZ1 C.st   EFF VMAG INASG QN--- FIEIK ALLH NQSGW PARVA DKLSF RYFV DLTEL IEAGY SASDV TITT ----N YN--- AGAKV TGLH PWNEA E
CelF  C.th   EFY VEAA VNAAG PG--- FVNIK ASII NKSGW PARGS DKLSA KYFV DISEA VAKGI TLDQI TVQS ----T TN--- GGAKV SQLL PWDPD N
CenB  C.fi   QLF VEAM LNQPP SGT-- FTEVK AMIR NQSAF PARSL KNAKV RYWF TTDGF ------ASDV TLSA ----N YSECA GAQSG KGVS AGGT- -
CelG  C.cl   EVI IKAG LNSTG PN--- YTEIK AVVY NQTGW PARVT DKISF KYFM DLSEV -AAGI DPLSL VTSS ----N YSEG- KNTKV SGVL PWDVS N
E4    T.fu   EIF VEAQ INTPG TT--- FTEIK AMIR NQSGW PARML DKGTF RYWF TLDE- ---GV DPADI TVSS ----A YNQCA TPEDV HHVS GD---  -

Family IIIa
CipA  C.th   -NA DTYL EISF- -TGGT LEPGA ---- HVQIQ GRFAK NDWSN -YTQ SNDYS FK--S ASQFV E--- WDQVT AYLNG VLVWG KEP
CbpA  C.cv   STY DTYV EFGFA SGRAT LKKGQ ---- FITIQ GRITK SDWSN -YTQ TNDYS FD-AS SSTPV V--- NPKVT GYIGG AKVLG TAP
CipC  C.cl   -NA DHYL EVALN SDAGS LPAGG ---- SIEIQ TRFAR NDWSN -FDQ SNDWS YT-AA GSYMD ---- WQKIS AFVGG TLAYG STP
Family IIIb
CelZ2 C.st   --G AYYL ETGFT DGAGY LQPNQ ---- SIEVQ NRFSK ADWTD -YIQ TNDYS FS--T NTSYG S--- NDRIT VYISG VLVSG IEP
CelA  B.la   --- DTYA ELGFS SGAGS IAEGG Q--- SGEIQ LRMSK ADWSN -FNE ANDYS FD-GA KTAYI D--- WDRVT LYQDG QLVWG IEP
CelV  E.ca   -KA NRYV LVTFA SG-GS LQPGA E--- TGEVQ VRIHA GDWSN -VNE TNDYS YG-PN ITSYT N--- WDKIT VHDKG TLVWG TEP
CelA  C.sa   --A DYYL EIGFK SGAGQ LQPGK D--- TGEIQ IRFNK DDWSN -YNQ GNDWS WI-QS MTSYG E--- NEKVT AYIDG VLVWG QEP
CelI2 C.th   -GA DTCL EVGFG SGAGT LDPGG ---- SVEVQ IRFSK EDWSN -YNQ SNDYS FN-PS ATDYT D--- WNRVT LYISN KLVYG KEP
Family IIIc
CelI1 C.th   --- IYYV EVDL- -SGTK IYPGG QSAY KKEVQ FRISA PEGTV -FNP ENDYS YQGLS AGTVV ---K SEYIP VYDAG VLVFG REP
CelZ1 C.st   --N IYYV NVDF- -TGTK IYPGG QSAY RKEVQ FRIAA PQNTN FWNN DNDYS FRDIK GVTSG NTVK TVYIP VYDDG VLVFG VEP
CelF  C.th   --H IYYV NIDF- -TGIN IFPGG INEY KRDVY FTITA PYGEG NWDN TNDFS FQGLE QGFTS K--K TEYIP LYDGN VRVWG KVP
CenB  C.fi   --- LGYV ELSC- -VGQD IHPGG QSQH RREIQ FRLTG PAG-- -WNP ANDPS YTGLT QTALA ---K ASAIT LYDGS TLVWG KEP
CelG  C.cl   --N VYYV NVIL- -TGEN IYPGG QSAC RREVQ FRIAA PQGTT YWNP ENDFS YDGLP TTSTV NTV- -TNIP VYDNG VKVFG NEP
E4    T.fu   --- LYYV EIDC- -TGEK IFPGG QSEH RREVQ FRIAG GPG-- -WDP SNDWS FQGIG NELAP ---- APYIV LYDDG VPVWG TAP
```

Figure 6 *Groupings of Family-III CBDs into 3 sub-families. Secondary structural elements (β strands) are indicated by arrows and are enumerated. Proposed cellulose-binding residues, as determined for the CipA scaffoldin from* C. thermocellum, *are shaded, as are the homologous residues in the other CBDs from this family. The "scaffoldin loop", which distinguishes subfamily-**a** CBDs from those of subfamily **b**, is denoted by the crosshatched region.*

In a very elegant work by Karplus, Wilson and colleagues,[61] the crystal structure of a subfamily-III**c** CBD, from *Thermomonospora fusca* cellulase E4, was recently determined together with the adjacent, intimately attached catalytic domain (Figure 7). The analysis of this structure by these authors has clarified many questions which have arisen concerning the sequence and biochemical properties of this type of CBD (see Wilson, et al., this volume). Additional evidence in understanding the biochemical importance of the subfamily III**c** CBD has been gained from the recent description of the cellulosomal CelG from *C. cellulolyticum*.[70]

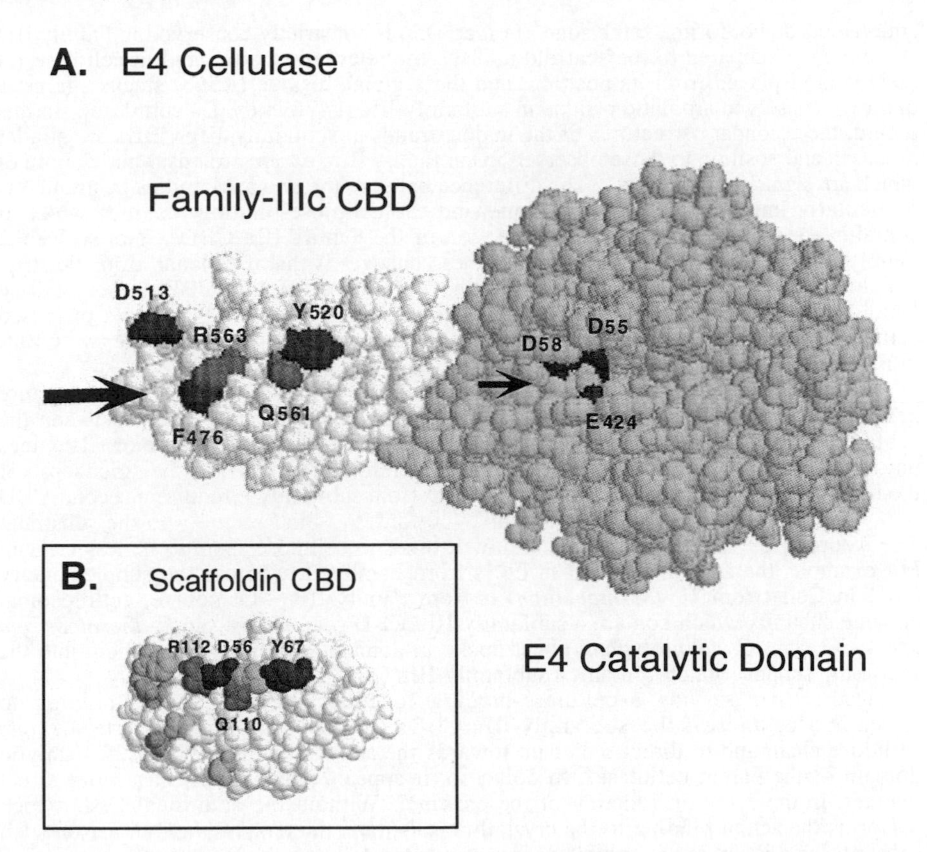

Figure 7 *Structural comparison of the binding faces of the Family III**a** and III**c** CBDs.* **A.** *The structure of the E4 cellulase from* T. fusca.[61] *The residues of the catalytic domain are shaded light gray, and those of the CBD are white. Four residues (D513, R563, Y520 and Q561), homologous to those of the other subfamily-III CBDs, are designated. In addition, a novel conserved aromatic residue (F476), characteristic of the subfamily-III**c** CBDs, is also labeled. A single cellulose chain appears to be bound by this CBD. Its proposed path is shown, starting from the interaction with the binding subsite on the CBD (large arrow) and continuing towards the active-site residues on the catalytic domain — D55, D58 and E424 (small arrow). For comparison, the same face of the scaffoldin CBD from the* C. thermocellum *cellulosome is given in* ***B*** *and the homologous binding residues are indicated. Note the altered distribution of the conserved residues. The path of the cellulose chain through the Family-III**c** CBD binding subsite en route to the active site of the catalytic domain is not equivalent to the planar strip residues of the other two subfamilies. D513 and Y520 are not believed to be involved directly in the interaction.*

Specifically, the lack of key cellulose-binding residues caused many research groups to suspect whether such CBDs would bind cellulose at all. Indeed, the lack of the full complement of planar strip residues would seem to preclude a strong cellulose-binding function. Structural evidence for the E4 CBD indicates that the D56 and R112 equivalents do not form a salt bridge. In fact, D56 is not strictly conserved in Family III**c** (Figure 6). Compared to the scaffoldin CBD, the relevant aspartic acid in cellulase E4 (D513) is displaced from its position, and the arginine residue (R563) stacks against a uniquely conserved aromatic residue in subfamily III**c** (F476 in the E4 cellulase). In this regard, the secondary structures of the major strands in subfamily-III**c** CBDs are similar in length and position to those observed in subfamily III**a**, *except* strands **4** and **5**, both of which are significantly shorter. The difference between the two subfamilies in strand **4** is particularly important, since this strand and the cellulose-binding residues which it contains are decisive to the binding function of the Family III**a** CBDs. Not so for the Family III**c** CBDs. The final outcome of these changes is that the planar aromatic strip, which appears to dominate the interaction between the scaffoldin CBD and crystalline cellulose, has been entirely dismantled in the subfamily III**c** CBDs, and its proposed cellulose-binding surface has been reorganized (compare the structures of the two CBDs in Figure 7).

Based on the new structural and biochemical information, the following picture begins to emerge. The functional significance of the difference observed between the subfamily-**c** CBDs and those of the other two subfamilies is that this type of CBD does not bind strongly to crystalline cellulose. It is therefore not a CBD in the strict sense of the definition. In fact, enzymes bearing a CBD from subfamily **c** require a second CBD — *i.e.*, a "true" cellulose-binding domain — in order to bind securely to the substrate. The nature of the additional CBD and how it binds to cellulose seem to be less critical. For example, the additional CBD in E4 is from Family II, whereas the supplementary CBD in CelI (from *C. thermocellum*) is from Family III**b**. Of course, cellulosomal enzyme subunits which contain a subfamily-III**c** CBD (*i.e.*, CelF from *C. thermocellum* and CelG from *C. cellulolyticum)* bear dockerin domains which integrate them into the scaffoldin subunit, which contains a subfamily-III**a** CBD.

Rather than serving a cellulose-binding function *per se* (*i.e.*, the binding to crystalline cellulose), the subfamily-III**c** CBDs seem to bind transiently to a *single* cellulose chain and to direct the chain towards the active-site cleft within the catalytic domain of the parent cellulase. In doing so, it appears to take part, in a more direct manner, in the catalytic function of the enzyme. Without the additional CBD which performs the actual binding to the crystalline substrate, the enzyme fails to act on such substrates and is limited to soluble or less structured cellulosics. Without the subfamily-III**c** CBD, the catalytic domain ceases to operate on soluble cellulosics. The CBDs of this subfamily may thus be considered to serve as a cellulose-binding "sub-site" of the catalytic domain.

6 THE COHESIN DOMAIN

A recombinant cohesin (the cohesin-2 domain from the cellulosome of *C. thermocellum)* has been crystallized[71] and its structure has been solved (PDB ID code 1ANU) by multiple isomorphous replacement (with the aid of two heavy atom derivatives).[72,73] The structure is the first of the highly conserved family of cohesin domains, and thus provides a model for other cohesin domains from this and other bacteria and fungi. The structure of a second cohesin (cohesin 7) from the scaffoldin subunit of this organism has also been crystallized,[74] and doubtless exhibits the same structure and fold.

Cohesin-2 is an elongated, conical molecule with approximate dimensions of 46 Å x 28 Å x 21 Å (Figure 8A). Numerous hydrogen bonds are formed between charged side-chain groups and main-chain nitrogens, notably in the loop regions. The H-bonding stabilization of the loops is particularly evident on loops **4/5**, **6/7** and **8/9** — all positioned on the same aspect of the cohesin domain.

Interestingly, the overall fold of the cohesin domain (Figure 8B) is remarkably similar to that of the CBD, despite the complete absence of sequence identity. It is thus interesting to speculate whether the jellyroll fold is a structural requirement for scaffoldin function and dynamics.

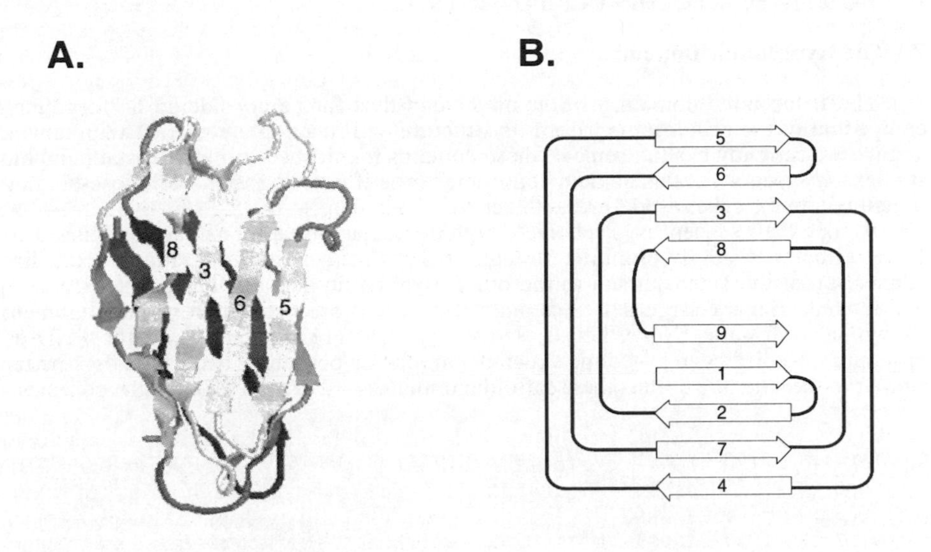

Figure 8 *Structure and topology of the cohesin domain.* ***A.*** *Ribbon diagram of the overall three-dimensional structure of cohesin-2.* ***B.*** *Topology diagram of the cohesin-domain showing the 9 strands, the two sheets and the connectivities which form the jellyroll fold.*

The various faces of the cohesin domain are shown in Figure 9. The side-chains of residues which form the β-strands are shaded and enumerated for clarity of orientation. Side chains of the loops are located at the two poles of the molecule. A greater familiarity with the structural elements which define the cohesin surface architecture is imperative to a greater understanding of its function, as will be detailed later in this chapter.

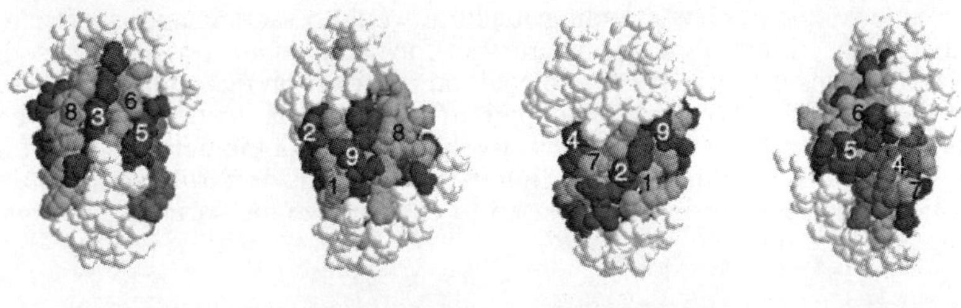
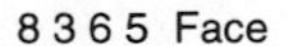

Figure 9 *"Mug shots" of the cohesin domain. Each face represents a 90° rotation of its neighbor along the vertical axis of the molecule. The diagrams are displayed in the spacefill mode. Compare the* **8365** *face with the ribbon diagram in Figure 8A.*

7 OTHER SCAFFOLDIN COMPONENTS

7.1 The Hydrophilic Domain

The hydrophilic domain, or domain X, can thus far be considered a "forgotten" component of the cellulosome. Both its structure and function are currently unknown, and we are currently cloning some of these domains for biochemical and crystallographic studies. It appears that this domain contributes little if anything to the cellulose-binding function or to the cohesin-dockerin interaction.

Among their sequences, a relatively high degree of similarity can be established — even with the CipA hydrophilic domain from *C. thermocellum* (Figure 10). The connection of the latter domain to the other hydrophilic domains has previously been overlooked. Several large deletions and/or insertions are evident from the alignment shown in the figure. Nevertheless, secondary structure analysis using the PHDsec program[75,76] indicates an all-β-type structure comprising between **8** and **9** strands, perhaps similar to the structure of the other scaffoldin domains — *i.e.*, the CBD and the cohesins.

```
                   bbb      bbb          bbbbbb   bbbb      bbbbb
CipA (Cth)  1703  YILPD------FSFDA--TVAPLVKAGFKVEIVGTELYAVTDANGYFEIT
CipC (Ccl)   198  TINPTSISAKAGSFAD-------TKIT--LTPNGNTF------NGISELQ
CbpA (Ccv)   197  IINPT-----SASFDKVNTKQADVKTT--MTLNGNTFKTITDANGTA-LN
CbpA (Ccv)   576  ILEPT-ISPVTASFDK--KAPADVATT--MTLNGYTF------NGITGLT
CbpA (Ccv)  1517  IVDST-VAPTAVSFDK--ANQADAAIT--MTLNGNTF-AIK--NGTATLV
CbpA (Ccv)  1608  IVNST-ITPVVATFEKTAAKQADVVVT--MSLNGNTFSAIK--NGTTTLV

                          bbbbbbb  bbb    bbbbbbb      bbbbbb
CipA (Cth)  1745  -GVPANASGYTLKISRATYLDRV--IANVVVTGDTSVSTS--QAPIMMWV
CipC (Ccl)   233  -SSQYTKGTNEVTLLAS-YLNTLPENTTKTLTFDFGVGTK--NPKLTITVL
CbpA (Ccv)   239  ASTDYSVSGNDVTISKA-YLAKQS-VGTTTLNFNFSAGN---PQKLVITVV
CbpA (Ccv)   615  -TSDYSISGNVVKISQA-YLAKQP-VGDLTLTFNFSNGNKTATAKLVVSI
CbpA (Ccv)  1560  KGTDYTVSENVVTISKA-YLAKQT--GTVTLEFVFDKGN---SAKVVVAV
CbpA (Ccv)  1653  KGTDYTISGSTVTISKA-YLATLA-DGSATLEFVFNQGA---SAKLRLTIV
```

Figure 10 *Sequence alignment of hydrophilic domains from clostridial scaffoldins. The parent scaffoldins (CipA from* C. thermocellum, *CipC from* C. cellulolyticum *and CbpA from* C. cellulovorans) *and the position of the first residue in the respective sequence are indicated. Similar or identical residues among the aligned sequences are shaded. The locations of proposed β strands are indicated by* "b".

It is tempting to speculate that the hydrophilic domains may assume a similar jellyroll topology. If so, then a binding function would be anticipated, similar to those observed for other proteins which share this topology. One possibility is that the hydrophilic domains might interact with each other, either on the same subunit (*i.e.*, for the scaffoldins which bear multiple copies of the hydrophilic domains) or between subunits of different cellulosomes. Alternatively, the hydrophilic domains might interact with neighboring cohesins or CBDs, filling in spatial voids or participating in dynamic conformational changes which are proposed to characterize the various activities of the cellulosome on cellulose.[21,77,78]

7.2 The Linkers

The determination of the structures for the CBD[60] and the cohesin domain,[73] together with alignment data for the hydrophilic domains (Figure 10), has provided us with a measure of the limits of these domains within the scaffoldin sequence. Residual portions of the scaffoldin sequences presumably represent linker segments which interconnect the various domains. Simple algebraic treatment of the three known scaffoldin species allows us to compare their linkers. By "subtracting" the portions which correlate to the CBD the cohesins and hydrophilic domains, we are left with the sequences listed in Table 1.

Table 1 *Representative linker sequences from the clostridial scaffoldins.*

C. thermocellum	*C. cellulovorans*	*C. cellulolyticum**
GSSVPTTQPNVPSDG	DTPVEA	DGGNPPPQDPT
GNATPTKGATPTNTATPTKSATATPTRPSVPTNTPTNTPANTPVSGNLK	NPIDNR	PKDIPGDS
GGSVVPSTQPVTTPPATTKPPATTKPPATTIPPSDDPNA	NIGDP	DPGTQPTKE (3)
GDTTVPTTPTTPVTTPTDDSNA	DAPKT	VPGIQPTKE (2)
GDTTEPATPTTPVTTPTTTDDLDA (4)	NAIGPVKT	DVGDVTPVNPT
GDTTVPTTSPTTTPPEPTITPNK	EPSQPVKT (3)	TPVVTG
TNKPVIEGYKVSG	AAADIKA	
	LPTVI	
	EIQ	
	PAVVDPVVTDF	

Numbers in parentheses indicate that the given linker occurs more than once in the corresponding scaffoldin sequence.
*Unpublished results, courtesy of J.-P. Belaich, *et al.*

It is immediately clear from the table that the linkers of the *C. thermocellum* scaffoldin are much longer than those of the other two species. It is also clear that, unlike the wealth of prolines and threonines which characterizes the CipA scaffoldin, the percentage of such residues contained in the other two species is comparatively low, particularly for the *C. cellulovorans* scaffoldin.

The Pro/Thr-rich linkers of the CipA scaffoldin from *C. thermocellum* have been likened to the plant cell-wall extensins which form extended rod-like structures.[23] The threonines of CipA are believed to be glycosylated with a distinctive D-galactofuranose-containing oligosaccharide,[79,80] and a second such oligosaccharide has also been described for a cellulosome-like entity from a non-clostridial bacterium.[81,82] A D-galactofuranosyl component (rare in nature) is characteristic of the extensins as well.[83]

The function of the linker segments and the oligosaccharide residues is currently unknown. It has been suggested that these structures may modulate cellulosome action in various ways.[23] It is clear that they are not simply of ornamental value, but that they fulfill an important functional or structural role. In the case of the *C. thermocellum* linkers, the high incidence of proline would suggest that the linkers form extended structures in the molecule,[84] which would separate the various scaffoldin domains. Such spatial separation of the cellulolytic subunits may facilitate intersubunit protein-protein interactions and promote dynamic synergistic interaction among the catalytic domains. In this context, proline-rich regions have been reported to act generally as "sticky arms", which serve to bind rapidly and "non-specifically" to other proteins.[85]

The differences in the linker segments of the three scaffoldins indicate that those of the two mesophilic strains would not be extensively glycosylated, due to the scarcity of threonines or serines. Their relatively short length would presumably preclude their major involvement in regulating the action of catalytic subunits, as suggested previously for the *C. thermocellum* linkers.[23] Insight into the action of the linker segments will undoubtedly result from additional sequences of scaffoldins from other species and from future research into the biophysical properties of the cellulosomes and scaffoldins.

8 THE DOCKERIN DOMAIN OF THE ENZYMATIC SUBUNITS

The crystallographic and NMR structures for the CBDs and cohesin domains have advanced greatly our knowledge concerning the function of these domains. On the other hand, the structure of the dockerin domain is not yet known. However, we can speculate as to a possible structure by analyzing the known dockerin sequences. For this purpose, a combined approach was employed, involving comparative biochemical evidence, intra- and inter-species sequence comparison, and homology modeling.[45]

The dockerin sequence can be subdivided into seven different parts (Figure 11A). All clostridial dockerins thus far described contain a duplicated sequence of about 22 amino acid residues, the first 12 of which (Figure 9, segments $\mathbf{b_1}$ and $\mathbf{b_2}$) are homologous to a known structure — the calcium-binding loop of the EF-hand motif. It is assumed that this 12-residue sequence in the dockerin structure would adopt the same fold when bound to calcium.

The potential calcium-binding characteristics of the duplicated sequence was first noted by Chauvaux, *et al.*[86] In this context, it has been demonstrated that the cohesin-dockerin interaction of *C. thermocellum* is indeed calcium dependent. In fact, a common feature of the dockerins is the conservation of the designated calcium-binding residues, notably the aspartic acid and asparagine or serine at positions **1**, **3**, **5**, **9** and **12** (Figure 11B). Indeed, the cohesin-dockerin interaction has been demonstrated to be calcium dependent.[36,43] Since calcium was not detected in the cohesin domain,[73] it follows that the calcium dependency of the interaction resides in the calcium-binding loop of the dockerin domain.

The second halves of the duplicated sequences (Segments $\mathbf{c_1}$ and $\mathbf{c_2}$) apparently form α-helices, according to secondary structure prediction (PHDsec).[76] These helices would be analogous to the F helix of the EF-hand motif.[87] The two duplicated sequences of the dockerin domain are connected by segment **d**, which is predicted to form a structured loop, according to PHDsec analysis. Finally, the dockerins bear connecting stretches (segments **a** and **e**) on the *N*- and *C*-termini of the duplicated sequences. A short β-strand is predicted for part of segment **a**. Since all bacterial dockerins thus far described are either internal or *C*-terminal, segment **a** invariably connects the dockerin to the other domains of the enzymatic subunit, *i.e.*, a CBD or a catalytic domain. If the dockerin is internal, segment **e** will connect it to supplementary domains (characteristically to an additional catalytic domain). Otherwise, the enzyme subunit is terminated by segment **e**.

Secondary structure prediction indicates that the E-helix equivalents (segment **a** and the concluding residues of segment **d**) may be replaced in the dockerin domain by alternative structural elements. The evidence thus suggests that, structurally, the dockerin domain would represent a variation of the EF-hand motif — or an F-hand motif.

Figure 11C shows a model of the dockerin domain, using the known structure of a typical EF-hand motif[88] as a prototype structure. The exact nature of the dockerin structure remains to be resolved by X-ray crystallography or NMR.

9 DIALOG BETWEEN SEQUENCE HOMOLOGY AND 3D STRUCTURE

In previous sections, we presented a background of the cellulosome field and reviewed recent structural advances concerning the non-catalytic domains of the cellulosome. Although the structures are nice and aesthetic, and they provide in themselves a basic picture of the given domain, we showed how sequence comparisons can yield a new dimension to these studies and provide supplementary information relevant to the function of the molecule. In the following sections, we will discuss how this dialog between sequence and structure can be extended further.

We are exceedingly lucky that so many different cellulosomal enzymes and associated subunits from different microbial strains have been sequenced during the past decade. Careful analysis of these sequences can provide considerable information regarding the structures and interactions of the various functional domains.

In the next several sections, we will discuss what kind of interpretations can be derived by such analyses and some of the conclusions we have reached. Although speculative, such an approach can also guide us in identifying amino acid residues of potential interest to both the structure and the function of the molecule. Based on the resultant information, mutants can be designed, which can be experimentally tested using site-directed mutagenesis, followed by comparative biochemical and biophysical analyses of the constructs. When necessary, the three-dimensional structure of the mutant of interest can be determined and compared with that of the wild type molecule.

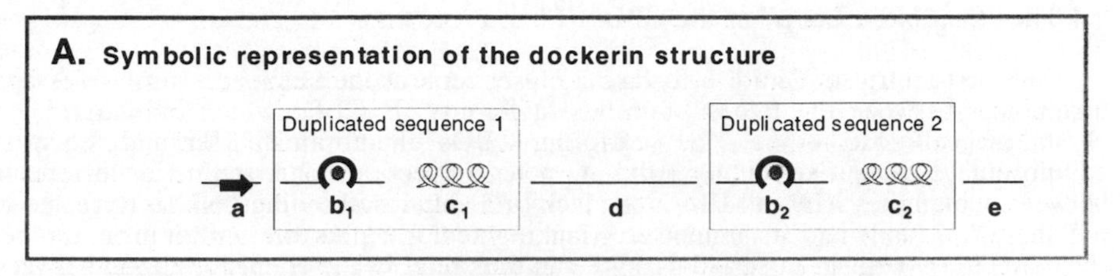

B. Conservation of calcium-binding residues

Troponin C	106	107	108	109	110	111	112	113	114	115	116	117
	D	K	N	A	D	G	F	I	D	I	E	E
	142	143	144	145	146	147	148	149	150	151	152	153
	D	K	N	N	D	G	R	I	D	F	D	E
CelS (C. th)	1		3		5		7		9	10	11	12
	D	V	N	D	D	G	K	V	N	S	T	D
	D	L	N	E	D	G	R	V	N	S	T	D
CelA (C. cl)	D	Y	N	N	D	G	N	V	D	A	L	D
	D	V	N	L	D	N	E	V	N	A	F	D

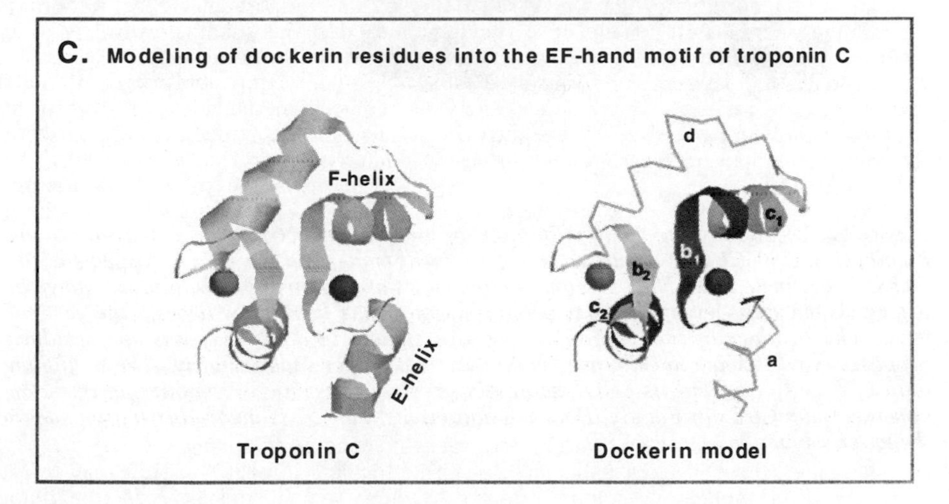

Figure 11 *Calcium-binding loop of the dockerin domain and proposed structural homology with the EF-hand motif. A. Symbolic representation of the dockerin structure. The sequences are partitioned into seven segments (designated* ***a*** *through* ***e****), and the 22-residue duplicated sequences are indicated. The calcium-binding motifs (*$\boldsymbol{b_1}$ *and* $\boldsymbol{b_2}$*) are indicated by a semicircle which encompasses the spherical calcium atom. Segments* $\boldsymbol{c_1}$ *and* $\boldsymbol{c_2}$ *are predicted to be α-helices and a short β-strand is predicted to form in segment* ***a*** *which precedes the initial duplicated sequence.* ***B.*** *Conservation of calcium-binding residues. Segments* $\boldsymbol{b_1}$ *and* $\boldsymbol{b_2}$ *of the dockerin domain exhibit strict sequence homology with the calcium-binding loop of troponin and similar calcium-binding proteins. The conserved shaded residues of the dockerin sequences either participate directly in the binding of calcium or are otherwise important for maintaining the structure of the motif.* ***C.*** *Homology modeling of the dockerin residues into the EF-hand motif of troponin C. The structurally homologous portions include the calcium-binding loop and the F helix. The E-helix equivalent may be replaced by alternative structural elements in segments* ***a*** *and* ***d*** *of the dockerin domain.*

9.1 The "Scaffoldin Loop" of the CBD

We first return to Figure 6 to take a closer look at the sequence similarities and differences between the two subfamilies of Family III CBD, which bind tightly to crystalline cellulose — *i.e.*, the scaffoldin CBDs of subfamily III**a** and the non-cellulosomal CBDs of subfamily III**b**. As noted in section 5.2.2, the major difference between subfamilies III**a** and III**b** is the lack of a cellulose-binding residue (tyrosine in the short **4'** strand) and its immediate flanking regions, plus the variation in another (histidine to tryptophan in strand **4**). We can thus refer to the connecting loop between strands **4** and **5** (which includes the **4'** strand) as the "scaffoldin loop". The location of the scaffoldin loop in the CipA CBD structure is shown in Figure 12. In subfamily III**a**, this loop would be very long whereas in subfamily III**b**, the connecting region would be very short. It is difficult to surmise precisely how the lack of this loop might affect the structure of the CBD. A structure for a member of subfamily III**b** CBD is clearly required.

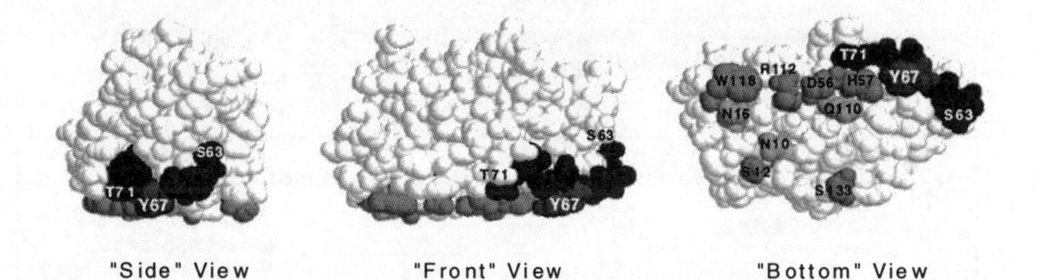

Figure 12 *Location of the "scaffoldin loop" of the Family III CBD. The structure of the Family III***a** *CipA CBD is displayed in the spacefill mode, and the various aspects of the molecule are indicated. The front and bottom views are synonymous with those shown in Figure 4; the side view represents a 90° rotation of the front view around the vertical axis. The residues of the scaffoldin loop are shaded in dark gray and the residues proposed to participate in cellulose binding are in a lighter shade of gray. The scaffoldin loop of Family III***a** *and its cellulose-binding residue, tyrosine 67, is missing from the non-scaffoldin CBDs of Family III***b**. *In addition, histidine 57 is invariably tryptophan in the latter subfamily.*

It is clear, though, that in the III**b** CBDs a major cellulose-binding residue (Y67) is missing from the linear aromatic strip of residues proposed to interact with the crystalline cellulose substrate. In addition, the absence of the entire scaffoldin loop would presumably reduce the surface area on the interacting face of the molecule. Although the exact nature of the alterations is currently unclear, the differences would be expected to translate into a slightly lower affinity for substrate in the CBDs from subfamily III**b**. A heightened affinity in the scaffoldin CBDs would be consistent with their increased burden in binding such a large complex, rather than a simple enzyme, to the substrate. It is intriguing to consider that such a small structural difference between the two subfamilies could account for the capacity to bind a 2-MDa complex versus a 100-kDa enzyme to the same substrate with similar efficiencies.

In order to examine the consequences of such differences, we can try to interconvert a member of one family to resemble another. For example, the scaffoldin loop can be deleted from the CipA CBD, and H57 can be replaced by a tryptophan. Conversely, the loop can be inserted into a member of Family III**b**, and a histidine can be substituted for the W57 equivalent. The binding activities and structures of the resultant mutant constructs can then be compared with those of the original CBD.

9.2 Intra-species Variance of Cohesin Sequences

It has long been noted that the cohesin domains from the scaffoldin subunit of the *C. thermocellum* cellulosome are highly conserved.[9] It has also been noted that the internal cohesins (cohesins 3 to 8) are very similar in this organism, whereas the external ones (cohesins 1,2 and 9) are less so.[37,43] Now that the structure of a representative cohesin is known,[73] we can map the divergent residues onto the surface of the molecule to gain further information regarding these differences.

Indeed, cohesins 3 through 8 are almost identical in their surface structure (Figure 13). In contrast, the external cohesins show a high degree of variance. The heterologous residues appear to be clustered, mainly on the **47219** face and on the flanking surfaces of cohesins 1, 2 and 9. The 3-residue deletion in cohesins 1 and 2 also occurs in a highly unconserved region, *i.e.*, the loop between strands **7** and **8**. On the other hand, the **8365** face and the crown of all cohesin domains in the *C. thermocellum* cellulosome are conserved.

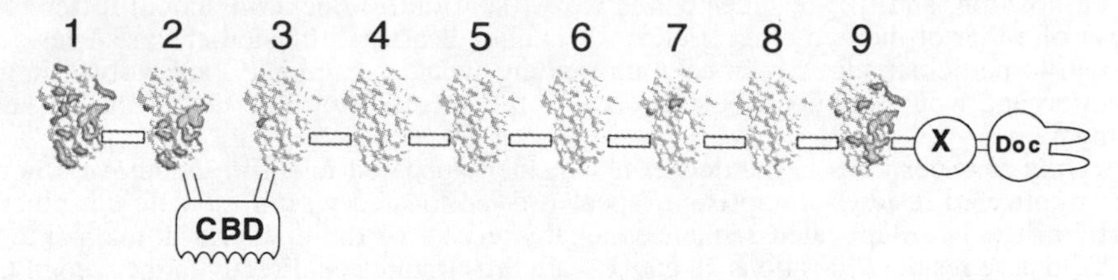

Figure 13 *Variant residues on the cohesins of* C. thermocellum. *The **47219** face of the molecule is shown (see Figure 9 for orientation), and the individual cohesin domains are numbered. The residues of the cohesin structure which differ from those of the consensus cohesin domains (cohesins 4 and 5) are mapped in gray on the surface of the molecule.*

It is interesting to speculate on the significance of the divergent cohesin surfaces. In this respect, the unconserved patches may reflect secondary interactions among neighboring scaffoldin components (*e.g.*, the CBD and linker segments), which are unrelated to dockerin binding. Alternatively, these regions could be related to the unique position of the individual divergent cohesins within the scaffoldin subunit. The interactions may modulate the packing of the cohesins with the cellulosome structure and may participate in conformational changes which have been suggested to take place upon binding to cellulose.

9.3 Recognition Codes of the Dockerin Domain

Earlier in section 8, we discussed how the structure of the duplicated sequence in the dockerin domain can correspond to a modification of the EF-hand motif, resulting in a proposed F-hand motif. We can now develop this finding further and gain more information by sequence comparison of dockerin domains which exhibit different specificities.

It was demonstrated previously, that the specificities of the dockerin domain of the enzymatic subunits of the *C. thermocellum* cellulosome are different from those of the dockerin on the scaffoldin subunit.[14,16] In this context, all of the dockerins from the enzyme subunits of this organism seem to exhibit the same specificity for the set of cohesins on the scaffoldin subunit.[37,43] As described earlier,[31] the scaffoldin dockerin is specific for the special type-II cohesin domains, located on at least three different cell surface proteins in this organism.

The question remained whether the dockerin domains would recognize the cohesins from different species of cellulosome (*i.e.*, from different microorganisms). The cross-species specificity of the cohesin-dockerin interaction was therefore investigated.[45] Cohesin-containing probes from the cellulosomes of two different bacterial species, *C. thermocellum* and *C. cellulolyticum,* were allowed to interact with cellulosomal enzymes from each species. In both cases, the cohesin domain of one bacterium interacted with enzymes from its own cellulosome only and failed to recognize enzymes from the other species. Thus, in the case of these two bacteria, the cohesin-dockerin interaction seems to be species specific.

Using this finding, we analyzed the dockerin sequences from the two species and also compared them with the dockerin from the scaffoldin subunit of *C. thermocellum* (Figure 14). In particular, we sought positions in the three sequences which would account for the different specificities. In this regard, strict intra-species conservation of residues coupled with cross-species variance would indicate a role in the biorecognition of the cohesin domain. Thus, in analyzing the sequences from the three groups of dockerin domains, we established the following criteria: (i) within a given species, the amino acid residues of all the dockerins should be conserved at each position, (ii) between the two species, the residues should show a high degree of dissimilarity at a given position, and (iii) residues of the CipA (scaffoldin) dockerin should differ from those of either of the two other groups. It is also clear that the side chains of residues known to participate directly in calcium-binding would presumably be involved in this function and would not contribute directly to the specificity properties of the dockerin domain.

Only four positions in the dockerin sequence appeared to fulfill the above criteria. The implicated residues comprise a repeated dyad located within the calcium-binding motif of the two duplicated sequences, characteristic of the dockerin domain (Figure 14A). These residues (positions **10** and **11**) are thus implicated directly in the recognition of the cohesins. According to the proposed model, these four residues do not participate in the binding of calcium *per se;* instead, they appear to serve as recognition codes in promoting the interaction with the cohesin domain. It is important to note that other dockerin residues, which do not fulfill the above criteria, might also serve as *contact residues* in the binding, but the four designated residues are considered critical for the *specificity* of the interaction.

Using the proposed structure of the dockerin domain (Figure 11C), we can predict where the four suspected residues might be located (Figure 14B). The two pairs of residues map at alternate poles of the calcium-binding loops (segments $\mathbf{b_1}$ and $\mathbf{b_2}$). We are currently performing site-directed mutagenesis of positions **10** and **11** from the dockerins of the two species, in attempts to interconvert the specificities and add further credence to the model. In modifying the suspected recognition codes, we hope to reverse the inherent specificities, such that the dockerin of one species will no longer recognize its own cohesins but will recognize those of the second species instead.

9.4 The Cohesin-Dockerin Interaction — A prediction

In the previous section, we showed how sequence comparisons can be combined with biochemical and structural information to predict which residues of the 70-residue dockerin domain might participate in the interaction with the cohesins. In this section, we will try to extend such an analysis and predict the complementary residues on the 140-residue cohesin domain which might similarly serve as specificity determinants in recognizing the dockerin domain.

Initial attempts at such an analysis compared the known cohesins of the scaffoldins from three species of bacteria.[73] This analysis yielded about 30 residues which fit the two required criteria: intra-species identity and cross-species variance. The labeled residues appeared to form at least three clusters over the cohesin molecule. It would be difficult to determine whether the selected areas would be important for dockerin binding, or whether these areas would be involved in protein-protein interactions, *e.g.*, for intra-subunit packaging of the cohesin domains. Clearly more information was necessary.

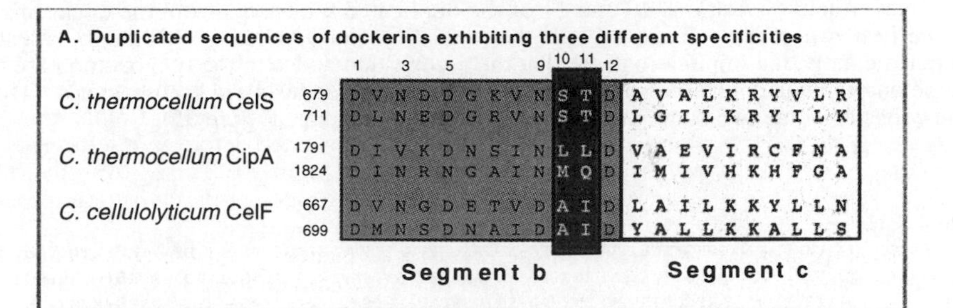

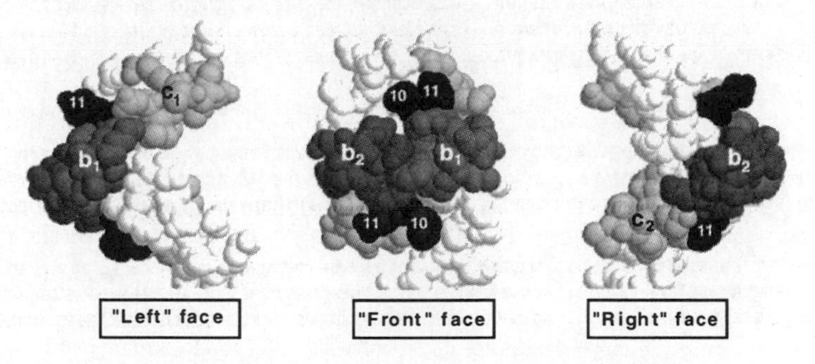

Figure 14 *Dockerin residues predicted to serve as recognition determinants in the cohesin-dockerin interaction.* **A.** *Duplicated sequences of dockerins exhibiting three different specificities. The corresponding portions of the dockerins from the CelS subunit of* C. thermocellum *and the CelF subunit of* C. cellulolyticum *are shown in the figure as representative of the two species. More than a dozen such dockerins are known from* C. thermocellum *and eight are currently known from* C. cellulolyticum. *The two domains shown were chosen from the analogous Family-48 cellulase component from the corresponding cellulosome. Segment* b *represents the calcium-binding loop, and segment* c *is predicted to be the F-helix equivalent (see Figure 11). In the case of the* C. thermocellum *dockerin, the two pairs of proposed recognition determinants are invariably ST or SS, whereas in the* C. cellulolyticum *dockerins, the two pairs are usually AI or AL. Both are different than the analogous residues (LL and MQ) from the only example of a scaffoldin dockerin presently known,* i.e., *the CipA dockerin from* C. thermocellum. ***B.*** *"Mug shots" of the proposed dockerin model. Chicken troponin C was used as a model for the spacefill structure shown in the figure. Segments* b *and* c *from the two duplicated segments are shaded (medium and light gray, respectively), and the positions of the two pairs of proposed specificity determinants are shaded in dark gray. The unshaded (white) residues represent areas of the molecule which are not considered to currently conform to the standard EF-hand motif. The final model (which will be based on crystallography or NMR) will presumably result in a distortion of the prototype.*

One problem in performing this analysis is the current deficiency of key sequences. For example, all nine scaffoldin cohesins are known for *C. thermocellum* and *C. cellulovorans* (the entire scaffoldin subunit has been sequenced in both bacteria),[9,10] but the sequences of only a few cohesins from the scaffoldin of *C. cellulolyticum* have been published.[34] On the other hand, many dockerin domains have been sequenced from cellulosomal subunits from *C. thermocellum* and *C. cellulolyticum,* but only a few dockerin domains have been reported for *C. cellulovorans.*[26] As more sequences are known, we will be able to refine the current analysis.

Nevertheless, positions 10 and 11 of the duplicated sequences from the dockerins of *C. cellulovorans* are consistent with those of *C. cellulolyticum.* Although untested experimentally, the implications are that the recognition codes in the two bacteria are the same and that the dockerins should cross-react with the cohesins in both bacteria. Thus, the cohesins of the two bacteria are placed in the same group (Figure 15).

```
          2° Struct  bb111bb      bbb222bbb      bb333bbb     bb444bbb           bb555b
GROUP 1
CipA2-ct    184  VVVEIGKVTGSVGTTVEIPVYFRGVPS-KGIANCDFVFRYDPNVLEIIGIDPGDIIVDPNPTKSFDTAIYPD
CipA5-ct    890  VRIKVDTVNAKPGDTVRIPVRFSGIPS-KGIANCDFVYSYDPNVLEIIEIEPGDIIVDPNPDKSFDTAVYPD
OlpA--ct     34  IEIIIGNVKARPGDRIEVPVSLKNVPD-KGIVSSDFVIEYDSKLFKVIELKAGDIVEN--PSESFSYNVVEK
GROUP 2
CbpA5-cv    952  VTATVGTATVKSGETVAVPVTLSNVP---GIATAELQVGFDATLLEVASITVGDIVLN--PSVNFSSVVNGS
CbpA8-cv   1377  VTATVGTATGKVGETVAVPVTLSNVP---GIATAEVQVGFDATLLEVASITAGDIVLN--PSVNFSSVVNGS
ClpC2-cl    435  LKVTVGTANGKPGDTVTVPVTFADVAKMKNVGTCNFYLGYDASLLEVVSVDAGPIVKN--AAVNFSSSASNG

          2° Struct   bb666bb             bbb777bbbb          bbbbb888bbbbb         bb999bb
GROUP 1
CipA2-ct    255  RKIIVFLFAEDSGTGAYAITKDGVFAKIRATVKSSAP----GYITFDEVGGFADNDLVEQ-KVSFIDGGVNV
CipA5-ct    961  RKIIVFLFAEDSGTGAYAITKDGVFATIVAKVKSGAPNGL-SVIKFVEVGGFANNDLVEQ-KTQFFDGGVNV
OlpA--ct    103  DEIIAVLYLEETGLGIEAIRTDGVFFTIVMEVSKDVKPGI-SPIKFESFGATADNDMNEM-TPKLVEGKVEI
GROUP 2
CbpA5-cv   1019  --TIKLLFLDDT-LGSQLISKDGVLATINFKAKTVTSKVT-TPVAVSGTPVFADGTLAEL-NMKTVAGSVTI
CbpA8-cv   1444  --TIKILFLDDT-LGSQLISKDGVFATVNFKVKSTATNSAVTPVTVSGTPVFADGTLAEL-KSESAAGRLTI
ClpC2-cl    505  --TISFLFLDNT-ITDELITADGVFANIKFKLFSVTAKTT-TPVTFKDGGAFGDGTMSKIASVTKTNGSVTI
```

Figure 15 *Prediction of cohesin residues which may be involved in binding to the dockerin domain. The sequences of the cohesin domains shown in the figure are aligned and grouped according to their proposed specificities. For the sake of simplicity, representative cohesins from the scaffoldins of the different bacteria are shown: scaffoldin cohesins 2 and 5 from* C. thermocellum *(CipA2-ct and CipA5-ct), non-cellulosomal cohesin from the latter bacterium (Orf3p-ct), and scaffoldin cohesins from* C. cellulovorans *(CbpA5-cv and CbpA8-cv) and* C. cellulolyticum *(CipC2-cl). The residues (numbered as they appear in the respective scaffoldin gene) in positions displaying combined group-specific identity and inter-group variance are emphasized in bold font. The positions of the β strands, based on the cohesin-2 crystal structure, are numbered and demarcated by* "b".

The breakthrough, which enabled a greater refinement of the data, was based on the sequence of a non-cellulosomal surface-borne cohesin (OlpA) from *C. thermocellum.*[52,54] Initially, it was postulated that this particular cohesin would bind to the dockerin on the scaffoldin subunit, thereby implanting the cellulosome into the cell surface. However, the OlpA cohesin was found to be a type-I cohesin in its structure and recognition properties. Like the cohesins on the scaffoldin subunit, this cohesin was shown to bind to the dockerin domains of the cellulosomal enzymes from this organism. Its sequence can thus be grouped with those of the scaffoldin cohesins from the same bacterium.

OlpA is sufficiently dissimilar, however, to allow us to reduce the number of suspected residues which might interact with the dockerin domain (Figure 15). These residues include S63, G84, A89, N122, and D123 of *C. thermocellum* (numbered according to the cohesin-2 convention),[73] which are strictly conserved in all of the cohesins in Group 1 (*i.e.*, OlpA and all nine scaffoldin cohesins) but are clearly different from those of Group 2 (which are largely identical among themselves).

Interestingly, all of these residues are located in the three highly stabilized loops **4/5**, **6/7** and **8/9**. The residues are mapped into a contiguous strip across the crown of the cohesin molecule (Figure 16). In addition, the neighboring residues of this cluster were inspected and one of them, E81, was also included in the list. The E→D replacement is usually considered to be a conservative one, but in this case, all of the cohesins of Group 1 show glutamic acid at this position, and all of the known cohesins of Group 2 show an aspartate residue. This fact, plus its proximity to the other residues, suggested that this residue should also be included in the recognition strip of the cohesin domain.

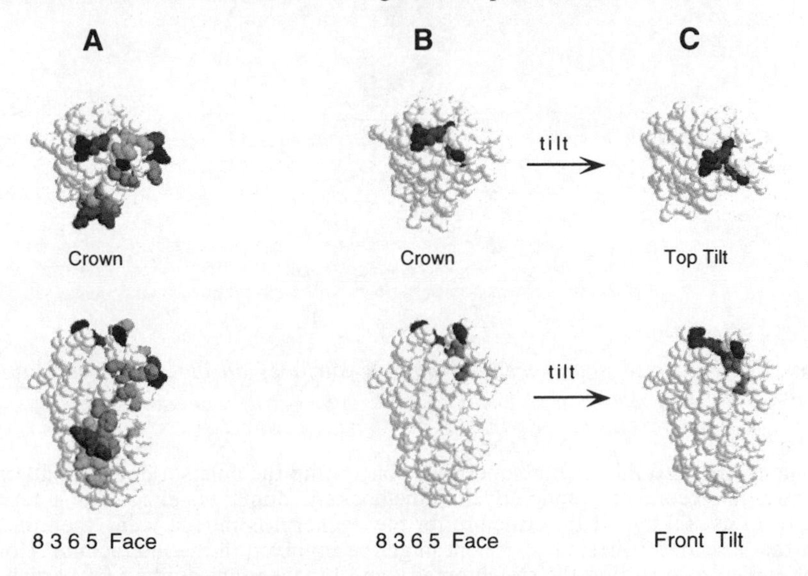

Figure 16 *Mapping of predicted cohesin residues, proposed to serve as recognition determinants for the dockerin domain.* ***A.*** *Analysis based on comparison of the cohesins from Group 1 — without OlpA — and those of Group 2.* ***B.*** *Analysis based on Group 1 including OlpA (see Figure 15).* ***C.*** *Tilted views of* ***B****, emphasizing the proposed "recognition strip".*

Another interesting difference between the cohesins of the two groups is the deletion of the glycine of the **6/7** loop in the cohesins of Group 2, rather than its replacement by another residue. Since this modification is consistent among all of the cohesins in the two groups, it is currently considered to be a true candidate for playing a role in dockerin recognition.

While sequence differences between the two groups of cohesin would indicate a difference in their recognition determinants, the differences between members within a given group would also be meaningful. For example, the differences between OlpA and the other cohesins of Group 1 may be relevant to a functional or structural difference between the cell-surface nature of OlpA and the location of the other cohesins in the scaffoldin subunit. In this regard, three extraneous patches of label can be discerned on the 8365 and the left faces of the cohesin domain (compare the structures in Figures 16A and B). This is a conserved face among all of the scaffoldin cohesins. One possibility is that these patches are involved in the interaction between the cohesins of the scaffoldin subunit. In this regard, this general region represents the complementary interface of the crystallographic dimer formed between two neighboring cohesin domains upon crystallization.[73]

Now that we have predicted portions of both the cohesin and dockerin domains which may serve as specificity determinants for each, we can examine whether the two structures fit. Using views of the two domains shown in Figures 14 and 16, we can see that one of the two recognition dyads of the dockerin domain seems to fit into a depression within the proposed recognition strip of the cohesin domain (Figure 17).

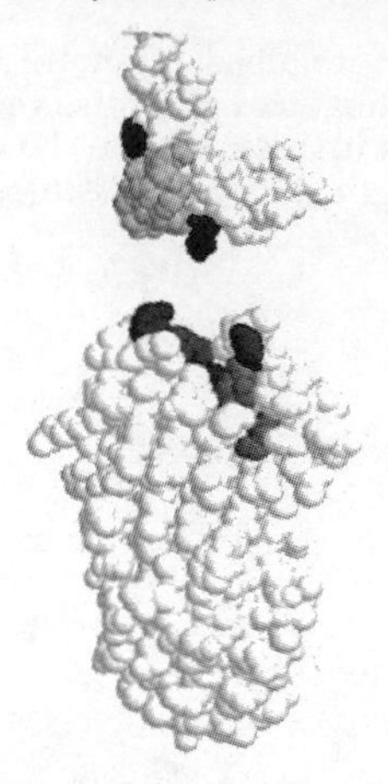

Figure 17 *Predicted complementary binding surfaces on the cohesin and dockerin domains.*

The model also raises some questions concerning the cohesin-dockerin interaction. As drawn, the second recognition dyad of the dockerin domain is directed at almost a 90° angle from its sibling. If the structure for the dockerin is indeed valid, then this could imply that a second cohesin component might be involved in the interaction. However, recent evidence regarding the stoichiometry and biochemical binding properties of the cohesin-dockerin interaction does not seem to support this notion.[89]

In any event, our analysis suggests that only six residues of the cohesin domain are predicted to participate in the recognition of the dockerin domain. Moreover, only four dockerin residues are considered to be critical to the specificity of the interaction. This is a testable hypothesis, and we are currently modifying, by site-directed mutagenesis, the relevant residues on both domains. Subsequent biochemical evidence should help us evaluate the accuracy of our prediction.

In the final analysis, the field awaits structure determination of the cohesin-dockerin complex, which will unequivocally define the interaction on the molecular level. We will then be able to judge, in retrospect, the legitimacy of our model and the validity of the approach.

10 CONCLUDING NOTE

In this chapter, we have documented the development of the cellulosome field, beginning with the early biochemical studies, continuing with the molecular biology and recent structure determinations. The information gained from these three disciplines can be combined with bioinformatics to provide novel structural and functional information, regarding the various components of the cellulosome.

As demonstrated in this chapter, the above combination allows us to focus on the hotspots of a given structure and to propose their possible functional implications. The identification of suspected residues is subject to experimental verification. Hence, based on our predictions, we can engage in a strategy of rational design and avoid the more random approach to mutagenesis. We can now select individual residues or groups of residues and test directly a well-defined hypothesis.

In an earlier review,[24] we suggested various uses of hybrid and chimeric forms of cellulosomes and their component parts for numerous types of applications. In one such application, different cohesin-dockerin pairs, which exhibit alternative specificities were suggested. The potential applications are very broad indeed. However, despite the progress and interest, the area is still in its infancy. The technical advances envisioned are slow to come, and we are dependent on further progress in our basic understanding of how the various cellulosomal components operate. The contribution of modern structural biology and homology modeling, combined with biochemical and genetic support will undoubtedly promote rapid strides in this direction.

Acknowledgments

We thank Dr. Miriam Eisenstein of the Weizmann Institute for initial docking studies on the cohesin-dockerin interaction and for productive discussions. Financial support is gratefully acknowledged from a contract from the European Commission (EU Contract BIO4-CT97-2303) and by grants from the Israel Science Foundation (administered by the Israel Academy of Sciences and Humanities, Jerusalem). Technical support was also provided by the Technion Otto Meyerhof Biotechnology Laboratories, established by the Minerva Foundation, Federal Republic of Germany.

References

1. R. Lamed, E. Setter, R. Kenig and E. A. Bayer, *Biotechnol. Bioeng. Symp.*, 1983, **13,** 163.
2. E. A. Bayer, R. Kenig and R. Lamed, *J. Bacteriol.*, 1983, **156,** 818.
3. R. Lamed, E. Setter and E. A. Bayer, *J. Bacteriol.*, 1983, **156,** 828.
4. P. Cornet, J. Millet, P. Béguin and J.-P. Aubert, *Bio/technology*, 1983, **1,** 589.
5. P. Béguin, P. Cornet and J.-P. Aubert, *J. Bacteriol.*, 1985, **162,** 102.
6. G. Joliff, P. Béguin and J.-P. Aubert, *Nucl. Acids Res.*, 1986, **14,** 8605.
7. G. P. Hazlewood, M. P. M. Romaniec, K. Davidson, O. Grépinet, P. Béguin, J. Millet, O. Raynaud and J.-P. Aubert, *FEMS Microbiol. Lett.*, 1988, **51,** 231.
8. P. Béguin, *Ann. Rev. Microbiol.*, 1990, **44,** 219.
9. U. T. Gerngross, M. P. M. Romaniec, T. Kobayashi, N. S. Huskisson and A. L. Demain, *Mol. Microbiol.*, 1993, **8,** 325.
10. O. Shoseyov, M. Takagi, M. A. Goldstein and R. H. Doi, *Proc. Natl. Acad. Sci. USA*, 1992, **89,** 3483.
11. N. R. Gilkes, B. Henrissat, D. G. Kilburn, R. C. J. Miller and R. A. J. Warren, *Microbiol. Rev.*, 1991, **55,** 303.
12. D. M. Poole, E. Morag, R. Lamed, E. A. Bayer, G. P. Hazlewood and H. J. Gilbert, *FEMS Microbiol. Lett.*, 1992, **99,** 181.
13. S. Salamitou, K. Tokatlidis, P. Béguin and J.-P. Aubert, *FEBS Lett.*, 1992, **304,** 89.
14. S. Salamitou, O. Raynaud, M. Lemaire, M. Coughlan, P. Béguin and J.-P. Aubert, *J. Bacteriol.*, 1994, **176,** 2822.
15. K. Tokatlidis, S. Salamitou, P. Béguin, P. Dhurjati and J.-P. Aubert, *FEBS Lett.*, 1991, **291,** 185.
16. K. Tokatlidis, P. Dhurjati and P. Béguin, *Protein Eng.*, 1993, **6,** 947.
17. R. Lamed and E. A. Bayer, *Adv. Appl. Microbiol.*, 1988, **33,** 1.
18. R. Lamed and E. A. Bayer, *in* Biochemistry and Genetics of Cellulose Degradation (J.-P. Aubert, P. Beguin and J. Millet, Eds.), pp. 101, Academic Press, London, 1988.
19. R. Lamed and E. A. Bayer, *in* Biosynthesis and biodegradation of cellulose and cellulose materials (C. H. Haigler and P. J. Weimer, Eds.), pp. 377, Marcel Dekker, New York, 1991.
20. P. Béguin, J. Millet and J.-P. Aubert, *FEMS Microbiol. Lett.*, 1992, **100,** 523.
21. E. A. Bayer and R. Lamed, *Biodegradation*, 1992, **3,** 171.
22. C. R. Felix and L. G. Ljungdahl, *Ann. Rev. Microbiol.*, 1993, **47,** 791.
23. R. Lamed and E. A. Bayer, *in* Genetics, biochemistry and ecology of lignocellulose degradation (K. Shimada, S. Hoshino, K. Ohmiya, K. Sakka, Y. Kobayashi and S. Karita, Eds.), pp. 1, Uni Publishers Co., Ltd., Tokyo, Japan, 1993.

24. E. A. Bayer, E. Morag and R. Lamed, *Trends Biotechnol.*, 1994, **12,** 378.
25. E. A. Bayer, E. Morag, Y. Shoham, J. Tormo and R. Lamed, *in* Bacterial adhesion: molecular and ecological diversity (M. Fletcher, Ed.), pp. 155, Wiley-Liss, Inc., New York, 1996.
26. P. Béguin and J.-P. Aubert, *FEMS Microbiol. Lett.*, 1994, **13,** 25.
27. P. Béguin and M. Lemaire, *Crit. Rev. Biochem. Molec. Biol.*, 1996, **31,** 201.
28. J. H. D. Wu, W. H. Orme-Johnson and A. L. Demain, *Biochemistry*, 1988, **27,** 1703.
29. K. Kruus, A. C. Lua, A. L. Demain and J. H. D. Wu, *Proc. Natl. Acad. Sci. USA*, 1995, **92,** 9254.
30. E. A. Bayer, E. Setter and R. Lamed, *J. Bacteriol.*, 1985, **163,** 552.
31. E. Leibovitz and P. Béguin, *J. Bacteriol.*, 1996, **178,** 3077.
32. B. R. Ali, M. P. Romaniec, G. P. Hazlewood and R. B. Freedman, *Enzyme Microbiol. Technol.*, 1995, **17,** 705.
33. T. Fujino, P. Béguin and J.-P. Aubert, *FEMS Microbiol. Lett.*, 1992, **94,** 165.
34. S. Pagès, A. Belaich, C. Tardif, C. Reverbel-Leroy, C. Gaudin and J.-P. Belaich, *J. Bacteriol.*, 1996, **178,** 2279.
35. M. Takagi, S. Hashida, M. A. Goldstein and R. H. Doi, *J. Bacteriol.*, 1993, **175,** 7119.
36. S. K. Choi and L. G. Ljungdahl, *Biochemistry*, 1996, **35,** 4906.
37. B. Lytle, C. Myers, K. Kruus and J. H. D. Wu, *J. Bacteriol.*, 1996, **178,** 1200.
38. M. M. Ahsan, T. Kimura, S. Karita, K. Sakka and K. Ohmiya, *J. Bacteriol.*, 1996, **178,** 5732.
39. K. Hon-nami, M. P. Coughlan, H. Hon-nami and L. G. Ljungdahl, *Arch. Biochem. Biophys.*, 1986, **145,** 13.
40. S. Kohring, J. Weigel and F. Mayer, *Appl. Environ. Microbiol.*, 1990, **56,** 3798.
41. B. Henrissat and A. Bairoch, *Biochem. J.*, 1996, **316,** 695.
42. A. Navarro, M.-C. Chebrou, P. Béguin and J.-P. Aubert, *Res. Microbiol.*, 1991, **142,** 927.
43. S. Yaron, E. Morag, E. A. Bayer, R. Lamed and Y. Shoham, *FEBS Lett.*, 1995, **360,** 121.
44. L. Gal, S. Pagès, C. Gaudin, A. Belaich, C. Reverbel-Leroy, C. Tardif and J.-P. Belaich, *Appl. Environ. Microbiol.*, 1997, **63,** 903.
45. S. Pagès, A. Belaich, J.-P. Belaich, E. Morag, R. Lamed, Y. Shoham and E. A. Bayer, *Proteins*, in press.
46. E. Lamed, J. Naimark, E. Morgenstern and E. A. Bayer, *J. Bacteriol.*, 1987, **169,** 3792.
47. E. Lamed, J. Naimark, E. Morgenstern and E. A. Bayer, *J. Microbiol. Methods*, 1987, **7,** 233.
48. R. Lamed, E. Morag (Morgenstern), O. Mor-Yosef and E. A. Bayer, *Curr. Microbiol.*, 1991, **22,** 27.
49. H. J. Flint, J. Martin, C. A. McPherson, A. S. Daniel and J. X. Zhang, *J. Bacteriol.*, 1993, **175,** 2943.
50. J. X. Zhang, J. Martin and H. J. Flint, *Mol. Gen. Genet.*, 1994, **245,** 260.
51. C. Fanutti, T. Ponyi, G. W. Black, G. P. Hazlewood and H. J. Gilbert, *J. Biol. Chem.*, 1995, **270,** 29314.
52. T. Fujino, P. Béguin and J.-P. Aubert, *J. Bacteriol.*, 1993, **175,** 1891.
53. A. Lupas, H. Engelhardt, J. Peters, U. Santarius, S. Volker and W. Baumeister, *J. Bacteriol.*, 1994, **176,** 1224.
54. S. Salamitou, M. Lemaire, T. Fujino, H. Ohayon, P. Gounon, P. Béguin and J.-P. Aubert, *J. Bacteriol.*, 1994, **176,** 2828.
55. M. Lemaire, H. Ohayon, P. Gounon, T. Fujino and P. Béguin, *J. Bacteriol.*, 1995, **177,** 2451.
56. E. A. Bayer and R. Lamed, *J. Bacteriol.*, 1986, **167,** 828.
57. F. Mayer, M. P. Coughlan, Y. Mori and L. G. Ljungdahl, *Appl. Environ. Microbiol.*, 1987, **53,** 2785.

58. P. Tomme, R. A. J. Warren, R. C. Miller, D. G. Kilburn and N. R. Gilkes, *in* Enzymatic degradation of insoluble polysaccharides (J. M. Saddler and M. H. Penner, Eds.), pp. 142, American Chemical Society, Washington, D.C., 1995.
59. E. Morag, A. Lapidot, D. Govorko, R. Lamed, M. Wilchek, E. A. Bayer and Y. Shoham, *Appl. Environ. Microbiol.*, 1995, **61,** 1980.
60. J. Tormo, R. Lamed, A. J. Chirino, E. Morag, E. A. Bayer, Y. Shoham and T. A. Steitz, *EMBO J.*, 1996, **15,** 5739.
61. J. Sakon, D. Irwin, D. B. Wilson and P. A. Karplus, *Nature Struct. Biol.*, 1997, **4,** 810.
62. G.-Y. Xu, E. Ong, N. R. Gilkes, D. G. Kilburn, D. R. Muhandiram, M. Harris-Brandts, J. P. Carver, L. E. Kay and T. S. Harvey, *Biochemistry*, 1995, **34,** 6993.
63. P. E. Johnson, P. Tomme, M. D. Joshi and L. P. McIntosh, *Biochemistry*, 1996, **35,** 13895.
64. P. J. Kraulis, G. M. Clore, M. Nilges, T. A. Jones, G. Pettersson, J. Knowles and A. M. Gronenborn, *Biochemistry*, 1989, **28,** 7241.
65. M.-L. Mattinen, M. Kontteli, J. Kerovuo, M. Linder, A. Annila, G. Lindeberg, T. Reinikainen and T. Drakenberg, *Protein Sci.*, 1997, **6,** 294.
66. M. R. Bray, P. E. Johnson, N. R. Gilkes, L. P. McIntosh, D. G. Kilburn and R. A. J. Warren, *Protein Sci*, 1996, **5,** 2311.
67. P. Tomme, A. L. Creagh, D. G. Kilburn and C. A. Haynes, *Biochemistry*, 1996, **35,** 13885.
68. P. Tomme, R. A. J. Warren and N. R. Gilkes, *Adv. Microb. Physiol.*, 1995, **37,** 1.
69. G. P. Hazlewood, K. Davidson, J. I. Laurie, N. S. Huskisson and H. J. Gilbert, *J. Gen. Microbiol.*, 1993, **139,** 307.
70. L. Gal, C. Gaudin, A. Belaich, S. Pagès, C. Tardif and J.-P. Belaich, *J. Bacteriol.*, in press, **179**.
71. S. Yaron, L. W. Shimon, F. Frolow, R. Lamed, E. Morag, Y. Shoham and E. A. Bayer, *J. Biotechnol.*, 1996, **51,** 243.
72. L. J. W. Shimon, F. Frolow, S. Yaron, R. Lamed, E. Morag, E. A. Bayer and Y. Shoham, *Acta Crystallogr., D.*, 1997, **53,** 114.
73. L. J. W. Shimon, E. A. Bayer, E. Morag, R. Lamed, S. Yaron, Y. Shoham and F. Frolow, *Structure*, 1997, **5,** 381.
74. P. Béguin, O. Raynaud, M.-K. Chaveroche, A. Dridi and P. M. Alzari, *Protein Sci.*, 1996, **5,** 1192.
75. B. Rost and C. Sander, *J. Mol. Biol.*, 1993, **232,** 584.
76. B. Rost and C. Sander, *Proteins*, 1994, **19,** 55.
77. E. Morag (Morgenstern), E. A. Bayer and R. Lamed, *Appl. Biochem. Biotechnol.*, 1991, **30,** 129.
78. E. Morag, E. A. Bayer and R. Lamed, *Appl. Biochem. Biotechnol.*, 1992, **33,** 205.
79. G. Gerwig, P. de Waard, J. P. Kamerling, J. F. G. Vliegenthart, E. Morgenstern, R. Lamed and E. A. Bayer, *J. Biol. Chem.*, 1989, **264,** 1027.
80. G. Gerwig, J. P. Kamerling, J. F. G. Vliegenthart, E. Morag (Morgenstern), R. Lamed and E. A. Bayer, *Eur. J. Biochem.*, 1991, **196,** 115.
81. G. Gerwig, J. P. Kamerling, J. F. G. Vliegenthart, E. Morag (Morgenstern), R. Lamed and E. A. Bayer, *Eur. J. Biochem.*, 1992, **205,** 799.
82. G. Gerwig, J. P. Kamerling, J. F. G. Vliegenthart, E. Morag, R. Lamed and E. A. Bayer, *J. Biol. Chem.*, 1993, **268,** 26956.
83. A. M. Showalter and J. E. Varner, *in* The biochemistry of plants. A comprehensive treatise (M. A., Ed.), Vol. 15, pp. 485, Academic Press, Inc., San Diego, 1989.
84. B. T. Nall, *Comments Mol. Cell. Biophys.*, 1985, **3,** 123.
85. M. P. Williamson, *Biochem. J.*, 1994, **297,** 249.
86. S. Chauvaux, P. Béguin, J.-P. Aubert, K. M. Bhat, L. A. Gow, T. M. Wood and A. Bairoch, *Biochem. J.*, 1990, **265,** 261.
87. C. I. Brändén and J. Tooze (1991) Introduction to Protein Structure, Garland Publishers, Inc., New York.
88. O. Herzberg and M. N. James, *J. Mol. Biol.*, 1988, **203,** 761.
89. I. Kataeva, G. Guglielmi and P. Béguin, *Biochem. J.*, 1997, **326,** 617.

THE CELLULOSOME: A VERSATILE SYSTEM FOR COUPLING CELLULOLYTIC ENZYMES AND ATTACHING THEM TO THE CELL SURFACE

Pierre Béguin, Sylvie Chauvaux, Marie-Kim Chaveroche, Gérard Guglielmi, Irina Kataeva, Emmanuelle Leibovitz and Isabelle Miras

Unité de Physiologie Cellulaire and URA CNRS 1300, Département des Biotechnologies, Institut Pasteur, 75724 Paris Cedex 15, France

1 INTRODUCTION

As discussed in other presentations (see Alzari *et al.* and Bayer *et al.*, this volume), the cellulase system of *Clostridium thermocellum* consists of high molecular weight exocellular complexes, termed cellulosomes, which comprise a variety of cellulolytic enzymes organised around a non-catalytic, cellulose-binding scaffolding subunit. Hydrolytic subunits bind to the scaffolding component by means of conserved, non-catalytic domains, termed dockerin domains of type I, which consist of two highly similar segments of about 22 residues each. The scaffolding component, termed CipA, comprises a cellulose-binding domain (CBD) and a set of nine highly similar modules, termed cohesin domains of type I, which serve as receptors for the dockerin domains borne by the catalytic subunits (for a review, see Béguin and Lemaire[1]).

The same principle is used to anchor the cellulosome to the cell surface. CipA also contains a COOH-terminal duplicated segment, termed dockerin domain of type II, since its sequence diverges from those harboured by the catalytic subunits. This domain fails to bind to the cohesin domains borne by CipA. By contrast, it binds to a specific set of cohesin domains, termed cohesin domains of type II, which are borne by several cell-bound polypeptides (Figure 1). Two of these polypeptides, OlpB and SdbA, have been characterised, as well as the corresponding genes.[2-4] Both were located in the cell envelope by immunocytochemical labelling and electron microscopy. Their NH_2-terminal region consists of four and one cohesin domain of type II, respectively. Their COOH-terminal region comprises three similar and conserved segments. These segments, termed SLH segments, are also found in S-layer proteins and exocellular enzymes of a variety of bacteria. Available evidence suggests that they serve to anchor polypeptides to the cell surface.

This presentation will focus on the reconstitution of simplified cellulase complexes, on improving the targeted integration of different polypeptides within artificial, cellulosome-derived complexes, and on the mode of attachment of the cellulosome to the surface of *C. thermocellum*.

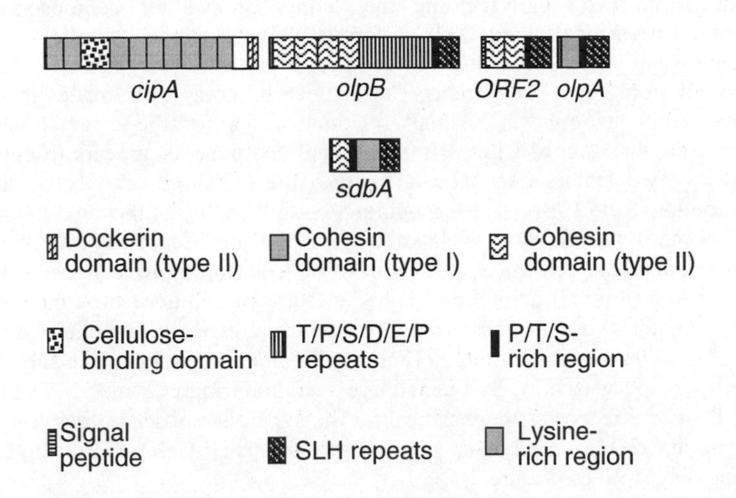

Figure 1 *Genes encoding surface proteins with cohesin domains of type II, which are probably involved in anchoring cellulosome and cellulosome components to the surface of* C. thermocellum. *The genes cipA, ORF2, and olpA are contiguous. sdbA maps in a different region. The genes are transcribed from left to right. The position of the segments encoding the various regions identified within each polypeptide are indicated with different patterns*

2 RECONSTITUTION OF SIMPLIFIED CELLULOLYTIC COMPLEXES

It has long been known that the cellulosome of *C. thermocellum* degrades crystalline cellulose with a high efficiency.[5,6] Activity is highly dependent on the association of the cellulolytic components, since dissociated subunits loose most of their activity against the crystalline substrate, although they retain to a large extent the capacity to hydrolyse amorphous and soluble cellulose derivatives.[7,8] However, little is known about the features of the cellulosome that are critical for activity. Indeed, not much is known concerning the structure and topology of the cellulosome besides that cohesin-dockerin interactions are responsible for anchoring catalytic subunits to CipA. In high resolution electron micrographs, cellulosomes appear in a variety of shapes, which may reflect different stages of maturation or decay.[9,10] Available evidence suggests that the stoichiometry and topological organisation of catalytic subunits along CipA is not unique. To begin with, there are at least 15 catalytic subunits containing a dockerin domain, and they cannot possibly fit on the 9 cohesin domains of CipA. Furthermore, most of the cohesin domains of CipA are extremely similar to one another, and even those that show the widest sequence divergence do not seem to discriminate between the different dockerin domains borne by the catalytic subunits.[11,12] Finally, "cellulosome" preparations appear to consist of a mixture of species with distinct physico-chemical and enzymatic properties.[13,14] However, binding of the catalytic subunits to the cohesin domains of CipA may be not entirely random, since interactions between the catalytic domains may force or rule out specific neighbourhood relationships between defined subunits.

With a view to understanding the factors critical for cellulosome activity, reconstitution experiments have been performed using mixtures of total[8,15] or selected cellulosome components.[16-20] Although good recovery of activity has been reported for reconstitution from total components, the activity of complexes made up of defined components amounts only to a small fraction of that of the original cellulosome. Nevertheless, association of CipA with individual components appears to enhance their activity.[16-20] Two factors may to contribute to the enhanced activity of cellulosome subunits complexed to CipA. Firstly, binding to CipA connects the catalytic subunits to the CBD borne by CipA. It is well known that the presence of a CBD enhances the activity of individual cellulases, which may be due to improved adsorption to the substrate or to a physical disruption of the structure of cellulose brought about by the CBD.[21,22] Another possible factor may be that the clustering of cellulolytic enzymes along CipA enables them to act more efficiently by attacking a site of the substrate whose structure is already weakened by the action of neighbouring enzymes. We investigated which of the two effects was more critical for the synergism observed between CipA and endoglucanase CelD.[20] Complexes were constructed using CelD and different forms of scaffolding polypeptides (Figure 2).[23]

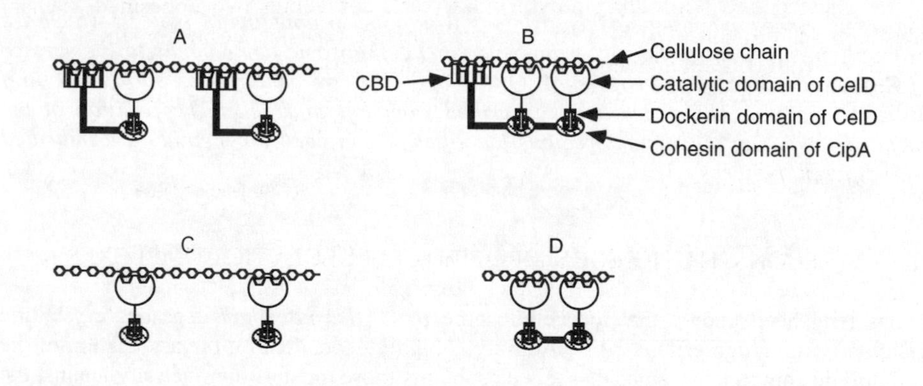

Figure 2 *Complexes of endoglucanase CelD with various scaffolding polypeptides. CelD is bound by means of its dockerin domain to polypeptides composed of a CBD linked to a single cohesin domain (A), a CBD linked to a double cohesin domain (B), a single cohesin domain with no CBD (C), or a double cohesin domain with no CBD (D)*

It was shown that a polypeptide consisting of a cellulose-binding domain plus a single cohesin domain was sufficient to yield maximum enhancement of the Avicelase activity of CelD: neither CipA nor a polypeptide consisting of a CBD linked to a cohesin domain repeated in tandem gave better results. In the latter case, it was shown by analytical ultracentrifugation analysis that two CelD molecules could indeed be bound simultaneously to the two cohesin domains linked in tandem. Polypeptides with one or two cohesin domains but no CBD failed to enhance cellulolytic activity.

Thus, in the case of the CelD-CipA synergism, linkage with a CBD, rather than clustering of the catalytic domains, appears to be the critical factor. Enhancement of CelD activity is not correlated with higher adsorption of the enzyme to the substrate, however. In the presence of sufficiently high cellulose concentrations, CelD can be bound almost quantitatively to the substrate, yet stimulation by CipA or by simplified scaffolding polypeptides bearing a CBD is still observed. The CBD + cohesin polypeptide may contribute to the activity of the complex by physically modifying the structure of cellulose. Alternatively, it may promote a more adequate presentation of the catalytic site than the spontaneously adsorbed enzyme.

3 IMPROVED TARGETED INTEGRATION OF DIFFERENT PROTEINS WITHIN CELLULOSOME-DERIVED COMPLEXES

As discussed above, cohesin and dockerin domains can be shuffled around and fused to foreign polypeptides, to which they confer their binding specificity. Thus, they lend themselves to the engineering of multiprotein complexes.[24,25] However, the general lack of binding specificity among cohesin and dockerin domains belonging to the same type would be expected to hamper the targeted integration of specific components. Combining cohesin-dockerin interactions of different types would provide a better way to specify where each component of the complex is to be integrated along the scaffolding polypeptide. We compared the formation of complexes integrating endoglucanases CelD and CelC using two scaffolding polypeptides, each containing two cohesin domains. One, termed Cip6, consisted of a tandem repeat of the type I cohesin domain derived from the seventh cohesin domain of CipA. The other, termed Cip20, was constructed by fusing the same cohesin domain to the type II cohesin domain of SdbA (Figure 3).

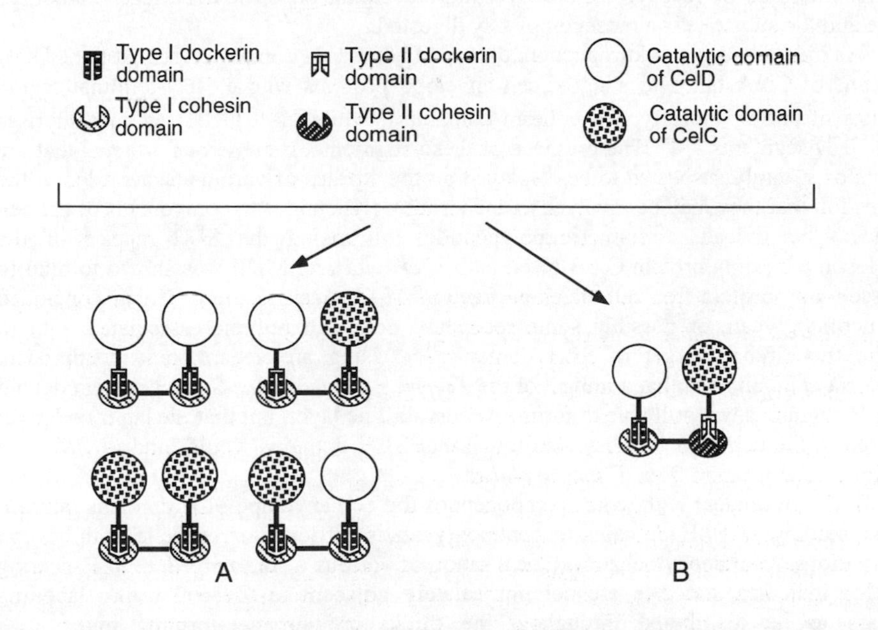

Figure 3 *Complexes containing the possible combinations that may form with Cip6, which contains two identical cohesin domains in tandem (A), and with Cip20, which contains a cohesin domain of type I linked to a cohesin domain of type II (B)*

Cip6 was mixed with varying proportions of CelD, bearing its natural type I cohesin domain, and CelC-DSCelD, in which the dockerin domain of CelD was fused to CelC (CelC, which is not a cellulosome component, does not possess a dockerin domain of its own). Likewise, Cip20 was mixed with different proportions of CelD and CelC-DSCipA, which carries the type II dockerin domain of CipA. The formation of complexes was assessed by non-denaturing gel electrophoresis and densitometry, and the polypeptide composition of each band was analysed by a second electrophoresis run under denaturing conditions. The yield of complexes containing both CelD and CelC was improved from 36 to 56 % by using Cip20 instead of Cip6 as a scaffolding protein. As expected, this was due to the forced integration of the two polypeptides harbouring different dockerin domains next to one another in Cip20, whereas the Cip6/CelD/CelC-DSCelD mixtures contained important quantities of the complexes Cip6/$(CelD)_2$ and Cip6/$(CelC\text{-}DSCelD)_2$ (Leibovitz and Béguin, submitted).

4 INTERACTION OF SLH SEGMENTS WITH COMPONENTS OF THE CELL ENVELOPE

Early investigations using immunocytochemical labelling and electron microscopy showed that cellulosomes are present in protuberances formed by the outer layer of the cell envelope of *C.* thermocellum.[26-28] Cell attachment is not permanent, though, and cellulosomes can be recovered from the culture medium once the cells reach stationary phase and the substrate is almost completely digested.

As mentioned in the Introduction, the type II dockerin domain present at the COOH terminus of CipA binds to a set of cell envelope proteins whose NH_2-terminal region consists of one or more type II cohesin domain(s), while the COOH region comprises three SLH segments.[3,4,29] The presence of these segments in numerous proteins that are known or strongly suspected to be displayed on the surface of various bacteria led to the suggestion that they may be involved in anchoring polypeptides to components of the cell surface.[29,30] Indeed, a chimeric polypeptide comprising the SLH repeats of the cellulosome-binding protein OlpB fused to *Escherichia coli* MalE was shown to bind to an essentially protein-free cell envelope fraction of *C. thermocellum*. It was concluded that peptidoglycan, or possibly some secondary cell wall polymer associated with it, maybe the binding target of SLH domains.[31] Using an elegant genetic approach, Olabarría *et al.* showed that a mutant of the *Thermus thermophilus* S-layer protein devoid of SLH segments was still able to form a paracrystalline layer, but that the latter no longer adhered to the cell wall.[32] They also found that SLH domains would bind *in vitro* to a peptidoglycan fraction from *T. thermophilus*.

It is still unclear with which component of the cell envelope SLH domains interact. Direct binding of SLH domains to peptidoglycan is difficult to reconcile with several observations. Immunocytochemical localisation of various *C. thermocellum* SLH-bearing proteins indicates that they are not immediately adjacent to the cell wall : labelling appears to be distributed throughout the thick, protuberance-forming outer layer surrounding the cells outside of the S-layer.[4,31,33] In addition, in *Bacillus stearothermophilus* PV72, the SLH-domain-containing S-layer protein SbsB fails to bind to the peptidoglycan fraction once it has been stripped of a secondary polysaccharide containing glucosamine and mannosamine (both presumably N-acetylated).[34] In *C. thermocellum*, a component capable to bind SLH repeats was identified after extracting cell envelopes with hot SDS. This component migrates like a 28 kDa polypeptide on

SDS-polyacrylamide gels.[31] However, it is resistant to protease treatment, remains in the aqueous phase upon phenol-chloroform extraction, and cannot be visualised with Coomassie blue nor with silver staining (unpublished data). Carbohydrate analysis of this component is in progress.

There is evidence that different SLH segments may have different binding specificities. For example, unlike the SLH segments of OlpB, the SLH segments from the cellulosome-binding protein SdbA fail to bind *in vitro* to SDS-extracted cell envelopes of *C. thermocellum*. Different specificities have also been observed between the SLH segments of the *C. thermocellum* S-layer protein SlpA, and the homologous segments present in the S-layer proteins EA1 and Sap from *Bacillus anthracis*. SLH segments from SlpA bind strongly to cell envelopes from *C. thermocellum*, but not at all to similarly prepared envelopes of *B. anthracis*. Conversely, the SLH segments of EA1 and Sap adsorb more strongly to envelopes from *B. anthracis* than to those from *C. thermocellum* (unpublished observations).

Elucidating the mode of interaction of SLH domains with the bacterial cell envelope will no doubt be important to understand how many bacteria manage to display a variety of exocellular proteins on their surface. Using rDNA technology, it may then be possible to engineer clones that would display various recombinant polypeptides. This opportunity may be worth exploring for practical applications such as the development of vaccines or immobilised enzymes.

5 CONCLUSION

Faced with cellulose, an abundant, but recalcitrant substrate, cellulolytic microorganisms have evolved a variety of polypeptide modules with a number of different functions, which can be combined in a number of different ways. In this respect, the tinkering chest of cellulosome-producing organisms such as *C. thermocellum* appears rather well stocked-up. To paraphrase Bayer and colleagues,[25] it is now up to the ingenuity of biotechnologists to turn it into a treasure trove for their own purposes.

References

1. P. Béguin and M. Lemaire, *Crit. Rev. Biochem. Molec. Biol.*, 1996, **31,** 201.
2. M. Lemaire and P. Béguin, *J. Bacteriol.*, 1993, **175,** 3353.
3. E. Leibovitz and P. Béguin, *J. Bacteriol.*, 1996, **178,** 3077.
4. E. Leibovitz, H. Ohayon, P. Gounon and P. Béguin, *J. Bacteriol.*, 1997, **179,** 2519.
5. E.A. Johnson, M. Sakajoh, G. Halliwell, A. Madia and A. L. Demain, *Appl. Environ. Microbiol.*, 1982, **43,** 1125.
6. R. Lamed, R. Kenig, E. Morag, J.F. Calzada, F. De Micheo and E.A. Bayer, *Appl. Biochem. Biotechnol.*, 1991, **27,** 173.
7. R. Lamed, E. Setter, R. Kenig and E.A. Bayer, *Biotechnol. Bioeng. Symp.*, 1983, **13,** 163.
8. K.M. Bhat and T.M. Wood, *Carbohydr. Res.*, 1992, **227,** 293.
9. R. Lamed, E. Setter and E.A. Bayer, *J. Bacteriol.*, 1983, **156,** 828.
10. F. Mayer, M.P. Coughlan, Y. Mori and L.G. Ljungdahl, *Appl. Environ. Microbiol.*, 1987, **53,** 2785.

11. S. Yaron, E. Morag, E.A. Bayer, R. Lamed and Y. Shoham, *FEBS Lett.*, 1995, **360,** 121.
12. B. Lytle, C. Myers, K. Kruus and J.H.D. Wu, *J. Bacteriol.*, 1996, **178,** 1200.
13. M. Pohlschröder, S.B. Leschine and E. Canale-Parola, *J. Bacteriol.*, 1994, **176,** 70.
14. B.R.S. Ali, M.P.M. Romaniec, G.P. Hazlewood and R.B. Freedman, *Enzyme Microb. Technol.*, 1995, **17,** 705.
15. L. Beattie, K.M. Bhat and T.M. Wood, *Appl. Microbiol. Biotechnol.*, 1994, **40,** 740.
16. J.H.D. Wu and A.L. Demain *in* FEMS Symposium No. 43 'Biochemistry and Genetics of Cellulose Degradation' (J.-P. Aubert, P. Béguin and J. Millet, eds), p. 117, Academic Press, London & New York, 1988.
17. S. Bhat, P.W. Goodenough and M.K. Bhat, *Int. J. Biol. Macromol.*, 1994, **16,** 335.
18. K. Kruus, A.C. Lua, A.L. Demain and J.H.D. Wu, *Proc. Natl. Acad. Sci. USA*, 1995, **92,** 9254.
19. M. Fukumura, K. Sakka, K. Shimada and K. Ohmiya, *Biosci. Biotechnol. Biochem.*, 1995, **59,** 40.
20. V. García-Campayo and P. Béguin, *J. Biotechnol.*, 1997, in press.
21. H. van Tilbeurgh, P. Tomme, M. Claeyssens, R. Bhikhabhai and G. Pettersson, *FEBS Lett.*, 1986, **204,** 223.
22. N. Din, N.R. Gilkes, B. Tekant, R.C. Miller Jr., R.A.J. Warren and D.G. Kilburn, *Bio/Technology*, 1991, **9,** 1096.
23. I. Kataeva, G. Guglielmi and P. Béguin, *Biochem. J.*, 1997, **326,** 617.
24. K. Tokatlidis, P. Dhurjati and P. Béguin, *Protein Eng*, 1993, **6,** 947.
25. E.A. Bayer, E. Morag and R. Lamed, *Trends in Biotechnol.*, 1994, **12,** 379.
26. E.A. Bayer, E. Setter and R. Lamed, *J. Bacteriol.*, 1985, **163,** 552.
27. E.A. Bayer and R. Lamed, *J. Bacteriol.*, 1986, **167,** 828.
28. A. Nolte and F. Mayer, *FEMS Microbiol. Lett.*, 1989, **61,** 65.
29. T. Fujino, P. Béguin and J.-P. Aubert, *J. Bacteriol.*, 1993, **175,** 1891.
30. A. Lupas, H. Engelhardt, J. Peters, U. Santarius, S. Volker and W. Baumeister, *J. Bacteriol.*, 1994, **176,** 1224.
31. M. Lemaire, H. Ohayon, P. Gounon, T. Fujino and P. Béguin, *J. Bacteriol.*, 1995, **177,** 2451.
32. G. Olabarría, J.L. Carrascosa, M.A. de Pedro and J. Berenguer, *J. Bacteriol.*, 1996, **178,** 4765.
33. S. Salamitou, M. Lemaire, T. Fujino, H. Ohayon, P. Gounon, P. Béguin and J.-P. Aubert, *J. Bacteriol.*, 1994, **176,** 2828.
34. W. Ries, C. Hotzy, I. Schocher, U.B. Sleytr and M. Sára, *J. Bacteriol.*, 1997, **179,** 3892.

CELLULOSOME ANALYSIS AND CELLULASES CELF AND CELG FROM *CLOSTRIDIUM CELLULOLYTICUM*

A. Belaich[1], J.-P. Belaich[1,2], H.-P. Fierobe[1], L. Gal[1], C. Gaudin[1], S. Pagès[1], C. Reverbel-Leroy[1] and C. Tardif[1,2]

[1] Laboratoire de Bioénergetique et Ingénierie des Protéines, IBSM, CNRS, 31 Chemin Joseph Aiguier, 13402 Marseille Cedex 20, France
[2] Université de Provence, Marseille, France.

1 INTRODUCTION

The mesophilic bacterium *Clostridium cellulolyticum* is a true cellulolytic organism able to grow on cellulose as the sole source of carbon and energy. A recent 16S rDNA analysis of polysaccharolytic clostridia has revealed that the mesophilic *C. cellulolyticum, Clostridium cellobioparum, Clostridium termitidis*, and *Clostridium papyrosolvens* are closely related (96.5-99.1 % of similarity)[1] and belong, with the thermophilic *Clostridium thermocellum* and *Clostidium stercorarium*, to the same cluster (group III). Surprisingly the mesophilic *Clostridium cellulovorans*, harbouring a cellulosome very similar to that of *C. cellulolyticum*, was found to be far from the other cellulolytic clostridia in the phylogenetic tree[1], which suggests a horizontal transfer of the overall cellulolytic system between quite distant species. The purpose of this paper is to review recent findings on the cellulolytic system of *C. cellulolyticum*, namely new data concerning the cellulosome analysis and the catalytic properties of cellulases CelF and CelG.

Six cellulase encoding genes (*celA, celD, celC, celF, celG,* and *celE*) have been cloned and sequenced[2,3,4,5,6]. All the deduced cellulase sequences contain a characteristic N-terminus signal sequence typical of secreted proteins, and a dockerin domain, which is a duplicated sequence of about 22 residues (DS), at the C-terminus, specific of the clostridial cellulases, which plays a major role in the cellulosome constitution (see below). Some cellulases (CelD, CelE and CelG) contain, in addition to the catalytic and the C-terminus dockerin domains, an extra domain, identified as a cellulose binding domain (CBD) on the basis of sequence comparisons.

The most striking feature in *C. cellulolyticum* was that, contrary to what has been reported for *C. thermocellum,* several genes were found to be clustered on the chromosome[3]. Such genetic organisation allowed walking on the chromosome from *celC*, to reveal *celF* encoding for the cellulase CelF[5], and *cipC* encoding for the scaffolding protein CipC[7,5]. The deduced amino acid sequence of CipC, revealed that the latter contains a type III Cellulose Binding Domain (CBD) at the N-terminus, at least two hydrophilic domains (HD) and probably 9 hydrophobic domains called cohesin domains (C). The number of cohesin domains has been estimated taking into account the fragments already sequenced and the size and the restriction map of the unsequenced fragment (Figure 1). This multi-domain protein is homologous to the scaffolding protein of two other cellulolytic clostridia: *C. thermocellum*[8]and *C. cellulovorans*[9]. Taking into account the scaffolding protein gene at the 5' end and the presence of a sixth gene, located downstream of *celE*, and encoding a putative endoglucanase (A. Bélaich personal communication), it can be assumed that the cluster is at least 16 kbp long.

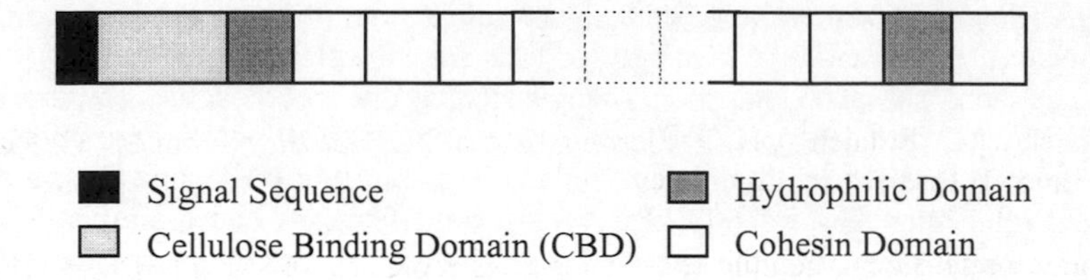

Figure 1 *Schematic representation of CipC. Solid line corresponds to sequenced parts, dotted line corresponds to unsequenced part.*

2 THE CELLULOSOME

2.1 Molecular reconstitution of cellulosome assemblage

The domains of the proteins (cellulases and scaffolding protein) which are involved in cellulosome assemblage were firstly identified using the following strategy. A recombinant truncated scaffolding protein, called miniCipC_1 containing the CBD, the first hydrophilic domain and the first cohesin domain of CipC was obtained[7]. Study of the interaction between this mini-scaffolding protein labelled by biotinylation and the endocellulase CelA was performed to identify the precise domains involved in the cellulosome assemblage.

CelA, as all the other cellulosomal catalytic subunits, contains a C-terminal dockerin domain and different forms of this protein were overproduced in *E. coli*: 1) the entire form containing the dockerin domain (CelA_2), 2) a truncated form devoid of the dockerin domain (CelA_3). A third form of the protein was generated by spontaneous proteolysis at the C-terminus of CelA2, leading to a protein containing only the first duplicated sequence of the dockerin domain. Only CelA_2 was found to interact with the mini-scaffolding polypeptide miniCipC_1. No binding was detected between any of the two other CelA forms. Furthermore, a mini-scaffolding protein called miniCipC_0 devoid of the cohesin domain did not bind any of the CelA forms[7].

It is therefore concluded that the cohesin domains of CipC act as receptors for the catalytic subunits. The entire dockerin domain of the cellulases was found to be necessary to interact with the cohesin domain. The CelA_2-miniCipC_1 interaction affinity constant, measured using the Plasmon Surface Resonance technique[7], was found to be 7.10^9 M^{-1}.

2.2 Analysis and properties of the cellulosomes from *Clostridium cellulolyticum*

2.2.1 Purification of the cellulosomes. The cellulosomes were prepared as described. After 5 days of growth on cellulose MN300, the culture was harvested by centrifugation. Ninety percent of the CMCase activity was located in the pellet (cells and cellulose). After extensive buffer washings, cells and non specifically adsorbed proteins were removed. The cellulosomal proteins were released from residual cellulose by extensive washing with distilled water and concentrated. The purification procedure led to the recovery of 50 mg of cellulosome per liter of culture with a specific CMCase activity of 2.6 IU/mg[10].

When analysed on SDS-PAGE, the cellulosomes were found to contain at least 14 proteins (Figure 2) which molecular masses range from 160 to 35 kDa. Among these proteins, the 160, 94 and 80.6 kDa polypeptides were the most abundant.

The apparent size and the homogeneity of the cellulosome preparation were analysed by gel filtration on CL-4B Sepharose. Two distinct peaks were obtained. The first one corresponded to an apparent molecular mass of 16 MDa, and the second to an apparent molecular mass of 600 kDa. Since both fractions had a similar SDS-PAGE pattern, it seems likely that the first peak corresponds to polycellulosomes composed by several (20-25) 600-kDa cellulosome units[10].

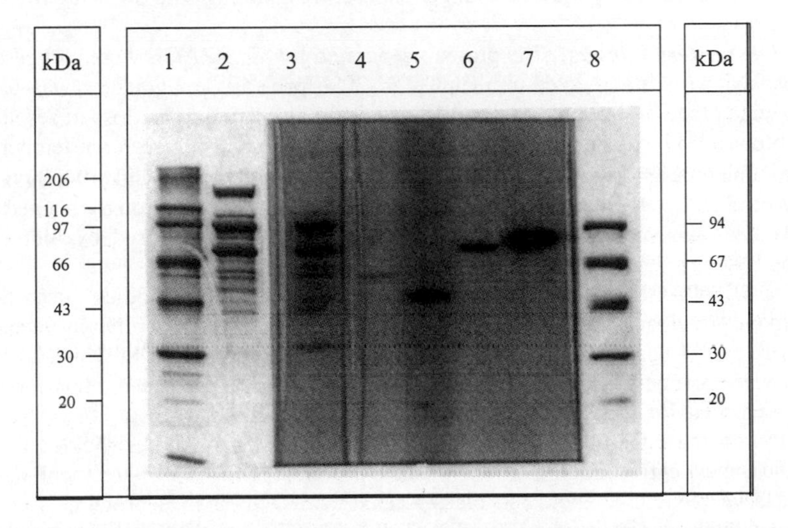

Figure 2 *Electrophoretic characterisation of components of the cellulosome from* Clostridium cellulolyticum. *Purified cellulosomes were analysed by SDS-PAGE. Lanes 1 and 8, MW markers (1: HMW and 8 : LMW calibration kits from Sigma and Pharmacia, respectively; the respective sizes (kDa) are indicated in columns on the right and left), lane 2, Coomassie blue-stained subunits of the cellulosome (14 bands were clearly visible, see table II for the respective sizes), lane 3, interaction of cellulosome subunits with biotinylated miniCipC1, lanes 4 to 7, Western blots showing patterns obtained with CelA, CelC, CelF and CBD-CelG antisera, respectively. In lane 7, the first band corresponding to 77.9 kDa is barely visible on the picture.*

2.2.2 Cellulose binding properties of the cellulosomes. The ability of the purified cellulosomes to bind cellulose was investigated. The apparent binding parameters were determined and found different from those of the miniCipC$_1$. The estimated dissociation constants (K_D = 1.4 10^{-8}M)[10] indicate that the cellulosome has a higher affinity (10-fold) for Avicel than miniCipC$_1$ for which a value of 1.4 10^{-7}M was obtained[11]. In addition, Avicel (Merck) bound 35-fold more molecules of miniCipC$_1$ (0.4 μmol/g) than of cellulosome (11.4 nmol/g). The parameters obtained with the CBD of the scaffolding protein CipA of *C. thermocellum* were very close to those measured with miniCipC$_1$, since a maximum number of sites of 0.54 μmol/g of cellulose and a K_D value of 4.0 10^{-7}M were reported for the CBD of CipA[12]. For each cellulosome, it is not possible to determine whether only the CBD of Cip, or the CBDs of the cellulases also interact with cellulose. Comparisons between the apparent binding parameters (smaller number of sites but stronger interaction) suggest that multiple interactions (involving several CBDs) may occur between the cellulosome units and cellulose.

2.2.3 Zymograms. Zymograms were performed on CMC (Carboxy Methyl Cellulose) with various quantities of pure cellulosomes[10]. The results showed the existence of 5 (or 5 groups of) proteins harboring CMCase activity: three major bands with an apparent *M*w of 89.6, 53 and 44.5 kDa and two minor bands corresponding to 35 and 29.5 kDa. Similar experiments were also carried out with swollen Avicel, xylan and laminarin. No activity was detected in these substrates, even with longer incubation times (36 h). With barley glucan, a single band corresponding to a 50 kDa protein was detected.

2.2.4 Western blotting. The proteins separated by SDS-PAGE were blotted before to be probed with biotinylated-miniCipC_1, (4-20% pre-cast gel gradient was used to detect a wide range of proteins, between 10 and 200 kDa, without any loss of resolution). All the blots were incubated under the same conditions (for 1 hour at room temperature) with the same quantity of biotinylated miniCipC_1 (7 μg). Thirteen bands (94, 89.6, 80.6, 77.9, 72.6, 67.7, 58.9, 54.2, 53, 49, 44.5, 43 and 29.5 kDa) were strongly stained. The 160 kDa band assumed to correspond to CipC was never recognised by biotinylated miniCipC1[10].

The SDS-PAGE blots were also incubated with various antibodies raised against recombinant proteins. The 80.6, 53 and 44.5 kDa proteins were specifically recognised by antisera raised against CelF, CelA and CelC, respectively. The 89.3 and 77.9 kDa proteins were recognised by CBD-CelG antibodies (Figure 2). The relative molecular masses measured for CelF, CelA and CelC were in agreement with the theoretical *M*w (Table 1). In the case of the CBD-CelG experiment, the antibodies yielded a barely detectable signal corresponding to a 77.9 kDa protein. This is in agreement with the expected *M*w of CelG. They also cross-reacted with a 89.3 kDa protein. In a third experiment the blot was incubated with CelA2 prior to antibodies raised against CelA. Two bands were revealed on the film: the first corresponded to CelA (49 kDa) and the second, to a large 160 kDa protein. When CelA3 was used, only the 49 kDa band was revealed. The 160 kDa form therefore appears to be the scaffolding protein CipC. It is worth noticing that no degradation of the scaffolding protein occurred during our purification procedure.

2.2.5 Glycosylation detections. The total proteins of the cellulosome were oxidised with metaperiodate to detect the glycosylated ones. Four proteins seemed to be glycosylated in the cellulosome[10]: CipC, CelE, CelF and the 67.7 kDa one. When less drastic oxidation was applied in order to detect only the sialic groups in the proteins, no signal was obtained.

2.2.6 N-terminal sequencing. N-terminal sequencing was carried out on the 80.6 kDa, 89.6 kDa, 94 kDa and 160 kDa bands[10]. The 80.6 kDa protein band gave the N-terminal sequence (ASSPAN) of CelF (34), which confirmed the results of the Western blotting experiments. As expected, the 160 kDa N-terminal (AGTGVV) matched the N-terminal sequence deduced from DNA sequence of *cipC* [7]. The N-terminal sequence (LVGAGD) of the 94 kDa protein corresponded to the CelE protein which gene is located downstream of *celG* in the cluster (1). The N-terminal sequence (QSTTPP) of the 89.6 kDa subunit did not correspond to any known cellulase. Table 1 summarises the results obtained.

Table 1 *Components of the cellulosome from* Clostridium cellulolyticum*: summary of the description and identification with the various methods used.*

Apparent molecular mass (kDa)	SDS-PAGE colored fractions	CMCase activity	CipC binding ability	Antibody identification	Glycosylated proteins	N-terminal sequence	Known components	
							name	Theoretical molecular mass (kDa)
160.0	S1				+	CipC	CipC	
114.4	S2							
94.0	S3		+		+	CelE	CelE	94
89.6	S4	+	+	CBD-CelG		QSTTPP		
80.6	S5		+	CelF	+	CelF	CelF	78
77.9	S6		+	CBD-CelG			CelG	76
72.6	S7		+					
67.7	S8		+		+		CelD	69
61.0	S9							
58.9	S10		+					
54.2			+					
53.0	S11	+	+	CelA			CelA	52
49.0	S12		+					
44.5		+	+	CelC			CelC	47
43.0	S13		+					
35.0	S14	+						
29.5		+	+					

3 CELLULOLYSIS

The global cellulolytic activity of the cellulosome is the result of synergism between the cellulases which are bound onto the scaffolding protein. The CBD of the scaffolding protein could be involved in the catalysis. In this section, cellulases CelF and CelG and some properties of the CBD belonging to the scaffolding protein are described.

3.1 The processive endocellulase CelF

3.1.1 CelF is a major component of the cellulosome. The CelF encoding sequence was isolated from *C. cellulolyticum* genomic DNA using the inverse PCR technique[5]. This gene lies between *cipC* and *celC* in the large *cel* cluster. Comparisons between the deduced amino acid sequence of the mature CelF (693 amino acids, Mr 77,626) and those of other β-glycanases showed that its N-terminal catalytic domain belongs to the family 48 of glycosyl hydrolases. CelF is composed of two domains: a N-terminal catalytic domain and a C-terminal dockerin domain. Immunoblotting analysis showed that CelF is one of the 3 major components of the cellulosome[13,10]. The ability of the entire form of CelF to interact with mini-CipC$_1$ was monitored by Western blotting and by Surface Plasmon Resonance technique. A K_D of 10^{-8} M was estimated for the dissociation of the complex mini-CipC$_1$-CelF. This value was quite similar to that found for CelA[13,7].

3.1.2 CelF is a cellulase. The recombinant form of the cellulase CelF, tagged by a C-terminal histidine tail, was overproduced in *Escherichia coli*. The fusion protein was purified by affinity chromatography on a Ni-NTA column. During storage at 4°C, a truncated form of CelF appeared, lacking the C-terminal duplicated segment, as previously observed with CelA and CelC. Truncated CelF and intact CelF were separated by performing chromatography on a Ni-NTA column, since only the intact form of the protein binds to the resin[13].

CelF is able to degrade PASC (Phosphoric Acid Swollen Cellulose), Avicel-cellulose and BMCC. Both entire and truncated purified forms degraded amorphous PASC (k_{cat} = 42 and 30 min^{-1}, respectively) and microcrystalline Avicel-cellulose (respective k_{cat} = 13 and 10 min^{-1}). On the other hand, only truncated CelF was able to weakly degrade CMC. No activity could be detected on xylan, laminarin or lichenan. The degradation of cellodextrins was studied by HPLC analysis. 10% of the G6 was found to be cleaved in 2 G3, and 90% into G4 + G2. G3 and G2 were the only hydrolytic products of G5, and G4 yielded only G2. Neither G3 nor G2 are substrate for CelF[13].

3.1.3 CelF is a processive endocellulase. The distribution of reducing sugars released by the truncated CelF and the truncated CelA (one endocellulase) from PASC between insoluble and soluble fractions were investigated and compared as described by Irwin *et al.*[14]. The sugars produced by truncated CelF were about 95% soluble and 5% insoluble, whereas those produced by truncated CelA were only 42% soluble and 58% insoluble. As expected for an endocellulase[14], CelA released a large quantity of insoluble reducing ends (Figure 3). Conversely, the high ratio of soluble reducing ends versus insoluble reducing ends released by truncated CelF from amorphous cellulose indicates that CelF is a processive enzyme[13].

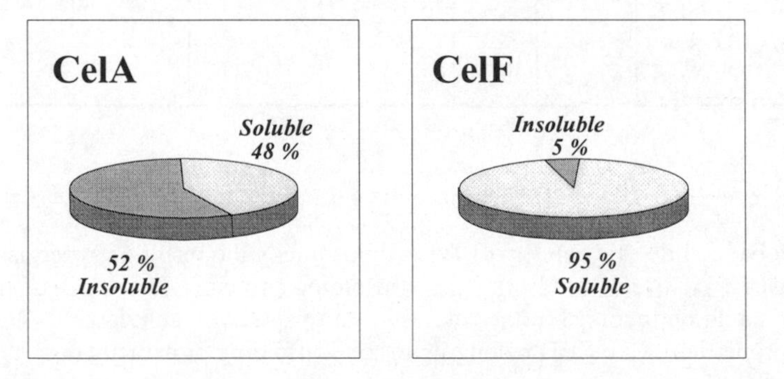

Figure 3 *Distribution of reducing ends between cellulose (insoluble fraction) and supernatant (soluble fraction) after hydrolysis of PAS-Cellulose by CelF and CelA (less than 1% of the cellulose had been digested by CelF or CelA).*

At the beginning of the enzymatic hydrolysis, a wide variety of soluble cellodextrins (from G2 to G6) was detected in the supernatant. After one hour, however, cellobiose was the major product (70%) and G3 and G4 amounted to 20% and 10%, respectively. No G5 and G6 were detected at this stage. The diversity of the cellodextrins released by truncated CelF from PASC at the beginning of the reaction indicated that the enzyme might hydrolyse internal β-1,4 bonds. In addition, although the truncated CelF generates very few reducing ends on CMC (Figure 4a), its action on this substrate leads to a large increase in CMC fluidity (Figure 4b).

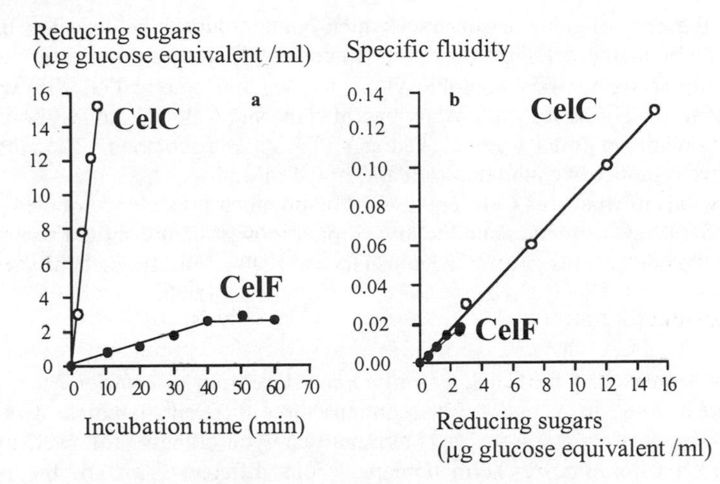

Figure 4 *CM-cellulose hydrolysis by CelF and by the endocellulase CelC. a) release of reducing sugars versus time. b) Increase in specific fluidity versus release of reducing sugars.*

It was previously hypothesised that CelF might initially hydrolyse cellulose molecules like an endocellulase and subsequently hydrolyse the substrate in a recurrent way. The hypothetical recurrent mode of action might be inhibited by the presence of carboxymethyl groups. The iterative gliding movement of the enzyme could be stopped by steric overcrowding. In order to check this second point, we tested the capacity of CMC to trap the enzyme. A significant decrease in the PASC activity was observed when truncated CelF was preincubated with CMC[13]. Part of the enzyme was in that case not available, the enzyme wedged into the substrate. The influence of the number of the extremities in the PASC on CelF activity was also tested. PASC was pretreated with truncated endocellulase CelA, washed, and was subsequently incubated with CelF. A decrease in the activity of truncated CelF was observed with increasing pretreatment times with CelA, showing that CelA hydrolyses potential CelF sites[13].

In view of the results, CelF can be classified as a processive endocellulase.

3.1.4 CelF has a low catalytic efficiency. If one compares kinetic parameters of CelF on its favourite substrate, PASC (K_M = 2.4 g/l, k_{cat} = 30 min^{-1}), with those of endocellulases CelA (K_M = 4 g/l, k_{cat} = 6644 min^{-1}) and CelC (K_M = 2.5 g/l, k_{cat} = 3900 min^{-1}) acting on CMC, it is obvious that the k_{cat} of CelF is about two orders of magnitude lower than those of endocellulases. It might be speculated that the recurrent mode of action of the enzyme along the cellulose chain is a slow process.

This low efficiency is quite surprising, since CelF is a major component of the cellulolytic system. Several hypothesis can be proposed: (i) some essential cofactors might be lost during the purification process; (ii) we possibly did not find or we were not able to obtain the best substrate for this enzyme (*e.g.* cellodextrins with a degree of polymerisation greater than 6); (iii) this protein might be of great importance during the synergistic hydrolysis of cellulose involving all the enzymes together. Davies and Henrissat[15] suggested that processivity is probably a key factor contributing to the efficient degradation of crystalline cellulose by cellulolytic systems.

3.1.5 Cocrystallisation of CelF with a thiooligosaccharide inhibitor. CelF degrades three soluble cellodextrins: cellohexaose, cellopentaose and cellotetraose. The latter is the only oligomeric substrate which contains a unique scissible bond[13]. The oxygen atom of the scissible bond was therefore replaced by a sulfur atom[16]. This compound, the methyl-4-*S*-β-cellobiosyl-4-thio-β-cellobioside (called IG4) was shown to be resistant to CelF hydrolysis. Activities of truncated CelF with or without 0.6 mM of the IG4 potential inhibitor were assayed using PASC as a substrate. The curves revealed that IG4 is a competitive inhibitor with a K_i of 0.4 mM.

Crystals of truncated CelF could only be obtained in presence of IG4. They were suitable for X-ray diffraction and the three dimentional structure should soon be available and will provide new insights on the unusual catalytic mechanism of family 48-cellulases.

3.2 The cellulase CelG

The gene coding for CelG, a family 9 cellulase from *C. cellulolyticum*, was cloned and overexpressed in *E. coli*. CelG contains three different domains identified on the basis of sequence comparisons, a N-terminal catalytic domain, followed by a type III CBD and a C-terminal dockerin domain. Four different forms of the protein were genetically engineered: CelGL (the entire form of CelG), CelGcat1 (the catalytic domain of CelG), CelGcat2 (CelGcat1+ 91 first amino-acids of the CBD) and GST-CBD_{CelG} (the CBD of CelG fusioned to Glutathione-S-transferase)[17].

3.2.1. Production of the different forms of CelG. To prevent the formation of inclusion bodies which commonly occurred when the entire form of CelG was produced by the recombinant strain of *E. coli* at 37°C with 1 mM of IPTG, wide ranges of growth and induction conditions were tested. The optimum conditions for the production of soluble recombinant CelG protein were obtained by inducing the culture at $OD_{600nm} = 2$ with 10 μM of IPTG and incubating the cells at 15°C for 17 h[17]. The same conditions were subsequently used to produce the various forms of CelG (CelGL, CelGcat1 and CelGcat2) in *E. coli*.

3.2.2. Purification of CelGL. Thirty grams wet weight of cells were used for purification which was performed as fast as possible in order to prevent the C-terminal degradation which usually occurs with *C. cellulolyticum* cellulases. From two liters of culture, 29 mg of active pure CelGL were obtained. The results of N-terminal micro-sequencing of this fraction (*i.e.* GTYNY) matches with the sequence deduced from the nucleotide sequence. The recombinant protein has an apparent mass of 76 kDa on SDS-PAGE which is in good agreement with its theoretical mass (76.109 kDa). The apparent pI of CelGL is 4.6. This protein is inactivated by storage at -70°C or -20°C. During storage at 4°C, and as previously observed in the case of CelA, CelC and CelF a partial hydrolysis occurs within 1-2 weeks, yielding to a mixture of the entire protein and a truncated protein which lacks the dockerin domain (CelGS). Since only the entire CelF contains the His-tag, the two forms were separated using a nickel column. Contrary to what was observed with CelA, CelC and CelF, the loss of the dockerin domain had no effect on the catalytic activity of CelG on the various substrates tested. Since no subsequent proteolysis was observed, CelGS was further chosen for comparisons with $CelA_3$, the dockerin lacking form of CelA.

3.2.3. Catalytic properties of CelGL. The specific activity of CelGL towards various substrates was measured. The highest specific activities were obtained with barley glucan and CMC. CelGL showed no activity towards laminarin, xylan, and chromogenic substrates such as *p*NP-glucose or *p*NP-cellobiose, but was able to degrade lichenan, Avicel, PASC, BMCC and natural cellodextrins with DP ranging from 6 to 3.

The apparent K_M and V_M of CelGL towards CMC and Avicel were found to be 8.7 g/l and 2000 IU/μmol and 17.8 g/l and 5 IU/μmol, respectively[17]. The K_M and V_M of

CelGL towards cellopentaose (DP 5) are 1.7 g/l and 670 IU/μmol. The optimum pH and temperature are pH 7.0 and 50°C. The final products of degradation of cellodextrins are cellobiose and glucose. The activity of CelGL towards cellodextrins increases with the DP of the substrate; the highest activity was observed on cellohexaose. The degradation of G4 and G3 was relatively slight compared to that of G6 or G5. The sugars released upon incubation of CelGL with PASC or Avicel PH101 were identical, *i.e.* G5 and G4 were first released (0-30 min), followed by accumulation of G3, G2 and G1 (2-3 hours).

3.2.4. Catalytic properties of CelGcat. The catalytic domain of CelG (CelGcat1) was produced. Surprisingly no catalytic activity was found against CMC, Avicel and PASC[17]. The catalytic domain, however, appeared to be correctly marked out in view of the only family 9 crystal structure published so far (3D-structure of CelD from *Clostridium* thermocellum)[18]. The *celG* gene was truncated to code for a protein containing the first 479 residues, being therefore 10 residues longer than the catalytic domain expected to lie from residu G_1 to $A_{469,}$ as suggested by the crystal structure of CelD. A longer peptide, containing in addition the first 91 amino acids of the putative CBD, CelGcat2, was produced and purified the same way. It was slightly active on CMC (0.5 IU/μmole of protein, corresponding to about 1/2000 of the activity of CelGL). It emerged clearly that the catalytic domain alone cannot efficiently hydrolyse CMC and that more than half of the second domain of the protein is required to obtain an active enzyme[17].

3.2.5. Cellulose binding assays. The binding properties of CelGL, CelGcat1 and GST-CBD_{CelG} were investigated on PASC, Avicel and BMCC[17]. Neither GST-CBD_{CelG} nor CelGcat1 were able to bind to these substrates although a large set of experimental conditions were tested. The entire protein CelGL was able to bind to Avicel. A reciprocal plot using a 5 % (final concentration) solution of Avicel indicated the existence of two classes of sites, characterised by the following parameters: K_D = 0.12 μM, 4 nmol of sites/g of Avicel and K_D = 7.4 μM, 95 nmol of sites/g. These properties seem to reflect a relatively weak binding ability in comparison with that of the CBD of miniCipC_1 which was found (on 0.5 % Avicel solution) to have a number of sites as high as 280 nmol/g and a K_D of 0.13 μM[17].

3.2.6. Comparison between the catalytic properties of CelGS and CelA$_3$. The activities of CelGS and CelA_3 were compared on CMC, PASC, Avicel and BMCC. The specific activity of CelA_3 on CMC was higher than that of CelGS[17]. Viscosimetric assays on CMC incubated for various times using high concentrations of CelGS or CelA_3 yielded similar profiles. It is clear that both enzymes have the same endo mode of action. Nevertheless, CelGS and CelA_3 exhibited quite different insoluble cellulose degradation activities, as shown in Figure 5. CelA_3 exhibited greater activity than CelGS on PASC, which is a low degree of crystallinity cellulose. The activities on Avicel and BMCC, which are more crystalline substrates than PASC, were also monitored. On Avicel as substrate, CelGS and CelA_3 had similar initial degradation rates (Figure 5A and Figure 6) but in the case of CelGS the degradation process continued, whereas that of CelA_3 stopped earlier. A study using a wide range of concentrations of the two enzymes and long incubation times confirmed this observation (Figure 6). With CelGS the number of sites on Avicel which were accessible for hydrolysis seems to be larger than in the case of CelA_3. When Avicel was incubated with both CelGS and Cel$A_{3,}$ no synergism was observed. Therefore none of these two enzymes seems to generate additionnal hydrolysable sites for the other.

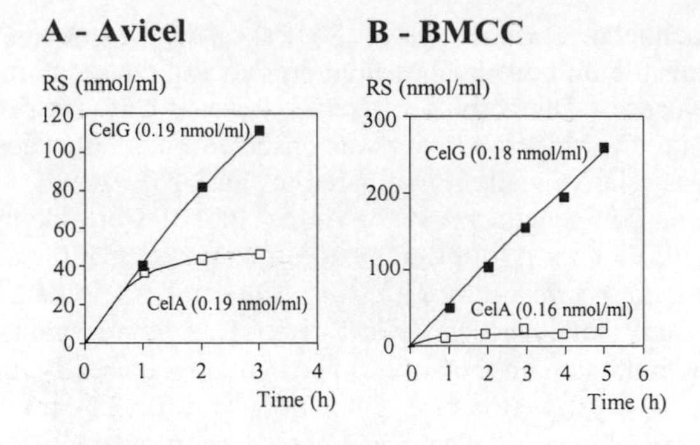

Figure 5 *Comparison between cellulase activities of CelGS (■) and CelA$_3$ (□) on: A - Avicel 8 g/l; B - BMCC 1.6 g/l.*

On BMCC, which is considered to be the most crystalline cellulose available, the activity of CelGS was 20-fold higher than that of CelA$_3$[17]. The sites present on BMCC were apparently not accessible or not easily accessible by CelA$_3$ whereas CelGS degraded this substrate efficiently and extensively (Figure 5B). Moreover, the rate of BMCC hydrolysis by CelG was faster than that of Avicel (Figure 5).

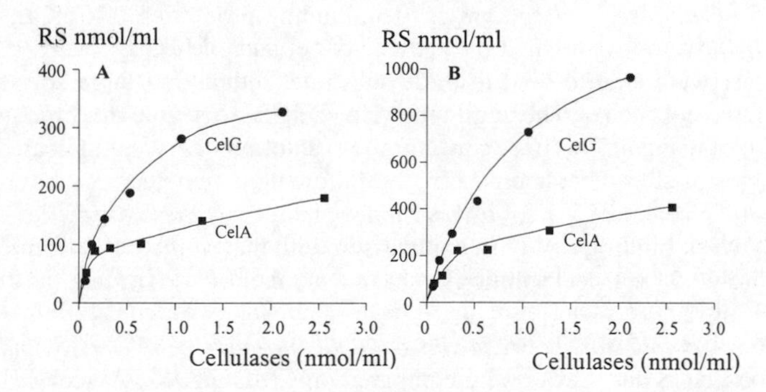

Figure 6 *Comparison between Avicel degradation observed at various concentrations of CelGS (●) and CelA$_3$ (■). RS: Reducing sugars released after A: 6h and B: 24h of incubation of enzymes with the substrate.*

3.3 Role of the scaffolding protein CipC in the cellulose degradation.

CipC possesses a family III CBD which might facilitate the anchoring of the cellulosome to the surface of the cellulose, increasing the local enzyme concentration.

The ability of the CBD of CipC to bind to cellulose was investigated. The binding parameters *i.e.* the dissociation constant (K_D) and the binding capacity (µmoles of miniCipC$_1$ bound per gram of cellulose) were determined with various insoluble substrates: BMCC, Avicel and PASC[11]. The CBD of CipC interacts with high affinity (10^8 or 10^7 M^{-1}, depending on the substrate) with all insoluble substrates tested. Moreover it has been observed that the binding of the CipC-CBD on cellulosic substrates leads to macroscopic changes in the cellulose particles flocculation. The presence of the

CipC-CBD prevents the flocculation of BMCC, PASC or Avicel (Figure 7). The mechanism responsible for this phenomenon remains unknown.

Figure 7 *Inhibition of the Avicel (10g/l in 25mM phosphate buffer pH 7) particles flocculation. Without (1) and with (2) pretreatment by the CBD of CipC (incubation at 20°C for 30 min with 20 μg of CBD/mg of cellulose followed by a removal of the CBD from cellulose using sterile water).*

Although the CBD of CipC alone does not have any detectable catalytic activity, it has been observed that the activity of the endoglucanase CelA, towards all the insoluble substrates tested, is enhanced by the presence of the CBD of CipC or by the presence of the mini-scaffolding protein miniCipC_1. In a same way, when the cellulose is pretreated with the CBD, as described above, the activity of CelA is strongly enhanced. These results suggest that the synergism observed between CelA and the CBD of CipC (or the miniCipC_1) is not exclusively due to the anchoring of the enzyme to the substrate but also that the CBD of CipC is able to generate new hydrolysable sites.

Grafting this CBD onto the catalytic domain of CelA (CelA_3) enhances the catalytic activity on all the insoluble substrates tested, but not toward soluble substrates such as CMC. The cellulose degradation profile of this chimeric protein (CelA_3-CBD-CipC) has been compared to that of CelA_3[11]. This additional domain involved modification of the cellulose kinetic degradation as shown in Figure 8.

The enhancement of the cellulolysis by this CBD varies from one enzyme to the other. The activities on PASC of CelGL and CelA_2 bound to miniCipC_1 have been compared. Two sets of experiments were carried out. First the specific activity of both enzymes bound to miniCipC_1 was monitored, and compared with that of the enzymes alone. Incubation with 0.02 μM of miniCipC_1 yielded a 1.3 fold enhancement of the specific activity of CelA_2, although it did not have any effect on the specific activity of CelGL. When a 500:1 ratio of miniCipC_1/enzyme was used, it drastically enhanced the specific activity of CelA_2 (6.7 fold), whereas with CelGL only a 1.7 fold increase was observed. The action of miniCipC_1 on this substrate was therefore not as beneficial for CelGL as for CelA_2.

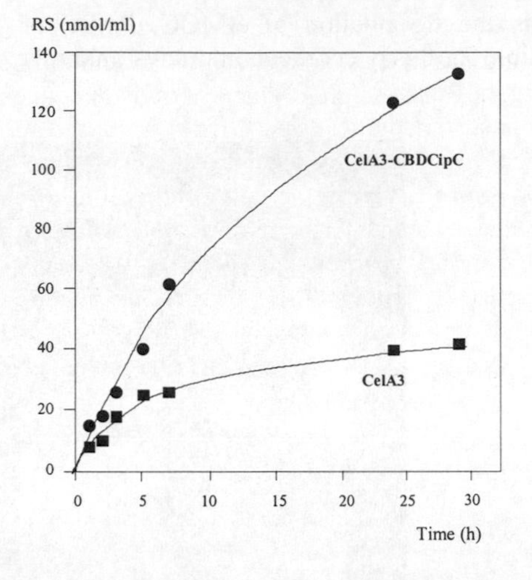

Figure 8 *Avicel degradation by $CelA_3$ and the chimeric protein $CelA_3$-CBD-CipC. The activity of CelA on Avicel, such as on the other insoluble cellulose tested, is rapidly stopped, probably by the lack of hydrolysable sites. In the case of the chimeric protein, the number of hydrolysable sites seems not a limiting factor during at least 30 hours.*

In order to check this assumption, the substrate was incubated with $miniCipC_1$ before addition of $CelA_2$ or CelGL. A 2.7-fold increase in the specific activity of $CelA_2$ was recorded, which was similar to that previously observed[11] and the increase was only 1.5 fold in the case of CelGL[17]. These results suggest that complex formation has various effects on cellulase activity, probably due to differences in the mode of action of the enzymes, and that the new sites exposed on the substrate by the action of miniCipC1 do not enhance the activity of CelGL as much as for $CelA_2$. One of the possible explanations of such a difference might be the presence of a typeIII-CBD in CelG.

4 CONCLUSION

The cellulolytic machinery of *C. cellulolyticum* is constituted of cellulosomes. Each cellulosome contains a multidomain-non enzymatic subunit of 170 kDa, CipC, and several cellulases. The strong interactions which occure between the cohesin domains of CipC and the entire dockerin domains of the cellulases are responsible of the cellulosome assemblage. Since the scaffolding protein contains nine cohesin domains, and since the analysis of pure preparation of cellulosomes showed thirteen proteins harboring a dockerin domain, it is obvious that all cellulosomes are not identical. Among these proteins, CelA, CelC, CelG, CelE and CelF were clearly identified. CipC, CelF and CelE are the three major components of the cellulosomes.

CipC contains also a Cellulose Binding Domain which anchors the cellulosome to the cellulose increasing the local enzyme concentration. This CBD has another role, by disrupting the cellulose and creating additional hydrolysable regions which can be a substrate for hydrolysing enzymes.

The two cellulases studied, CelF and CelG, differ from the cellulases already studied and probably play an important role in crystalline cellulose degradation. CelF is so far the first family 48 enzyme to be extensively studied. It is one of the major proteins of the cellulosome and has been characterised as an endoprocessive cellulase. CelF has been crystallised and the 3D structure of the protein is almost established.

CelG, which is an endoglucanase, differs from the three others already studied (CelA, CelC and CelD) by its ability to hydrolyse crystalline cellulose. CelG is a three domains protein containing a family 9 catalytic domain, a type III-CBD and a dockerin domain. It appears that, contrary to the majority of multidomain-cellulases, the catalytic domain and the CBD are not independent. The recent results obtained by Sakon *et al.*[19] studying the 3D structure of the cellulase E4 from *Thermomonospora fusca* which is very homologous to CelG, clearly showed that the catalytic domain and the CBD interact.

The cellulosome of *C. cellulolyticum* is very active on crystalline cellulose. The endoprocessive mechanism of CelF, the properties of CelF and CelG which efficiently hydrolyse crystalline cellulose, and the properties of the CBD of the scaffolding protein probably have a key role in the hydrolysis of natural cellulose.

Acknowledgments

This research was supported by grants from the Centre National de la Recherche Scientifique, the University de Provence, the EEC (BIOTECH contract BIO-CT-94-3018), and the Région Provence-Alpes-Côte d'Azur.

References

1. F. A. Rainey and E. Stackebrant, *FEMS Microbiol. Lett.*, 1993, **113**, 125.
2. E. Faure, A. Belaich, C. Bagnara, C. Gaudin and J.P. Belaich, *Gene*, 1989, **84**, 39.
3. C. Bagnara-Tardif, C. Gaudin, A. Belaich, P. Hoest, T. Citard and J.P. Belaich, *Gene*, 1992, **119,** 17.
4. S. Shima, Y. Igarashi and T. Kodama, *Gene*, 1991, **104**, 33.
5. C. Reverbel-Leroy, A. Belaich, A. Bernadac, C. Gaudin, J.P. Belaich and C. Tardif, *Microbiol.*, 1996, **142,** 1013.
6. J.P. Belaich, C. Tardif, A. Belaich and C. Gaudin, *J. Biotechnol.*, 1997, in press.
7. S. Pagès, A. Belaich, C. Tardif, C. Reverbel-Leroy, C. Gaudin and J.P. Belaich, *J. Bacteriol.*, 1996, **178**, 2279.
8. U.T. Gerngross, M.P. Romaniec, T. Kobayashi, N.S. Huskisson and A.L. Demain, *Mol. Microbiol.*, 1993, **8**, 325.
9. O. Shoseyov, M. Takagi, M.A. Goldstein and R.H. Doi, *Proc. Natl. Acad. Sci. USA*, 1992, **89**, 3483.
10. L. Gal, S. Pagès, C. Gaudin, A. Belaich, C. Reverbel-Leroy, C. Tardif and J.P. Belaich, *Appl. Environ. Microbiol.*, 1997, **63**, 903.
11. S. Pagès, L. Gal, A. Belaich, C. Gaudin, C. Tardif and J.P. Belaich, *J. Bacteriol.*, 1997, **179**, 2810.
12. E. Morag, A. Lapidot, D. Govorko, R. Lamed, M. Wilchek, E.A. Bayer and Y. Shoham, *Appl. Environ. Microbiol.*, 1995, **61**, 1980.
13. C. Reverbel-Leroy, S. Pagès, A. Belaich, J.P. Belaich and C. Tardif, *J. Bacteriol.*, 1997, **179**, 46.
14. D.C. Irwin, M. Specio, L.P. Walker and D.B. Wilson, *Biotechnol. Bioeng.*, 1993, **42**, 1002.
15. G. Davies and B. Henrissat, *Structure*, 1995, **3**, 853.

16. C. Reverbel-Leroy, G. Parsiegla, V. Moreau, M. Juy, C. Tardif, H. Driguez, J.P. Belaich and R. Haser, *Acta Crystall.*, *section D*, 1997, in press.
17. L. Gal, C. Gaudin, A. Belaich, S. Pagès, C. Tardif and J.P. Belaich, *J. Bacteriol.*, 1997, in press.
18. M. Juy, A.G. Amit, P. Alzari, R. Poljak, M. Claeyssens, P. Béguin and J.P. Aubert, *Nature*, 1992, **357**, 89.
19. J. Sakon, D. Irwin, D.B. Wilson and P.A. Karplus, *Nature, Struct. Biol.*, 1997, **4**, 810.

EGZ (CEL5) OF *ERWINIA CHRYSANTHEMI*: TWO DOMAINS AND THREE POSSIBILITIES

F. Barras, I. Bortoli-German, E. Brun, M. El Hassouni, A. Marteau and B. Py

LCB-CNRS, IBSM, 31 Chemin Joseph Aiguier, 13402 Marseille, Cedex 20, France

1 INTRODUCTION

Most plant pathogens, bacteria or fungi, produce extracellular enzymes, their substrates being present in the plant cell wall. Likewise, *Erwinia chrysanthemi*, a Gram-negative bacterium, causing soft-rot symptoms of a large number of dicotyledons, secretes a battery of depolymerising enzymes, including pectinases, proteases and an endoglucanase referred to as EGZ[1]. Soft-rot essentially stems from the pectinolytic activity and the precise role of cellulase remains to be clearly assessed[2,3]. However, the occurrence of cellulases in all soft-rot erwinia strains, *i.e. E. chrysanthemi* and *E. carotovora*, analysed to date, suggests that they provide an advantage to the bacterium.

Despite a large diversity in biochemical and structural characteristics, pectinases (comprising pectate lyases, polygalacturonases, pectin methyl esterases, pectin acetyl esterase) and cellulases are secreted across the bacterial envelope by the same pathway referred to as type II (or GSP)[4,5]. This secretion system includes over 20 proteins which are thought to form two distinct machineries, namely the Sec and the Out machineries allowing the crossing of the cytoplasmic and the outer membrane, respectively. Targeting information on pectinases and cellulases allowing their interaction with the Sec machinery lies certainly within the leader peptide which is processed upon crossing the cytoplasmic membrane. The nature of the second targeting motif, if any, permitting interaction of these enzymes with the Out machinery remains mysterious. The identification of the second Out specific targeting information is a main goal of our research. For this purpose, EGZ was used as a protein model.

EGZ is a typical modular cellulase since it exhibits a catalytic domain fused to a cellulose binding domain (CBD) via a Ser/Thr-rich linker region[6]. Hence, EGZ possesses both the ability to catalyse cleavage of β1,4 glycosidic bonds in soluble cellulose and to bind onto microcrystalline cellulose. As we shall see below, this two-domains view, classically used to describe cellulase/cellulose interactions, cannot be used to describe the interactions between EGZ and the Out secretion machinery.

This text aims at summarising the biochemical, genetic and structural information available on EGZ. This information stems from initial biochemical studies in J. Cattanéo's group in Marseille[7,8] and is currently the site of a collaborative effort with NMR spectroscopists P. Gans and D. Marion in Grenoble and with crystallographers M. Czjzeck and R. Haser in Marseille.

2 HYDROLYSING β1,4 GLYCOSIDIC BONDS

2.1 Substrates

EGZ acts upon carboxymethylcellulose as well as on small substrates (*para*-nitrophenyl-β-cellobioside). ^{1}H NMR proved that it catalyses the cleavage of cellotetraose to cellobiose with retention of configuration at the anomeric carbon[9].

2.2 Genetic analysis

The catalytic domain of EGZ belongs to family 5/2 of β-glycosyl hydrolases. In the absence of a 3D-structure, a thorough genetic analysis was undertaken with two main goals: (i) to unravel the role of a series of conserved positions throughout family 5, and (ii) to identify structural positions, to be exploited in our studies on secretion (see below). The use of informational suppression to protein analysis, as proposed by J.H. Miller[10], allowed us to obtain a pattern of 13-14 substitutions per position studied[11]. A simple protocol, coupling a test for enzymatic activity (by plate assay) and *in vivo* stability (by immunoblot analysis) was used to study each mutated protein. Overall, 16 positions were studied, yielding 208 mutated proteins. Analysis of pattern substitutions allowed us to categorise the mutated positions as follows: (i) neutral, *i.e.* any of the substitutions is tolerated without altering the studied property (catalysis, stability), (ii) restricted, *i.e.* only a subset of the substitutions is compatible with the studied property, (iii) constrained, *i.e.* the wild type residue only is compatible with the studied property. For those positions classified as « restricted », comparison of properties of the « allowed vs. non allowed » residues pointed to the type of physical or chemical properties that was needed at these positions. This genetic study led us to formulate the following predictions[11]: (i) Glu133 and Glu220 are the catalytic residues; (ii) His192 is important for catalysis but might also have a structural role; (iii) His98 has a substrate binding role; (iv) Arg155 has a structural role, forms a buried salt bridge and is located far from the catalytic centre; (v) Arg57 has a central role, intervening in both catalysis and structure, forming a buried salt bridge, possibly with one of the catalytic Glu residues; (vi) Arg203 and Arg208 have a structural role, and possibly belong to an α-helix. These predictions proved to be correct after the three-dimensional structure of two catalytic domains belonging to family 5, namely *Clostridium cellulolyticum* CelCCA and *C. thermocellum* CelC, became available[12,13]. Yet, the low level of primary sequence similarity between EGZ and the two others (about 20%) makes precise interpretation of the behaviour of the 208 variants impossible. Solving the structure of a family 5/2 member is therefore awaited.

3 BINDING CRYSTALLINE CELLULOSE

Binding of EGZ onto microcrystalline cellulose (Avicel) is mediated by a C-terminal cellulose binding domain (CBD). Sequence analysis revealed that this CBD exhibits very peculiar features as compared with other known CBDs[14]. It is 62 residues in size, which is intermediate between the so-called fungal CBDs (approx. 33-36 residues) and the so-called bacterial ones (110 to 155 residues). Moreover, it shares no sequence similarity with any of the known CBDs and it was proposed to make a family (*i.e.* family V) on its own[14].

The three-dimensional structure of the CBD was solved by two-dimensional proton nuclear magnetic resonance spectroscopy[15,16]. The overall shape resembles a skiing boot. It is made of a triple anti-parallel β-sheet fused to a less-ordered region. It has a well-organised, compact hydrophobic core, made of 7 residues. Sticking out from this ordered region are two aromatic residues (Trp43 and Tyr44) that are likely to be instrumental in binding cellulose (see below). The less-ordered region is made of two very rigid regions (residues 3-14 and residues 27-30) which bracket a large, disordered loop (residues 15-26). The first rigid region (residues 3-14) includes both Cys3 (covalently linked to Cys61) and an aromatic residue belonging to the hydrophobic core of the molecule. The disulphide bond lies in the vicinity of the linker region and is exposed to the solvent; this bond does not appear to be essential in maintaining the stability, although in its absence, the CBD becomes a very good target for intracellular proteases. A remarkable feature of the 3-14 region is the presence of a *cis*-Pro, that might have a role in the secretion of EGZ (see below). The disordered loop (residues 15-26) includes a Trp18 residue that sticks out of the structure and aligns with both the Trp43 and Tyr44 residues (see above). All three form a flat surface that is reminiscent of cellulose binding faces identified in other CBDs. Moreover, the distance (10.5 Å) between the aromatic centroids of the Trp18 and Tyr44 residues matches the distance between the glucose moieties in the cellulose chain and the CBD is predicted to cover 5-6 glucose residues.

As mentioned above, sequence analysis failed to reveal any convincing similarity between CBD of EGZ and others. Actually, a few sequences were repeatedly extracted from the data bank that shared a few conserved positions, but the level of similarity was too low to be taken as significant. Surprisingly, availability of the three-dimensional structure revealed that the few conserved residues were those forming the hydrophobic core and the cellulose binding face in the CBD of EGZ. Most of the other sequences correspond to modules of unknown function, belonging to bacterial chitinases or other proteins. Our current effort aims at characterising biochemically and structurally those modules and to further investigate the interaction of EGZ and chitin.

4 GOING OUT OF *ERWINIA CHRYSANTHEMI*

4.1 Evidences against a discrete, sequential secretion motif

In order to reach the plant cell wall, EGZ must cross the double membrane envelope of *E. chrysanthemi.* In most protein targeting processes, secretion information lies within the structure of the targeted polypeptide as a discrete segment of residues. The best studied secretion motif is the leader peptide found in all proteins which eventually cross the bacterial membrane or the rough endoplasmic reticulum membrane. Other examples include mitochondrial targeting peptide or C-terminal secretion motifs of bacterial toxins or proteases. Hence, most studies converge to suggest that secretion information lies within a sequential motif.

Taking advantage of the modular organisation of EGZ we used a deletion domain approach in order to reduce the size of the target to be searched for secretion information[17]. The net result of this was that all regions of EGZ, *i.e.* catalytic domain, CBD and linker region need to be present for the polypeptide to be secreted. Since modifying the size of the linker region, or deleting it, impaired secretion as well, it was proposed that the two-dimensional view of EGZ with two functionally and structurally independent domains was not able to account for the secretability[17]. Rather, we proposed that secretability reflects interactions between the two domains. A simple possibility was that both domains get together in close vicinity such that residues from each domain can participate to a « patch » signal. The linker region would have a guiding role. This was rather a heretic view of protein translocation since it implied that the EGZ has to acquire some level of structural organisation before crossing the outer membrane, at odds with the dogmatic view of protein folding and protein translocation that are considered as mutually exclusive. Preliminary data had been obtained with the pullulanase secretion system which were consistent with the view that the type II secretion system is compatible with folding of the passenger protein[18]. In fact, *Vibrio cholera* toxin had been shown, in 1987, to be secreted as a folded pentamer but at that time, nobody knew that cholera toxin goes out of *Vibrio* the same way as EGZ goes out of *E. chrysanthemi*[19].

4.2 Structure/secretion relationships within the CBD

In order to test the hypothesis that EGZ folded prior to secretion, we made use of the disulphide bond present within the CBD. Using both reducing agent and site directed mutagenesis, we showed that the disulphide bond is indeed formed within the periplasm[20]. Disulphide bond formation was shown to be catalysed by DsbA protein both in *E. coli* and *E. chrysanthemi*[21]. Moreover, behaviour of disulphide bond-less mutants supported the view that the disulphide bond was not only tolerated by the translocation process but is required for it. A possibility is that disulphide bond formation allows maintenance of the putative patch motif.

Recently, secretability of EGZ variants that have substitutions in one of the hydrophobic core forming residues were analysed. Interestingly, a mutation (Ile changed in Tyr) resulted in an apparent heterogeneity in the population. Steady-state analysis showed that half of the molecules was found to be secretable and half remained stuck in the periplasm. Analysis of the secreted forms showed that they bind to cellulose like the wild type, suggesting that they possess a correctly folded CBD. In contrast, those periplasmic forms were found to be reduced in their efficiency in binding cellulose. Though very preliminar, these observations fit in the view that the CBD must fold in its native form for EGZ to be secreted. Interestingly, a graphic modelling representation of the mutation predicts a clash with the *cis*-Pro centred region. Hence, an attractive hypothesis is that the mutation impairs isomerisation of this peptide bond, yielding essentially two populations, with *trans*- or *cis*-Pro residues, of which only the *cis*-Pro containing forms would be able to be secreted and bind efficiently to cellulose.

4.3 Structure/secretion relationships within the catalytic domain and the role of Arg57

If the « folding for secretability » rule applies in the catalytic domain as well, any conformation-modifying mutation might affect secretability of EGZ. Such a structural position was uncovered by our genetic study (see above). Mutated proteins in which Arg57 residue was changed to Lys, His, Gln or Ser were studied for both their *in vivo* stability and secretability. The efficiency of secretability and the susceptibility to proteolysis varied in opposite directions. An interpretation is that modifying the position 57 invariably affected the conformation, and as a consequence, secretability was affected. This was taken as pointing to the existence of a very tight link between structure and secretability in the catalytic domain.

An alternative interpretation is that Arg57 makes contact directly with a residue belonging to the secretion patch. Based upon models of *Clostridium* enzymes (see above), Arg57 was predicted to be hydrogen bonded with at least 4 residues. Each of them was therefore mutagenised and the resulting protein assayed for secretability, activity and stability. With the exception of mutation at Glu133, the acid/base residue, none of the mutants were found to be affected in activity or secretability.

Hence, any type of substitution at position 57 affects activity and destabilises the protein (with the exception of Lys substitution which is conservative for stability). Yet, none of the four residues contacted by Arg57 makes an important contribution on its own. Our *in vivo* characterisation of mutations at position Arg57 suggests that this residue has a general (pleiotropic) role in the functioning of EGZ, since abilities to fold to a protease-resistant form, to be catalytically active and to be secreted out of the cell, all can be compromised by a modification of the lateral chain at this position. Solving the 3D-structure of EGZ catalytic domain will hopefully shed some light on this.

4.4 Is the secretion motif lying in the loops?

E. carotovora secretes a similar set of enzymes (cellulases and pectinases) as *E. chrysanthemi* and produces soft-rot disease[1,5]. Pectinases of both species share above 60% sequence identity and cellulases are about 40% identical. Both the Out machineries of *E. chrysanthemi* and *E. carotovora* share sequence similarities (30 to 80%) throughout the 14 *out* encoding genes. Yet, heterologous secretion is not possible[22,23]. Hence, while within one bacterial species a large array of biochemically and structurally dissimilar enzymes are secreted by the same pathway, an exogenous enzyme exhibiting about 80% identity with an endogenous one (such as PelC enzymes) is not secreted. This « species specificity » is still eluding us.

This « species specificity » was used in an attempt to locate the secretion motif as follows. *E. carotovora* cellulase, referred to as CelV, has a catalytic domain 46% identical to that of EGZ, and a family III cellulose binding domain[24]. A hybrid cellulase was therefore constructed that has the *E. carotovora* CelV catalytic domain (and the cognate linker region) with the EGZ CBD fused to it. The resulting hybrid exhibits about 80% identity with EGZ. This chimera, active and binding on cellulose, was secreted neither by *E. chrysanthemi* nor by *E. carotovora*. While deceptive, this result suggested that the secretion information present within the catalytic domain lies in those parts that diverge between EGZ and CelV. Sequence alignments and use of structural informations available from the study of the *Clostridium* enzymes showed that the more dissimilar regions are, the more they tend to form the exposed loops. Mutagenesis studies focusing on these regions might allow to uncover part of the secretion motif.

5 CONCLUSION AND PERSPECTIVES

On one hand, studying the *Erwinia* cellulase EGZ at first suffered from two handicaps: (i) *Erwinia* is not a cellulolytic organism and (ii) EGZ is a classic endoglucanase. On the other hand, it offered two originalities: (i) the biological role of EGZ has to be appreciated within the context of a plant destroying program, acting in concert with biochemically differentiated enzymes and (ii) secretability provides both a complex biological process to decipher and an additional biochemical « marker » to studying structure/function relationships. Moreover, the *Erwinia* system offered the amenability to genetic tools.

Work ahead includes, as a priority, the solving of the three-dimensional structure of the full length EGZ. This stands now as a prerequisite to hope to understand secretion. Yet, even knowing the domain/domain interaction within the wild type EGZ will not be sufficient; solving the three-dimensional structure of variants impaired in secretion will be required for comparison. Hybrid cellulases and cellulases modified in the linker region might be the best candidates.

Studies of secretion suggested that some domain/domain interactions might be underlying the motion of these « typical modular proteins ». Questions related to folding of EGZ, and to protein/protein interactions taking place between EGZ and the Out proteins, are next on the priority list.

All soft-rot *Erwinia sp.* produce one or two cellulases. Pathogenic power of cellulase-negative mutants of *Erwinia* are slightly reduced, if at all, depending upon the strain. What is the actual role of those cellulases? Are they making the way for pectinases to go through and reach the pectin middle lamella? If so, a more serious reduction in pathogenicity would be expected in mutated strains devoid of cellulase activity. Are cellulases intervening as weapons *de luxe* to be used in special occasions?

6 ACKNOWLEDGEMENTS

I. Bortoli-German, E. Brun, M. El Hassouni, A. Marteau and B. Py all contributed to our knowledge on EGZ as parts of their PhD programs in the laboratory; their names are listed in alphabetical order. The solving of the CBD structure was part of an enjoyable collaboration with NMR spectroscopists P. Gans and D. Marion (IBS, Grenoble). We are grateful to B. Henrissat (CERMAV-CNRS, Grenoble), J. Haiech (LCB-CNRS, Marseille), M. Czjzeck and R. Haser (IBSM, Marseille) for their help. Thanks are also due to M. Chippaux for his interest in initial part of this work. Fruitful discussion with T. Teeri (Stockholm) was essential for the writing of this text. This work was made possible by grants from the CNRS, the Ministère of Research and Technology and the University of Aix-Marseille II.

References

1. F. Barras, F. van Gijsegem and A.K. Chatterjee, *Ann. Rev. Phytopathol.*, 1994, **32**, 201.
2. M. Bocara, J.L. Aymeric and C. Camus, *J. Bacteriol.*, 1994, **176**, 1524.
3. V. Cooper and G.P.C. Salmond, in prep.
4. A.P. Pugsley, *Microbiol. Rev.,* 1993, **57**, 50.
5. G.P.C. Salmond, *Ann. Rev. Phytopathol.*, 1994, **32**, 181.
6. B. Py, I. Bortoli-German, J. Haiech, M. Chippaux and F. Barras, *Prot. Eng.*, 1991, **4**, 825.
7. M.H. Boyer, J.P. Chambost, M. Magnan and J. Cattaneo, *J. Biotechnol.*, 1984, **1**, 241.
8. M.H. Boyer , B. Cami, J.P. Chambost, M. Magnan and J. Cattaneo, *Eur. J. Biochem.*, 1987, **162**, 311.
9. F. Barras, I. Bortoli-German, M. Bauzan, J. Rouvier, C. Gey, A. Heyraud and B. Henrissat, *FEBS Lett.*, 1992, **300**, 145.
10. J.H. Miller, *Meth. Enzymol.*, 1991, **208**, 543.

11. I. Bortoli-German, J. Haiech, M. Chippaux and F. Barras, *J. Mol. Biol.*, 1995, **246**, 82.
12. V. Ducros, M. Czjzek, A. Belaich, C. Gaudin, H.P. Fierobe, J.P. Belaich, G.J. Davies and R. Haser, *Structure*, 1995, **3**, 939.
13. R. Dominguez, H. Souchon , S. Spinelli, Z. Dauter, K.S. Wilson, S. Chauvaux, P. Beguin and P.M. Alzari, *Nature Struct. Biol.*, 1995, **2**, 569.
14. P. Tomme, R.A.J. Warren and N.R. Gilkes, *Adv. Microb. Physiol.*, 1995, **37**, 1.
15. E. Brun, P. Gans, D. Marion and F. Barras, *Eur. J. Biochem.*, 1995, **231**, 142.
16. E. Brun, F. Moriaud, P. Gans, M.J. Blackledge, F. Barras and D. Marion, *Biochemistry*, in press.
17. B. Py, M. Chippaux and F. Barras, *Mol. Microbiol.*, 1993, **7**, 785.
18. A.P. Pugsley, I. Poquet and M.G. Kornacker, *Mol Microbiol,* 1991, **5**, 865.
19. T.R. Hirst and J. Holmgren, *Proc. Natl. Acad. Sci.* U.S.A., 1987, **84**, 7418.
20. I. Bortoli-German, E. Brun, B. Py, M. Chippaux and F. Barras, *Mol. Microbiol.*, 1994, **11**, 545.
21. V. Shevchik, I. Bortoli-German, J. Robert-Baudouy, S. Robinet, F. Barras and G. Condemine, *Mol. Microbiol.*, 1995, **16**, 745.
22. S.Y. He, M. Lindeberg, A.K. Chatterjee and A. Collmer, *Proc. Natl. Acad. Sci.* U.S.A., 1991, **88**, 1079.
23. B. Py, G.P. Salmond, M. Chippaux and F. Barras, *FEMS Microbiol. Lett.*, 1991,**79**, 315.
24. V.J. Cooper and G.P.C. Salmond, *Mol. Gen. Genet.*, 1993, **241**, 341.

INTERACTION OF ENDO-β-1,4-XYLANASES WITH COMPOUNDS CONTAINING D-GLUCOPYRANOSYL RESIDUES

Peter Biely[a], Mária Vršanská[a] and M. K. Bhat[b]

[a]Institute of Chemistry, Slovak Academy of Sciences, 842 38 Bratislava, Slovakia; [b]Food Macromolecular Science Department, Institute of Food Research, Reading laboratory, Early Gate, Whiteknights Road, Reading RG6 6BZ, UK

1 INTRODUCTION

Endo-β-1,4-xylanases (EXs, EC 3.2.1.8) occur in nature in two families, family 10 (formerly F) and 11 (formerly G) (Henrissat and Bairoch, 1993). The enzymes of the two families differ in physicochemical properties, arrangement of hydrophobic clusters, conserved amino acids and in the overall tertiary structures (Wong *et al.*, 1988; Gilkes *et al.*, 1991; Prade, 1995). In spite of limited knowledge on the differences in catalytic properties of EXs in the two families, it is certain that the larger enzymes, belonging to family 10, exhibit greater catalytic versatility or lower substrate specificity than enzymes of family 11 (Biely *et al.*, 1997). For instance, EXs of family 10 catalyse hydrolysis of artificial substrates of cellulolytic enzymes, aryl β-cellobiosides (van Tilbeurgh and Claeyssens, 1985), at the agluconic linkage (Grepinet *et al.*, 1988; Gilbert *et al.*, 1988; Lüthi *et al.*, 1990; Shareck *et al.*, 1991; Hass *et al.*, 1992). EX of *Dichomitus squalens* (Rouau and Odier, 1986) and EXII of *Schizophyllum commune* (Biely, 1987) which catalyse the same reaction, are apparently members of the family 10. The low-molecular-mass EXs, the family 11 enzymes, do not possess this catalytic property. Common features of both EX families are their endo-character demonstrated viscosimetrically and the double displacement mechanism of the hydrolysis of glycosidic bond, which means that both types of enzymes are retaining glycanases (Withers *et al.*, 1986; Gebler *et al.*, 1992; Biely *et al.*, 1994).

Recently, a comparison of amino acid sequences and tertiary structures of various glycosyl hydrolases has indicated that the two EXs families evolved from different ancestors. EXs of family 10 are closely related to endo-β-1,4-glucanases of family 5 (formerly A) (Dominguez *et al.*; 1995; Jenkins *et al.*, 1995), while EXs of family 11 show weak, but significant similarity with low molecular mass endo-β-1,4-glucanases of family 12 (formerly H) (Törrönen *et al.*, 1993). In this connection it was of interest to examine the tolerance of substrate binding sites of EXs of family 10 and family 11 for replacement of D-xylopyranosyl residues by D-glucopyranosyl residues in some artificial substrates.

Several compounds containing D-glucopyranosyl residues were tested with four different EXs (belonging to two EX families) as substrates for hydrolysis and glycosyl transfer reactions. They were used as both glycosyl donors and acceptors, to gain information as to which subsites of the substrate binding sites of EXs have the ability to accommodate D-glucopyranosyl residues.

2 MATERIALS AND METHODS

2.1 Enzymes

Family 10 EXs: *Cryptococcus albidus* EX (48 kDa, pI 5.0) and XlnA of *S. lividans* (47 kDa, pI 5.2) were purified as described (Biely and Vršanská, 1986; Morosoli *et al.*, 1986). Family 11 EXs: EX from *Thermomyces lanuginosus* ATCC 46882 (26 kDa, pI 3.7) and XlnC of *S. lividans* (22 kDa, pI 10.25) were purified from culture filtrates as described in previous papers (Bennett *et al.*, 1997; Kluepfel *et al.*, 1992). One unit of EX activity is defined as the amount of enzyme liberating 1 μmol of xylose equivalents in 1 min (Somogyi-Nelson procedure).

2.2 Substrates

4-Nitrophenyl β-D-xylopyranoside (NPh-Xyl), 4-nitrophenyl β-D-glucopyranoside (NPh-Glc), 4-nitrophenyl β-cellobioside (NPh-Cel), 4-nitrophenyl β-lactoside (NPh-Lac), cellooligosaccharides and β-1,4-xylooligosaccharides were either synthesised or obtained from commercial sources.

2.3 Enzyme-substrate mixtures and their analysis

A 150 mM solution of NPh-Xyl or NPh-Glc, 50 mM solution of NPh-Cel and 25 mM solution of NPh-Lac were incubated with enzymes (3 U/ml) in 0.02 M acetic acid buffer (pH 5.4) and the products of hydrolysis were analysed by TLC on microcrystalline cellulose (Merck) in the solvent system ethyl acetate/acetic acid/water (3:2:2, by vol.). 4-Nitrophenol (NPh-OH) and NPh-glycosides were visualised under UV light and reducing sugars were detected with the aniline/hydrogen phthalate reagent. The compounds were identified on the basis of standards. Quantitative results were obtained by determination of liberated NPh-OH (410 nm) after termination of the reactions with addition of two volumes of saturated sodium tetraborate solution.

3 RESULTS

3.1 NPh-Xyl and NPh-Glc

NPh-Xyl was attacked by both EXs of family 10. Thus both EXs of family 10 exhibit aryl β-xylosidase activity. The rate of liberation of NPh-OH as a function of NPh-Xyl concentration does not obey the Michaelis-Menten kinetics and shows sigmoidal dependence (Figure 1), particularly with the *C. albidus* enzyme. The reaction rate is extremely slow at substrate concentrations below 10 mM, but increases considerably in going to higher values. At high substrate concentration (50-150 mM) the liberation of NPh-OH was accompanied by the formation of xylose and xylooligosaccharides which showed chromatographic mobility of β-1,4-xylooligosaccharides. At higher substrate concentrations, the predominant reaction is not the hydrolysis of the substrate but the glycosyl transfer reactions leading to the synthesis of oligosaccharides. The degradation of NPh-Xyl by EXs of family 10 represents a complex reaction pathway that involves a series of slow and more rapid hydrolytic and glycosyl transfer reactions as shown below in Scheme 1.

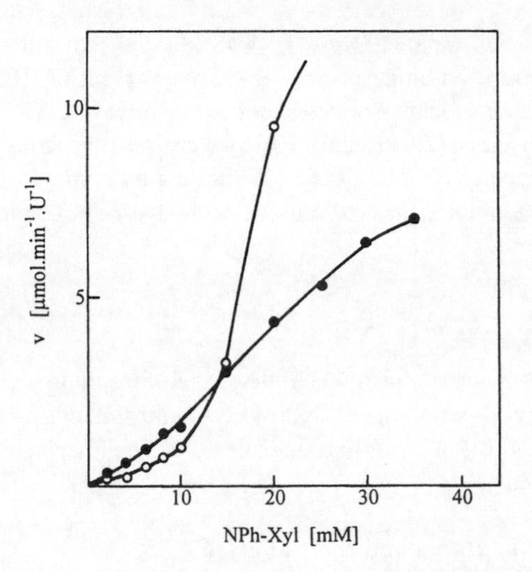

Figure 1 *The rate of liberation of NPh-OH from NPh-Xyl by XlnA from* Cryptococcus albidus (**O**) *and EX from* Streptomyces lividans (●) *as a function of substrate concentration*

Slow reactions:

NPh-Xyl → Xyl + NPh-OH

NPh-Xyl + NPh-Xyl → NPh-Xyl_2 + NPh-OH

NPh-Xyl_2 + NPh-Xyl → NPh-Xyl_3 + NPh-OH

Faster reactions:

NPh-Xyl_2 → Xyl_2 + NPh-OH

NPh-Xyl_3 → Xyl_3 + NPh-OH

NPh-Xyl + Xyl_3 → Xyl_2 + NPh-Xyl_2

Scheme 1 *The degradation of NPh-Xyl by EXs of family 10*

At the same enzyme concentration in terms of units per ml, none of the tested EXs of family 11 hydrolysed the xyloside or catalysed the synthesis of NPh-glycosides of xylooligosaccharides. An evidence for extremely slow hydrolysis of NPh-Xyl was obtained only after 4 day incubation with EX from *T. lanuginosus*. The rate of NPh-OH liberation was estimated to be by several orders slower than that with EXs of family 10.

Neither EXs of family 10 nor of family 11 attacked NPh-Glc at an appreciable rate. The addition of the C-6 hydroxymethyl group to the xyloside, changes dramatically the ability of the substrate to form a productive complex with EXs of family 10. However, qualitative evidence has been obtained by TLC that NPh-Glc (150 mM) served as xylosyl acceptor during degradation of 40 mM Xyl_3 with the *C. albidus* EX. The character of the formed linkage has not been determined. In this connection, the interaction of the enzymes with NPh-β-glycosides of cellobiose and lactose is interesting.

3.2 NPh-Cel and NPh-Lac

Both glycosides were slowly cleaved only by EXs of family 10. TLC product analysis demonstrated that the cleavage took place exclusively at the agluconic linkage to afford NPh-OH and cellobiose and lactose, respectively (Scheme 2).

Glcβ1-4Glc-NPh → Glcβ1-4Glc + NPh-OH

Galβ1-4Glc-NPh → Galβ1-4Glc + NPh-OH

Scheme 2 *Cleavage of NPh-Cel and NPh-Lac by EXs of family 10*

K_m values were determined in substrate concentration range 20 to 50 mM for NPh-Cel and 2 to 20 mM for NPh-Lac. With NPh-Cel the reaction at substrate concentrations below 20 mM does not obey the Michaelis-Menten kinetics. The upper concentration limit was given by solubility of the substrates in the reaction buffer. For both XlnA from *S. lividans* and EX from *C. albidus*, the K_m values were found to be similar and of very high value; around 80 mM for NPh-Cel and 100 mM for NPh-Lac. 4-O-Glycosylation of the non-hydrolysed NPh-Glc, resulting in glycosides of disaccharides (NPh-Cel, NPh-Lac), increased greatly the chance of the formation of productive enzyme-substrate complexes with EXs of family 10. The reason of this fact is apparently a strong affinity of subsite -II towards the non-reducing D-glucopyranosyl or D-galactopyranosyl residues.

One important difference was observed between the two EXs of family 10. The EX from *C. albidus*, in contrast to XlnA of *S. lividans*, catalyses, in addition to hydrolysis, cellobiosyl or lactosyl transfer to another substrate molecule. TLC analysis of the reaction mixtures showed the presence of low amounts of NPh-glycosides with mobility of NPh-glycosides of tetrasaccharides. The type of the newly formed linkage in such products have not been established, however, in analogy with 6'-O-xylosylation of cellobiose by *C. albidus* EX reported earlier (Biely and Vršanská, 1983), we assume that the cellobiosyl transfer to NPh-Cel, or lactosyl transfer to NPh-Lac leads also to formation of 6'-O-β-glycosidic linkage (Scheme 3). One could not expect that the 4'-O-β-transfer reaction takes place with both cellobioside and lactoside, because they are each other's 4'-epimers.

Glcβ1-4Glc-NPh + Glcβ1-4Glc-NPh → Glcβ1-4Glcβ1-6Glcβ1-4Glc-NPh + NPh-OH

Galβ1-4Glc-NPh + Galβ1-4Glc-NPh → Galβ1-4Glcβ1-6Galβ1-4Glc-NPh + NPh-OH

Scheme 3 *Cellobiosyl and lactosyl transfer catalysed by family 10 EX from* C. alibidus

Under the same experimental conditions 4-nitrophenyl glycosides of disaccharides were resistant towards EXs of family 11. The fact, that both EX families cleaved easily 4-nitrophenyl β-xylobioside (Biely *et al.*, 1997; Vršanská *et al.*, 1997), suggests that a possible repulsion of the aryl aglycon at the subsite +I in substrate binding site of EXs of family 11 is apparently not the main reason of the resistance of NPh-Xyl, NPh-Cel and NPh-Lac towards action of EXs of family 11. One of the subsites (-I or -II of the binding sites of these enzymes) is probably unable to accommodate glucopyranosyl or galactopyranosyl residues.

In a different set of experiments, we examined the ability of EXs to accommodate D-glucopyranosyl residues in those subsites of the substrate binding sites which are at the right of the catalytic groups, *i.e.* the subsites +I and +II. Cellobiose and NPh-Cel were examined as xylosyl acceptors in the glycosyl transfer reactions which are catalysed by EXs of family 10 at high concentration of xylotriose or NPh-Xyl, and by EXs of family 11 at high concentration of xylooligosaccharides. The only enzyme which utilised cellobiose or NPh-Cel as xylosyl acceptor was EX of *C. albidus*. This observation, confirming earlier results (Biely and Vršanská, 1983), pointed again to considerable differences in the construction of the substrate binding sites even within one enzyme family. For example, none of the following compounds which could be synthesised and hydrolysed by EX from *C. albidus* at the β-1,6-linkage (arrows in Scheme 4) will be attacked by XlnA of *St. lividans* and EXs of family 11.

↓ ↓ ↓

Xylβ1-6Glcβ1-4Glc Xylβ1-6Glcβ1-4Glc-NPh Xylβ1-6Galβ1-4Glc-NPh

↓ ↓

Glcβ1-4Glcβ1-6Glcβ1-4Glc-NPh Glcβ1-4Glcβ1-6Glcβ1-4Glc-NPh

Scheme 4 *Compounds recognised by EX from* C. albidus *but not by XlnA of* St. lividans *nor by EXs of family 11*

4 DISCUSSION

In agreement with the suggestion that EXs evolved from a common ancestor with endo-β-1,4-glucanases of family 5 (formerly A) (Dominguez *et al.*, 1995; Jenkins *et al.*, 1995), EXs of family 10 contain in their substrate binding sites at least two subsites that can accommodate β-D-glucopyranosyl residues. As shown in Figure 2, this evolutionary heritage applies to the subsites left of the catalytic groups designated as -I and -II. They can accommodate cellobiosyl and lactosyl moieties, and probably any mixed-type dimers that contain one β-1,4-linked D-glucopyranosyl and one D-xylopyranosyl moiety. Kitaoka *et al.* (1993) have also demonstrated that some high molecular mass EXs can attack the agluconic bond in 4-nitrophenyl glycosides of cellotriose and cellotetraose which indicates that EXs of family 10 have the ability to tolerate the replacement of a D-xylopyranosyl residue by a D-glucopyranosyl residue at positions that would correspond theoretically to subsites -III and -IV (Figure 2). However, the resistance of NPh-Glc and the extremely high K_m values for NPh-Cel and NPh-Lac of the two family 10 EXs suggest that the decisive affinity in productive binding of these two glycosides is exhibited by subsite -II, although this affinity is much lower for β-D-gluco- and β-D-galactopyranosyl residues than for β-D-xylopyranosyl residues. This is supported by the data on 50-times higher catalytic efficiency of a family 10 EX of *Cellulomonas fimi* for NPh-Xyl_2 than for NPh-Cel (Bedarkar *et al.*, 1992; Gebler *at al.*, 1992).

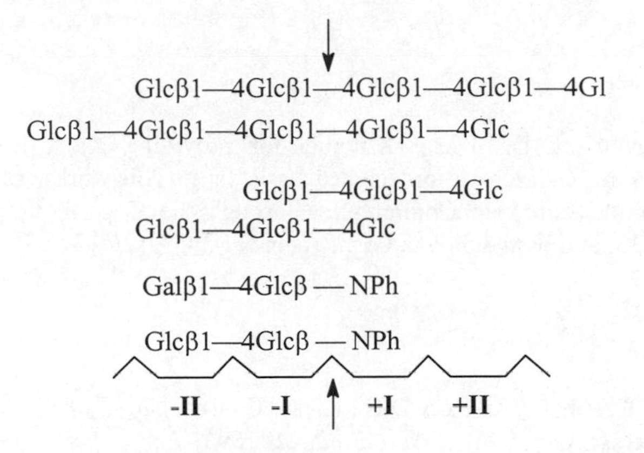

Figure 2 *Possible productive enzyme-substrate complexes of EXs of family 10 with compounds containing β-D-glucopyranosyl residues. The indentations in the blocks and roman numerals represent the sugar-binding subsites; the arrow in the block represents the catalytic groups of the enzymes.*

The subsites of the substrate binding sites of EXs of family 10 located at the right of the catalytic groups are more specific for β-D-xylopyranosyl residues. However, an evidence for hydrolysis of cellotriose or cellopentaose has been obtained at 20 mM substrate concentration and 50-times higher enzyme concentration as those used for NPh-Cel or NPh-Lac. This means that under extreme conditions there is some probability of the formation of productive enzyme-substrate complexes of EXs of family 10 with compounds that contain three or more consecutive β-1,4-linked D-glucopyranosyl residues. Interesting is the finding that the *C. albidus* EX was found to be the only EX that can use cellobiose, NPh-Cel or NPh-Lac as glycosyl acceptors in glycosyl transfer reactions catalysed at high concentration of NPh-Xyl or Xyl_3. This serves as an evidence that some gluco-compounds can fit into subsites of the substrate binding site right of the catalytic groups (Figure 2). However, the orientation of cellobiose, NPh-Cel and NPh-Lac in subsites +I and +II seems to be distorted in comparison with the binding of xylobiose, because xylosylation of cellobiose takes place at position 6 (Biely and Vršanská, 1983). The EX of *C. albidus* was also capable of xylosylation of NPh-Cel and NPh-Lac, and moreover, the transfer of a cellobiosyl or a lactosyl moiety by the enzyme was observed during degradation of NPh-Cel and NPh-Lac to another substrate molecule, respectively. This property of the *C. albidus* EX was not observed with XlnA from *S.*

lividans, also a family 10 EX, thus it can not be considered as a general feature of EXs of family 10. However, we are aware that final conclusions on the non-specificity of EXs of family 10 will require quantitative kinetic data.

EXs of family 11 do not tolerate the replacement in their substrates of a D-xylopyranosyl residue with a D-glucopyranosyl residue. They do not hydrolyse cellobiosides, lactosides, cellooligosaccharides and do not use these compounds as xylosyl acceptors in glycosyl transfer reactions at high concentration of xylooligosaccharides. They also do not exhibit significant aryl β-xylosidase activity. On the basis of these considerations, one can speculate that endo-β-1,4-glucanases of family 12, which show certain resemblance to EXs of family 11 (Törrönen *et al.*, 1993), probably evolved from EXs and not *vice versa*.

Acknowledgements

The authors thank Dr. Dieter Kluepfel for providing XlnA from *Streptomyces lividans* and Mrs. M. Cziszárová for technical assistance. This work was supported by the Commission of the European Communities through the Copernicus program (project CIPA-CT94-8232) and by the Slovak Grant Agency VEGA (2/4147/97).

References

Bedarkar, S., Gilkes, N.R., Kilburn, D.G., Kwan, E., Rose, D.S., Miller Jr., R.C., Warren, R.A.J., Withers, S.G. (1992), *J. Mol. Biol.*, **228**, 693.

Bennett, N.A., Ryan, J., Biely, P., Vršanská, M., Kremnický, L., Macris, B.J., Kekos, D., Kakopodis, P., Claeyssens, M., Nerinckx, W., M.K. Bhat (1997), *Carbohydr. Res.*, submitted for publication.

Biely, P. (1987), 'Differentiation of glycanases of microbial cellulolytic systems using chromogenic and fluorogenic substrates', *in*: Extracellular Enzymes of Microorganisms (Chaloupka, J. and Krumphanzl, V., eds.) pp.187-192, Plenum Press, New York and London.

Biely, P., Kluepfel, D., Morosoli, R., Shareck, F. (1993), *Biochim. Biophys. Acta*, **1162**, 246.

Biely, P., Kremnický L., Alföldi, J., Tenkanen, M. (1994), *FEBS Letters*, **356**, 137.

Biely, P. and Vršanská, M. (1983), *Carbohydr. Res.*, **123**, 97.

Biely, P., Vršanská, M. (1986), *Methods Enzymol.*, **160**, 638.

Biely, P., Vršanská, M., Tenkanen, M., Kluepfel, D. (1997), *J. Biotechnol.*, in press.

Christakopoulos, P., Nerinckx, W., Klarskov, K., Van Beeumen J., Kekos, B., Macris, B., Claeyssens, M. (1995), *Med. Fac. Landbouww. Univ. Gent*, **60/4a**, 2005.

Dominguez, R., Souchon, H., Spinelli, S., Dauter, Z., Wilson, K.S., Chauvaux, S., Beguin, P., Alzari, P.M. (1995), *Nature Struct. Biol.*, **2**, 569.

Gebler, J., Gilkes, N.R., Claeyssens, M., Wilson, D.B., Beguin, P., Wakarchuk, W.W., Kilburn, D.G., Miller-Jr, R.C., Warren, R.A., Withers, S.G. (1992), *J. Biol. Chem.*, **267**, 12559.

Gilbert, H.J., Sullivan, D.A., Jenkins, G., Kellett, L.E., Minton, N.P., Hall, J. (1988), *J. Gen. Microbiol.*, **134**, 3239.

Gilkes, N.R., Henrissat, B., Kilburn, D.G., Miller Jr., R.C., Warren, R.A.J. (1991), *Microbiol. Rev.*, **55**, 303.
Grepinet, O., Chebrou, M.-C., Beguin, P. (1988), *J. Bacteriol.*, **170**, 4576.
Hass, H., Herfurth, E., Stöffer, G., Rendl, B. (1992), *Biochim. Biophys. Acta*, **1117**, 279.
Henrissat, B., Bairoch, A. (1993), *Biochem. J.*, **293**, 781.
Jenkins, J., Lo Leggio, L., Harris G., Pickersgill, R. (1995), *FEBS Letters*, **362**, 281.
Kitaoka, M., Haga, K., Kashiwagi, Y., Sasaki, T., Taniguchi, H., Kusakabe, I. (1993), *Biosc. Biotech. Biochem.*, **57**, 1987.
Kluepfel, D., Daigneault, N., Morosoli, R., Shareck, F. (1992), *Appl. Microbiol. Biotechnol.*, **36**, 626.
Lüthi, E., Love, D.R., McAnulty, J., Wallace, C., Caughey, P.A., Saul, D., Bergquist, P.L. (1990), *Appl. Environ. Microbiol.*, **56**, 1017.
Morosoli, R., Bertrand, J.-L., Mondou, F., Shareck, F., Kluepfel, D. (1986), *Biochem. J.*, **239**, 587.
Rouau, X., Odier, E. (1986), *Carbohydr. Res.*, **9**, 25.
Shareck, F., Roy, C., Yaguchi, M., Morosoli, R., Kluepfel, D. (1991), *Gene*, **107**, 75.
Prade, R.A. (1995), *Biotechnol. Genet. Engin. Rev.*, **13**, 101.
Törrönen, A., Kubicek, C.P., Henrissat, B. (1993), *FEBS Letters*, **321**, 135.
van Tilbeurgh, H., Claeyssens, M. (1985), *FEBS Letters*, **187**, 283.
Vršanská, M., Nerinckx, W., Biely, P., Claeyssens, M. (1997), 'Fluorogenic substrates for endo-β-1,4-xylanases'. Abstracts of 8th Bratislava Symposium on Saccharides, Smolenice, Slovakia, p. 80.
Withers, S.G., Dombrowski, D., Berven, L.A., Kilburn, D.G., Miller, RC. Jr., Warren, R.A.J., Gilkes, N.R. (1986), *Biochem. Biophys. Res. Commun.*, **139**, 487.
Wong, K.K.Y., Tan, L.U.L., Saddler, J.N. (1988), *Microbiol. Rev.*, **52**, 305.

BIOCHEMICAL CHARACTERISATION OF CELLULASES AND XYLANASES FROM *THERMOASCUS AURANTIACUS*

M.K. Bhat[1], N.J. Parry[1*], S. Kalogiannis[1**], D.E. Beever[2], E. Owen[2], W. Nerinckx[3] and M. Claeyssens[3]

[1]Food Macromolecular Science Department, Institute of Food Research, Reading Laboratory, Earley Gate, Whiteknights Road, Reading, RG6 6BZ, UK. [2]Department of Agriculture, The University of Reading, Earley Gate, PO Box 236, Reading, RG6 6AT, UK. [3]Department of Biochemistry, University of Gent, K.L. Ledeganckstraat 35, B-9000, Gent, Belgium. Present addresses: *Unilever Research, Colworth Laboratory, Colworth House, Bedford, MK44 1LQ, UK. **Hellenic Sugar Industry, Sindos, 57400, Thessaloniki, Greece.

1 INTRODUCTION

Cellulose and hemicellulose are the major plant cell wall polysaccharides accounting for approximately 70% of the plant biomass and represent the major renewable energy source. The initial speculation that cellulosic wastes can be converted to soluble sugars using cellulolytic enzymes appeared very attractive both from economic and ecological view points. However, research on cellulolytic enzymes and their evaluation for hydrolysing plant biomass to soluble sugars indicated that improved cellulases and hemicellulases are necessary to make such conversion economically viable[1,2]. Alternatively, research on cellulases and hemicellulases revealed that these enzymes have potential applications in food, animal feed, paper pulp and related industries[3]. Furthermore, recent research from our laboratory and other laboratories indicated that thermostable cellulases and hemicellulases with broad substrate specificities could be useful for many industrial applications[4,5].

The thermophilic fungus *Thermoascus aurantiacus* produces a complete cellulase and hemicellulase system when grown on cellulosic carbon sources such as Avicel, corn cob and cereal straw[6,7]. However, the level and number of cellulase and hemicellulase components produced varies depending on the carbon source used[6,7]. We have recently demonstrated that *T. aurantiacus* IMI 216529 produces relatively high levels of all cellulase and hemicellulase components when grown on corn cob[6-8]. In the present paper, the purification of cellulases and xylanases from *T. aurantiacus* IMI 216529 is presented along with details of their biochemical properties, thermostability, substrate specificity and mode of action. The data presented clearly indicate that the cellulases and xylanases from *T. aurantiacus* have considerable commercial potential.

2 MATERIALS AND METHODS

2.1 Materials

T. aurantiacus IMI 216529 was obtained from the International Mycological Institute, Surrey, UK. The crude cellulase and hemicellulase rich fractions of *T. aurantiacus* IMI 216529 were kindly supplied by Prof. Macris, NTUA, Athens, Greece and Prof. Tiraby, Cayla, France. All chemicals used were analytical grade and purchased either from Sigma or BDH.

2.2 Methods

2.2.1 Enzyme assays. The endoglucanase, exoglucanase and β-D-glucosidase activities were measured as described[9], whereas the xylanase activity was measured according to the method of Bailey *et al*[10].

2.2.2 Purification of cellulase components. The cellulase rich fraction from *T. aurantiacus* was desalted on Bio-Gel P6DG column (2.6 × 100 cm) and fractionated on a DEAE-Sepharose column (2.6 × 90 cm) at pH 5.0. A major xylanase and a minor β-D-glucosidase fraction were eluted in the buffer wash. The bound endoglucanases, exoglucanases and β-D-glucosidase were eluted using a NaCl gradient (0 - 0.5 M) and further separated using an Ultrogel ACA 44 column (XK 1.6 × 100 cm). The major β-D-glucosidase was completely separated from endo- and exo-glucanase activities. The endo- and exoglucanases were purified using an affinity matrix, *p*-aminobenzyl 1-β-thio-cellobioside coupled to Sepharose 4B, and a novel principle, which involved the use of 0.1 M sodium acetate buffer at pH 5.0, containing 0.1 M glucose and 0.1 M cellobiose for the elution of endoglucanase and exoglucanases, respectively. Also, the separation of the major and minor endoglucanases as well as the high and low affinity exoglucanases was achieved using the above mentioned affinity column. Final purification of the major endoglucanase and two exoglucanases was achieved using a Mono-Q column.

2.2.3 Purification of the major and minor xylanases (XLN I and II). The hemicellulase rich fraction from *T. aurantiacus* was desalted on a Bio-Gel P6DG column and was fractionated using a Q-Sepharose fast flow column (20x3.4 cm) at pH 8.7. The XLN I was eluted in the buffer wash and established to be pure. The bound XLN II, β-D-xylosidase, acetylesterase and α-L-arabinofuranosidase were eluted using a 0 - 0.4 M NaCl gradient. XLN II was subsequently purified to homogeneity using a Hi trap SP column at pH 3.5.

2.2.4 Characterisation of the cellulases and XLN I. The molecular mass, pI, extent of glycosylation, effect of protease and amino acid composition were determined as described[11-14].

2.2.5 Determination of pH and temperature optima, and of stability. The pH optima of the purified β-D-glucosidase, the endoglucanase and both exoglucanases were determined using a citrate / phosphate buffer in the pH range of 2.8 - 6.8. The temperature optima were determined by measuring the activity between 30 - 90°C at optimum pH. The thermostabilities of the β-D-glucosidase and the endoglucanases were determined for 48 h at different temperatures (50 - 80°C) as a function of pH (2.8 - 6.8) by measuring the residual activity. The pH and temperature optima of XLN I were determined by measuring the activity in the pH region 2.2 - 10.2 and in the temperature range 50 - 90°C. Also, the stability of XLN I was determined at different pH (3.0 - 9.8) and temperatures (50 - 80°C) for six days by measuring the residual activity.

2.2.6 Determination of substrate specificities. The substrate specificities of the β-D-glucosidase, the endoglucanase and both exoglucanases were determined by measuring the activity towards a variety of *p*-nitrophenyl glycosides, disaccharides, oligosaccharides and polysaccharides. Also, the substrate specificities of XLN I and II were determined by measuring their activity towards various natural polysaccharides and pNP-glycosides.

2.2.7 Determination of modes of action. The modes of action of the β-D-glucosidase, the endoglucanase and both exoglucanases were determined by analysing the hydrolysis products released from cellooligosaccharides (Glc_n, n = 2-6), *p*-nitrophenyl cellooligosaccharides ($pNPGlc_n$, n = 1-5) and 4-methylumbelliferyl cellooligosaccharides ($MeUmbGlc_n$, n = 1-6) as a function of time. In addition, the K_m, k_{cat} and bond cleavage frequencies were determined for the endo- and exo-glucanases using $MeUmbGlc_n$. The mode of action of XLN I on xylooligosaccharides (Xyl_n, n=3-6) was determined by analysing the hydrolysis products released as a function of time using HPLC.

3 RESULTS AND DISCUSSION

3.1 Purification of the cellulases and xylanases from *T. aurantiacus*

The cellulase and hemicellulase rich fractions of *T. aurantiacus* were used to obtain pure cellulases and xylanases. Using mainly ion-exchange, gel filtration and affinity chromatography techniques, a major endoglucanase, a major β-D-glucosidase and two major exoglucanases were purified in mg quantities from the cellulase rich fraction of *T. aurantiacus*[15,16]. In addition, a minor endoglucanase, a minor β-D-glucosidase and an exoglucanase were purified in small quantities. The two major exoglucanases were designated as high and low affinity exoglucanases, based on their affinity towards *p*-aminobenzyl 1-β-thio-cellobioside.

The hemicellulase rich fraction from *T. aurantiacus* contained mainly xylanases and xylan debranching enzymes with low level of cellulase activities. One major xylanase (XLN I), one minor xylanase (XLN II), one β-D-xylosidase and one acetyl esterase were isolated and purified. The purification protocol adopted in the present investigation was relatively simple and yielded mg quantities of pure XLN I and II. In addition, obtaining mg quantities of pure enzymes resulted in the crystallisation and 3D-structure determination of the major endoglucanase and of XLN I from *T. aurantiacus*[17,18].

3.2 General properties of the cellulases, XLN I and XLN II from *T. aurantiacus*

The general properties of the purified endoglucanase, both exoglucanases (low and high affinity), β-D-glucosidase, XLN I and II are presented in Table 1. Both exoglucanases and the β-D-glucosidase were highly glycosylated, whereas the endoglucanase was less glycosylated. All cellulase components from *T. aurantiacus* were resistant against the protease from *Aspergillus sojae* for 6 h between pH 6.0 - 8.0. Determination of amino acid composition revealed that all cellulase components and the XLN I were rich in acidic amino acids.

3.3 Temperature and pH optima, and stabilities of the cellulases, XLN I and XLN II from *T. aurantiacus*

The endoglucanase and exoglucanases were optimally active at 70°C, compared with 80°C for the β-D-glucosidase and XLN I. In contrast, XLN II was optimally active around 60°C. The endoglucanase, the β-D-glucosidase and XLN I had optimal activities at pH 3.5, 4.0 and 4.4 respectively, whereas both exoglucanases were optimally active between pH 4.4 - 5.2.

Table 1 *General properties of the cellulases, XLN I and II from* T. aurantiacus.

Property	Endoglucanase	Exo-HA	Exo-LA	β-D-glucosidase	XLN I	XLN II
M_r (kDa)	34	52 - 55	52 - 55	120	32	17
pI (pH)	3.5 - 3.7	3.6 - 3.8	3.6 - 3.8	4.1 - 4.5	7.0	ND
Sugar content (%)	2.7	15.5	10.5	19.6	ND	ND

Exo-HA, exoglucanase - high affinity; Exo-LA, exoglucanase - low affinity; M_r, molecular mass; pI, isoelectric point; ND, not determined.

The stability of the endoglucanase and of the β-D-glucosidase at pH 5.2 at different temperatures is shown in Figures 1 A and B. The endoglucanase was completely stable up to 60°C, at pH 5.2 for 48 h (Figure 1 A). At 70°C and at pH 5.2, the endoglucanase retained around 60% of its original activity for 48 h, but at 80°C under the same conditions, the enzyme lost all of its activity sharply (Figure 1A). However, the endoglucanase retained its full activity between pH 2.8 - 6.8 at 50°C for 48 h (data not shown). In contrast, the β-D-glucosidase retained its full activity up to 70°C at pH 5.2 for 48h (Figure 1B) but it was less stable at pH values above and below 5.0 (data not shown).

The stability of XLN I was determined at different pH and temperatures as a function of time. The enzyme retained its full activity up to 80°C and between pH 4.4 - 6.2 for six days. The stability of XLN I at 60°C at different pH values is shown in Figure 2. XLN I retained 90% of its original activity at 60°C between pH 4.4 - 8.0, and 50% of its original activity at 60°C at pH 3.0 and 9.8 for 2 days and 2 h, respectively.

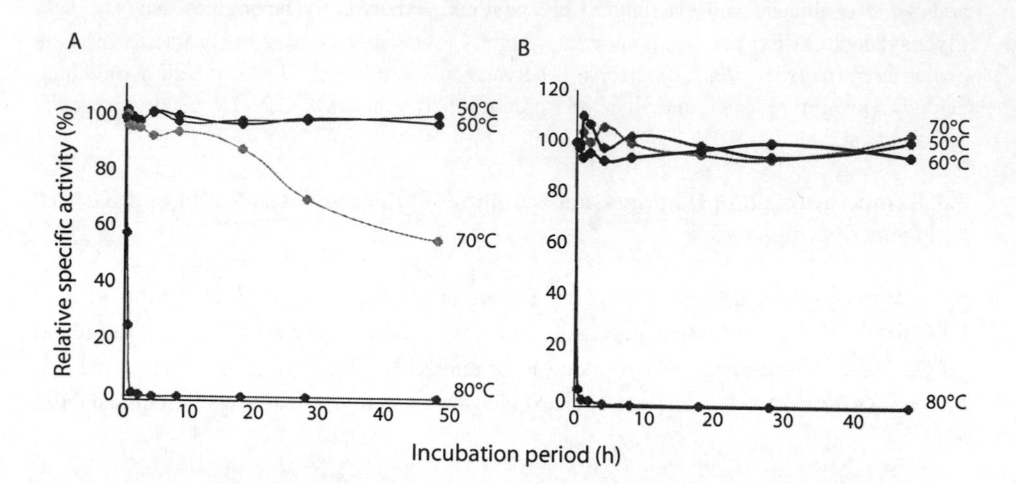

Figure 1 *Stabilities of the endoglucanase (A) and the β-D-glucosidase (B) from* T. aurantiacus *at pH 5.2.*

3.4 Substrate specificities of cellulases, XLN I and XLN II from *T. aurantiacus*

The endoglucanase hydrolysed polysaccharides such as carboxymethylcellulose (CMC), barley β-glucan and lichenan. It also hydrolysed substituted and unsubstituted Glc_n derivatives (*e.g.* pNP-glycosides and $pNPGlc_2$), xylan and Avicel, although to a limited extent. The enzyme was inactive on carboxymethyl pachyman, laminarin, β(1→3) linked glycosides and different cyclodextrins. This demonstrates its specificity for the β(1→4) glycosidic bond.

The β-D-glucosidase showed its highest activity towards pNPGlc. However, the enzyme was active towards β-linked glycosides, β-linked disaccharides and oligosaccharides. Furthermore, the enzyme hydrolysed laminarin and CMC to a limited extent but it was inactive with respect to Avicel, xylan and lichenan. Also, it was inactive on α(1→4), α(1→6) and α(1→1) linked glucopyranosyl compounds, revealing its strict specificity for β-linked glucopyranosides.

Both exoglucanases demonstrated a high activity towards $pNPGlc_2$ and pNP-lactoside, and were active towards substituted and unsubstituted Glc_n, Avicel, barley β-glucan, lichenan and to a limited extent on CMC. Also, the high affinity exoglucanase demonstrated a higher activity towards all polysaccharides compared to the low affinity exoglucanase with its highest activity towards lichenan. However, both exoglucanases were inactive towards laminarin, xylan, CM-pachyman, neosugar and cyclodextrins.

XLN I from *T. aurantiacus* showed the same level of activity towards different xylans. However, this xylanase showed higher affinity towards substituted xylans, and was active against $pNPGlc_2$, $MeUmbXyl_2$ and MeUmbXylGlc. However, the activity of XLN I towards pNP-α-L-arabinopyranoside and pNP-α-L-arabinofuranoside was higher than towards pNP-β-D-xylopyranoside. These and other biochemical properties presented elsewhere suggest that XLN I from *T. aurantiacus* belongs to family 10 of glycosyl hydrolases[7,17].

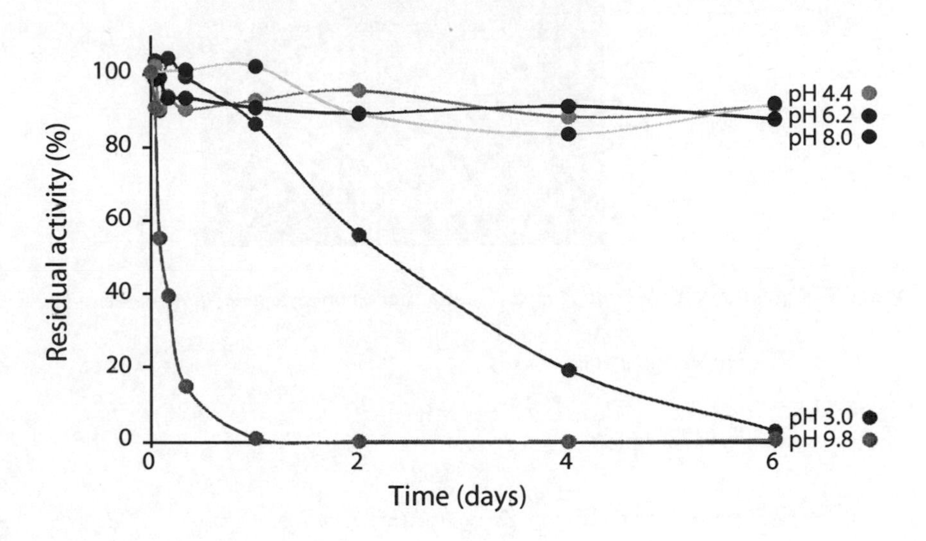

Figure 2 *Stability of XLN I from* T. aurantiacus *at 60°C.*

The substrate specificities of XLN I and II from *T. aurantiacus* were compared using a number of MeUmb derivatives as shown in Figure 3. Although XLN I was active on $MeUmbXyl_2$ and MeUmbXylGlc, XLN II was only active only towards $MeUmbXyl_2$. These results are in agreement with Kitaoka *et al*[19], who suggested that $pNPXyl_2$ and pNPXylGlc could be used to biochemically distinguish the xylanases from family 10 and 11. Also, the results presented in Figure 3 revealed that XLN I and II from *T. aurantiacus* belong to families 10 and 11 of glycosyl hydrolases respectively. This was corroborated by the 3D-structure determination of XLN I[17].

3.5 Modes of action of the cellulases, the β-D-glucosidase and XLN I from *T. aurantiacus*

The mode of action of the endoglucanase was determined using $MeUmbGlc_n$. The initial product ratios and the reaction kinetics were determined and are summarised in Figure 4. The endoglucanase preferentially cleaved the glycosidic bond adjacent to MeUmb of glycosides up to $MeUmbGlc_3$. In addition, the turnover number of the endoglucanase towards the preferential site of $MeUmbGlc_3$ was ten times higher than that towards the preferential site of $MeUmbGlc_2$. The enzyme showed preference towards the internal glycosidic bonds of $MeUmbGlc_4$ and $MeUmbGlc_5$. The action of the endoglucanase on $MeUmbGlc_5$ was random, indicating that it is a true endoglucanase and similar to the endoglucanases characterised from *Penicillium pinophilum*[20].

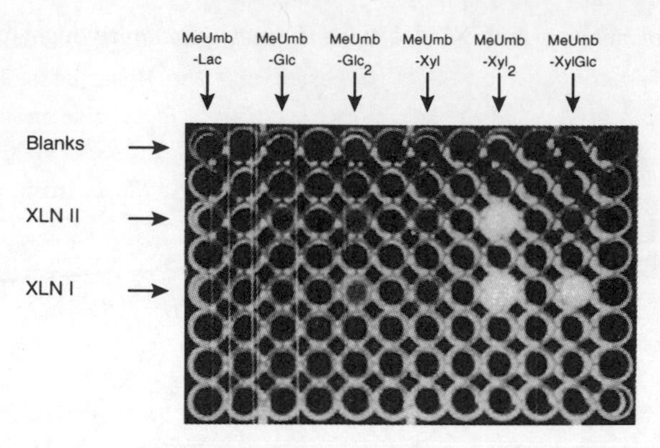

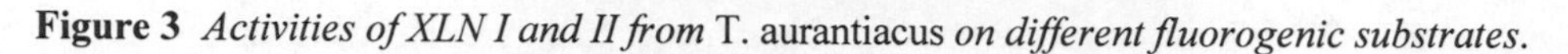

Figure 3 *Activities of XLN I and II from* T. aurantiacus *on different fluorogenic substrates.*

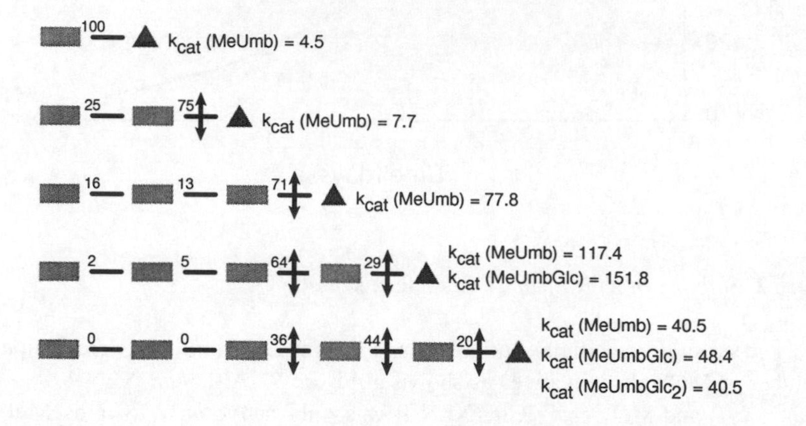

Figure 4 *Bond cleavage frequencies and turnover numbers (sec.$^{-1}$) for the hydrolysis of MeUmbGlc$_n$ by the endoglucanase from* T. aurantiacus.

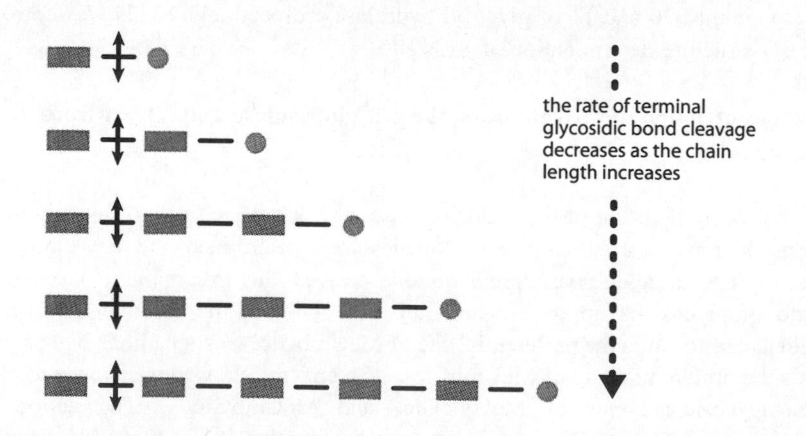

Figure 5 *Mode of action of the β-D-glucosidase from* T. aurantiacus *on pNPGlc$_n$.*

The mode of action of the β-D-glucosidase was initially determined using unsubstituted Glc_n. Although the results suggested that the β-D-glucosidase cleaves the terminal glycosidic bond, it was not possible to establish the precise mode of action on Glc_n. However, using $pNPGlc_n$ and subsequent analysis of the hydrolysis products by HPLC, it was possible to demonstrate that the β-D-glucosidase from *T. aurantiacus* cleaved one glucose unit at a time from the non-reducing end of $pNPGlc_n$ (Figure 5). The rate of hydrolysis decreased with increasing chain length of $pNPGlc_n$. This showed that the β-D-glucosidase from *T. aurantiacus* is specific for disaccharides and trisaccharides rather than for higher cellooligosaccharides.

The modes of action of the exoglucanases were determined using $MeUmbGlc_n$ and analysis of the hydrolysis products by HPLC. Based on the initial product ratios, the reaction kinetics were determined; the results are summarised in Figure 6. The pattern of hydrolysis of $MeUmbGlc_n$ by both exoglucanases was identical. However, differences in the k_{cat} values were observed (Figure 6). Unfortunately, the hydrolysis of $MeUmbGlc_5$ by both exoglucanases could not be followed accurately besause of the very fast rate of substrate hydrolysis. Both exoglucanases preferentially cleaved the second glycosidic bond distal from the aglycon of $MeUmbGlc_2$ and $MeUmbGlc_3$. Furthermore, when $MeUmbGlc_4$ was used, both exoglucanases showed a higher preference towards the first holosidic bond near MeUmb (Figure 6). Thus, both exoglucanases showed an identical mode of action but with small differences in their turnover number towards various $MeUmbGlc_n$.

The mode of action of XLN I from *T. aurantiacus* was studied using Xyl_n with DP 3 up to 6. The enzyme released equimolar amounts of Xyl and Xyl_2 from Xyl_3 with concentrations up to 2 mM (Figure 7). Interestingly, with higher concentration of Xyl_3, the ratio of Xyl_2 to Xyl was higher than one (Figure 7). This suggested that, at higher concentrations of Xyl_3, the enzyme catalysed both the hydrolysis and transglycosylation reactions.

Xyl_4 was found to be a better substrate for this xylanase than Xyl_3. The enzyme released Xyl, Xyl_2 and Xyl_3 from Xyl_4, with Xyl_2 as the main product. However, the concentration of Xyl_3 was considerably higher than Xyl when the substrate (Xyl_4) concentration was greater than 2 mM. This again indicated that this xylanase catalysed both hydrolysis and transglycosylation reactions simultaneously at high substrate concentration.

It was reported that three xylanases from *Streptomyces lividans* hydrolysed Xyl_4 to Xyl_2 and Xyl_3, through a series of hydrolytic and transfer reactions[21]. Similar observations have been made with the xylanase from *Thermomyces lanuginosus*[22].

When Xyl_5 was used as a substrate, the enzyme released mainly Xyl_2 and Xyl_3. Also, from Xyl_6, the enzyme released mainly Xyl_3, and equimolar amounts of Xyl_2 and Xyl_4. These results clearly suggested that XLN I from *T. aurantiacus* is an endoxylanase.

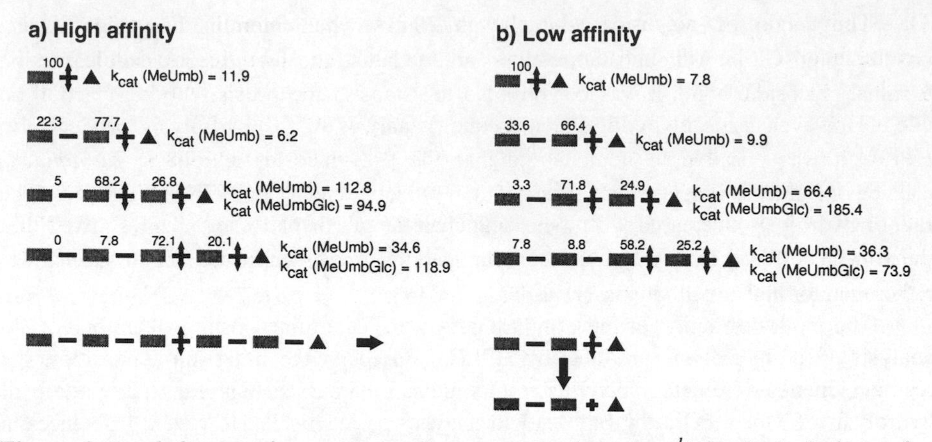

Figure 6 *Bond cleavage frequencies and turnover numbers (sec.$^{-1}$) for the hydrolysis of MeUmbGlc$_n$ by the exoglucanases from* T. aurantiacus.

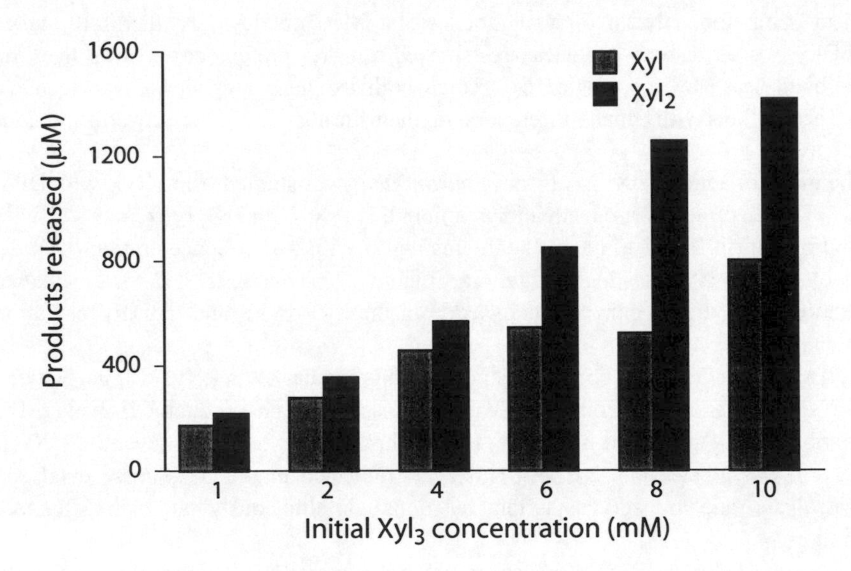

Figure 7 *Ratio of Xyl and Xyl_2 released from different concentrations of Xyl_3 by XLN I from* T. aurantiacus.

4 CONCLUSIONS

We have developed relatively simple procedures for the purification of several cellulases, β-D-glucosidases and xylanases from *T. aurantiacus*, and one endocellulase, two exocellulases, one β-D-glucosidase and two xylanases have been characterised. The molecular mass of the cellulases and xylanases from *T. aurantiacus* ranged from 17 - 55 kDa except that of β-D-glucosidase, which was 120 kDa when determined by SDS-PAGE. Determination of the pH and temperature optima and the stabilities of cellulases and xylanases from *T. aurantiacus* showed some interesting properties and suggested that these enzymes could be of commercial interest. Substrate specificity and mode of action studies revealed that the cellulase components from *T. aurantiacus* are highly specific for β-linked glycosidic bonds and exhibit the mode of action as expected for typical endoglucanases, exoglucanases and β-D-glucosidases. However, XLN I from *T. aurantiacus* showed a broad substrate specificity and was more active on pNP-α-L-arabinopyranoside and pNP-α-L-arabinofuranoside than on pNPXyl. Furthermore, the determination of the substrate specificities of XLN I and II from *T. aurantiacus* using various model substrates revealed that they belong to the families 10 and 11 of glycosyl hydrolases respectively.

Acknowledgements

We thank Prof. Macris, NTUA, Athens, Greece and Prof. Tiraby, Cayla, Toulouse, France for their kind gift of enzyme samples. Also, we acknowledge the Commission of the European Communities (AIR2- CT93-1272) and BBSRC for financial support.

References

1. D.D. Ryu and M. Mandels, *Enzyme Microb. Technol.*, 1980, **2**, 91.
2. A.A. Klyosov, *Biochemistry,* 1990, **29**, 10577.
3. M.K. Bhat and S. Bhat, *Biotechnol. Adv.*, 1997, **15**, 583.
4. C. Sudgen and M.K. Bhat, *World J. Microbiol. Biotechnol.*, 1994, **10**, 444.
5. J.G. Zeikus, C. Lee, Y.-E. Lee and B.C. Saha, *in* "Enzymes in Biomass Conversion" edited by G.F. Leatham and M. E. Himmel, ACS Symp. Vol. **460**, American Chem. Soc., Washington, DC, 1991, Chapter 4, 37.
6. N.J. Parry, S. Kalogiannis, E. Owen, D.E. Beever and M.K. Bhat, Poster presented at "PIRA International Conference on Non-Wood Fibres for Industry", Silsoe Research Institute, 1994.
7. S. Kalogiannis, E. Owen, D.E. Beever and M.K. Bhat, *Med.Fac.Landbouww. Univ. Gent*, 1995, **60/4a**, 1995.
8. S. Kalogiannis, E. Owen, D.E. Beever, M. Claeyssens, W. Nerinckx and M.K. Bhat, SCI Lecture papers number 78.
9. T.M. Wood and M.K. Bhat, *Methods Enzymol.*, 1988, **160**, 87.
10. M.J. Bailey, P. Biely and K. Poutanen, *J. Biotechnol.*, 1992, **23**, 257.
11. B. Lugtenberg, J. Meijers, R. Peters, P. Van Der Hoek and L. Van Alphen, *FEBS Lett.*, 1975, **58**, 254.
12. M.K. Bhat and T.M. Wood, *Biotechnol. Bioeng.*, 1989, **33**, 1242.
13. M. Dubois, K.A. Gilkes, J.K. Hamilton, P.A. Reberts and F. Smith, *Anal. Chem.*, 1956, **28**, 350.

14. P. Bohlen and R. Schroeder, *Anal. Biochem.*, 1982, **126**, 144.
15. N.J. Parry, D.E. Beever, E. Owen and M.K. Bhat, Poster presented at "23rd meeting of the Federation of European Biochemical Societies", Basel, Switzerland, 1995.
16. N.J. Parry, E. Owen, D.E. Beever, M. Claeyssens and M.K. Bhat, Poster presented at "British Society of Animal Science Meeting", Scarborough, UK, 1996.
17. L. Lo Leggio, S. Kalogiannis, N.J. Parry, M.K. Bhat and R.W. Pickersgill, Poster presented at "Understanding Protein Structure Determination", Huddinge, Sweden, 1996.
18. L. Lo Leggio, N.J. Parry, J. Van Beeumen, M. Claeyssens, M.K. Bhat and R.W. Pickersgill, *Acta Cryst.*, 1997, **D53**, 599.
19. M. Kitaoka, K. Haga, Y. Kashiwagi, T. Sasaki, H. Taniguchi and I. Kusakabe, *Biosci. Biotech. Biochem.*, 1993, **57**, 1987.
20. M.K. Bhat, A.J. Hay, M. Claeyssens and T.M. Wood, *Biochem. J.*, 1990, **266**, 371.
21. P. Biely, D. Kluepfel, R. Morosoli and F. Shareck, *Biochim. Biophys. Acta*, 1993, **1146**, 246.
22. N.A. Bennett, J. Ryan, P. Biely, M. Vrsanská, L. Kremnicky, B.J. Macris, D. Kekos, P. Christakopoulos, P. Katapodis, M. Claeyssens, W. Nerinckx, P. Ntarima and M.K. Bhat, *Carbohydr. Res.*, 1997 (in press).

Structure/Function of Carbohydrases

STRUCTURE AND FUNCTION IN β–1,4-GLYCANASES

R.A.J. Warren

Department of Microbiology & Immunology and Protein Engineering Network of Centres of Excellence, University of British Columbia, Vancouver BC, Canada V6T 1Z3

1 INTRODUCTION

As more and more cellulase systems are examined, it is clear that those from different organisms, including both bacteria and fungi, are similar in the types and the activities of the enzymes comprising them. The same types of enzymes from different organisms have similar efficiencies. This has broadened the focus of research on the hydrolysis of the polysaccharides in plant cell walls. Much has been learned about the structures, catalytic mechanisms, and interactions of the enzymes, and of the general characteristics of the systems from which they are obtained. An overview of some aspects of the field is presented here. More comprehensive coverage of most of the material is found elsewhere in this volume.

2 STEREOSPECIFICITY

All glycoside hydrolases can be divided into two large groups: those that hydrolyse the glycosidic bond with retention of conformation at the anomeric centre, and those that cause inversion of the conformation[1]. The stereospecificity of an enzyme is readily determined by NMR spectrometric analysis of the initial products of the reaction before significant mutarotation has occurred. The β-1,4-glycanases include both types. The retaining enzymes catalyse hydrolysis by a two-step mechanism involving a covalent enzyme-substrate intermediate; the inverting enzymes use a concerted mechanism that does not involve a covalent intermediate. In general, both types of reaction involve two catalytic carboxyl groups on the enzyme[1]. In retaining enzymes, one carboxyl group acts as a nucleophile, the other as an acid/base catalyst. In inverting enzymes, one acts as an acid , the other as a base catalyst. An important consequence of the two different mechanisms is that retaining but not inverting enzymes can transglycosylate. Retaining enzymes can be used to synthesise glycosides under appropriate conditions.

3 MODULES AND FAMILIES

Domains are independently folding, spatially distinct structural units in proteins. The sequence need not be contiguous. Modules are a subset of domains that are contiguous in sequence and that are used repeatedly as domains in functionally diverse proteins[2]. The majority of β-1,4-glycanases are modular proteins, with the number of modules generally ranging from two to six (Table 1). The modules retain their functions when separated by proteolysis or by manipulation of the DNA sequences encoding them[3,4]. It is customary to refer to the modules in β-1,4-glycanases as domains.

Table 1 *Domain organisation in β-1,4-glycanases from Cellulomonas fimi*

Enzyme	*Organisation*
Endoglucanase CenA	CBD2/CD6
Endoglucanase CenB	CD9/CBD3/FN3/FN3/FN3/CBD2
Endoglucanase CenC	CBD4/CBD4/CD9/?/?
Endoglucanase CenD	CD5/FN3/FN3/CBD2
Cellobiohydrolase CbhA	CD6/FN3/FN3/CBD2
Cellobiohydrolase CbhB	CD48/FN3/FN3/FN3/CBD2
Xylanase XynB	CD10/CBD2
Xylanase XynC	NodB/TST/CD10/CBD9/CBD9/?
Xylanase XynD	CD11/CBD2/NodB/CBD2

CD: catalytic domain with family number; CBD: cellulose-binding domain with family number; FN3: fibronectin type III repeat; TST: thermostabilising domain; NodB: acetyl xylan esterase domain related to NodB from *Rhizobium* spp.; ?: domain of unknown function.

All of the enzymes comprise at least a catalytic domain (CD) or module. The commonest ancillary module is a substrate-binding domain. Where present, the substrate-binding domain is usually a cellulose-binding domain (CBD). CBDs occur in enzymes other than cellulases, such as xylanases and mannanases. A number of other types of module also occur in the enzymes (Table 2). It is striking that several types of three-dimensional structures occur in the CDs, whereas all of the ancillary domains characterised to date are folded β-sheet structures (Table 2).

Glycoside hydrolases are classified into families of related amino acid sequences, using the sequences of the CDs only[5,6]. Where analysed, all CDs in a family have similar three-dimensional structures and exhibit the same stereospecificity of hydrolysis. A family may contain enzymes with different specificities and modes of action (Table 3). Such differences between enzymes in a family are determined by the fine details of their structures, not by their global folds[7,8]. Several of the families are part of a superfamily, designated Clan GH-A, in which all of the enzymes are retaining and all of their CDs are 4/7$(\beta/\alpha)_8$ barrels in which the catalytic carboxy amino acids are adjacent to the carboxy termini of β-strands four and seven[8-11]. The entire clan comprises at least 11 different families of glycoside hydrolases exhibiting at least 18 different specificities.

CBDs are also classified into families of related amino acid sequences (Table 4).

Table 2 *Types of Modules in β–1,4–Glycanases*

Module	*Families* #	*Types of Structure* *
Catalytic domains	20	Barrels: $(\beta/\alpha)_8$; α_{12}; $(\alpha/\alpha)_6$; β_6 β-sandwiches
CBDs	13	β-sandwiches
Fibronectin Type III repeats	1	β-sandwiches
Dockerins	2	
Cohesins	2	β–sandwiches
S-layer homology (SLH) domains	1	
Thermostabilising domains	1	
NodB sequences	1	

Number of families of related amino acid sequences at the time of writing.
* Different types of fold reported at the time of writing

Table 3 *Families of β–1,4-Glycanases*

Family	*Stereospecificity*	*Substrate Specificity*	*Organisms*
3	Retaining	β-Glucosidases Cellodextrinases	Fungi; Bacteria
5	Retaining	Endoglucanases Mannanases Cellodextrinases	Fungi; Bacteria
6	Inverting	Endoglucanases Cellobiohydrolases	Fungi; Bacteria
7	Retaining	Endoglucanases Cellobiohydrolases	Fungi; Bacteria
8	Inverting	Endoglucanases	Bacteria
9	Inverting	Endoglucanases	Bacteria
10	Retaining	Xylanases	Fungi; Bacteria
11	Retaining	Xylanases	Fungi; Bacteria
12	Retaining	Endoglucanases	Fungi; Bacteria
16	Retaining	Lichenases Laminarinases	Fungi; Bacteria
26	Retaining	Mannanases Endoglucanases	Bacteria
43	Inverting	Xylanases	Fungi; Bacteria
48	Inverting	Cellobiohydrolases Endoglucanases	Fungi; Bacteria

Table 4 *Families of Cellulose-Binding Domains*

Family	*Amino Acids* #	*Number in Family**	*Organisms* +
I	33-37	45	Fungi
II	83-108	48	Bacteria
III	84-172	34	Bacteria
IV	125-148	6	Bacteria
V	63	1	Bacterium
VI	78-92	7	Bacteria
VII	240	1	Bacterium
VIII	152	1	Dictyostelium
IX	117-187	15	Bacteria
X	51-55	7	Bacteria
XI	116-180	4	Bacteria
XII	49-56	12	Bacteria
XIII	41-47	23	Bacteria

The number of amino acids comprising members of the family

* The number of different CBDs in the family

+ The organisms in which members of the family are found

4 SPECIFICITY AND ACTIVITY

The hydrolytic activities of the enzymes can be analysed on several levels. Oligosaccharides can be used to determine the preference or specificity for a glycoside, and for defining the number of sugar-binding subsites within the active site. Similarly, chromogenic substrates can yield useful data on the kinetics of enzymes but not necessarily on the hydrolysis of the corresponding polysaccharides. The enzymes in general hydrolyse glycosidic bonds efficiently; the limiting factor in the hydrolysis of the polysaccharides is probably the initial entry of a substrate molecule into the active site of the enzyme.

In crystalline cellulose, for example, individual molecules are held closely together by hydrogen bonds and van der Waals forces; before an enzyme can hydrolyse a glycosidic bond, it must sequester a single molecule of cellulose and insert it into its active site. Kinetic analyses are difficult with the true substrates of the enzymes. Analysis of the insoluble products of cellulose hydrolysis can be done by converting those products to soluble derivatives and determining their sizes by size-exclusion chromatography[12-14]. Such analyses are important in differentiating between endo-acting and exo-acting enzymes.

Cellulases are differentiated into exoglucanases, or cellobiohydrolases, and endoglucanases. The three-dimensional structures of enzymes of both types within a family suggest that cellobiohydrolases have tunnel-shaped active sites whereas those of endoglucanases are open clefts[15-18]. This could explain the difference in the modes of action of the two groups. It seems unlikely, however, that the distinction between exo- and endo-acting enzymes is absolute. Some cellobiohydrolases have low endoglucanase activity[19,20], and some endoglucanases appear to hydrolyse the substrate progressively from the initial site of hydrolysis[21-23]. It is probably more accurate to say that some enzymes are preferentially exo- and others preferentially endo-acting. Hydrolysis would be more efficient if cellobiohydrolases did not always have to find the ends of molecules but could create them if necessary. Furthermore, it would be easier for them to find ends created by endoglucanases if the latter removed a few sugar residues at the initial site of hydrolysis. This may be the role of the progressive endoglucanases. Some cellobiohydrolases act from the reducing ends, others from the non-reducing ends of molecules[20,23,24]. This also increases the efficiency of hydrolysis.

It was believed for some time that cellobiohydrolases but not endoglucanases could solubilise crystalline cellulose. Some endoglucanases can, however, hydrolyse crystalline cellulose extensively[13].

It must be emphasised that in nature the enzymes may encounter pure cellulose rarely, if at all. Ideally the enzyme activities should be compared on plant cell walls.

5 MECHANISMS

An understanding of the mechanism of a glycanase requires identification of its catalytic amino acid residues. In virtually all instances these are carboxy amino acids. When possible, a combination of methods rather than a single method should be used to identify these residues.

As more enzymes are added to a family of catalytic domains, the fully conserved amino acids will be those essential for activity. Fully conserved carboxy amino acids are then targets for site-directed mutation. In general, mutation of such a residue should lower the activity of an enzyme several orders of magnitude, depending on the substrate employed[25-27].

Determination of the three-dimensional structure of an enzyme, especially of an enzyme-substrate or enzyme-inhibitor complex, should also pinpoint the catalytically important residues. Again, however, site-directed mutation of the carboxy amino acids targeted by this approach should also lower activity significantly.

These approaches are applicable to both retaining and inverting enzymes. Other approaches are applicable only to retaining enzymes. The nucleophile in a retaining enzyme can be identified by labelling it with a mechanism-based inhibitor that forms a relatively stable, covalent enzyme-inhibitor complex[28,29]. The enzyme remains catalytically competent because inactivation is reversible by transglycosylation of the inhibitor to an appropriate acceptor sugar. Again, mutation of the carboxy amino acid so identified should lower the activity significantly. If the carboxy amino acid is mutated to alanine, activity can be restored virtually to wild-type levels by small nucleophiles such as azide and formate, providing the aglycone of the substrate is a good leaving group, such as 2,4-dinitrophenyl. The product is the α-glycosyl azide or formate, so the enzyme is inverting rather than retaining under these conditions[26]. Formation of the α-glycoside confirms that the mutated carboxy amino acid is the nucleophile.

Although inhibitors specific for the acid/base catalyst in a retaining enzyme are unavailable, it can be identified. Amino acid sequence conservation and/or three-dimensional structure determination are used to target particular carboxy amino acids.

Mutation of the acid/base catalyst should lower activity significantly[25]. The activity of the alanine mutant can be restored virtually to wild-type levels by azide or formate. In this case, however, the product is the β-glycoside, confirming that the mutated carboxy amino acid is the acid/base catalyst[25].

Identification of its catalytic carboxy amino acids opens the way to a detailed analysis of the catalytic mechanism of an enzyme. For example, the pK_a of the acid/base catalyst in a retaining xylanase cycles between pH 6.7 and pH 4.2 in the course of the reaction[30].

6 ENZYME SYSTEMS

The systems of enzymes produced by different micro-organisms for hydrolysis of the polysaccharides in the cell walls of plants can be compared in two ways: either the numbers of enzymes from different families or the numbers of enzymes with particular specificities. Bearing in mind that not all of the enzymes produced by a particular organism may be known, both comparisons reveal that the enzyme systems produced by a number of different micro-organisms are quite similar.

All micro-organisms from which multiple β–1,4–glycanases have been characterised produce enzymes from several families (Table 5). Obviously, there has been considerable horizontal transmission of the genes encoding the enzymes in most families within and between bacteria and fungi. At present family 7 appears to be confined to fungi, families 9 and 48 to bacteria.

The similarities between the systems are much more apparent from the range of specificities found in each of them (Table 6). These similarities serve to emphasise that, contrary to earlier opinion, bacteria can degrade crystalline cellulose[13].

Table 5 *Families of β-1,4-Glycanases Produced by Selected Organisms*

Organism	*Enzyme Family #*
Humicola insolens	5(1); 10(1); 6(2); 7(2); 11(1); 12(1); 45(1)
Trichoderma reesei	3(1); 5(2); 6(1); 7(2); 11(2); 12(1); 45(1)
Butyrivibrio fibrisolvens	3(1); 5(2); 9(1); 10(1); 43(1)
Fibrobacter succinogenes	5(2); 9(2); 11(2); 16(1)
Caldocellulosiruptor saccharolyticus	5(3); 9(2); 10(3); 48(1)
Clostridium cellulolyticum	5(2); 8(1); 9(2);43(1); 48(1)
Clostridium thermocellum	3(1); 5(5); 8(1); 9(3); 10(3); 16(2); 26(1); 48(1)
Cellulomonas fimi	5(1); 6(2); 9(2); 10(2); 11(1); 26(1); 48(1)
Thermomonospora fusca	5(1); 6(2); 9(2); 11(1);48(1)
Streptomyces lividans	5(2); 10(1); 11(2); 12(1)
Pseudomonas fluorescens subsp. *cellulosa*	3(1); 5(1); 9(1); 10(2); 26(1); 45(1)

The number is the family; the number in parentheses is the number of enzymes produced in that family.

Table 6 *Specificities of the β-1,4-Glycanases Produced by Selected Organisms*

Organism	*Type of Enzyme*#						
	Cd	*Eg*	*Cbh*	*Xyn*	*Man*	*Lic*	*Lam*
Humicola insolens		5	2	2			
Trichoderma reesei	4	2	2	1			
Butyrivibrio fibrisolvens		3		1			
Fibrobacter succinogenes	1	3		2		1	
Caldocellulosiruptor saccharolyticus	3	1	3	2			
Clostridium cellulolyticum		6		1			
Clostridium thermocellum		10	1	3		1	1
Cellulomonas fimi	4	2	3	1			
Thermomonospora fusca		4	2	1			
Streptomyces lividans		2		3	1		
Pseudomonas fluorescens subsp. *cellulosa*	2	2		2	1		

Cd, cellodextrinase; Eg, endoglucanase; Cbh, Cellobiohydrolase; Xyn; xylanase; Man, mannanase; Lic, lichenase; Lam, laminarinase. Numbers are the enzymes of each type produced by the organism.

7 CONCLUSIONS

Molecular genetics, chemistry, biochemistry and X-ray crystallography are revealing in exquisite detail the catalytic mechanisms of a variety of β-1,4 glycanases, thereby allowing an understanding of the determinants of substrate specificity and mode of action. Identification of the enzymes comprising systems for the hydrolysis of the polysaccharides in plant cell walls is showing that different micro-organisms are very similar in the ways in which they hydrolyse this recalcitrant substrate. In spite of this, relatively little is known of the ways in which the enzymes sequester and hydrolyse molecules of their polysaccharide substrates, of the processivity of some enzymes, and of the exact manner in which enzymes in a particular system interact in the hydrolytic events. It is hoped that progress in these areas will be presented at the next Tricel conference.

Acknowledgements

I am indebted to all present and past members of and collaborators with "The UBC Cellulase Group" for their friendship, insights, suggestions and industry, and to *Cellulomonas fimi* for providing such an interesting enzyme system.

References

1. J.D. McCarter and S.G. Withers, *Curr. Opin. Struct. Biol.*, 1994, **4**, 885.
2. P. Bork, A.K. Downing, B. Kieffer and I.D. Campbell, *Quart. Rev. Biophys.*, 1996, **29**, 119.
3. N. R. Gilkes, B. Henrissat, D.G. Kilburn, R.C. Miller, Jr., and R.A.J. Warren, *Microbiol. Rev.*, 1991, **55**, 305.
4. P. Tomme, R.A.J. Warren and N.R. Gilkes, *Adv. Microb. Physiol.*, 1995, **37**, 1.
5. B. Henrissat, *Biochem. J.*, 1991, **280**, 309.
6. B. Henrissat and A. Bairoch, *Biochem. J.*, 1993, **293**, 781.
7. G. Davies and B. Henrissat, *Structure*, 1995, **3**, 853.
8. B. Henrissat, I. Callebaut, S. Fabrega, P. Lehn, J.-P. Mornon and G. Davies, *Proc. Nat. Sci. USA*, 1995, **92**, 7090.
9. J. Jenkins, L. Lo Leggio, G. Harris and R. Pickersgill, *FEBS Lett.*, 1995, **362**, 281.
10. R. Dominguez, H. Souchon, S. Spinelli, Z. Dauter, K.S. Wilson, S. Chauvaux, P. Béguin and P. Alzari, *Nature Struct. Biol.*, 1995, **2**, 569.
11. V. Ducros, M. Czjzek, A. Belaich, C. Gaudin, H.-P. Fierobe, J.P. Belaich, G.J. Davies and R. Haser, *Structure*, 1995, **3**, 939.
12. K. Kleman-Leyer, E. Agosin, A.H. Conner and T.K. Kirk, *Appl. Environ. Microbiol.*, 1992, **58**, 1266.
13. K.M. Kleman-Leyer, N.R. Gilkes, R.C. Miller, Jr., and T.K. Kirk, *Biochem. J.*, 1994, **302**, 463.
14. K.M. Kleman-Leyer, M. Siika-Aho, T.T. Teeri and T.K. Kirk, *Appl. Environ. Microbiol.*, 1996, **62**, 2883.
15. J. Rouvinen, T. Bergfors, T. Teeri, J.K.C. Knowles and T.A. Jones, *Science*, 1990, **249**, 380.
16. M. Spezio, D.B. Wilson, *Biochemistry*, 1993, **32**, 9906.
17. C. Divne, J. Ståhlberg, T. Reinikainen, L. Ruohonen, G. Pettersson, J.K.C. Knowles, T.T. Teeri and T.A. Jones, *Science*, 1994, **265**, 524.
18. G.J. Kleywegt, J.-Y. Zou, C. Divne, G.J. Davies, I. Sinning, J. Ståhlberg, T. Reinikainen, M. Srisodusk, T.T. Teeri and T.A. Jones, *J. Mol. Biol.*, 1997, **272**, 383.
19. J. Ståhlberg, G. Johansson and G. Pettersson, *Biochim. Biophys. Acta*, 1994, **1157**, 107.
20. H. Shen, N.R. Gilkes, D.G. Kilburn, R.C. Miller, Jr., and R.A.J. Warren, *Biochem. J.*, 1995, **311**, 67.
21. P. Tomme, E. Kwan, N.R. Gilkes, D.G. Kilburn and R.A.J. Warren, *J. Bacteriol.*, 1996, **178**, 4216.
22. C. Reverbel-Leroy, S. Pages, A. Belaich, J.-P. Belaich and C. Tardif, *J. Bacteriol.*, 1997, **179**, 46.
23. B.K. Barr, Y.-L. Hsieh, B. Ganem and D.B. Wilson, *Biochemistry*, 1996, **35**, 586.
24. A. Meinke, N.R. Gilkes, E. Kwan, D.G. Kilburn, R.A.J. Warren and R.C. Miller, Jr., *Mol. Microbiol.*, 1994, **12**, 413.
25. A.M. MacLeod, T. Lindhorst, S.G. Withers and R.A.J. Warren, *Biochemistry*, 1994, **33**, 6371.
26. A.M. MacLeod, D. Tull, K. Rupitz, R.A.J. Warren and S.G. Withers, *Biochemistry*, 1996, **35**, 13165.

27. H.G. Damude, S.G. Withers, D.G. Kilburn, R.C. Miller, Jr. and R.A.J. Warren, *Biochemistry*, 1995, **34**, 2220.
28. S.G. Withers, *Pure & Appl. Chem.*, 1995, **10,** 1673.
29. S.G. Withers and R. Aebersold, *Protein Sci.*, 1995, **4**, 361.
30. L.P. McIntosh, G. Hand, P.E. Johnson, M.D. Joshi, M. Korner, L.A. Plesniak, L. Ziser, W.W. Wakarchuk and S.G. Withers, *Biochemistry*, 1996, **35,** 9958.

STRUCTURE-FUNCTION RELATIONSHIPS IN CELLULASES: THE ENZYMATIC DEGRADATION OF INSOLUBLE CELLULOSE

C. Boisset, S. Armand, S. Drouillard, H. Chanzy, H. Driguez and B. Henrissat*

Centre de Recherches sur les Macromolécules Végétales†
Centre National de la Recherche Scientifique
BP 53
F-38041 Grenoble cedex 9 (France)

1 INTRODUCTION

Cellulases often have a modular structure with a catalytic domain attached to one or several non-catalytic cellulose-binding domains.[1] Biochemically, a classical view of cellulases is their subdivision into two categories, namely endoglucanases which would attack cellulose chains at random, and cellobiohydrolases which would degrade the chains in an exo-fashion from the chain ends created by the endoglucanases. Although the three-dimensional structure of a large number of cellulase catalytic and cellulose-binding domains is now known,[2-4] the mechanism by which cellulases degrade crystalline cellulose is still poorly understood. One of the most intriguing features of cellobiohydrolases is their tunnel-shaped active sites which suggest that the cellulose chains are recognised from their ends and that the resulting mode of action is exo.[5-7] It has been proposed that cellobiohydrolases fall into one of two main types: those attacking cellulose chains from their non-reducing ends, and those selectively hydrolysing the chains from the reducing ends.[7-10] This very useful distinction has led to a better understanding of the fundamental differences between cellobiohydrolases. However, even with this subdivision, there are aspects of cellobiohydrolase action which are still difficult to interpret. In particular, (i) the degree of synergy between endoglucanases and cellobiohydrolases appears lower than that expected for a strict endo/exo cooperation, (ii) cellobiohydrolases sometimes display significant activity on carboxymethylcellulose, and (iii) what is the origin of the synergy observed between the two classes of cellobiohydrolases? Recently, it has been hypothesised that the loops which make the lid of the active site of cellobiohydrolases could perhaps sometimes "open" to allow an occasional initial endo-attack.[2]

To cast some light on these aspects, the present work attempts to identify the sites of initial attack of cellulases on glucan chains. Two approaches have been followed: (i) the comparative adsorption and degradation features of cellulose I (native cellulose) and cellulose III, a different polymorph obtained by solid-state swelling/de-swelling cellulose I with ethylene diamine and, (ii), the preparation and assay of synthetic oligosaccharides carrying non-carbohydrate substituents at either or both extremities.

* Corresponding author
† Affiliated with the Joseph Fourier University, Grenoble

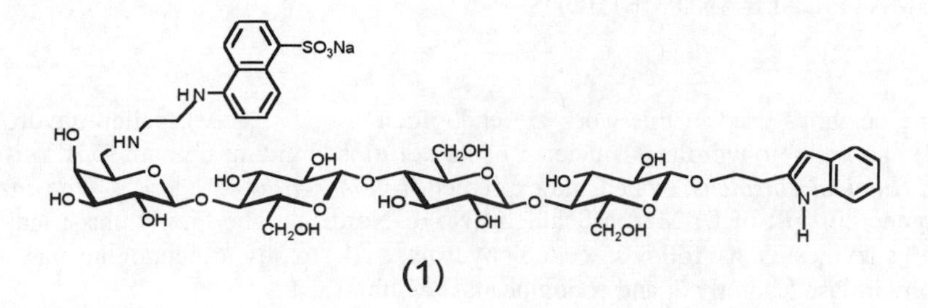

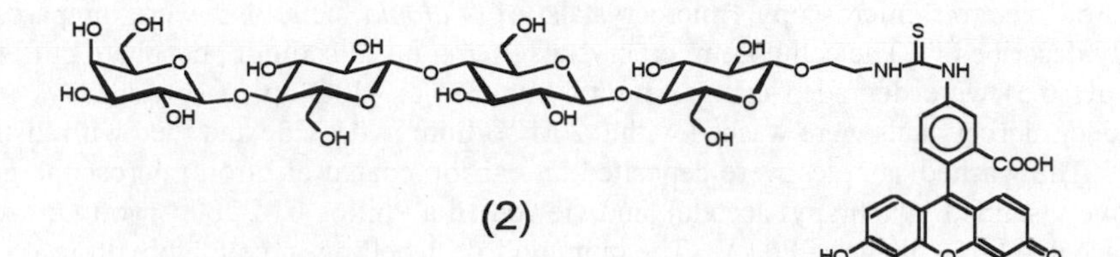

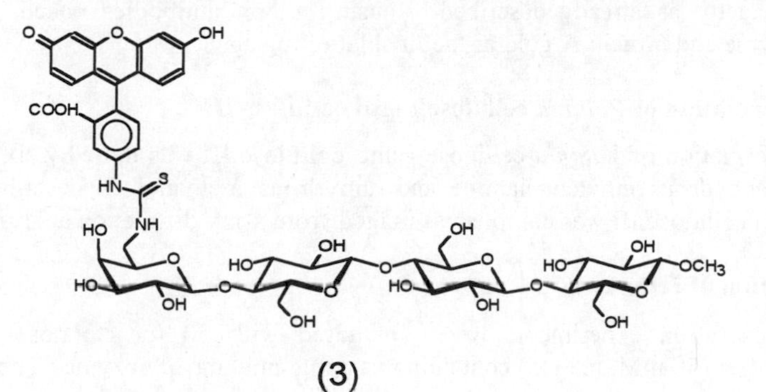

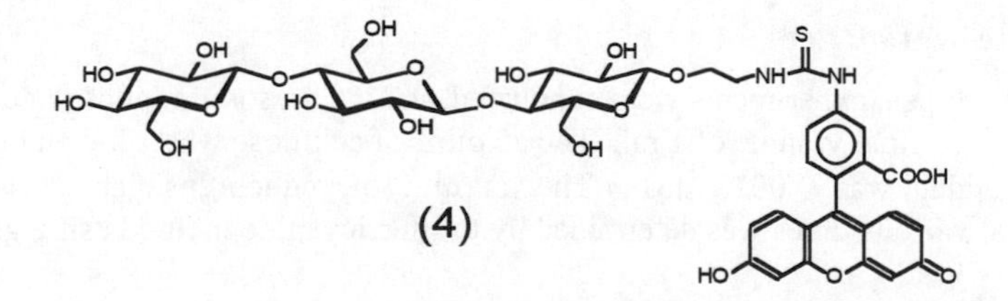

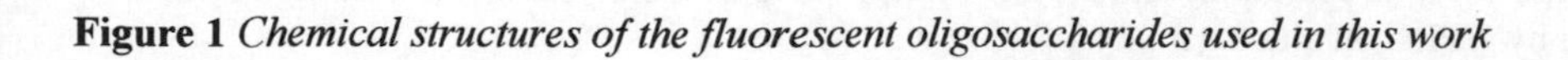

Figure 1 *Chemical structures of the fluorescent oligosaccharides used in this work*

2 MATERIALS AND METHODS

2.1 Enzymes

The enzymes used in this work are endoglucanase V (EG V), cellobiohydrolase I (CBH I) and cellobiohydrolase II deleted of its cellulose-binding domain (CBH II-core), all from *Humicola insolens*, cloned and expressed in *Aspergillus oryzae.*[11,12] The enzymes were a generous gift of Dr Martin Schülein (Novo-Nordisk). The family classification[13-15] of these enzymes is as follows: cellobiohydrolase II, family 6; endoglucanase I and cellobiohydrolase I, family 7; and endoglucanase V, family 45.

2.2 Transmission electron microscopy

For electron microscopy, microcrystals of *Valonia* cellulose were prepared as already described.[16] The cellulose microcrystals (1 mg/ml in sodium phosphate buffer, 50 mM, pH 6.5) were degraded using 1 mg/ml enzyme at 41°C up to several days. The degraded microcrystals were washed with 0.2 M sodium hydroxide and then with distilled water. The washed samples were deposited on carbon-coated electron microscope grids, negatively stained (1% uranyl acetate), and viewed in a Philips CM 200 Cryo microscope at an accelerating voltage of 80 kV. The immunogold labelling of cellobiohydrolase I was conducted exactly as already described[11,17] using rabbit antibodies raised against the purified enzyme and protein A-gold as the final labelling agent.

2.3 Transformation of *Valonia* cellulose I into cellulose III

Transformation of *Valonia* cellulose I into cellulose III was done by six successive soaking in anhydrous ethylenediamine and anhydrous methanol as described.[18] The conversion in cellulose III was complete as judged from x-ray diffraction analysis.

2.4 Adsorption of cellulases

The adsorption experiments were conducted with 50 μg cellulose in sodium phosphate buffer (50 mM, pH 6.5) containing varying amounts of enzyme. The amount of proteins adsorbed on cellulose was determined after 20 min at 4°C by measuring the concentration of proteins in the supernatant by the Bradford method using known concentrations of each enzyme for calibration.

2.5 Cellulose degradation

The hydrolysis experiments were conducted at 41°C in sodium phosphate buffer (50 mM, pH 6.5) in a final volume of 1 ml. The amount of cellulose was 1 mg and the amount of enzyme added was 0.0028 μM. The amount of reducing sugar released in the supernatant at various times was determined by the ferricyanide method using glucose as a standard.

2.6 Preparation of synthetic oligosaccharides

A bifunctionalised tetrasaccharide (Figure 1) was prepared as described.[19] Three oligosaccharides carrying a fluorescein group either at the reducing or at the non-reducing end have also been prepared (Armand, Drouillard, Schülein, Driguez & Henrissat, manuscript in preparation) and are shown in Figure 1. Analysis of the cleavage sites was done by HPLC on silica gel.

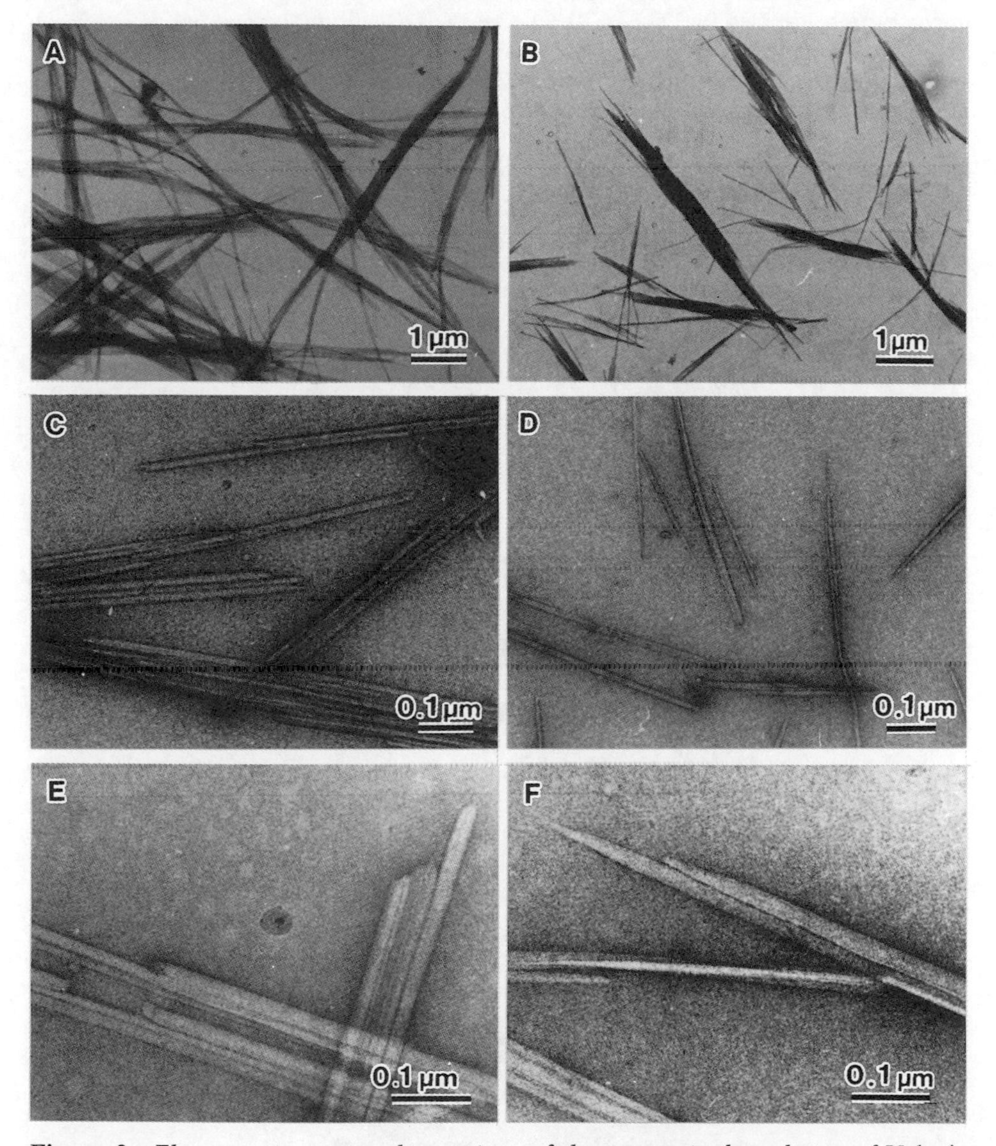

Figure 2 *Electron microscopy observations of the enzymatic degradation of* Valonia *cellulose microcrystals. The micrographs on the left (*2A, 2C and 2E*) show the initial microcrystals and those on the right show the microcrystals degraded by* H. insolens *EG V (*2B*), CBH I (*2D*) and CBH II-core (*2F*). The low magnification images (*2A, 2B*) are unstained. The specimen in* 2C-2F *were negatively stained with uranyl acetate.*

3 RESULTS

3.1 Electron microscopy observations

Transmission electron microscopy allows the visualisation of the consequences of the enzymatic degradation of crystalline cellulose. The morphology of *Valonia* cellulose I microcrystals degraded by EG V, CBH I and CBH II-core of *H. insolens* is shown in Figure 2. Upon treatment by EG V, the initial microcrystals (Figure 2A) are degraded into shorter microcrystals (Figure 2B), presumably through the selective hydrolysis of the kinks created in cellulose during the acido-mechanical treatment used to prepare the microcrystals. In the case of the degradation by the cellobiohydrolases, higher magnifications had to be used to reveal the morphological changes occurring to the crystals. In the case of CBH I, the initial 200 Å-wide microcrystals (Figure 2C) are eroded into thinner microcrystals (Figure 2D) as was previously observed using *Trichoderma reesei* CBH I.[20] The morphological change accompanying degradation by CBH II-core is the classical "pointed tip" morphology (Figure 2F) already observed in several instances.[21,22]

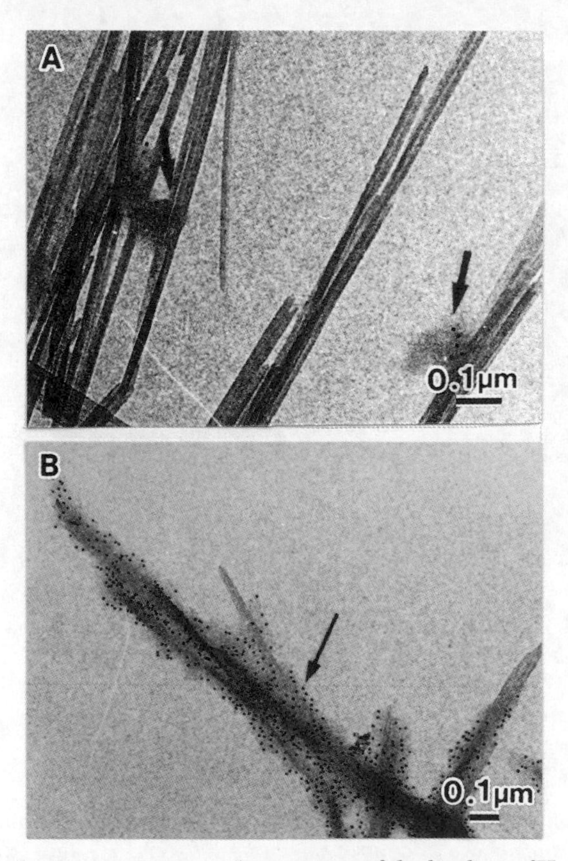

Figure 3 *Immunoelectron microscopy observations of the binding of* H. insolens *CBH I to* Valonia *cellulose I (3A) and cellulose III (3B) microcrystals. Arrows show some of the gold particles. The specimen were unstained.*

The binding of *H. insolens* CBH I to *Valonia* cellulose I and cellulose III microcrystals was followed by immunogold labelling using a previously established double sandwich method.[17] In the case of cellulose I, the gold labelling was low and irregular (Figure 3A), while it was considerably higher and more homogeneous when cellulose III was employed (Figure 3B). The binding of EG V gave very similar pictures (data not shown). CBH II-core, lacking a cellulose-binding domain, did not show any strong labelling.

3.2 Adsorption of cellulases on *Valonia* cellulose I and cellulose III

The adsorption of *H. insolens* CBH I and EG V on *Valonia* cellulose I and cellulose III is presented in Figure 4. The transformation of cellulose I into cellulose III increases the adsorption capacity of cellulose by a factor of x5 in the case of EG V and more than 5 in the case of CBH I (the exact value cannot be estimated since saturation was not reached in our experimental conditions).

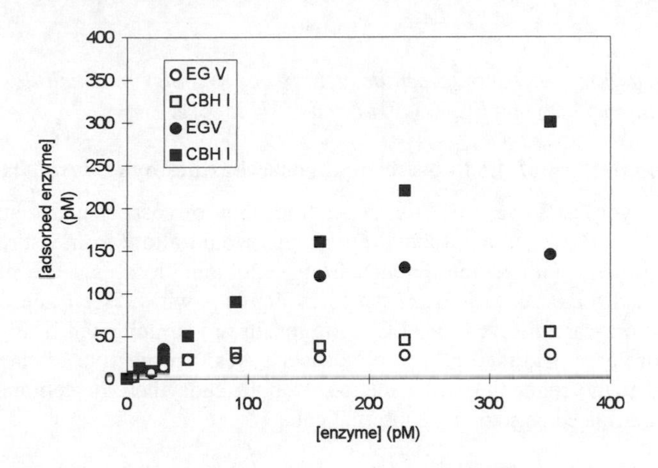

Figure 4 *Adsorption of* Humicola insolens *EG V and CBH I on* Valonia *cellulose I (open symbols) and cellulose III (filled symbols).*

3.3 Degradation of *Valonia* cellulose I and cellulose III by *H. insolens* cellulases

Figure 5 shows that the transformation of cellulose I into cellulose III increases the amount of reducing sugar produced at extended times by a factor of x5 for both EG V and CBH IIc, and by a factor of almost 10 in the case of CBH I. The digestibility of cellulose therefore appears to correlate with its adsorption capacity for the enzymes. The rapid swelling-unswelling of the crystals which is used to transform cellulose I into cellulose III generates a significant increase in the disordered cellulose surface chains and cannot reduce significantly its degree of polymerisation. This suggests that disorder in the surface chains thus controls both the adsorption capacity and the overall digestibility of cellulose by cellulases, irrespective of their classification as endoglucanases or cellobiohydrolases. In other words, disordered chains probably represent valid initial sites of attack for all cellulases. To study if cellobiohydrolases are indeed able to make internal cuts in the chains, several oligosaccharides substituted at the extremities by non-carbohydrate groups were tested (see next section).

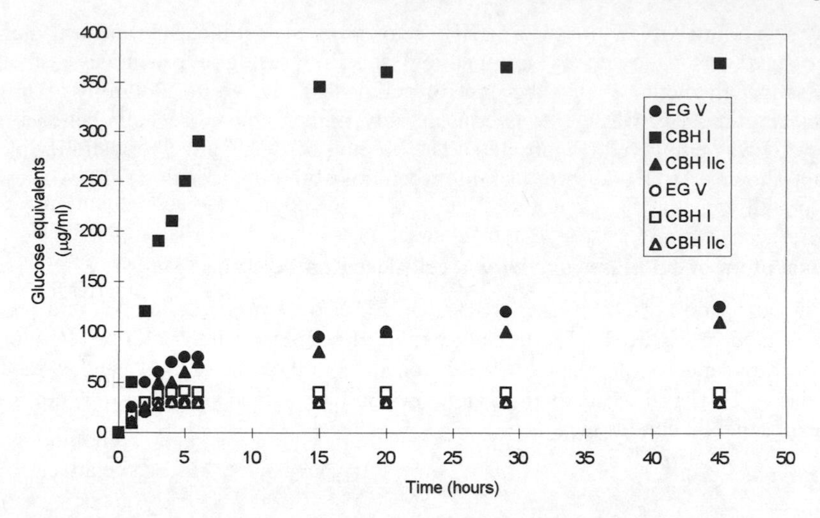

Figure 5 *Degradation of* Valonia *cellulose I (open symbols) and cellulose III (filled symbols) by* Humicola insolens *EG V, CBH I and CBH II core.*

3.4 Degradation pattern of the fluorescent oligosaccharides by *H. insolens* cellulases

If cellobiohydrolases were strictly exo-acting, then oligosaccharides substituted at each or both extremities by non-carbohydrate groups would allow to measure selectively endoglucanase activity without interference from cellobiohydrolases. We have assayed the four substituted oligosaccharides presented in Figure 1 with CBH I and CBH II-core and the degradation was followed by TLC. The results are summarised in Figure 6. The two cellobiohydrolases degraded all the oligosaccharides, irrespective of the location of the substituents at the reducing, non-reducing or both ends thereby demonstrating that these enzymes are indeed able to make internal cuts.

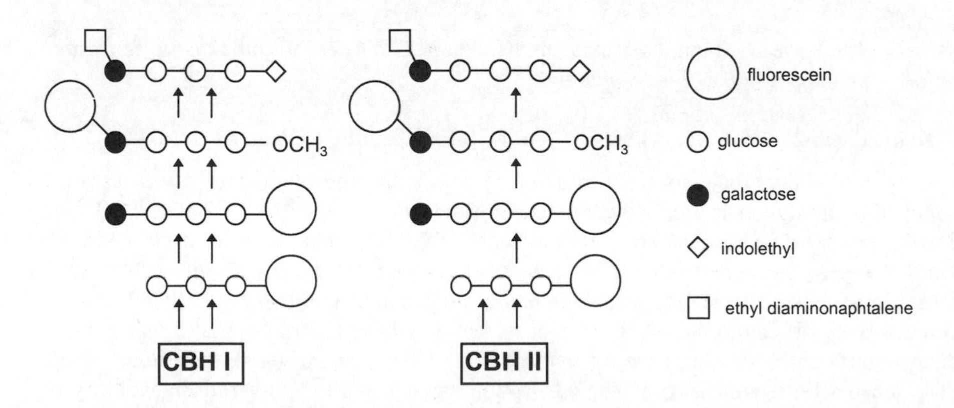

Figure 6 *Degradation pattern of the various fluorescent oligosaccharides by* H. insolens *cellobiohydrolase I (CBH I) and cellobiohydrolase II (CBH II) as determined by TLC analysis. The arrows show the observed cleavage points.*

4 DISCUSSION

The fact that cellobiohydrolases can make internal cuts in all the fluorescent oligosaccharides is consistent with the faster degradation of cellulose III where the only change compared to cellulose I is the amount of disordered chains and the packing of the chains. The mechanism by which cellobiohydrolases make those internal cuts is not directly established. However, given the size of the fluorescent substituents attached to the oligosaccharides, it is likely that this occurs by an opening of the loops which form the lid of the tunnel active site of cellobiohydrolases. Cellobiohydrolases therefore could have a mechanism featuring (i) occasional loop opening allowing internal cuts, (ii) followed by a processive mechanism by which cellobiose is released from one of the ends created by the internal cut.

How does this -still hypothetical- mechanism explain the morphology of cellulose during enzymatic hydrolysis is not known. It is likely that there are disordered chains prone to enzymatic attack at both ends of the cellulose microcrystals. Cellobiohydrolases can perhaps initiate their attack at these sites. The subsequent processive attack has then very different consequences depending on which end of the microcrystal is being degraded and on the directionality of the processive action (Figure 7). Attack at the "wrong" end would release the enzyme after a very limited number of cuts without generating any noticeable morphological change, whereas attack at the "good" end would allow a larger number of cuts. Depending on the degree of processivity of the enzyme, the processive action can generate pointed tips or thinner microcrystals (Figure 7).

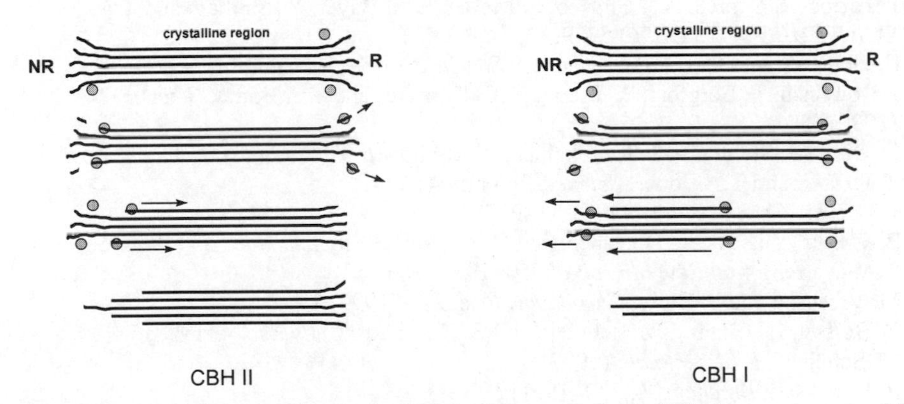

Figure 7 *Putative mechanism by which different cellobiohydrolases generate different morphologies during the degradation of cellulose microcrystals. In both cases, the initial sites for attack are the disordered chains at the ends of the microcrystals. From these sites, CBH II (left) then proceeds in the non-reducing-to-reducing end direction. When the initial site was close to the reducing end, then a few processive cuts do not change the morphology of the tip of the crystal, and the enzyme is rapidly released in solution. At the other end, the processive degradation can continue for a number of times before the enzyme eventually desorbs. This generates a pointed tip morphology. For CBH I (right), everything is the same except that the directionality of the attack is reversed. The difference in the morphology of the crystals, here thinner, is tentatively attributed to a higher processivity of CBH I.*

5 CONCLUSION

Although considerable progress has been made through recent 3D-structure determinations of cellulases, we are still far from fully understanding the mechanism of enzymatic degradation of cellulose. It is possible that processivity, to varying extents, is a general feature of cellulases. The particular behaviour of cellobiohydrolases is one extreme of this processive mechanism, and the role of the loops that cover their active site is clearly significant. The most likely role for these loops is perhaps not so much to ensure a chain-end recognition, but to prevent the dislodged cellulose chain from re-adhering to the crystal surface as the enzyme processes.[2,5,6] Several aspects remain to be studied, such as the ways to measure the frequency of loop opening and the origin of the synergy of cellulases.

Acknowledgements: We wish to thank Dr. M. Schülein (Novo-Nordisk) and Dr. G.J. Davies (University of York) for their help and support. This work was funded by the European Community BIOTECHNOLOGY Programme. C.B. was the recipient of a joint fellowship from ADEME and Novo-Nordisk.

References

1. N.R. Gilkes, B. Henrissat, D.G. Kilburn, R.C. Miller and R.A.J. Warren, *Microbiol. Rev.*, 55 (1991) 303.
2. G. Davies and B. Henrissat, *Structure*, 3 (1995) 853.
3. J. Tormo, R. Lamed, A.J. Chirino, E. Morag, E.A. Bayer, Y. Shoham and T.A. Steitz, *EMBO J.*, 15 (1996) 5739.
4. B. Henrissat and G. Davies, *Curr. Op. Struct. Biol.*, 7 (1997) 637.
5. J. Rouvinen, T. Bergfors, T. Teeri, J.K.C. Knowles and T.A. Jones, *Science*, 249 (1990) 380.
6. C. Divne, J. Stahlberg, T. Reinikainen, L. Ruohonen, G. Pettersson, J.K.C. Knowles, T.T. Teeri and T.A. Jones, *Science*, 265 (1994) 524.
7. T.T. Teeri, *Trends Biotechnol.*, 15 (1997) 160.
8. B.K. Barr, Y.L. Hsieh, B. Ganem and D.B. Wilson, *Biochemistry*, 35 (1996) 586.
9. R.A. Warren, *Ann. Rev. Microbiol.*, 50 (1996) 183.
10. M. Vrsanska and P. Biely, *Carbohydr. Res.*, 227 (1992) 19.
11. C. Boisset, H. Chanzy, M. Schülein and B. Henrissat, *Cellulose*, 4 (1997) 7.
12. M. Schülein, *J. Biotechnol.*, (1997)
13. B. Henrissat, *Biochem. J.*, 280 (1991) 309.
14. B. Henrissat and A. Bairoch, *Biochem. J.*, 293 (1993) 781.
15. B. Henrissat and A. Bairoch, *Biochem. J.*, 316 (1996) 695.
16. H. Chanzy and B. Henrissat, *Carbohydr. Polym.*, 3 (1983) 161.
17. N.R. Gilkes, D.G. Kilburn, R.C. Miller Jr., R.A.J. Warren, J. Sugiyama, H. Chanzy and B. Henrissat, *Int. J. Biol. Macromol.*, 15 (1993) 347.
18. H. Chanzy, B. Henrissat, R. Vuong and J.F. Revol, *Holzforschung*, 40 (1986) 25.
19. S. Armand, S. Drouillard, M. Schülein, B. Henrissat and H. Driguez, *J. Biol. Chem.*, 272 (1997) 2709.
20. H. Chanzy, B. Henrissat and R. Vuong, *FEBS Lett.*, 172 (1984) 193.
21. H. Chanzy and B. Henrissat, *FEBS Lett.*, 184 (1985) 285.
22. M. Koyama, W. Helbert, T. Imai, J. Sugiyama and B. Henrissat, *Proc. Natl. Acad. Sci. USA*, 94 (1997) 9091.

THERMOMONOSPORA FUSCA CELLULASE E4: A PROCESSIVE ENDOCELLULASE

David B. Wilson, Diana Irwin, Joshua Sakon*, P. Andrew Karplus
Section of Biochemistry, Molecular and Cell Biology, Cornell University
Ithaca, New York 14853
*Present address: Department of Chemistry and Biochemistry, University of Arkansas
Fayetteville, AR 72701

1 INTRODUCTION

Thermomonospora fusca is a filamentous soil bacterium that degrades most plant cell wall polymers including cellulose. It secretes six different cellulases (E1-E6) including two exocellulases, E3 and E6, three endocellulases, E1, E2, and E5, and one unusual endocellulase, E4. The activities of these enzymes on several cellulosic substrates are shown in Table 1.[1]

We have cloned and sequenced all six structural genes and find that their catalytic domains belong to four families, 5, 6, 9, and 48 (Table 2).[2,3,4](Irwin, unpublished results) Surprisingly, the mesophilic soil bacterium *Cellulomonas fimi* produces a set of six cellulases[5] that are equivalent to those of *T. fusca* in their activities and catalytic domains (Table 2). The two sets do not appear to have come from a common ancestor since the level of sequence identity in the catalytic domains ranges from 21 to 80% and the family II cellulose binding domains are at opposite ends of the molecule in four of the five *C. fimi* cellulases that contain a cellulose binding domain (CBD). The only enzymes that show a high level of amino acid sequence identity and similar domain orders are Cen B, and E4.

Table 1 *Activities of* Thermomonospora fusca *Cellulases*

PROTEIN	MW (kD)	SPECIFIC ACTIVITIES (µmoles Cellobiose/min.-µmol) CMC	Swollen Cellulose	Filter Paper	BMCC	% insoluble reducing sugar (filter paper hydrolysis)
E1	101.2	5,410.0	363.0	0.18 *	0.4 *	40
E2	43.0	433.0	758.0	0.77	6.5	32
E3	59.6	0.3 *	3.9	0.13 *	2.0	8
E4-90	90.4	475.0	202.0	1.03	19.1	13
E4-68	68.0	488.0	54.0	0.24 *	6.1	22
E4-51	51.4	108.0	6.3	0.13 *	0.2 *	63
E4-74	74.0	121.0	23.2	0.27 *	0.3 *	44
E5	46.3	2,840.0	90.0	0.83	5.0	31
E6	104.0	0.4 *	0.7 *		0.6 *	4

*target digestion could not be achieved
The activities were measured as described in reference 1.

Table 2 *Relationship between catalytic domains of* T. fusca *and* C. fimi *hydrolases.*

Family	**9**	**6**	**6**	**9**	**5**	**48**
T. fusca	E1	E2	E3	E4	E5	E6
C. fimi	Cen C	Eg A	CBH A	Cen B	Cen D	CBH B
% Identity	59	42	55	80	21	33
Family II CBD locations						
T. fusca	C	C	N	C	N	C
C. fimi	---	N	C	C	C	C

2 PROPERTIES OF E4

E4 differs from all the other *T. fusca* cellulases in that its initial soluble product from cellulose is cellotetraose[6]. Later in the reaction this cellotetraose is cleaved to about 50% cellobiose and 25% each of glucose and cellotriose.[7] E4 also produces much less insoluble reducing sugar (13%) than do other endocellulases (30 - 40%)[6] (Table 1). Finally E4 is the only enzyme that can synergise with endocellulases and both classes of exocellulases in crystalline cellulose degradation (Table 3).[1,6]

E4 contains two different CBDs, one is in family II and is located at the C-terminus (Figure 1), while the other is in family IIIc and is next to the catalytic domain. Limited proteolysis of E4-90 removes the family II CBD, a short linker and a fibronectin III like domain to produce E4-68. We have produced two other E4 species by PCR , E4-74 which lacks the internal family IIIc CBD and E4-51 which only contains the family 9 catalytic domain.[6]

Table 3 *Synergism between cellulases during filter paper hydrolysis.*

Cellulases	Activity (µmoles cellobiose/min.-µmol)
E4	1.0
E4 + E3	2.0
E4 + E5	2.1
E4 + E1	0.7
E3 + E5	3.8
E4 + CBHII	2.6
E4 + E5 + E6	4.6
E3 + E5 + CBHI	9.2
E4 + E3 + E5 + CBHI	12.4
E3 + E5	4.1
E4 + E3 + E5	5.4

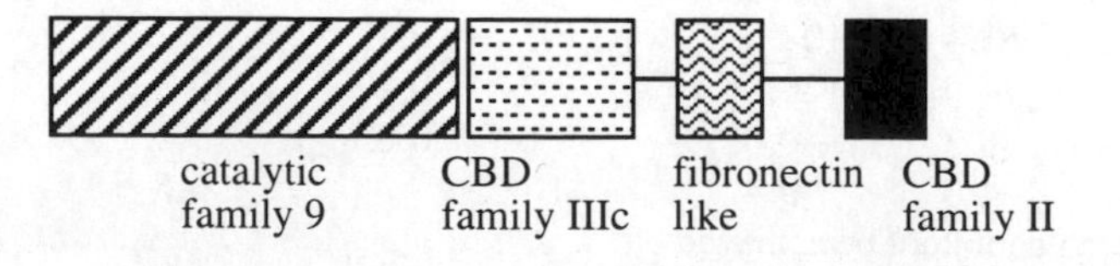

Figure 1 *Domains of* Thermomonospora fusca *E4*

Only the forms that contain the family II CBD, E4-90 and E4-74, bind tightly to cellulose while E4-68 has barely detectable binding and E4-51 has none at all.[6] However, E4-68 D55C, which contains a mutation that reduces activity on swollen cellulose to 9% of wild type E4-68, has significant linear binding to BMCC.[6] The activities of E4 and its derivatives on several forms of cellulose are shown in Table 1.[6] Clearly the family IIIc CBD is very important for activity on BMCC and filter paper, while the family II CBD is very important for activity on filter paper and also affects activity on BMCC. It is notable that although E4-74 binds very well to BMCC, it has almost no activity on crystalline substrates.

Analysis of the percentage of reducing ends found in the supernatant (soluble) versus the filter paper (insoluble) after hydrolysis shows a clear difference between endo and exo cellulases (Table 1).[1,6] E4-90 produces only 13% insoluble reducing ends, while E4-74, which is only missing the family IIIc CBD, behaves like an endocellulase in that it produces 44% insoluble reducing ends on filter paper. This indicates that it is the family IIIc CBD which is responsible for the unusual properties of E4.

3 CRYSTAL STRUCTURE OF E4-68

E4-68 was crystallised and its 3-dimensional structure was determined at 1.9Å resolution by X-ray crystallography.[8] A model of the structure is shown in Figure 2. The α-barrel catalytic domain is rigidly joined to the β barrel family IIIc CBD. The catalytic domain has a shallow open active site cleft which is aligned with the flat face of the family IIIc binding domain in such a way that a cellulose chain could bind to both sites (Figure 3). Tormo *et al.*[9] have solved the crystal structure of a *Clostridium thermocellum* family III CBD from the scaffoldin protein Cip and have proposed that it is this flat face which binds to cellulose. There are two calcium ions, one is near the cellulose binding cleft of E4cd and the other is buried in the CBD.

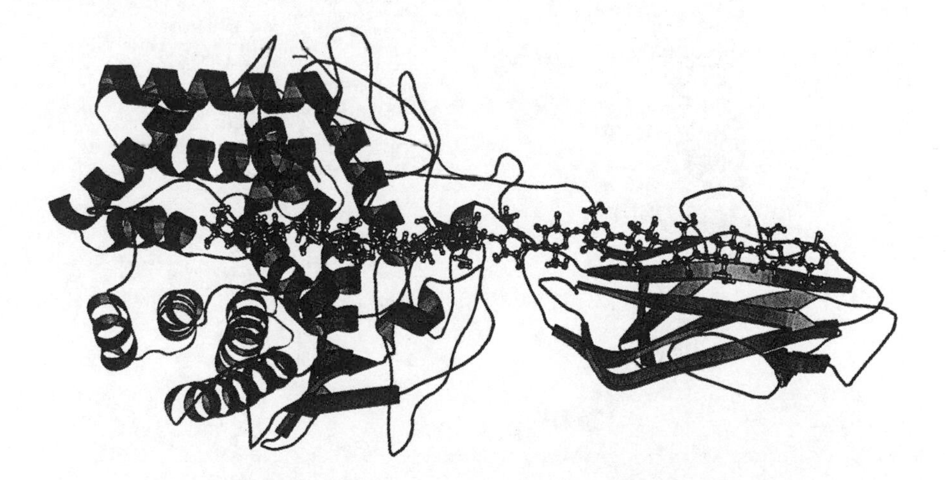

Figure 2 *A ribbon representation of the structure of E4-68. The catalytic domain of the enzyme exhibits an (α/α)6 barrel fold and the binding domain exhibits an antiparallel β–barrel fold. The cellulase strand was modelled onto the structure.*

Figure 3 *Molecular surface of E4-68 showing the proposed interaction of an extended cellulose chain with the catalytic domain and the family IIIc CBD. Reprinted with permission, Nature Structural Biology.*[9]

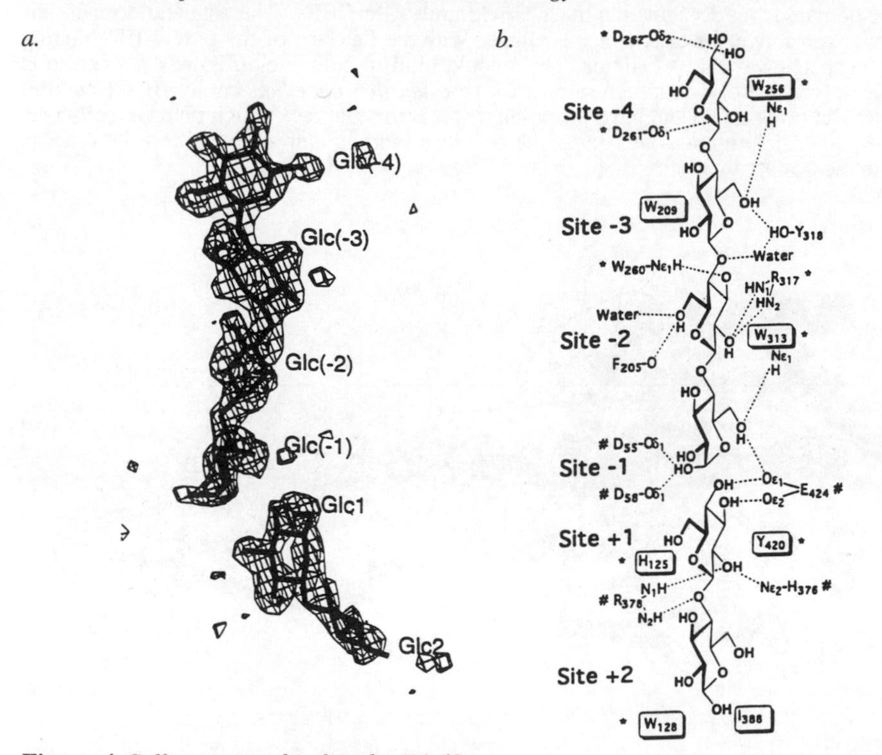

Figure 4 *Cellopentaose binding by E4-68.*
*(a) The conformation of "cellohexaose" and its electron density. (b) Schematic representation of the protein-ligand interactions. Residue names for the aromatic platforms are boxed. * indicates conservation in E4 -68 like enzymes; # indicates conservation in family 9. Reprinted with permission, Nature Structural Biology.*[8]

The structure of E4-68 crystals that were soaked in cellopentaose showed the products formed in the active site cleft, the orientation of the non-reducing end of the cellulose chain, and the location of the scissile bond (Figure 4a,b). From the relative densities of the bound sugars, it appears that there are two binding modes for cellopentaose on E4 that are shifted from each other by one glucose residue. Thus, six glucose residues were seen and those at each end had about one half the density of the four middle residues. This is reasonable, since if an oligosaccaride is rotated 180° along its long axis, it is equivalent to the original after a displacement of one residue. The cleavage product has an α configuration at its reducing end as would be expected for an inverting enzyme. The E4 active site cleft is blocked by a loop at the non reducing end (Figure 3) in such a way that a longer cellulose chain must bend to exit the cleft. This blockage could explain why cellotetraose units are cleaved from swollen cellulose.

A model of the active site configuration (Figures 5) shows a water molecule in almost the identical position of the 1 hydroxyl group in the product. This water is bound to Asp55 and Asp58, with Asp58 being best positioned to function as the catalytic base, while Glu424 appears to be the catalytic acid.

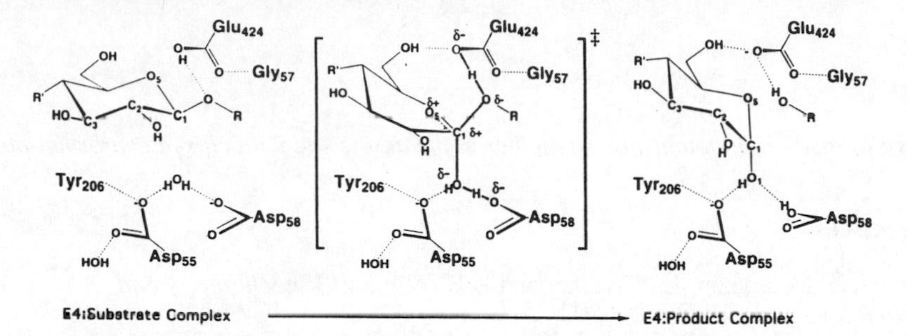

Figure 5 *Proposed strucurally detailed catalytic mechanism, reprinted with permission, Nature Structural Biology.*[8]

A comparison of the liganded and unliganded structures (closed and open configurations) shows that binding of substrate causes a conformational change in the enzyme (Figure 6). There are three amino acid residues that show significant shifts between the two structures, Tyr 420, Trp 313 and Glu 424. After the shift Glu 424 is positioned to form a hydrogen bond to the scissile oxygen of the polysaccaride. The position of several of the bound sugars is changed from that seen in a native cellulose chain (Figure 4a, b). Glc(-2) and Glc(-1) are forced toward the nucleophilic water and there is a smaller shift of Glc(+1).

4 CONCLUSIONS

From these studies it seems likely that after an initial endocellulolytic cleavage, E4 will processively cleave cellotetraose molecules from the nonreducing end of the cellulose chain until it is either completely hydrolysed or dissociates. After each cleavage the non reducing end cellotetraose would be expected to dissociate while the fragment which is bound to sites +1 and +2 and to the family IIIc CBD appears to be fed into the active site cleft in a processive manner. E4 is probably able to give synergy with exocellulases because it can bind and cleave internal sites, while it is able to give synergy with other endocellulases because of its processive activity.

These experiments show the power of combining structural studies with recombinant DNA techniques and that E4 is a member of a new group of family 9 cellulases which at present only contains bacterial enzymes (E4, CenB and several clostridal cellulases including CelG from *C. cellulolyticum*[10]).

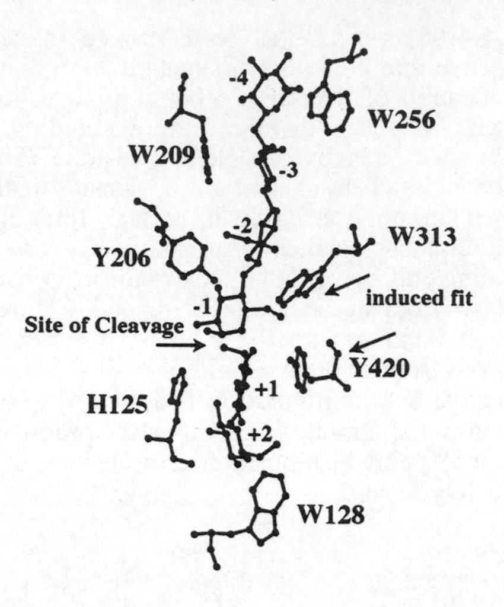

Figure 6 *Position of bound sugar residues in the active site with closed conformation.*

References

1. D. Irwin, M. Spezio, L.P. Walker and D.B. Wilson, *Biotechnology and Bioengineering*, 1993, **42**, 1002.
2. E.D. Jung, G. Lao, D. Irwin, B.K. Barr, A. Benjamin and D.B.Wilson, *Appl. Environ. Microbiol.* 1993, **59**, 3032.
3. S. Zhang, G. Lao and D.B. Wilson, *Biochemistry* 1995, **34**, 3386.
4. G. Lao, G. Ghangas, E.D. Jung and D.B. Wilson, *J. Bacteriol.*, 1991, **173**, 3397.
5. H. Shen, A. Meinke, P. Tomme, H.G. Damude, E. Kwan, D.G. Kilburn, R.C. Miller, Jr., R.A.J. Warren, and N.R. Gilkes, 'Enzymatic Degradation of Insoluble Substrates', eds. J.N. Saddler and M.H. Penner, American Chemical Society, Washington D.C., 1995, Chapter 12, p. 177.
6. D. Irwin, D.-H. Shin, S. Zhang, B. Barr, J. Sakon, A. Karplus, D.B. Wilson, *J. Bacteriol.*, submitted.
7. B.K. Barr, Y.L. Hsieh, B. Ganem and D.B. Wilson, *Biochemistry*, 1996, **35**, 586.
8. J. Sakon, D. Irwin, D.B. Wilson, P.A. Karplus, *Nature Structural Biology,* 1997, **4**, 810.
9. J. Tormo, R. Lamed, A. Chirino, E. Morag, E.A. Bayer, Y. Shoham and T.A. Steitz, *EMBO J.* 1996, **15**, 5739.
10. L. Gal, C. Gaudin, A. Belaich, S. Pages, C. Tardif and J.-P. Belaich, *J. Bacteriol.*, 1997, **179**, 21.

THE STRUCTURE AND FUNCTION OF CELLULOSE BINDING DOMAINS

A. Boraston[1,3], M. Bray[3], E. Brun[2], A.L. Creagh[1,4], N.R. Gilkes[1,3], M.M. Guarna[1], E. Jervis[1,4], P. Johnson[2], J. Kormos[1,3], L. McIntosh[2], B.W. McLean[1,3], L.E. Sandercock[1,3], P. Tomme[1,3], C.A. Haynes[1,4], R.A.J. Warren[3] and D.G. Kilburn[1,3]

Protein Engineering Network of Centres of Excellence
Biotechnology Laboratory[1], Departments of Biochemistry and Molecular Biology[2], Microbiology & Immunology[3], Chemical & Bio-Resource Engineering[4], University of British Columbia, Vancouver Canada V6T 1Z3

1 INTRODUCTION

Many enzymes involved in the hydrolysis of insoluble polysaccharides contain a structurally and functionally independent domain which mediates adsorption to the substrate. Such enzymes include numerous cellulases (β-1,4-glucanases) as well as certain β-1,3-glucanases, β-1,4-xylanases, chitinases and amylases. In several instances, it has been shown that removal of the substrate-binding domain significantly reduces the ability of the enzyme to hydrolyse insoluble substrates while the hydrolysis of small, soluble substrates is unchanged.

More than 180 putative cellulose-binding domains (CBDs) have been identified and grouped into 13 families based on amino acid sequence similarities (1,2). They vary in length from 33 to 240 amino acids. The structures of CBDs from five different families (Families I to V) are now available. All are anti parallel β strand polypeptides. Our group has studied the properties of Family II and Family IV CBDs of β-glycanases from the cellulolytic bacterium *Cellulomonas fimi* in detail.

2 *CELLULOMONAS FIMI* CBDS

We have now cloned and expressed genes for four endoglucanases, CenA, CenB, CenC and CenD, two cellobiohydrolases, CbhA and CbdB, and a mixed function exoglycanase, Cex, from *Cellulomonas fimi*. Each of these enzymes is modular with at least two distinct domains, one (or more) of which mediates binding to cellulose (Figure 1). CBDs occur at the N terminus, C terminus or internally in these enzymes. With the exception of CenC, each of the enzymes has a terminally located Family II CBD.

Family II CBDs are about 110 amino acid β-barrel structures which bind to crystalline and to amorphous cellulose. The binding involves three conserved solvent exposed aromatic residues (3,4) linearly arranged on a surface ridge or planar face (Figure 2). These residues presumably interact hydrophobically with the cellulose chain by stacking onto the glucosidic rings. In CBD_{Cex}, chemical oxidation (5) or mutation (M. Bray and B. McLean, unpublished data) of these tryptophan residues dramatically reduces the binding affinity.

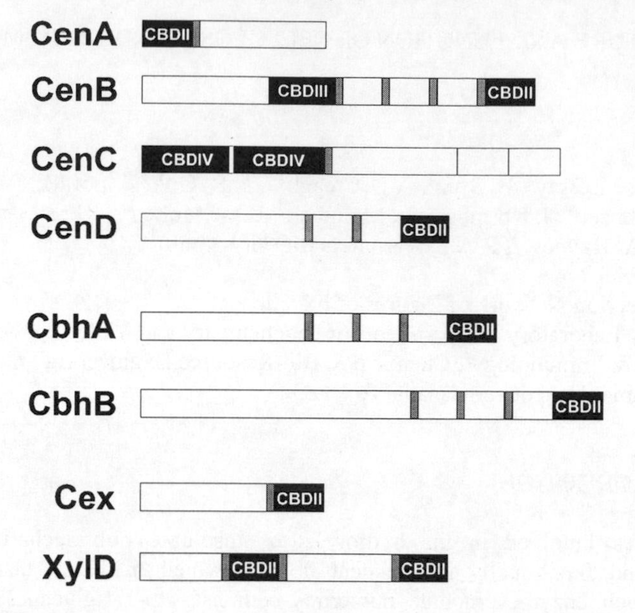

Figure 1 *Graphic representation of the cellulases of* Cellulomonas fimi. *Each protein is shown as a linear map to outline their domain organisation. CenA, CenB, CenC and CenD are β-1,4-glucanases, CbhA and CbhB are cellobiohydrolases, Cex (XynB) is a mixed exoglucanse/xylanase.* CBD_{II}, CBD_{III} *and* CBD_{IV} *refer to the family II, III and IV cellulose binding domains respectively.*

When bound to cellulose, the accessibility of these residues to oxidation is decreased. However, they are not completely protected indicating that the interaction of the CBD with the cellulose surface is dynamic (Figure 3). Thermodynamic analysis of the binding of CBD_{Cex} to crystalline cellulose by microcalorimetry indicates that the driving force for adsorption is entropic with a small enthalpic component (6). This points to the dominance of dehydration effects in the formation of the CBD-cellulose complex. We propose that binding involves an initial event in which a number of hydrophobic residues along the CBD binding face make sufficient contact to dehydrate both the binding face and the underlying cellulose surface. This then facilitates the formation of hydrogen bonds between the protein and the cellulose surface accounting for the enthalpic component of binding.

The CBDs from Families I, II and III all have planar binding faces and all bind crystalline cellulose with similar affinities (K_a in the micromolar range). The Family IV CBD from *C. fimi* CenC binds to amorphous cellulose and soluble cellooligosaccharides ($K_a \cong 10^{-4}$M) but not to crystalline cellulose (7,8). Unlike other known CBDs, it does not have a binding face but a 5-stranded binding cleft on one side with a central strip of hydrophobic residues flanked on both sides by polar H-bonding residues (Figure 4). The cleft binds single cellulose molecules whereas the binding faces of the CBDs in Families I, II and III bind to the parallel, close packed molecules at the surface of crystalline cellulose.

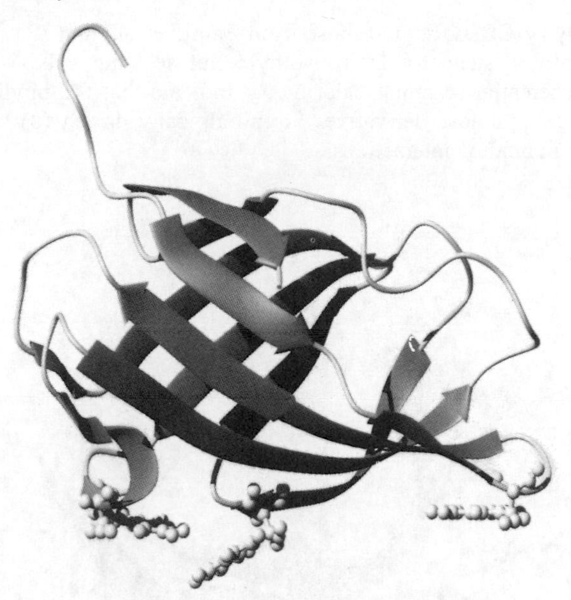

Figure 2 *Schematic illustration of the three dimensional structure of CBD_{Cex}. Three solvent exposed tryptophan residues involved in binding are shown in a ball and stick representation.*

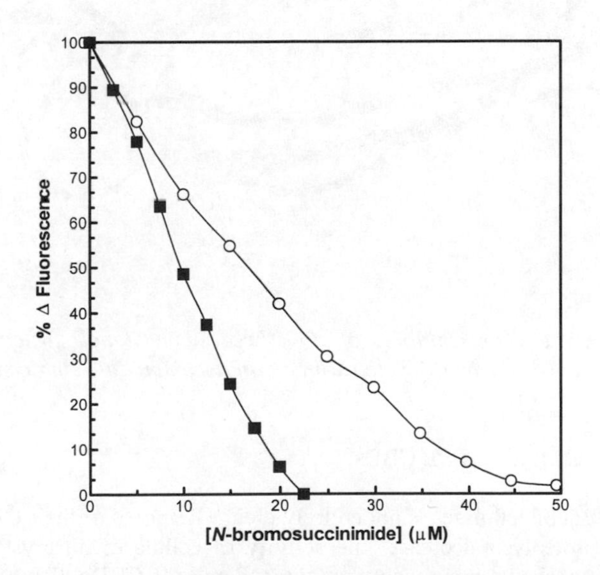

Figure 3 *Chemical modification of the tryptophans of CBD_{Cex}. Chemical oxidation of the tryptophans of CBD_{Cex} with increasing concentrations of N-bromosuccinimide was monitored by the loss of fluorescence emission at 342 nm after excitation at a wavelength of 295 nm. Closed squares represent measurements for CBD in the absence of cellulose, open circles represent those for CBD in the presence of cellulose.*

The Family IV CBDs, but not those from Families I, II and III, also bind to soluble cellulose derivatives such as hydroxyethylcellulose and soluble oligosaccharides. Titration and differential scanning calorimetry indicate that the binding of CBD_{CenC} to soluble sugars and cellulose derivatives is enthalpically driven (8), analogous to most other sugar-protein binding interactions.

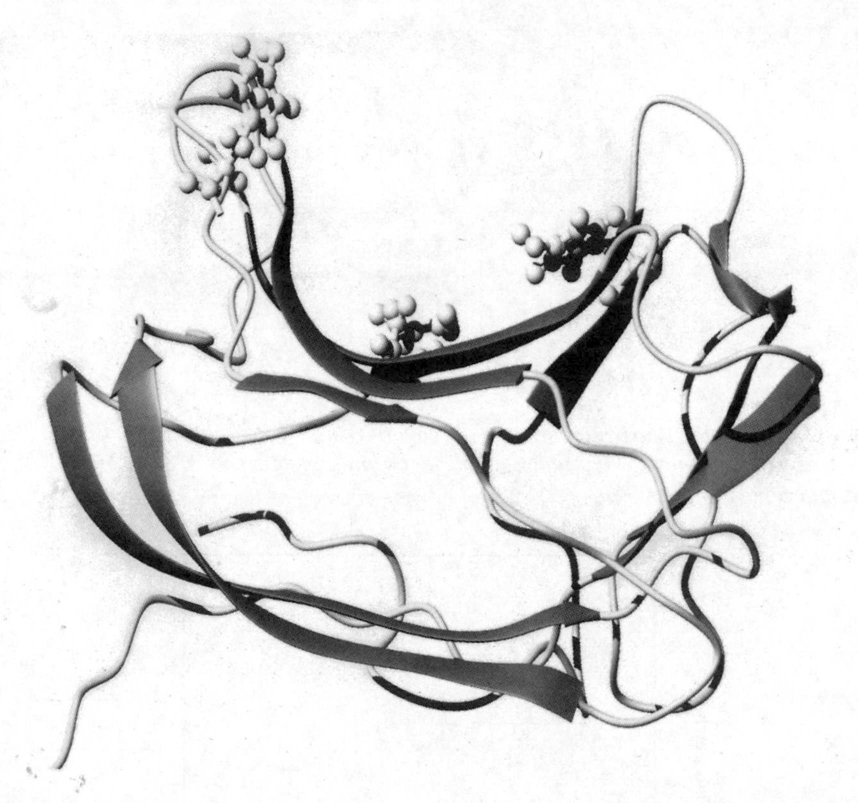

Figure 4 *Schematic representation of the three dimensional structure of CBD_{IV}. Illustrated in this view is the binding cleft which can accommodate single sugar strands.*

3 BIOLOGICAL ROLE OF CBDs

The role of the CBD of cellulases is not entirely clear. Removal of their CBDs by genetic manipulation or proteolysis decreases the activity of cellulases on crystalline cellulose (9,10,11). We have shown previously that some Family II CBDs disrupt the surface of cellulose fibres and release fine particles from cotton or Avicel. The mechanism responsible for this phenomenon is as yet unknown but must rely on the disruption of non-covalent bonds between adjacent clusters of fibrils or particles since the CBDs demonstrate no measurable hydrolytic activity. When added separately, a Family II CBD and its cognate catalytic domain synergise in the degradation of crystalline cellulose (12). Family I and IV CBDs have not been shown to disrupt the cellulose structure.

The essentially irreversible binding of Family II CBDs to crystalline cellulose is a paradox (6). Although adsorption appears to follow an equilibrium Langmuir isotherm (Figure 5), bound CBD does not dissociate if the concentration of free CBD is reduced by dilution. Immobilising an enzyme to its substrate would appear to restrict its access to susceptible bonds. However, we know that the present of a CBD enhances enzyme activity on crystalline cellulose. This suggests that the bound enzyme must be free to move on the surface of the cellulose.

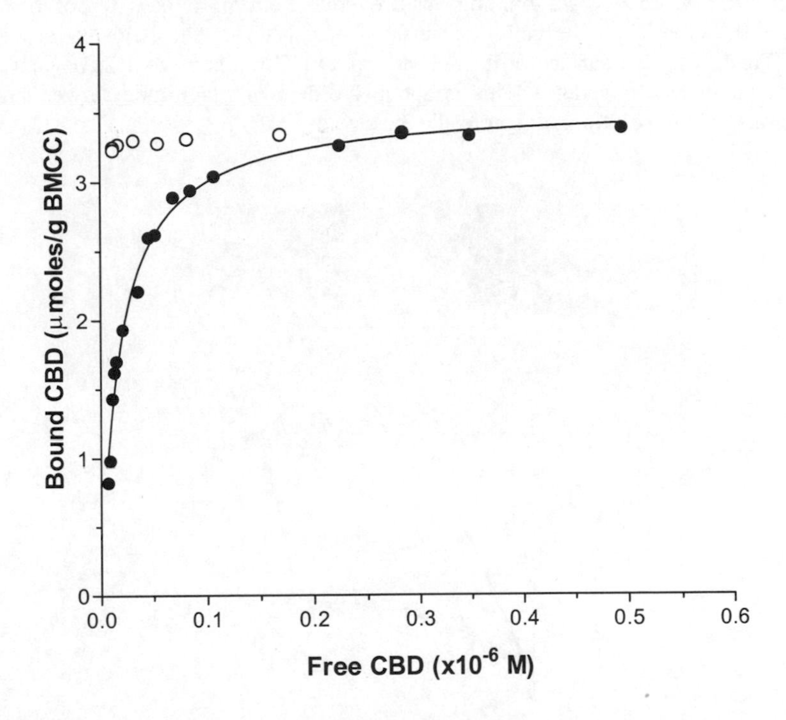

Figure 5 *Irreversible binding of CBD_{Cex}. Ascending (closed circles) and descending (open circles) isotherms of CBD_{Cex} binding on bacterial microcrystalline cellulose are shown. For the ascending isotherm, CBD remaining in solution after adsorption was determined by absorbance at 280 nm. Bound CBD was determined by the difference of the measured free CBD from a protein only control. The descending isotherm was measured by re-equilibration of CBD bound to cellulose with buffer and CBD in solution quantified by absorbance at 280 nm. Binding appears to follow a Langmuir isotherm equilibrium, however, this is not re-established upon dilution of free CBD.*

The movement of enzymes along the surfaces of biopolymers containing enzyme-susceptible sites can be characterised on the basis of recovery of fluorescence after photobleaching (FRAP). Fluorescent-labelled enzyme bound to substrate is irreversibly bleached by a burst of laser light directed to a sharply defined area of the surface. After bleaching, the non-fluorescent enzyme in the observation region may be replaced by fluorescent molecules by diffusion across the bleached interface and from the liquid phase. The pattern of recovery, *i.e.* uniform or progressing from the unbleached interface, indicates if movement along the surface occurs. The rate relates to the way in which the enzyme interacts with the surface. Using this technique it has been shown that the motion of fluorescently labelled amylase on a starch surface occurs by both lateral diffusion along the surface (over micron distances) and exchange between bound and free enzyme molecules in the solution (13).

We used FRAP to examine the surface diffusion of an exoglucanase and an endoglucanase from *Cellulomonas fimi* (Cex and CenA) and their respective Family II CBDs on sheets of crystalline cellulose microfibrils prepared from the cell walls of *Valonia ventricosa.* In all cases, binding was irreversible but more than 70% of the bound molecules was mobile on the cellulose surface. Analysis of the recovery curves for CBD_{Cex} (Figure 6) gave surface diffusion rates of $2x10^{-11}cm^2sec^{-1}$ to $1.2x10^{-10}cm^2\ sec^{-1}$ depending on surface coverage. This is about 4 orders of magnitude slower than the diffusion rate, free in solution, of a molecule the size of CBD_{Cex}.

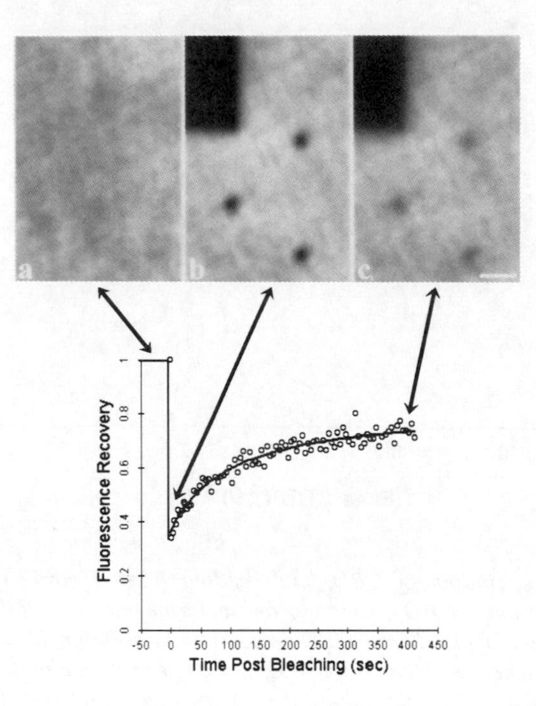

Figure 6 *Recovery of fluorescence after photo-bleaching CBD_{Cex} bound to Valonia cellulose. CBD_{Cex} bound to the cellulose sheet and examined by confocal laser scanning microscopy. The image sequence at the top of the figure shows the cellulose surface prior to bleaching, immediately after bleaching, and 7 minutes after photobleaching. High intensity laser light was used to bleach test spots and a large control rectangle. Over the recovery period of 7 minutes fluorescence in the centre of the test spots recovered to 70% of the original value. During this time there was less than 5% recovery in the centre of the control region indicating that there was no contribution from free CBD in solution and that diffusion was confined to the surface.*

There were only small differences in the surface diffusion rates of the enzymes and their respective CBDs. The mobile fraction of molecules was consistently higher for the endoglucanase CenA and its CBD.

The targeting and irreversible binding of a cellulase to its substrate could provide an extremely efficient means of maximising and retaining activity in the proximity of the bacterial cell secreting the enzyme. This would be true only if the surface diffusion rate did not limit access to cleavable sites. Although we do not know the spacing of cleavage sites on the cellulose surface, it seems unlikely that the surface diffusion rate limits enzyme activity. Based on our measurements, bound CenA will traverse several hundred lattice units in one minute. The turnover rate of CenA is about 0.25 moles of reducing sugar per mole of enzyme per minute, several orders of magnitude less than the transit rate of the enzyme on the surface. It is of interest that at least some of the fungal Family I CBDs bind cellulose reversibly (14). This difference in the properties of fungal and bacterial CBDs may relate to their modes of attack. Bacteria secrete cellulases into their environment and in some cases adhere to their substrate. Binding and retention of the enzyme close to the cell would thus be important. Fungal hyphae penetrate the substrate and release enzymes within the confines of the target. Under these conditions a smaller, reversible CBD may be preferable. During the course of hydrolysis, bacterial cellulases may be released from the substrate as a consequence of its degradation. Alternatively, for an organism like *C. fimi*, the CBD may be cleaved from the catalytic domain by the controlled secretion of a protease by the bacterium, generating a second class of non-adherent cellulases active on soluble products (15). This example of economy of operation provides *C. fimi* with the advantages of enzyme targeting and retention on cellulose as well as the ability to deal with any subsequent supply of soluble cellooligosaccharides.

References

1. P. Tomme, R.A.J. Warren and N.R. Gilkes, *Advances Microb. Physiol.,* **37** (1995).
2. P. Tomme, R.A.J. Warren, R.C. Miller Jr., D.G. Kilburn and N.R. Gilkes, *in* 'Enzymatic Degradation of Insoluble Carbohydrates', J.N. Saddler and M.H. Penner (Editors), American Chemical Society Symposium Series, 1995, p. 142.
3. D.M. Poole, G.P. Hazlewood, N.S. Huskisson, R. Virden, R. and H.J. Gilbert (1993), *FEMS Microbiol. Lett.*, **106**, 77.
4. N. Din, I.J. Forsythe, L.J. Burtnick, N.R. Gilkes, R.C. Miller Jr., R.A.J. Warren and D.G. Kilburn (1994), *Mol. Microbiol.*, **11**, 747.
5. M.R.. Bray, P.E. Johnson, N.R. Gilkes, L.P. McIntosh, D.G. Kilburn and R.A.J. Warren (1996), *Protein Sci.*, **5**, 2311.
6. L. Creagh,E. Ong, E. Jervis, D.G. Kilburn and C.A. Haynes (1996), *Proc. Nat. Acad. Sci. USA*, **93**, 12229.
7. J.B. Coutinho, N.R. Gilkes, R.A.J. Warren, D.G. Kilburn and R.C. Miller Jr. (1992), *Mol. Microbiol.*, **6**, 1243.
8. P. Tomme, A.L. Creagh, D.G. Kilburn and C.A. Haynes (1996), *Biochemistry*, **35**, 13885.
9. N.R. Gilkes ,R.A.J. Warren, R.C. Miller Jr. and D.G. Kilburn (1988), *J. Biol. Chem.*, **263**, 10401.
10. J. Hall, G.W. Black, L.M.A. Ferreira, S.J. Millward-Sadler, B.R.S. Ali, G.P. Hazlewood and H.J. Gilbert (1995), *Biochem. J.*, **309**, 749.

11. G. Maglione,O. Matsushita, J.B. Russell and D.B. Wilson (1992), *Appl. Environ. Microbiol.*, **58**, 3593.
12. N. Din, H.G. Damude, N.R. Gilkes, R.C. Miller Jr., R.A.J. Warren and D.G. Kilburn (1994) Proc. Nat. Acad. Sci., USA 91, 11383-11387.
13. E. Katchalski-Katzir, J. Rishpon, E. Sahar, R. Lamed and Y.I. Henis (1985) Biopolymers 24, 257-277.
14. M. Linder and T. Teeri (1996) Proc. Nat. Acad. Sci. USA 93, 12251-12255.
15. L.E. Sandercock, A. Meinke, N.R. Gilkes, D.G. Kilburn and R.A.J. Warren (1996) FEMS Microbiol. Lett. 143, 7-12.

STRUCTURE-FUNCTION RELATIONSHIPS IN THE CELLULASE-HEMICELLULASE SYSTEM OF ANAEROBIC FUNGI

G. P. Hazlewood and H. J. Gilbert

Laboratory of Molecular Enzymology
The Babraham Institute
Cambridge CB2 4AT

Department of Biological and Nutritional Sciences
The University of Newcastle upon Tyne
Newcastle upon Tyne NE1 7RU

1 INTRODUCTION

Anaerobic chytridiomycete fungi constitute a relatively recently discovered group of microorganisms. They occur widely as endosymbionts in the gastrointestinal tracts of herbivores and, together with other members of the gut microflora, they play an important role in the digestive physiology of the animal by colonising and digesting the large quantities of ingested plant biomass. To fulfill this role, anaerobic fungi synthesise and secrete a comprehensive repertoire of hydrolytic enzymes, including some of the most active cellulases and hemicellulases known.

The first full description of these fungi is attributed to Orpin who showed in 1975 that zooflagellates which had been known for many years to occur widely in the sheep rumen were, in fact, zoospores of the obligately anaerobic fungus *Neocallimastix frontalis*[1]. He subsequently identified two further species of anaerobic fungi in the sheep rumen, each of which had a motile stage (the zoospore) and a non-motile thallus supporting the zoosporangium[2,3]. Since these early studies anaerobic fungi have been found in the gastrointestinal tracts of many different species of ruminant and nonruminant herbivores, but to date none have been found in other likely anaerobic environments such as landfill sites or muds.

In the past, taxonomy of the anaerobic fungi has relied heavily on zoospore morphology and fine structure[4]; currently, some 17 species have been described and assigned to five genera. In terms of their life cycle, three genera are monocentric and two polycentric; in monocentric species (*e.g.*: *Neocallimastix, Piromyces, Caecomyces*), the cell body, which contains the developing zoospores and eventually becomes the zoosporangium, is attached to a radial array of anucleate rhizoids. In the polycentric species (*e.g.*: *Orpinomyces, Anaeromyces*), the vegetative thallus consists of hyphae which are multinucleate and multiple sporangia develop at many points throughout the thallus.

For further details of the biology of anaerobic fungi, the reader is referred to recent review articles [5-8].

2 ENZYMIC DEGRADATION OF PLANT CELL WALLS

Almost without exception, anaerobic fungi thus far isolated and studied share the capacity to utilize some and, in most cases, all of the structural polysaccharides contained in plants. Enzymic dissolution of plant cell walls is a unique and complex process; it involves a highly heterogeneous substrate[9] comprising interlocking matrices of cellulose, hemicelluloses and pectic polysaccharides, and occurs at the interface between liquid and solid phases. In general, microorganisms able to carry out this process, synthesise and secrete enzyme systems composed of large numbers of polysaccharide hydrolases with complementary specificities, which act cooperatively to degrade cell wall polysaccharides, liberating soluble sugars that can be taken up to provide energy and carbon (Table 1)[10,11].

Table 1 *Enzymes that digest plant cell wall polysaccharides**

Cellulases	*Xylan-degrading enzymes*
endoglucanase	endoxylanase (xylanase)
exoglucanase (cellobiohydrolase)	β-xylosidase
β-glucosidase	α-arabinofuranosidase
cellodextrinase	α-glucuronidase
	acetylxylan esterase
Mannan-degrading enzymes	phenolic acid esterase
endomannanase (mannanase)	
mannosidase	
α-galactosidase	

*For a detailed account of the composition and structures of polysaccharide substrates and the biochemistry of the enzymes, the reader is referred to recent review articles[9-11].

3 BIOCHEMICAL PROPERTIES OF CELLULASES AND HEMICELLULASES FROM ANAEROBIC FUNGI

Biochemical analysis of the extracellular enzymes produced by anaerobic fungi in culture has shown that these organisms produce the full repertoire of cellulase and hemicellulase activities. Most of these studies have been conducted with strains belonging to the genera *Neocallimastix*, *Piromyces* and *Orpinomyces*, but it is probably reasonable to assume that the same will prove true for all known genera. Results obtained independently by two groups indicate that the specific activities of cellulases produced by anaerobic fungi are at least as high as those reported for the cellulases of aerobic fungi belonging to the genus *Trichoderma*[12,13].

3.1 *Piromyces equi*

Studies carried out with a strain of *Piromyces equi* isolated from the cecum of a pony, have revealed that the fungus produces a range of extracellular cellulase and hemicellulase activities when cultured in medium containing either cellulose, xylan or cellobiose as carbon source. The cellulase and hemicellulase activities can be recovered from the culture supernatant by adsorbing onto either amorphous or substantially crystalline cellulose. Strong binding of the enzymes to cellulose was not easily reversed although some activity was recovered by repeated washing of the polysaccharide at low ionic strength or with increasing concentrations of ethylene glycol. The bound proteins accounted for up to 90% of the total cellulase and hemicellulase activities of the culture (Table 2) and were shown by activity staining of SDS-PAGE gels to include multiple endoglucanases, xylanases and mannanases. Once released from cellulose by washing with water, the bound proteins could not be further fractionated by gel filtration chromatography[14]. Taken together, these results indicate that the cellulase/hemicellulase system of *P. equi* constitutes a large multiprotein complex which has high affinity for cellulose, is not easily fractionated and contains cellulases as well as xylan- and mannan-degrading enzymes.

Table 2 *Enzyme activities of the* P. equi *cellulose-binding complex*

Enzyme	*Substrate*	*Specific activity (units/mg)*	*Percentage of total activity*
cellulase	carboxymethylcellulose	5.26	88
	acid-swollen cellulose	1.51	85
	crystalline cellulose	0.53	86
xylanase	soluble oat xylan	8.75	90
mannanase	carob galactomannan	1.43	81
cellobiosidase	*p*-nitrophenyl cellobioside	0.07	65
β-glucosidase	*p*-nitrophenyl glucoside	0.50	51
α-galactosidase	*p*-nitrophenyl galactoside	-	-

3.2 Other species

Results obtained elsewhere have confirmed that the above is true for other *Piromyces* strains[15-17] and for *Neocallimastix* species also[18].

4 ARCHITECTURE OF MICROBIAL CELLULASE/HEMICELLULASE SYSTEMS

It is now widely acknowledged that microbial cellulase/hemicellulase systems conform to two basic models[11]. Aerobic organisms like the filamentous fungus *Trichoderma reesei* and the soil bacteria *Pseudomonas fluorescens* subsp. *cellulosa* and *Cellulomonas fimi* produce multiple polysaccharide hydrolases which act cooperatively, but do not aggregate with each other. In general, the individual enzymes of these non-complexed systems are modular in structure and contain polysaccharide-binding domains, chiefly cellulose binding domains (CBD) which, at the first level, enable the enzymes to recognise and bind to their native cellulosic substrates. The second model is prevalent in anaerobic bacteria like, for example, *Clostridium thermocellum*[11,19]. It is characterised by the formation of highly-ordered multienzyme complexes or cellulosomes in which individual enzymes bind to a cellulosome-integrating protein (Cip) via the interaction of their highly-conserved docking domains or dockerins with receptor domains or cohesins on Cip. The high affinity of such complexes for cellulose can be ascribed to a CBD carried by Cip; their catalytic efficiency depends in large part on the structural integrity of the complex being maintained.

Based on the results of biochemical analysis, it is clear that there are marked similarities between the cellulase/hemicellulase systems of anaerobic fungi and the cellulosome-like complexes produced by anaerobic bacteria. In order to understand more about the way that the individual enzymes of the fungal complex interact with each other and with their native substrates, and to define the molecular mechanisms mediating assembly of the complex, we have used rDNA techniques to isolate and characterise genes coding for the individual components of the cellulase/hemicellulase systems of two species of anaerobic fungi, *Neocallimastix patriciarum* and *Piromyces equi*.

5 STRUCTURE-FUNCTION RELATIONSHIPS OF THE CELLULASES AND HEMICELLULASES FROM ANAEROBIC FUNGI

5.1 *Neocallimastix patriciarum*

Nucleotide sequences of xylanase (*xynA*)[20] and endoglucanase (*celB*)[21] cDNAs from *N. patriciarum* revealed open reading frames (ORF) of 1821 nucleotides (nt) and 1422 nt respectively, encoding polypeptides with predicted sizes of 66192 daltons and 53070 daltons. Both enzymes were modular in structure (Figure 1). Xylanase A (XylA) contained duplicated glycosyl hydrolase Family 11 catalytic domains of 220 residues, joined in tandem by a linker sequence. Downstream of the second catalytic domain was a repeated 40-residue sequence which was not required for catalytic activity and displayed no close homology with the sequences of any known proteins. Both catalytic domains encoded a functional xylanase[20].

Endoglucanase B (CelB) was also modular and was divided into two parts by a threonine-rich linker of 20 residues (Figure 1). The 365-residue domain upstream of the linker folded into an active endoglucanase, and displayed significant sequence identity with the catalytic domains of Family 5 endoglucanases from several anaerobic rumen bacteria. Downstream of the linker there was a repeated 40-residue sequence which was very similar to the 40-residue reiterated noncatalytic domain found at the *C*-terminus of XylA[21].

5.2 *Piromyces equi*

In order to establish whether *N. patriciarum* represents a good model for the molecular architecture of the cellulases and hemicellulases from anaerobic fungi in general, we have undertaken detailed molecular analysis of the polysaccharide hydrolase enzymes from a second fungus, namely *Piromyces equi*. A *P. equi* cDNA library comprising 2 x 10^7 clones was screened in two ways. First for expression of enzyme activity, resulting in the isolation of multiple cellulase$^+$, xylanase$^+$ and mannanase$^+$ clones[14]. The second approach used an antibody raised against affinity-purified fungal cellulase/hemicellulase complex and resulted in more than 300 positive clones. *P. equi* xylanase A (XylA), encoded by the *xynA* cDNA, had a predicted size of 68057 daltons and was modular in structure[22]. The enzyme comprised three distinct regions separated by linkers (Figure 1); the two largest domains of 220 residues displayed 54% sequence identity with each other and 30% identity with the Family 11 catalytic domains of XylA from *N. patriciarum*. Between the putative catalytic domains was located the same 40-residue repeat seen previously in XylA and CelB from *N. patriciarum*. Deletion of part of the *C*-terminal catalytic domain from *P. equi* XylA inactivated the whole enzyme, indicating that only the second catalytic domain formed an active enzyme.

Other cDNAs from the *P. equi* library have been sequenced and ORF identified, reinforcing this general view of the molecular architecture of the fungal polysaccharide hydrolases. For example, *P. equi* mannanase A (ManA), a 68000 dalton protein encoded by the *manA* cDNA, was also modular in structure[22]. The *N*-terminal signal peptide of ManA was followed by a 340-residue catalytic domain, homologous to the catalytic domains of other microbial mannanases belonging to glycosyl hydrolase Family 26. Downstream of the catalytic domain, and separated from it by a linker, were three repeats of the 40-residue noncatalytic domain seen previously in other fungal enzymes (Figure 1). Interestingly, gene duplication in *P. equi* has resulted in two other manannase isoforms which differ only slightly from ManA[23].

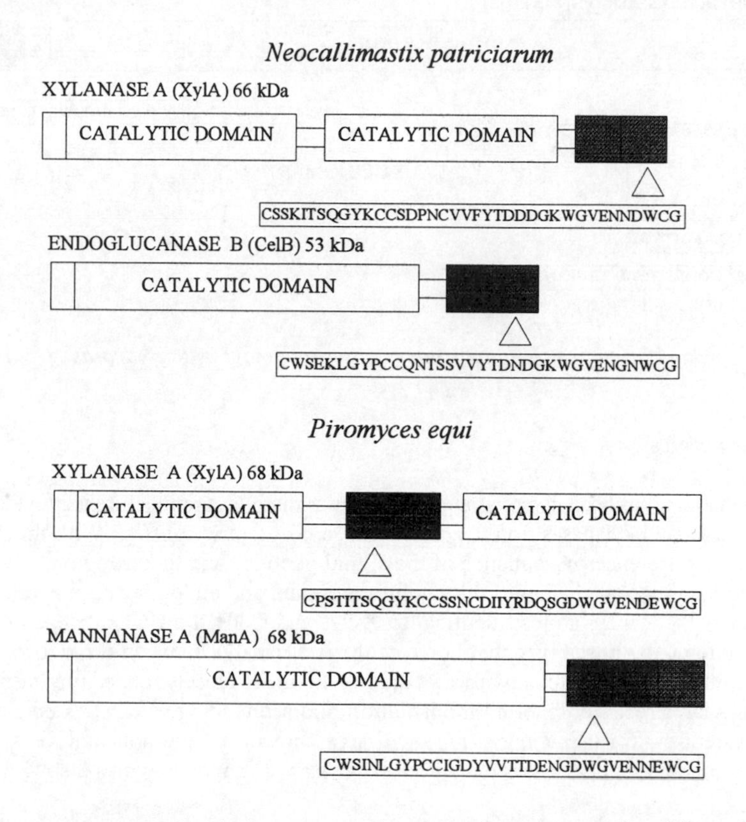

Figure 1 *Architecture of endoglucanase, xylanase and mannanase from* N. patriciarum *and* P. equi

Other examples of *P. equi* enzymes studied include a cinnamoyl ester hydrolase (Ceh) and two cellobiohydrolases (Cbh) (I.J. Fillingham and A.C. Freelove - unpublished). All three were modular in structure and contained the 40-residue noncatalytic domain common to the other anaerobic fungal cellulases and hemicellulases (Figure 2). In Ceh, an enzyme that removes esterified ferulic and *p*-coumaric acids from plant cell walls, a single copy of the 40-residue domain was located at the *N*-terminus and was followed by 12 full repeats of a 13-residue motif NQGGGMPWGDFGG; immediately downstream was an esterase catalytic domain comprising 270 residues. *P. equi* CbhA and CbhB contained catalytic domains that displayed significant sequence identity with cellobiohydrolases belonging to glycosyl hydrolase Family 6 and Family 48 respectively. The reiterated noncatalytic 40-residue domains were located at the *N*-terminus of CbhA and at the *C*-terminus of ChbB, and in both examples were separated from the catalytic domain by a typical hinge or linker (Figure 2).

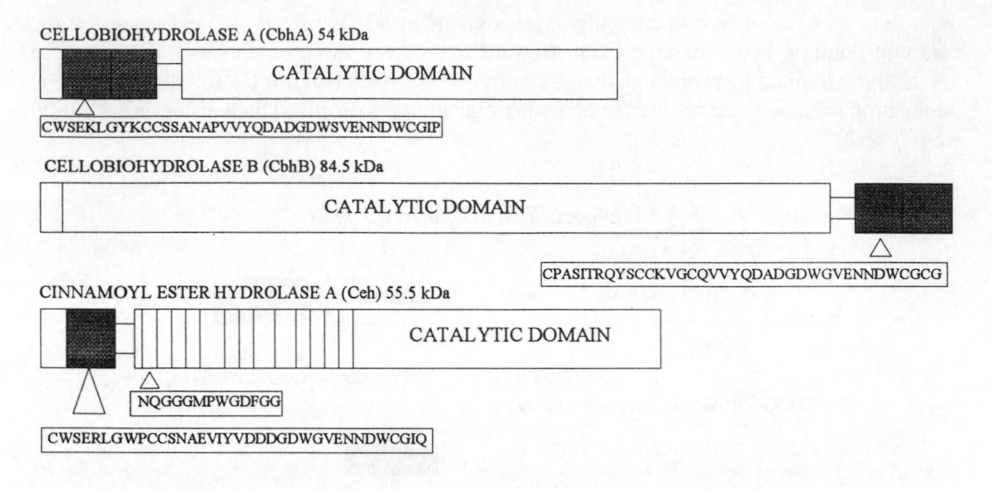

Figure 2 *Architecture of cellobiohydrolases and cinnamoyl ester hydrolase from* P. equi

5.3 Other species

Elsewhere, it has been demonstrated that similar modular architecture is prevalent in the endoglucanases and xylanases from *Orpinomyces*, a polycentric species. With the exception of some small differences in sequence of the signal peptides and linker regions, xylanase A (XynA), containing a single Family 11 catalytic domain, and endoglucanase B (CelB) from *Orpinomyces* PC-2 were almost identical to XylA and CelB from *N. patriciarum*[24]. This raises the interesting possibility that horizontal gene transfer between the two species has occurred quite recently or alternatively that the two fungi are closely related phylogenetically.

Others have recently reported the isolation and sequencing of cDNAs encoding four different esterases from *N. patriciarum*, and have shown that the noncatalytic 40-residue repeated domain is highly-conserved within this class of fungal enzymes also[25].

5.4 Summary

Based on the results summarised above, it is possible to draw together the following conclusions regarding the plant cell wall degrading enzymes of anaerobic fungi: (I) they exhibit a modular architecture, being composed of discrete catalytic and noncatalytic domains joined by hinges or linkers in which N S Q T and G residues predominate; (ii) on the whole, their catalytic domains appear more closely related to bacterial rather than fungal enzymes, raising the interesting possibility that gene transfer between prokaryotes and lower eukaryotes has taken place within the closed rumen ecosystem; (iii) anaerobic fungal xylanases provide the first example of such enzymes with identical duplicated catalytic domains; (iv) cellulose-binding domains which are prevalent in polysaccharide hydrolases from aerobic microbes are absent from the fungal enzymes; (v) a 40-residue noncatalytic domain (Figure 3), present in one, two or three copies, is highly conserved in the fungal enzymes and is not restricted in its location within the enzyme.

```
Np CelB1    C F S - - V N L G Y S C C - N - G C E V E Y T D S D G E W G V E N G N W C G I K
Np CelBII   C W S - - E K L G Y P C C Q N - T S S V V Y T D N D G K W G V E N G N W C G I Y
Np XylAI    C S A R I T A Q G Y K C C S D P N C V V Y Y T D E D G T Q G V E N N D W C G C G
Np XylAII   C S S K I T S Q G Y K C C S D P N C V V F Y T D D D G K W G V E N N D W C G C G
Pe XylAI    C P S T I T S Q G Y K C C S S - N C D I I Y R D Q S G D W G V E N D E W C G C G
Pe XylAII   C P S S I K N Q G Y K C C S D - S C E I V L T D S D G D W G I E N D E W C G C G
Pe ManAI    C W S - I - N L G Y P C C I G - D Y - V V T T D E N G D W G V E N N E W C G I V
Pe ManAII   C W S - - E P L G Y P C C V G - N T - V I S A D E S G D W G V E N N E W C G I V
Pe ManAIII  C W A - - E F L G Y P C C V G - N T - V I S T D E F G D W G V E N D D W C G I L
Pe CbhAI    C W S - - E K L G Y K C C S S A N A P V V Y Q D A D G D W S V E N N D W C G I P
Pe CbhAII   C W S - - E K L G Y P C C K - S T S A V V Y Q D A D G D W G V E N N D W C G I S
Pe CbhBI    C P A S I T R Q G Y S C C K - V G C Q V V Y Q D A D G D W G V E N N D W C G C G
Pe CbhBII   C P T S I T N Q G Y S C C S - S C G P V Y Y Q D A D G D W G V E N G D W C G M P
Pe Ceh      C W S - - E R L G W P C C S D S N A E V I Y V D D D G D W G V E N N D W C G I Q
```

Figure 3 *Conserved noncatalytic 40-residue domain in cellulases and hemicellulases from* N. patriciarum (Np) *and* P. equi (Pe)

6 ASSEMBLY OF THE MULTIENZYME CELLULOSE-BINDING CELLULASE/HEMICELLULASE COMPLEX FROM ANAEROBIC FUNGI

Conservation of the noncatalytic 40-residue domain in numerous different cellulase and hemicellulase enzymes from at least three different species of anaerobic fungi suggests that the domain fulfills an important function. Since none of the enzymes that contain it bind strongly to cellulose, it is clearly not a CBD. By analogy with the aggregated cellulase complexes of anaerobic bacteria like *C. thermocellum,* it is possible that the conserved 40-residue domain is a dockerin that mediates assembly of the multienzyme cellulase complex by binding to a receptor carried on a scaffolding protein. To test this hypothesis DNA encoding the reiterated 40-residue fungal domain was fused to the glutathione-S-transferase (GST) gene to produce a GST-dockerin hybrid. A variety of approaches were used to show that the hybrid binds specifically to 97- and 116-kDa polypeptides that are components of the multienzyme cellulase complexes from *P. equi* and *N. patriciarum* respectively[22]. No binding of the hybrid to *E. coli* proteins or to cellulosomal proteins form *C. thermocellum* was observed.

Based on these results, we believe that the highly-conserved noncatalytic 40-residue domain shown in Figure 3 and widely reported in cellulases and hemicellulases from anaerobic fungi is an assembly domain that promotes formation of the multiprotein fungal cellulase complex by binding to a scaffolding protein. Fungal enzymes like xylanase B (XylB)[26] and cellulase A (CelA)[27] from *N. patriciarum* (Figure 4) which do not carry the 40-residue repeated sequence are probably not part of the aggregated complex and could be regarded as being analogous to the "extracellulosomal" cellulases and xylanases produced by cellulolytic anaerobic bacteria. It is interesting to note that this proposed mode of assembly of the fungal complex is quite similar to that established for the cellulosome of *C. thermocellum*[19], the main difference being that the putative fungal dockerin occurs as one, two or three copies and shares no significant sequence similarity with the dockerins of the clostridial enzymes. This suggests that a similar mechanism for degrading plant cell walls has evolved in two taxonomically distinct organisms, and may be considered further evidence for the efficiency of the aggregated cellulosome-like complexes.

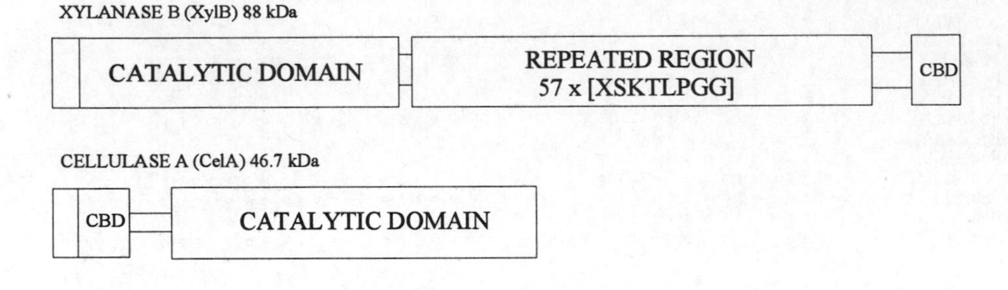

Figure 4 *Architecture of xylanase B and cellulase A from* N. patriciarum

7 CONCLUSIONS

Biochemical analysis of the cellulase/hemicellulase enzyme systems of anaerobic fungi, coupled with structure-function analysis of individual glycosyl hydrolases, strongly suggests that the multiprotein cellulase complex characteristic of these organisms is formed by the binding of multiple enzymes to a putative scaffolding protein of about 100 kDa in size. Binding is apparently mediated by a highly-conserved 40-residue noncatalytic domain present in one, two or three copies in each of the catalytic components of the complex. Strong binding of the complex to cellulose is not mediated by CBD borne on the catalytic components, but probably depends on a CBD which is either part of the scaffolding protein or a polypeptide which is always associated with it.

The main objectives of future work will be to characterise the scaffolding protein and its interaction with individual enzymes, and to define the molecular basis for the strong binding of the fungal complex to cellulose.

References

1. C. G. Orpin, *J. Gen. Microbiol.*, 1975, **91**, 249.
2. C. G. Orpin, *J. Gen. Microbiol.*, 1976, **94**, 270.
3. C. G. Orpin, *J. Gen. Microbiol.*, 1977, **99**, 107.
4. J. Li, B. Heath and L. Packer, *Can. J. Bot.*, 1993, **71**, 393.
5. J. Li and B. Heath, *Can. J. Microbiol.*, 1993, **39**, 1003.
6. D. A. Wubah, D. E. Akin and W. S. Borneman, *Crit. Rev. Microbiol.*, 1993, **19**, 99.
7. A. P. J. Trinci, D. R. Davies, K. Gull, M. I. Lawrence, B. B. Nielsen, A. Rickers, and M. K. Theodorou, *Mycol. Res.*, 1994, **98**, 129.
8. D. O. Mountfort and C. G. Orpin (eds.), "Anaerobic Fungi: Biology, ecology and function", Marcel Dekker, Inc., New York, 1994.
9. N. C. Carpita and D. M. Gibeaut, *The Plant Journal*, 1993, **3**, 1.
10. M. P. Coughlan and G. P. Hazlewood (eds.), "Hemicellulose and hemicellulases", Portland Press, London, 1993.
11. P. Tomme, R. A. J. Warren and N. R. Gilkes, *Adv. Microbial Physiol.*, 1995, **37**, 1.

12. T. M. Wood, C. A. Wilson, S. I. McCrae and K. N. Joblin, *FEMS Microbiol. Letts*, 1986, **34**, 37.
13. G. Beldman, M. F. Searle-Van Leeuwen, F. M. Rombouts and F. G. J. Voragen, *Eur. J. Biochem.*, 1985, **146**, 303.
14. B. R. S. Ali, L. Zhou, F. M. Graves, R. B. Freedman, G. W. Black, H. J. Gilbert and G. P. Hazlewood, *FEMS Microbiol. Letts*, 1995, **125**, 15.
15. M. J. Teunissen, G. V. M. De Kort, H. J. M. Op Den Camp and J. H. J. Huis in't Veld, *J. Gen. Microbiol.*, 1992, **138**, 1657.
16. M. J. Teunissen, J. M. H. Hermans, J. H. J. Huis in't Veld and G. D. Vogels, *Arch. Microbiol.*, 1993, **159**, 265.
17. R. Dijkerman, M. B. F. Vervuren, H. J. M. Op Den Camp and C. Van Der Drift, *Appl. Environ. Microbiol.*, 1996, **62**, 20.
18. C. A. Wilson and T. M. Wood, *Appl. Microbiol. Biotechnol.*, 1992, **37**, 125.
19. P. Béguin and M. Lemaire, *Crit. Rev. Biochem. Mol. Biol.*, 1996, **31**, 201.
20. H. J. Gilbert, G. P. Hazlewood, J. I. Laurie, C. G. Orpin and G.-P. Xue, *Mol. Microbiol.*, 1992, **6**, 2065.
21. L. Zhou, G.-P. Xue, C. G. Orpin, G. W. Black, H. J. Gilbert and G. P. Hazlewood, *Biochem. J.*, 1994, **297**, 359.
22. C. Fanutti, T. Ponyi, G. W. Black, G. P. Hazlewood and H. J. Gilbert, *J. Biol. Chem.*, 1995, **270**, 29314.
23. S. J. Millward-Sadler, J. Hall, G. W. Black, G. P. Hazlewood and H. J. Gilbert, *FEMS Microbiol. Letts.*, 1997, **141**, 183.
24. X.-L. Li, H. Chen and L. G. Ljungdahl, *Appl. Environ. Microbiol.*, 1997, **63**, 628.
25. B. P. Dalrymple, D. H. Cybinski, I. Layton, C. S. McSweeney, G.-P. Xue, Y. J. Swadling and J. B. Lowry, *Microbiology*, 1997, **143**, 2605.
26. G. W. Black, G. P. Hazlewood, G.-P., Xue, C. G. Orpin and G. P. Hazlewood, *Biochem. J.*, 1994, **299**, 381.
27. S. Denman, G.-P. Xue and B. Patel, *Appl. Environ. Microbiol.*, 1996, **62**, 1889.

STRUCTURE AND FUNCTION ANALYSIS OF *PSEUDOMONAS* PLANT CELL WALL HYDROLASES

Geoffrey P. Hazlewood[1] and Harry J. Gilbert[2]

[1]Laboratory of Molecular Enzymology, The Babraham Institute, Babraham, Cambridge CB2 4AT, U.K. and [2]Department of Biological & Nutritional Sciences, University of Newcastle upon Tyne, Newcastle upon Tyne NE1 7RU, U.K.

1 *PSEUDOMONAS FLUORESCENS* SUBSP. *CELLULOSA*

Pseudomonas fluorescens subsp. *cellulosa* (NCIMB 10462) can hydrolyse crystalline cellulose, xylan, arabinan, galactan and mannan, and can use all but the last polysaccharide as sole source of carbon and energy. Conversion of these plant cell wall polysaccharides to metabolisable sugars by strain NCIMB 10642 is mediated by extracellular polysaccharide hydrolases, which are induced by culturing with cellulose, carboxymethylcellulose (CMC) and xylan, repressed during growth with cellobiose, glucose or xylose and secreted without apparently forming aggregates or associating with the cell surface[1]. The complexity of the *P. fluorescens* subsp. *cellulosa* system has become apparent through the cloning of genes in *Escherichia coli*, a bacterium devoid of plant cell wall hydrolases. These studies have revealed that the capacity of *Pseudomonas* strain NCIMB 10642 to digest plant cell wall polysaccharides can be attributed to the production of multiple extracellular cellulase and hemicellulase enzymes all of which are the products of distinct genes belonging to multigene families[2,3]. Knowledge of the primary structures of each gene, coupled with gene sectioning and functional analysis of the encoded protein have provided a detailed picture of structure/function relationships and have established the cellulase-hemicellulase system of *P. fluorescens* subsp. *cellulosa* as a paradigm for the plant cell wall degrading enzyme systems of aerobic cellulolytic bacteria[1].

2 ARCHITECTURE OF *PSEUDOMONAS* PLANT CELL WALL HYDROLASES

2.1 Cellulases

The cellulase system of *P. fluorescens* subsp. *cellulosa* is composed of at least three different activities - endoglucanase, cellodextrinase and β-glucan glucohydrolase. Three different genes encoding endoglucanase have been isolated from *P. fluorescens*, *celA*[4], *celB*[5] and *celE*[6]; in each case the encoded enzyme is modular in structure and contains a distinct catalytic domain and two other domains, joined together by serine-rich linker sequences (Figure 1). Based on hydrophobic cluster analysis (HCA), the catalytic domains of endoglucanases CELA, CELB and CELE belong to glycosyl hydrolase families 9, 45 and 5 respectively[7]. Functional analysis of the polypeptides encoded by gene fragments has shown that the highly-conserved 100 residue domain located at the *C*-terminus of CELE and CELA and at the *N*-terminus of CELB is a Type II cellulose-binding domain (CBD).

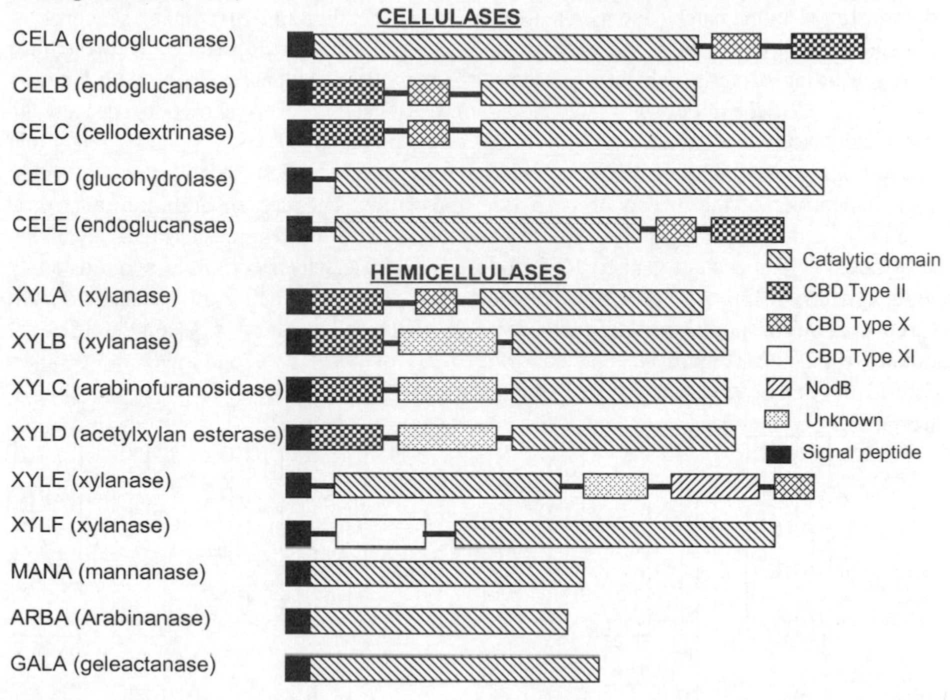

Figure 1 *Molecular architecture of P. fluorescens subsp. cellulosa polysaccharidases*

A similar Type II CBD (Figure 2) is present in the *N*-terminal 100 residues of the cellodextrinase (CELC) which has a catalytic domain belonging to glycosyl hydrolase family 5[8]. The middle domain of each of these cellulase enzymes contains a novel CBD (Figure 3) of about 30 residues, first described in xylanase E (XYLE) from *P. fluorescens* subsp. *cellulosa*[9]. The fifth component of the cellulase system is a 1,4-β-D-glucan glucohydrolase D (CELD) belonging to family 3. CELD is not modular, and its association with the cell envelope of *P. fluorescens* subsp. *cellulosa* is apparently mediated by a hydrophobic *N*-terminal region linked to the catalytic domain by a short sequence rich in hydroxyamino acids[10] (Figure 1).

2.2 Xylan-degrading enzymes

Xylan-degrading activity of *P. fluorescens* subsp. *cellulosa* is attributable to six proteins encoded by multiple genes[3]. Four genes (*xynA*, *B*, *E* and *F*) coding for endoxylanases have been isolated[9,11,12]. All four enzymes are modular in structure (see Figure 1), and contain three or more distinct domains joined together by linkers rich in hydroxyamino acids. The catalytic domains of XYLA, B and F belong to glycosyl hydrolase family 10, while that of XYLE is in family 11. The *N*-terminal 100 residues of XYLA and XYLB comprise a Type II CBD, the middle domain of XYLA and the *C*-terminal domain of XYLE contain the same small novel CBD previously identified in the middle region of the four *Pseudomonas* cellulases. A second novel CBD, with no homology to sequences contained in the protein sequence databases, has been demonstrated immediately downstream of the signal peptide and first linker sequence of XYLF[9]. XYLE contains a fourth domain, homologous with the NodB protein of nitrogen-fixing bacteria, which is present also in modular xylanases from *Cellulomonas fimi*[13] and *Cellvibrio mixtus*[9]. This domain has recently been shown to deacetylate chemically acetylated xylan[14]. *P. fluorescens* subsp. *cellulosa* also contains genes that encode enzymes which attack the side-chains of xylans; the first, *xynC*, maps to within 150 nucleotides of the 3′ end of *xynB* (see above) and encodes an arabinofuranosidase (XYLC) active against polymeric arabinoxylan but not against small synthetic substrates[12]. The second gene *xynD* encodes an esterase which combines synergistically with xylanase to release acetic and phenolic acids from plant cell wall material[15-17]. All three enzymes contain an *N*-terminal Type II CBD followed by a second domain of unknown function (Figure 1). Nucleotides 114-931 of *xynB, xynC* and *xynD* and residues 39-311 of XYLB, XYLC and XYLD were identical[12,17].

```
CELA  865  C Q Y V V T N Q W N N G F T A V I R V R N N G S S A I N R W S V N W S Y S D G S R I T N S W N A N V
CELB   32  C E Y R V T N E W G S G F T A S I R I T N N G S S T I N G W S V S W N Y T D G S R V T S S W N A G L
CELC   39  C E Y V V T N S W G S G F T A A I R I T N S T S S V I N G W N V S W Q Y N - S N R V T N L W N P N L
CELE  473  C S Y T V T N Q W S N G F T A S I R I A N N G T S P I N G W N L S W S Y S D G S R V T N S W N A N V
XYLA   31  C S Y N I T N E W N T G Y T G D I T I T N R G S S A I N G W S V N W Q Y A - T N R L S S S W N A N V
XYLB   39  C T Y T I D S E W S T G F T A N I T L K N D T G A A I N N W N V N W Q Y S - S N R M T S G W N A N F
XYLC   39  C T Y T I D S E W S T G F T A N I T L K N D T G A A I N N W N V N W Q Y S - S N R M T S G W N A N F
XYLD   38  C T Y T I D S E W S T G F T A N I T L K N D T G A A I N N W N V N W Q Y S - S N R M T S G W N A N F

CELA       T G N N P Y A A S A L G W N A N I Q P G Q T A E F G F Q G T K G A G S R Q V P A V T G S V C Q  961
CELB       S G A N P Y S A T P V G W N T S I P I G S S V E F G V Q G N N G S S R A Q V P A V T G A I C G  128
CELC       S G S N P Y S A S N L S W N G T I Q P G Q T V E F G F Q G V T N S G T V E S P T V N G A A C T  134
CELE       S G N N P Y T V S N L G W N G S I Q P G Q A V E F G F Q G T K N N S A A A I P T L S G N V C N  569
XYLA       S G S N P Y S A S N L S W N G N I Q P G Q S V S F G F Q V N K N G G S A E R P S V G G S I C S  126
XYLB       S G T N P Y N A T N M S W N G S I A P G Q S I S F G L Q G E K N G S T A E R P T V T G A A C N  134
XYLC       S G T N P Y N A T N M S W N G S I A P G Q S I S F G L Q G E K N G S T A E R P T V T G A A C N  134
XYLD       S G T N P Y N A T N M S W N G S I A P G Q S I S F G L Q G E K N G S T A E R P T V T G A A C N  133
```

Figure 2 *The primary structure of P. fluorescens subsp. cellulosa Type II cellulose binding domains*

2.3 Other hydrolases

P. fluorescens subsp. *cellulosa* also produces enzymes which attack the smaller quantities of more accessible polysaccharides found associated with the cell wall matrix; these include mannans, arabinans and galactans. Genes coding for enzymes active against each of these polysaccharides have been isolated and sequenced. Mannanase A (MANA) encoded by the *manA* gene is a family 26 glycosyl hydrolase active against galactomannan and mannoligosaccharides, but with little activity against other cell wall polysaccharides[18].

Arabinanase activity of *Pseudomonas* was attributed to an extracellular enzyme, ArbA, encoded by the *arbA* gene. ArbA belonged to glycosyl hydrolase family 43, was homologous with arabinanase from *Aspergillus niger* and hydrolysed linear but not branched arabinan and arabinooligosaccharides. Initial attack of arabinan by ArbA was random but subsequent release of exclusively arabinotriose indicated a processive mechanism and suggested that ArbA combines endo- and exo- modes of action[19]. Galactanase A (GalA) an endo-β-,1,4-galactanase encoded by the *ganA* gene from *P. fluorescens* subsp. *cellulosa* attacked only galactan and galactooligosaccharides[20]. It exhibited significant sequence identity with a galactanase from *Aspergillus aculeatus* and, on that basis, belongs to the small glycosyl hydrolase family 53. Each of the three enzymes described above differed significantly from the other glycosyl hydrolases of *P. fluorescens* subsp. *cellulosa.* First, each occurred as a single copy. Second, none displayed the modular architecture typical of the cellulases and xylanases (Figure 1). Finally, CBD which are prevalent in cellulases and hemicellulases from *P. fluorescens* subsp. *cellulosa* are absent from MANA, ArbA and GalA, perhaps reflecting the greater accessibility of their target substrates in plant cell walls and a reduced requirement for polysaccharide-binding domains to enhance catalytic efficiency.

```
XYLE  Q C N W W G T F Y P L C Q T Q T S G W G W E N S R S C I S T S T C N S Q G T  653
XYLB  Q C N W W G T R Y P L C T N T A S G W G W E N N T S C I T T S T C N S Q G A  649
XYLA  Q C N W Y G T L Y P L C V T T T N G W G W E D Q R S C I A R S T C A A Q P A  282
CELA  R C N W Y G T L Y P L C V T T Q S G W G W E N S Q S C I S A S T C S A Q P A  836
CELC  Q C N W Y G T L Y P L C V S T T S G W G Y E N N R S C I S P S T C S A Q P A  327
CELB  - C N W Y G T L T P L C N N T S N G W G Y E D G R S C V A R T T C S A Q P A  317
```

Figure 3 *The primary structure of P. fluorescens subsp. cellulosa Type X cellulose binding domains*

3 FUNCTION OF NONCATALYTIC DOMAINS

The domains of cellulases and hemicellulases from *P. fluorescens* subsp. *cellulosa* which do not hydrolyse glycosidic bonds are of two types: cellulose-binding domains and NodB domains.

3.1 Cellulose-binding domains and linker sequences

As discussed above, all cellulases and xylanases from *P. fluorescens* subsp. *cellulosa* contain one or more CBDs. Recent work has shown that two CBDs, contained within a single polypeptide, can interact synergistically resulting in higher affinity binding than observed for a single CBD[21]. Whether this is true for the paired CBD of *P. fluorescens* subsp. *cellulosa* cellulases and hemicellulases remains to be seen. The high frequency with which CBDs occur in the cellulases and hemicellulases from *P. fluorescens* subsp. *cellulosa* suggests that the selective advantage conferred by these domains is so great that it has led to evolutionary subduction of less well adapted enzymes.

A unifying role for CBD and linker sequences has yet to be established. There is evidence that CBDs play a major role in potentiating the activity of cellulases against the more resistant forms of insoluble cellulose[6,22,23]. For example, Din *et al.*[24] demonstrated that the isolated Type II CBD from *C. fimi* endoglucanase A (CenA) could open up the structure of highly crystalline cellulose making it more accessible to enzyme attack. Similarly, removal of the CBD from CenA diminished activity against bacterial microcrystalline cellulose (BMCC), but not amorphous cellulose, while full-length CELE from *P. fluorescens*, containing a *C*-terminal Type II CBD, displayed 4 times higher activity against Avicel than a truncated form lacking the CBD[8]. Recent studies focusing on XYLA and XYLC from *Pseudomonas,* both of which contain a Type II CBD, have provided evidence for the biological roles of both CBD and linkers. The CBD, when covalently attached to these enzymes increased their activity against cellulose/hemicellulose complexes. Addition of the hemicellulase CBDs, to the respective catalytic domain, *in trans*, did not enhance the catalytic activity of the enzymes. The CBD alone elicited no detectable change in the structural integrity of either crystalline cellulose or plant cell walls when assessed using the constant-load extension assay (creep assay) originally developed for evaluating the biological properties of expansins. These latter plant proteins are active in catalytic amounts and loosen the structure of plant cell walls by physically disrupting the close-packed structure of microfibrils at the cellulose/hemicellulose interface. Taken together, these data indicate that the hemicellulase CBD enhances catalytic activity not by disrupting the complex interactions between plant structural polysaccharides in the manner of expansins, but simply by promoting close contact between the catalytic domain of the enzyme and its target substrate. Removal of the linker sequences from XYLA did not affect catalytic activity against soluble xylan or its ability to bind to cellulose. However, the activity of the linkerless enzyme against its target substrate contained in cellulose-xylan complexes derived from plant cell walls was significantly lower than full-length XYLA. A possible explanation for this observation is that during hydrolysis of the xylan component by, for example, XYLA, the CBD restricts substrate availability by anchoring the enzyme at a fixed location. Structural flexibility conferred on the enzyme by the linker may increase the number of glycosidic bonds accessible to the catalytic domain.

4 STRUCTURES AND CATALYTIC MECHANISMS OF *PSEUDOMONAS* PLANT CELL WALL HYDROLASES

Enzymatic hydrolysis of the glycosidic bonds contained in plant structural polysaccharides proceeds via general acid catalysis and has a critical requirement for amino acid residues to act as proton donor and nucleophile/base respectively[25]. Based on similarities in their amino acid sequences, the current total of almost 1000 known glycosyl hydrolases can be fitted into a classification comprising some sixty different families[7]. This sequence-based classification has proved to be enormously beneficial in reflecting common structural features and evolutionary relationships and as a tool for deducing mechanistic information. Crystal structures have been determined for twenty or so glycosyl hydrolases[26], but there are no structural paradigms for more than half of the sixty families. Below is a description of some of our work on the structure/function relationships of the catalytic domains of *Pseudomonas* glycosyl hydrolases.

4.1 Xylanase A

4.1.1 Catalytic Domain. The crystal structure of one of the *P. fluorescens* subsp. *cellulosa* family 10 xylanases (XYLA) has been determined at 1.8Å resolution and a crystallographic *R* factor of 0.166[27,28]. These studies showed that the catalytic core of XYLA and, by inference, those of other family 10 xylanases, has the architecture of the 8-fold α/β barrel or TIM barrel, with the nucleophilic glutamate (E246) located at the carboxyl end of β-strand 7 of the barrel and the catalytic acid-base (E127) within 5.5Å at the end of β-strand 4. Similar 8-fold α/β architecture has been demonstrated for representatives of glycosyl hydrolase families 1, 2, 5, 26, 30, 35, 39 and 42; for each of the nine members of this superfamily or clan (GH-A) the catalytic residues and catalytic mechanism are strictly conserved, indicating that all have diverged from a common ancestor[26]. Compared with other 8-fold α/β barrels within the superfamily, XYLA has one additional α-helix and an atypically long loop after strand 7 which appeared by X-ray crystallography to be stabilised by calcium. Xylopentaose soaked into crystals of the inactive enzyme occupied an active-site cleft formed by the longer loops that follow β-strands 4 and 7; five substrate-binding subsites -1 to +4 were identified within the cleft, with the cleaved bond located between subsites -1 and +1, indicative of an exo-mode of action. A cluster of conserved residues in the substrate-binding cleft adjacent to subsite -1 was indicative of an additional subsite (subsite -2) which was obscured from xylopentaose by contacts within the crystal. The presence of subsite -2 was confirmed by biochemical analysis which showed that (i) X_2 and X_3 were the main products formed from xylan and xylooligosaccharides (X_3 to X_6) and (ii) activity against X_6 was higher than against X_5 or any other xylooligosaccharides. These results indicated that XYLA has an endo-mode of action and at least six substrate-binding subsites, with at least two xylose-binding sites on each side of the catalytic residues[29]. Recent experiments (unpublished) using end-labelled xylooligosaccharides have established that the XYLA substrate binding cleft consists of four subsites aglycone and three subsites glycone of the two catalytic residues (Figure 4).

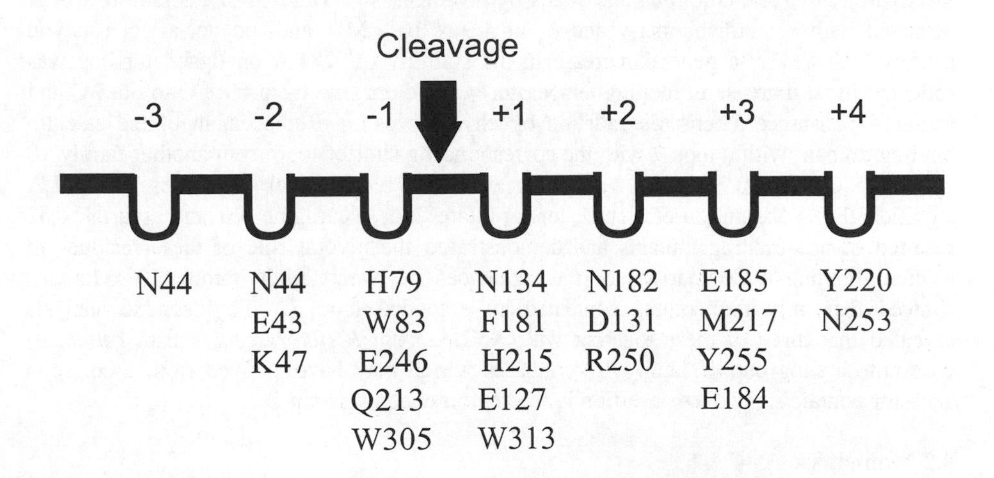

Figure 4 *Residues on the surface of the substrate binding cleft of xylanase A*

The functional importance of subsite -2 and the roles of other conserved residues located on the surface of the active-site cleft of family 10 xylanases, was investigated by site-directed mutagenesis, involving the substitution of alanine (A) for glutamate 43 (E43A), asparagine 44 (N44A), lysine 47 (K47A) and methionine 46 (M46A)[29]. The results showed that binding to the -2 site is essential for efficient hydrolysis of xylooligosaccharides. Mutants N44A and E43A were much less active than the wild-type enzyme against X_3, X_4 and X_5 but their activity against xylan was unaltered. K47A was less active against xylan and had little activity against xylooligosaccharides. These data were consistent with a reduction in the capacity of the -2 site to bind substrate, leading to random binding of oligosaccharides along the cleft at sites +4 to +1, and the production of dead-end complexes; activity against polymeric xylan remained unchanged or only slightly reduced because binding of xylan at sites +4 to +1 would place adjacent xylose residues in sites -1 and -2. Thus the primary role of the -2 subsite of XYLA is to prevent small oligosaccharides from forming non-productive enzyme-substrate complexes.
We have also investigated the role of amino acids at other xylose-binding pockets within the extended substrate-binding cleft of *Pseudomonas* XYLA. Of particular interest was residue N182 which is located at the boundary between subsites +2 and +3. In xylanase A from *Streptomyces lividans*, an enzyme which also belongs to glycosyl hydrolase family 10, the equivalent residue, N173, is clearly important for the subsite +3 xylose-binding pocket. However, N182A, N182R and N182D mutants of *Pseudomonas* XYLA generated xylotriose and xylose from xylotetraose and not xylobiose as produced by the wild type enzyme. These data suggest that in XYLA, N182 plays an important role in the function of subsite +2, but not subsite +3, and provide evidence that residues which are highly conserved in xylanases belonging to the same glycosyl hydrolase family do not necessarily play equivalent roles in enzyme function.

4.1.2 Calcium-binding site. Other mutagenesis studies conducted with XYLA have shown that the calcium-binding site, which is unique among family 10 xylanases, stabilises the structure of the extended loop 7 and protects XYLA from thermal inactivation, thermal unfolding and attack by proteinases[30]. Binding of calcium to XYLA occurred with 1:1 stoichiometry and K_a of 4.9 x 10^4 M^{-1}, and did not affect catalytic activity. However, a general increase in the stability of XYLA on ligand binding was reflected in an increase in melting temperature of the enzyme from 60.8°C to 66.5°C and acquired resistance against degradation by chymotrypsin. Replacement of the calcium binding domain within loop 7 with the corresponding shorter loop from another family 10 xylanase (Cex from *C. fimi*)[31] did not alter the biochemical properties of XYLA significantly. Substitution of alanine for aspartate 256, asparagine 261 and aspartate 262 resulted in non-binding mutants and demonstrated the pivotal role of these residues in calcium binding. Comparison of the sequences of twenty eight family 10 xylanases showed that only four others contained an extended loop 7. Phylogenetic analysis revealed that three of these together with XYLA from *P. fluorescens* subsp. *cellulosa*, constitute a subgroup of family 10 xylanases which may have evolved from a common ancestor containing a DNA insertion in the region encoding loop 7.

4.2 Mannanase A

Apart from X-ray crystallographic determination of the tertiary structures of enzyme-substrate or enzyme-inhibitor complexes, a number of other approaches have been used to facilitate identification of the key catalytic residues of glycosyl hydrolases. These include carrying out detailed kinetic analysis on mutant enzymes in which the putative catalytic carboxylate residues have been altered by site-directed mutagenesis. This latter approach has recently been used for two *P. fluorescens* subsp. *cellulosa* enzymes belonging to glycosyl hydrolase families for which no structural paradigms exist.

The first enzyme, mannanase A (MANA), belongs to family 26. Analysis of the stereochemical course of mannotetraose hydrolysis by purified recombinant MANA showed that the configuration of the anomeric carbon was retained on cleavage of the middle glycosidic bond, indicating a double displacement mechanism[32]. Hydrophobic cluster analysis (HCA) revealed that two glutamate and two aspartate residues were conserved in all members of family 26, and identified sequence motifs which indicated that MANA was related to glycosyl hydrolases of the GH-A superfamily, all of which have 8-fold α/β barrel architecture. Site-directed mutagenesis was carried out to test the prediction that E320 and E212 are the catalytic nucleophile and acid-base respectively. Replacement of the conserved aspartates with alanine and glutamate had no dramatic effect on catalytic activity. In contrast, substituting alanine or aspartate for either of the conserved glutamates substantially decreased activity against galactomannan, mannotetraose and 2,4-dinitrophenyl β-mannobioside (2,4-DNPM). The apparent K_m of E320A was similar to that of wild-type MANA, but a large reduction in K_m was seen for E212A. Analysis of the pre-steady state kinetics of 2,4-DNPM hydrolysis by E212A revealed a rapid initial rate of 2,4-dinitrophenol (2,4-DNP) release, as would be predicted for the hydrolysis by MANA of a substrate with a good leaving group, followed by a rapid decay leading to a slow steady state release of 2,4-DNP as deglycosylation became rate limiting. During hydrolysis of 2,4-DNPM by E320A, glycosylation was apparently rate limiting and there was no significant pre-steady state burst of 2,4-DNP release. These results were consistent with the view that E212 and E320 respectively are the catalytic acid-base and nucleophile residues of the retaining glycosyl hydrolase MANA.

4.3 Galactanase A

A second *Pseudomonas* enzyme for which the catalytic mechanism and key catalytic residues have been identified in the absence of detailed structural information is the endo-β-1,4-galactanase, GalA, which belongs to glycosyl hydrolase family 53[32]. Analysis of the stereochemical course of 2,4-dinitrophenyl-β-galactobiose (2,4-DNPG_2) hydrolysis by GalA in the presence of D_2O revealed that the β-conformation of the anomeric carbon was retained on cleavage of the aglycone bond, suggesting that the hydrolysis of glycosidic bonds by GalA proceeds by a double displacement mechanism. HCA indicated that GalA and, by inference, other members of family 53, are related to the GH-A superfamily, members of which have an 8-fold α/β barrel structure. Furthermore, it predicted that E161 and E270 were the key catalytic acid-base and nucleophile residues respectively. GalA mutants in which E161 and E270 had been replaced by alanine or aspartate were unaltered in conformation, as evidenced by circular dichroism spectroscopy, but were virtually inactive against galactan. E161A exhibited a much lower K_m for 2,4-DNPG_2 than wild type GalA and elicited a rapid initial pre-steady state burst of 2,4-DNP release; no such pre-steady state burst was seen during hydrolysis of 2,4-DNPG_2 by mutant E270A. These data are consistent with the view that E161 and E270 are the catalytic acid-base and nucleophile residues of the retaining enzyme GalA.

Acknowledgements

The authors gratefully acknowledge the continuing financial support of the Biotechnology and Biological Sciences Research Council, and the substantial contributions made to this research by their colleagues and coworkers.

References

1. G.P. Hazlewood, J.I. Laurie, L.M.A. Ferreira and H.J. Gilbert, *J. Appl. Bact.*, 1992, **72**, 244.
2. H.J. Gilbert, G. Jenkins, D.A. Sullivan and J. Hall, *Mol. Gen. Genet.*, 1987, **210,** 551.
3. H.J. Gilbert, D.A. Sullivan, G. Jenkins, L.E. Kellett, N.P. Minton and J. Hall, *J. Gen. Microbiol.*, 1988, **134**, 3239.
4. J. Hall and H.J. Gilbert, *Mol. Gen. Genet.*, 1988, **213,** 112.
5. H.J. Gilbert, J. Hall, G.P. Hazlewood and L.M.A. Ferreira, *Mol. Microbiol.*, 1990, **4**, 759.
6. J. Hall, G.W. Black, L.M.A. Ferreira, S.J. Millward-Sadler, B.R.S. Ali, G.P. Hazlewood and H.J. Gilbert, *Biochem. J.*, 1995, **309,** 749.
7. B. Henrissat and A. Bairoch, *Biochem. J.*, 1996, **316**, 695.
8. L.M.A. Ferreira, G.P. Hazlewood, P.J. Barker and H.J. Gilbert, *Biochem. J.*, 1991, **279**, 793.
9. S.J. Millward-Sadler, K. Davidson, G.P. Hazlewood, G.W. Black, H.J. Gilbert and J.H. Clarke, *Biochem. J.*, 1995, **312**, 39.
10. J.E. Rixon, L.M.A. Ferreira, A.J. Durrant, J.I. Laurie, G.P. Hazlewood and H.J. Gilbert, *Biochem. J.*, 1992, **285**, 947.
11. J. Hall, G.P. Hazlewood, N.S. Huskisson, A.J. Durrant and H.J. Gilbert, *Mol. Microbiol.*, 1989, **3**, 1211.
12. L.E. Kellett, D.M. Poole, L.M.A. Ferreira, A.J. Durrant, G.P. Hazlewood and H.J. Gilbert, *Biochem. J.*, 1990, **272**, 369.
13. J.H. Clarke, K. Davidson, H.J. Gilbert, C.M.G.A. Fontes and G.P. Hazlewood, *FEMS Microbiol. Letts*, 1996, **139**, 27.
14. J.I. Laurie, J.H. Clarke, A. Ciruela, C.B. Faulds, G. Williamson, H.J. Gilbert, J.E. Rixon, J. Millward-Sadler and G.P. Hazlewood, *FEMS Microbiol. Letts* 1997, **148,** 261.
15. L.M.A. Ferreira, T.M. Wood, G. Williamson, C. Faulds, G.P. Hazlewood, G.W. Black and H.J. Gilbert, *Biochem. J.*, 1993, **294**, 349.
16. C.B. Faulds, M.-C. Ralet, G. Williamson, G.P. Hazlewood and H.J. Gilbert, *Biochim. Biophys. Acta*, 1995, **1243**, 265.
17. B. Bartolomé, C.B. Faulds, P.A. Kroon, K. Waldron, H.J. Gilbert, G.P. Hazlewood and G. Williamson, *Appl. Environ. Microbiol.*, 1997, **63**, 208.
18. K.L. Braithwaite, G.W. Black, G.P. Hazlewood, B.R.S. Ali and H.J. Gilbert, *Biochem. J.*, 1995, **305**, 1005.
19. V.A. McKie, G.W. Black, S.J. Millward-Sadler, G.P. Hazlewood, J.I. Laurie and H.J. Gilbert, *Biochem. J.*, 1997, **323**, 547.

20. K.L. Braithwaite, T. Barna, T. Spurway, S. Charnock, G.W. Black, N. Hughes, J.H. Lakey, G.P. Hazlewood, B. Henrissat and H.J. Gilbert, *Biochemistry* (in press) (1997).
21. M. Linder, I. Salovuori, L. Ruohonen and T.T. Teeri, *J. Biol. Chem.*, 1996, **271**, 21268.
22. P. Tomme, H. van Tilbeurgh, G. Pettersson, J. van Damme, J. Vandekerckhove, J. Knowles, T. Teeri and M. Claeyssens, *Eur. J. Biochem.*, 1988, **170**, 575.
23. J.B. Coutinho, N.R. Gilkes, R.A.J. Warren, D.G. Kilburn and R.C. Miller, Jr., *FEMS Microbiol. Lett.*, 1993, **113**, 211.
24. N. Din, N.R. Gilkes, B. Tekant, R.C. Miller Jr., R.A.J. Warren and D.G. Kilburn, *Bio/technology*, 1991, **9**, 1096.
25. D. Kafetzopoulos, G. Thireos, J.N. Vournakis and V. Bouriotis, *Proc. Natl. Acad. Sci. USA*, 1993, **90**, 8005.
26. G. Davies and B. Henrissat, *Structure*, 1995, **3**, 853.
27. G.W. Harris, J.A. Jenkins, I. Connerton, N. Cummings, L. LoLeggio, M. Scott, G.P. Hazlewood, J.I. Laurie, H.J. Gilbert and R.W. Pickersgill, *Structure*, 1994, **2**, 1107.
28. G.W. Harris, J.A. Jenkins, I. Connerton and R.W. Pickersgill, *Acta Cryst.*, 1996, **D52**, 393.
29. S.J. Charnock, J.H. Lakey, R. Virden, N. Hughes, M.L. Sinnott, G.P. Hazlewood, R. Pickersgill and H.J. Gilbert, *J. Biol. Chem.*, 1997, **272**, 2942.
30. T.D. Spurway, C. Morland, A. Cooper, I. Sumner, G.P. Hazlewood, A.G. O'Donnell, R.W. Pickersgill and H.J. Gilbert, *J. Biol. Chem.* 1997 **272,** 17523.
31. D. Tull, S.G. Withers, N.R. Gilkes, D.G. Kilburn, R.A.J. Warren and R. Aebersold, *J. Biol. Chem.*, 1991, **266**, 15621.
32. D.N. Bolam, N. Hughes, R. Virden, J.H. Lakey, G.P. Hazlewood, B. Henrissat, K.L. Braithwaite and H.J. Gilbert, *Biochemistry*, 1996, **35**, 16195.

SUBSTRATE DISTORTION IN GLYCOSIDE HYDROLYSIS: THE USE OF MODIFIED OLIGOSACCHARIDES TO STUDY CELLULASE STRUCTURE AND FUNCTION

Gideon J. Davies

Department of Chemistry
University of York
York, Y01 5DD
U.K.

1 INTRODUCTION

Our understanding of the mechanisms of catalysis displayed by various cellulases is derived essentially from two sources, traditionally envisaged as mutually exclusive pursuits. Historically, *X-ray crystallography* has been used as a tool to study protein three-dimensional structure, whilst details on the kinetics, substrate specificity and mechanism have come from the *chemical* disciplines. Recent X-ray structure determinations with specifically designed and synthesised oligosaccharides has provided the first exciting glimpses of unhydrolysed sugars bound in the active-sites of these enzymes. This work clearly illustrates the potential of an integrated inter-disciplinary approach to structure-function studies. In this paper I review some of the recent advances in the field of glycoside hydrolase structure and function, in particular the observations of unhydrolysed oligosaccharides in enzyme active sites.

2 HISTORICAL PERSPECTIVE

It is thirty years since the work of David Phillips and colleagues provided the first glimpse of a distorted saccharide bound to the active site of a retaining glycoside hydrolase, hen egg-white lysozyme (HEWL)[1]. The foundations of the lysozyme mechanism were the distortion of the bound sugar towards the transition-state, coupled with the stabilisation of a long-lived oxocarbenium ion intermediate. Whilst we now realise that the vast majority of retaining glycoside hydrolases actually performs catalysis *via* a double displacement mechanism, in which the catalytic intermediate is covalently linked (and thus is not a long-lived oxocarbenium ion), it is with this pioneering work on HEWL that the role of substrate distortion was first proposed.

The double-displacement mechanism, in which a covalent glycosyl-enzyme intermediate is formed and subsequently hydrolysed *via* oxocarbenium ion transition states, was first clearly outlined by Koshland in 1953 (Scheme 1)[2]. Evidence for such a mechanism comes primarily from secondary deuterium isotope effects, where $k_H/k_D > 1$.[3] Observation of such an isotope effect demands that the reaction intermediate displays tetrahedral (sp^3) geometry and is thus wholly inconsistent with a long-lived oxocarbenium ion intermediate. The covalent intermediate may also be trapped through the use of 2-deoxy-2-fluoroglycosides[4]. The 2-position fluorine substituent inductively destabilises the transition state and under favourable leaving-group conditions, causes the accumulation of the covalent glycosyl-enzyme intermediate. Indeed, the structure of the covalent glycosyl-enzyme intermediate has been revealed by the X-ray crystallographic analyses on a number of systems[5,6]. Thus, the catalytic mechanism of HEWL, whilst often described as the "paradigm" for retaining glycoside hydrolases, is far more paradox than paradigm. If it functions literally as described by Phillips *via* the stabilisation of a long-lived oxocarbenium ion *intermediate*,[7] then it probably possesses a unique mechanism.

It is, however, with the early work on complexes of HEWL that the importance of substrate distortion during catalysis was first developed[1]. Analysis of structure of HEWL with a modified tetrasaccharide lactone derivative, together with an appreciation of the concepts of transition-state theory (*i.e.*, that an enzyme binds the transition state with increased affinity compared to the substrate) revealed the probable role of substrate distortion in glycoside hydrolysis. These ideas were later confirmed by higher resolution analysis of both a NAM-NAG-NAM complex[7] and an oligosaccharide product complex of a partially inactive mutant of HEWL[8]. Unfortunately, these and many other studies, represented product or pseudo-product complexes, and did not contain an intact glycosidic bond linked to a sugar in the leaving-group site. An additional consequence of this is that since these distorted sugars were inevitably terminal, they were often associated with poor electron density and subsequently high crystallographic temperature factors.

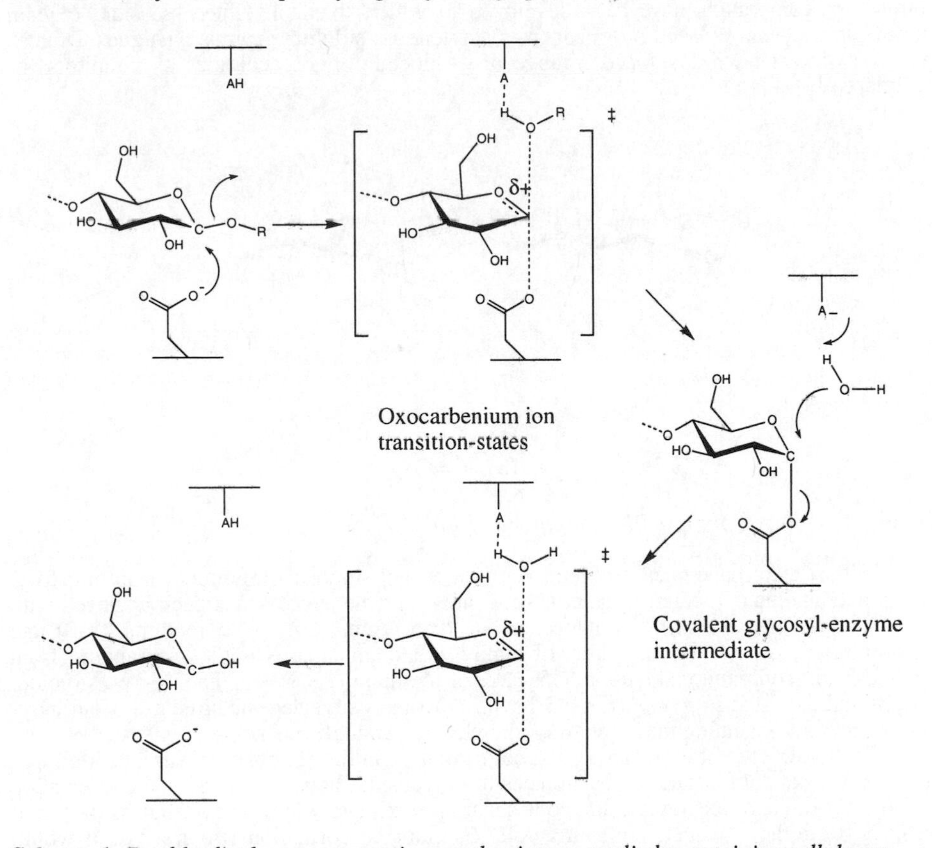

Scheme 1 *Double-displacement reaction mechanism as applied to retaining cellulases*

Recently, there has been an explosion of structures for the many sequence families of glycoside hydrolases[9,10]. Representatives for over 26 of the families have been described at the three-dimensional level. Many of these are cellulases. The pursuit of oligosaccharide complexes for these structures, embracing the disciplines of synthetic chemistry and X-ray crystallography, has revealed new insights into substrate binding and catalysis by these enzymes.

3 STRUCTURAL STUDIES ON CELLULASES AND RELATED SYSTEMS

The fundamental problem to address when attempting to obtain complexes of non-hydrolysed oligosaccharides bound to enzyme active-sites is how to prevent enzymatic hydrolysis long enough for an X-ray diffraction experiment. So far, three classes of study have been successful: the use of chemically non-hydrolysable analogues, the manipulation of crystal pH and temperature (in combination with more slowly hydrolysable substrates) and rapid soaking followed by immediate data collection.

3.1 The use of non-hydrolysable substrate analogues

In order to obtain non-hydrolysable substrate and transition-state analogues, a number of compounds have been developed in which the labile interglycosidic oxygen atom(s) have been replaced by nitrogen, methylene or sulphur groups. Hugues Driguez and co-workers have developed a range of 4-thiocello-oligosaccharides as inhibitors for cellulases[11,12] (Figure 1).

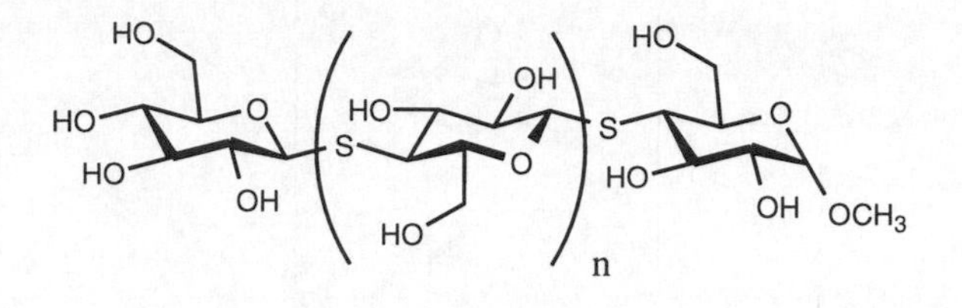

(1) n = 1
(2) n = 2
(3) n = 3

Figure 1 *Non-hydrolysable thio-oligosaccharides*

Whilst there have been problems with these and similar compounds, resulting from weakened binding compared to natural substrates, notable success has been achieved with the *Fusarium oxysporum* endoglucanase I from family 7. Compounds (1-3) are predominantly competitive inhibitors of family 7 endoglucanases with K_i's ranging from 330 μM for compound (1) up to 35 μM for compound (3)[11]. The X-ray structure determination of the *F. oxysporum* EG I with (3) finally revealed the structure of an intact oligosaccharide spanning the active site of a retaining endoglucanase[13].

In this structure, the pyranose ring conformation in the -1 subsite[14] was distorted into a $^1S_3$ skew-boat conformation. This places the glycosidic bond in a quasi-axial orientation and has the effect of placing the aglycone sugar approximately 8 Å away from its predicted position were the -1 subsite ring in a full $^4C_1$ chair conformation (Figure 2). It would appear that substrate distortion is driven, at least in part, by the interactions in the +1 binding site. Sulzenbacher and co-workers have obtained complexes with both the unhydrolysed, distorted thio-oligosaccharide and the corresponding product complex (with cellobiose bound in the -2 / -1 subsites). The cellobiose product complex displays no distortion in the -1 subsite, as one might expect[15].

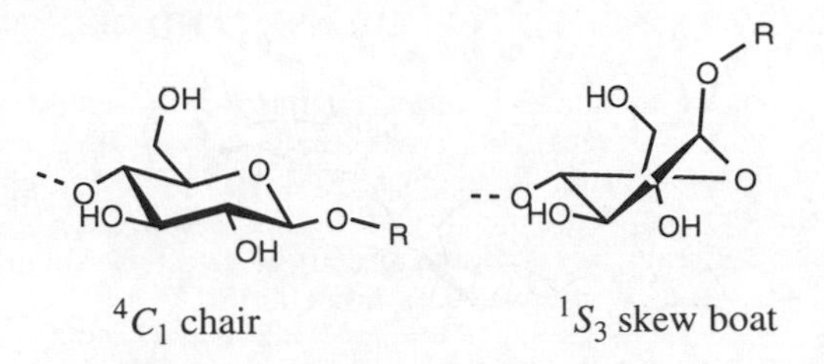

Figure 2 *The $^4C_1$ chair and $^1S_3$ skew-boat conformations. Note the huge effect that a conformational interconversion has on the position of the leaving-group "R"*

3.2 The use of crystal pH and modified disaccharides

Recently, we have trapped a family 5 endoglucanase with an unhydrolysed *O*-linked substrate. In this case, the pH of the protein crystals was lowered to pH 5, a pH at which the alkalophilic *Bacillus agaradherans* cellulase Cel5A is inactive due to protonation of its catalytic nucleophile. The reduced activity at low pH, presumably coupled with the use of a 2-fluoro-2-deoxy-cellobioside derivative (compound 5, Figure 3), prevents hydrolysis of the substrate in the crystal (Davies, Brzozowski and Schülein, unpublished observations). In addition, cryocrystallographic techniques at 120 K were used in order to further reduce the likelihood of enzymatic action. The crystal structure revealed a remarkable distortion in the -1 subsite. The pyranoside ring conformation is again clearly a $^1S_3$ skew-boat, with the axial glycosidic bond and leaving-group orientation described above.

It is relevant that in this study, as in the Sulzenbacher study on the family 7 enzyme[13,15] the analogous product complex (in this case β-cellotriose bound in the -3, -2 and -1 subsites) shows no distortion (Davies, Brzozowski, Dauter, Mackenzie, Withers and Schülein, manuscript in preparation). As with the Sulzenbacher study on the family 7 enzyme, this is indicative of a major role for the +1 subsite interactions in substrate distortion.

3.3 Observations on related β-retaining enzyme systems

1996 Also saw the publication of another unhydrolysed sugar in the active site of a glycoside hydrolase. Tews and co-workers presented the structure of the family 20 chitobiase, in complex with its natural substrate chitobiose[16] (compound 6, Figure 3). For reasons that are unclear, but which may relate to the very short soaking times involved, the natural substrate remains unhydrolysed. What is important is that the -1 subsite ring is distorted to present the axial leaving group with essentially an identical conformation to those seen in both the EG I thio-oligosaccharide study and in the *Bacillus* Cel5A structure (Figure 3). Although this conformation is described by the authors of the chitobiase work as a "4-sofa" with five ring atoms lying in a plane, this may well be an artefact of the refinement. This work is particularly interesting, since although chitobiase is a retaining enzyme, the catalytic nucleophile is donated not by the enzyme, but is instead provided by the C2 *N*-acetamido group of the *N*-acetylglucosamine sugar itself. Indeed, when overlapped with the family 5 and family 7 cellulase complexes, described above, the enzymatic nucleophile in the cellulase structures overlaps perfectly with the substrate-donated nucleophile in the chitobiose structure[13,15].

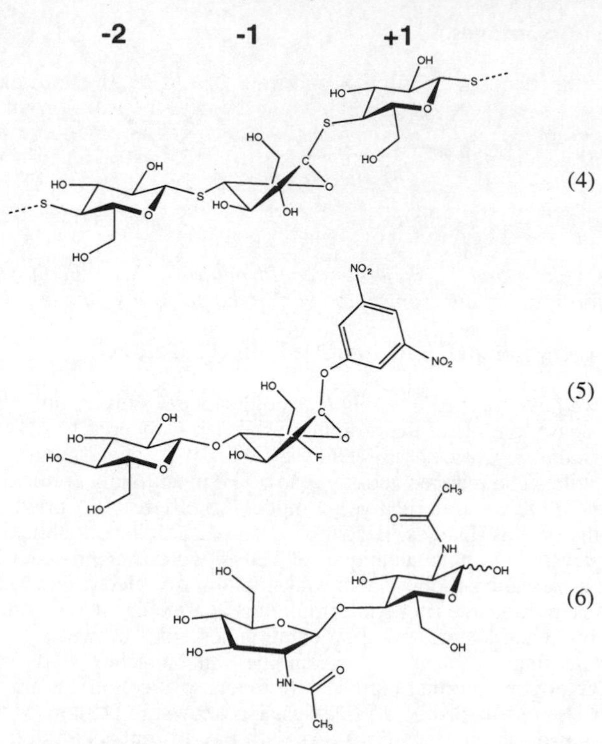

Figure 3 *Distorted sugars in β-retaining glycoside hydrolases*

4 PYRANOSIDE RING DISTORTION

4.1 Similar distortions in unrelated structures

The three independent analyses, described above (3.1 - 3.3), all reveal unhydrolysed saccharides spanning the active-sites of β-retaining enzymes. Notably, they have been trapped on-enzyme by three separate means: non-hydrolysable sulphur-containing substrate analogues[13], utilisation of an acid pH at which the enzyme is inactive due to protonation of its catalytic nucleophile and by short-crystallographic soaking times to kinetically trap an otherwise thermodynamically unfavoured species[16]. Structurally these pyranoside ring contortions cause the position of the adjacent (+1 subsite) sugar to move by over 8 Å compared to a regular β-linkage between two $^4C_1$ ring sugars which has enormous consequences for the correct modelling of oligosaccharides into enzyme active sites. We propose that distortion is primarily driven by the extra favourable interactions in the +1 (aglycone) subsite. The sugar binding sites of these structurally unrelated enzymes all undergo a marked "kink" at the point of cleavage, such that a regular polymer of β-D-sugars in $^4C_1$ conformation would be unable to bind across the point of cleavage. In the absence of a covalent link to the +1 subsite sugar, there would appear to be no driving force for the distortion.

4.2 Catalytic advantages

What are the benefits of such a substrate distortion ? Both the classic double-displacement with an enzymatic nucleophile and the neighbouring-group participation with a substrate nucleophile, require nucleophilic attack at the anomeric carbon. This attack, concomitant with leaving group departure, must take place *via* an in-line approach. The problem facing the enzyme is that nucleophilic attack at an anomeric C1 with an equatorial leaving group (such as is found in all β-D or α-L linked oligosaccharides) is sterically hindered, primarily by the presence of the hydrogen atom bound to C1, but also by the H3 and H5 protons (Figure 4). This concept is well documented in organic chemistry where Bentley and Schleyer, amongst others, have demonstrated a lack of nucleophilic attack on, conformationally restricted, 2-adamantyl substrates (compound 7, Scheme 2) due to an analogous steric hindrance[17], Scheme 2. Thus, in order for the enzyme to make a nucleophilic substitution at C1, the sugar must first be distorted away from its ground-state $^4C_1$ chair conformation in which the leaving group is equatorial, towards a boat or skew-boat form in which the glycosidic-bond and leaving group become axial. This is exactly what is observed in the EG I, Cel5A and chitobiase complexes.

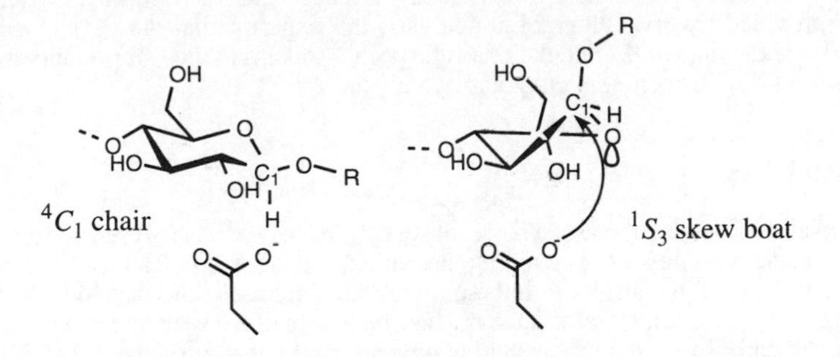

Figure 4 *Presentation of the anomeric carbon for in-line attack in the $^1S_3$ skew-boat conformation*

Although this situation arises for nucleophilic substitution at a β-linked sugar with an equatorial leaving group, an analogous situation can be envisaged for the deglycosylation of the covalent enzyme intermediate formed by enzymes hydrolysing α-glycosidic bonds with net retention of configuration. In these cases, the glycosyl-enzyme intermediate is equatorially β-linked and direct attack by the hydrolysing water molecule may be similarly hindered. This leads one to speculate that such a β-linked covalent enzyme intermediate may exist in a conformation other than the "preferred" $^4C_1$ chair conformation.

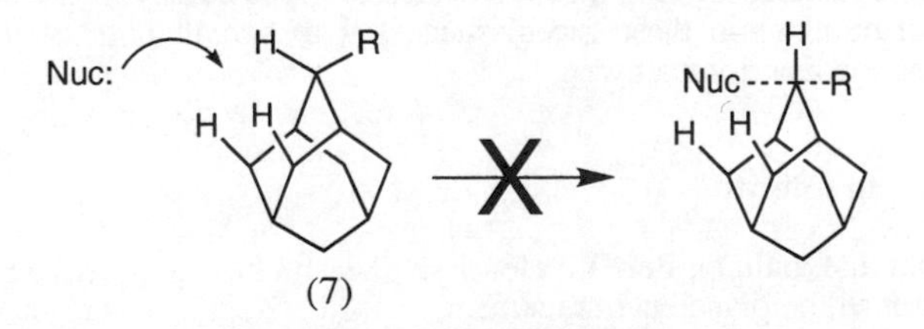

Scheme 2 *Conformationally restricted adamantane systems display steric hindrance to in-line nucleophilic attack*

Catalytically, such a substrate distortion provides enormous advantages for the enzyme. Primarily this comes from the appropriate presentation of the anomeric carbon for nucleophilic attack which is otherwise sterically hindered by C1-H, as discussed above.

But additionally, the axial glycosidic-bond is now closer to the conformation required in the oxocarbenium ion transition-state, consistent with the dictates of transition-state theory. This is reflected in the fact that the transition-state $^4H_3$ half-chair conformation lies between the $^4C_1$ chair and the observed $^1S_3$ skew-boat on the pyranoside ring interconversion pathway. In the case of both the EG I thio-oligosaccharide structure and the Cel5A complex, the distortion also presents the glycosidic oxygen in an appropriate position for protonation by the enzymatic Brønsted acid, which was not possible with the substrate in a ground-state $^4C_1$ chair conformation with the glycosidic bond equatorial to the ring plane. A final, but perhaps less significant effect, is that the axial glycosidic bond orientation is consistent with a stereoelectronic role in glycoside catalysis. According to the dictates of the anti-periplanar lone pair hypothesis, glycosidic bond hydrolysis should be enhanced by the "antiperiplanar" overlap of one of the ring oxygen lone-pairs with the anti-bonding orbitals of the glycosidic bond[18,19]. The role of ring distortion in the enzyme catalysed hydrolysis of equatorial glycosides has also recently been demonstrated by a series of *ab initio* molecular orbital calculations on a model system, 2-oxanol[20]. A clear role for ring-distortion is demonstrated in this case, through reduction of the energy required to stretch the glycosidic bond and delay of the transition-state, both of which reduce the energy barrier for bond cleavage. Hence, it can clearly be seen that the empirical observations of distortion provided by crystallographic analysis, the experimental studies on small model compounds and *ab initio* calculations all point towards the importance and wide applicability of ring distortion in enzymatic catalysis.

5 SUMMARY

Whilst these studies represent the first observations of unhydrolysed oligosaccharides spanning the active sites of β-retaining enzymes, others have trapped non-hydrolysed sugars across the active-sites of the family 13 α-amylases and cyclodextrin glycosyl transferases. In particular, great success has been obtained with the acarbose series of inhibitors. Acarbose is a tetrasaccharide whose inhibitory properties benefit from an unsaturated ring, whose $^2H_3$ half-chair conformation mimics the distortion expected at the transition-state. This group is linked *via* a non-hydrolysable *N*-glycosidic linkage to the adjacent sugar. A large number of acarbose (and extended acarbose-derived oligosaccharides) have now been described[21,22] and have allowed dissection of the protein-ligand interactions at the (pseudo) transition-state for these enzymes. Sadly, suitable active-site spanning *transition-state* analogues for β-retaining enzymes are not yet available. It is hoped that the recent structure determinations described here will assist in the design and synthesis of appropriate transition-state mimics for cellulase systems too.

It is 30 years since the ground-breaking work of Sir David Phillips and co-workers provided the first glimpse of substrate distortion in glycoside hydrolysis. Recent collaborations between synthetic chemists and protein crystallographers have revealed the full nature of substrate distortion in the active sites of glycoside hydrolases. It is hoped that future work will build upon these recent successes to reveal more of the protein-ligand interactions along the reaction pathway.

Acknowledgements

The author would like to thank Bernard Henrissat, Martin Schülein, Mike Sinnott and Steve Withers for continued help and encouragement.

References

1. C.C.F. Blake, L.N. Johnson, G.A. Mair, A.C.T. North, D.C. Phillips, and V.R. Sarma, *Proc. Roy. Soc. ser. B*, 1967, **167**, 378.
2. D.E. Koshland, *Biol. Rev.*, 1953, **28**, 416.
3. G. Davies, M.L. Sinnott, and S.G. Withers, in 'Comprehensive Biological Catalysis', ed. M.L. Sinnott, Academic Press, London, 1997, p.119.
4. S.G. Withers, I.P. Street, P. Bird, and D.H. Dolphin, *J. Am. Chem. Soc*, 1987, **109**, 7530.
5. A. White, D. Tull, K. Johns, S.G. Withers, and D.R. Rose, *Nat. Struct. Biol.*, 1996, **3**, 149.
6. W.P. Burmeister, S. Cottaz, H. Driguez, S. Palmieri, and B. Henrissat, *Structure*, 1997, **5**, 663.
7. N.C.J. Strynadka and M.N.G. James, *J. Mol. Biol.*, 1991, **220**, 401.
8. A.T. Hadfield, D.J. Harvey, D.B. Archer, D.A. MacKenzie, D.J. Jeenes, S.E. Radford, G. Lowe, C. M. Dobson, and L. N. Johnson, *J. Mol. Biol*, 1994, **243**, 856.
9. G. Davies and B. Henrissat, *Structure*, 1995, **3**, 853.
10. B. Henrissat and G.J. Davies, *Curr. Op. Struct. Biol.*, 1997, **7**, 637.
11. C. Schou, G. Rasmussen, M. Schülein, B. Henrissat, and H. Driguez, *J. Carbohydr. Chem.*, 1993, **12**, 743.
12. H. Driguez, *in* 'Carbohydrate Bioengineering', ed. S.B. Petersen, B. Svennson, and S. Pedersen, Elsevier, Amsterdam, 1995, p. 113.
13. G. Sulzenbacher, H. Driguez, B. Henrissat, M. Schülein, and G.J. Davies, *Biochemistry*, 1996, **35**, 15280.
14. G.J. Davies, K.S. Wilson, and B. Henrissat, *Biochem. J.*, 1997, **321**, 557.
15. G. Sulzenbacher, M. Schülein, and G.J. Davies, *Biochemistry*, 1997, **36**, 5902.
16. I. Tews, A. Perrakis, A. Oppenheim, Z. Dauter, K.S. Wilson, and C.E. Vorgias, *Nat. Struct. Biol.*, 1996, **3**, 638.
17. T.W. Bentley and P.R. Schleyer, *Adv. Phys Org. Chem.*, 1977, **14**, 1.
18. P. Deslongchamps, 'Stereoelectronic effects in Organic Chemistry', Pergamon Press, 1983.
19. A.J. Kirby, *Acc. Chem. Res.*, 1984, **17**, 305.
20. B.J. Smith, *J. Am. Chem. Soc.*, 1997, **119**, 2699.
21. B. Strokopytov, D. Penninga, H.J. Roseboom, K.H. Kalk, L. Dijkhuizen, and B.W. Dijkstra, *Biochemistry*, 1995, **34**, 2234.
22. A.M. Brzozowski and G.J. Davies, *Biochemistry*, 1997, **36**, 10837.

STRUCTURAL STUDIES OF THE CELLULOSOME

G. A. Tavares, H. Souchon, D. M. Guérin, M.-B.Lascombe and P. M. Alzari*

Unité d'Immunologie Structurale
Institut Pasteur
75724 Paris, France

1 INTRODUCTION

The thermophilic anaerobe *Clostridium thermocellum* produces a high molecular weight multi-enzymatic complex, the cellulosome, which is very active against crystalline forms of cellulose[1-4]. The quaternary organisation of the glycosidases within the complex appears to play an essential role in potentiating their activity against an insoluble substrate[5]. The cellulosome includes, in addition to several hydrolytic components, various non-catalytic protein modules which ensure the integration of the enzymes within the quaternary complex and the attachment of the complex to the bacterial cell membrane.

1.1 The catalytic domains

Several *C. thermocellum* enzymes have been cloned from different bacterial strains and characterised[4,6]. Their catalytic domains belong to at least 12 different glycosidase families according to a classification based on amino acid sequence similarities[7]. The crystal structures of representatives of four of these families have been determined at high resolution in our laboratory[8-11].

Endoglucanases CelA (family 8) and CelD (family 9) are inverting glycosidases that fold into an $(\alpha/\alpha)_6$ barrel formed by six inner and six outer α-helices. Endoglucanase CelD also contains an N-terminal immunoglobulin-like domain tightly packed against one side of the barrel. This β-barrel domain of unknown function is missing in some members of family 9 cellulases and is not required for catalysis. In both family 8 and family 9 cellulases, the active site cleft is a long acidic groove defined by the loops connecting the C-terminal end of the external α-helices with the N-terminal end of the internal helices. Despite their overall structural similarity, however, CelA and CelD hydrolyse the β-1,4 glycosidic linkage by a different reaction mechanism. CelD is able to cleave short cellooligosaccharides (DP3), whereas CelA requires at least five bound D-glucosyl residues for efficient hydrolysis.

Two other *C. thermocellum* enzyme structures, family 5 endoglucanase CelC and family 10 xylanase XynZ, fold into $(\beta/\alpha)_8$ barrel domains similar to that initially found in triose phosphate isomerase (TIM). These enzymes are part of a larger class of retaining TIM-barrel glycosidases which also includes family 1 β-glucosidases, family 2 β-galactosidases and family 17 β-glucanases, among other enzymes[12,13]. Independently of their distinct substrate specificity, all glycosidases in this class share a conserved catalytic machinery with the proton donor Asp residue located at the end of strand 4 and the nucleophilic Glu residue at the end of strand 7. The structure of CelC revealed an additional helical subdomain of 54 residues inserted between the sixth strand and the sixth helix of the$(\beta/\alpha)_8$ barrel. This insertion gives rise to a deep substrate binding site and could explain the conformational changes observed in CelC upon substrate binding[10], which is not observed in other family 5 cellulases lacking the helical subdomain[14].

The structural diversity revealed by these four *C. thermocellum* glycosidases (and others whose structure can be inferred by homology with known related enzymes) demonstrate that cellulolytic bacteria have evolved an efficient mechanism to integrate a variety of enzymes displaying quite distinct molecular architectures within a single stable macromolecular aggregate.

1.2 The scaffolding protein CipA

The general principle of cellulosome assembly appears to be relatively simple. In addition to the catalytic domain, each hydrolytic component of the complex comprises a conserved 'dockerin' domain of about 65 amino acid residues that serves to anchor the enzyme to a non-catalytic scaffolding polypeptide called CipA (for Cellulosome Integrating Protein). The amino acid sequence of CipA includes a series of nine similar modules of 150 residues - the cohesin domains - separated by short peptide linker segments[15]. Each cohesin domain serves as a receptor subunit that specifically binds to the dockerin domains of the hydrolytic components of the cellulosome[16,17].

In addition to the enzyme-binding domains, CipA contains a family III cellulose-binding domain (CBD_{CipA}) that mediates binding of the complex to cellulose fibers and a C-terminal dockerin domain similar to those carried by cellulosomal enzymes. However, the dockerin domain of CipA does not bind to cohesin domains of the same polypeptide[18], but it specifically recognises a different type of cohesin domain found in some membrane-associated proteins. Based on differences in amino acid sequences and binding specificities, two types of cohesin and dockerin domains can thus be distinguished[19]. Type I cohesin domains of the scaffolding protein CipA bind to the cognate dockerin domains borne by the catalytic components, whereas type II cohesin domains (found in the N-terminal regions of at least three different membrane-associated proteins from *C. thermocellum*) recognise the dockerin domain of CipA.

2 COHESIN DOMAINS

Type I cohesin domains have been isolated or cloned from the scaffolding proteins of three bacteria: *C. thermocellum* CipA[15,20,21], *C. cellulovorans* Cbpa[22] and *C. cellulolyticum* CipC[23]. In general, the central cohesin domains from the same scaffolding protein are highly conserved (90% identical residues), whereas the domains located towards the N- and C-terminal regions tend to show a lower similarity. Amino acid sequence conservation is still weaker between cohesin domains from different scaffolding proteins (typically 30 - 50% identity), suggesting that the fine specificity of cohesin-dockerin interactions may differ from one species to another.

We have cloned the seventh cohesin domain of CipA from *C. thermocellum* (CipA_7) and determined its three-dimensional structure at 1.7 Å resolution[21,24]. The structure is closely similar to that of the second cohesin domain from CipA (CipA_2) which has been determined independently at 2.15 Å resolution[25] and shows only minor topological differences.

2.1 Overall fold

The type I cohesin domain folds into a compact, conically-shaped structure composed of 143 amino acid residues. Nine major β-strands are arranged in three distinct sheets forming together a flattened β-barrel with a typical jelly roll topology (Figure 1a). The β-strands are connected by peptide loops of various lengths at both extremities of the molecule, although the connecting loops are systematically longer (10-15 residues) at one end of the barrel.

The protein core of the cohesin domain is highly hydrophobic, including several aromatic and aliphatic side chains from the internal face of the β-sheets and connecting loops. The hydrophobic character of amino acid residues at these positions is highly conserved in all known type I cohesin domains[24], indicating that hydrophobic interactions play a primary role in protein folding and stability.

2.2 Cohesin-cohesin interactions in the crystal

Crystallisation of CipA_7 leads to extensive intermolecular surface contacts between neighbouring molecules. The single largest contact involves the formation of a crystallographic 'dimer' with internal two-fold symmetry which results from the face-to-face packing of the four-stranded β-sheets (Figure 1a). This interaction buries a contact surface of 1040 $Å^2$ and is characterised by extended hydrophobic contacts as well as several direct and water-mediated hydrogen bonding interactions. A majority of the hydrophobic and polar residues forming the 'dimer' interface are conserved in all nine cohesin domains from the scaffolding protein CipA (and to a lesser extent in other type I cohesin domains), suggesting that similar dimers could form between any pair of CipA domains[24,25]. Although ultracentrifugation experiments failed to reveal stable oligomeric species of CipA_7 in solution, cohesin-cohesin interactions similar to those observed in the crystal might also occur within the scaffolding protein CipA where the cohesin domains are covalently linked by short peptide linkers. Indeed, binding of dockerin domains to a pair of covalently-linked cohesin domains was found to be highly cooperative[26], suggesting that domain-domain interactions within the same polypeptide could mask the respective dockerin-binding sites.

2.3 Type II cohesin domains

Three genes, encoding membrane-associated polypeptides containing one to four N-terminal type II cohesin domains each, have been cloned and sequenced[19,27]. The COOH terminus of these proteins consists of triplicated S-layer homology (SLH) domains, which are commonly found in exocellular, cell-bound proteins. This suggests that type II cohesin domains, which specifically bind to the dockerin domain of CipA, could serve to anchor the cellulosome to the bacterial cell surface. The amino acid sequences of type II cohesin domains display 50-60% of invariant amino acid positions. In contrast, the sequence similarity between type I and type II domains is low, typically 15 - 25% amino acid identity. In spite of this low similarity, however, key structural residues are conserved between the two families, strongly suggesting that type II cohesin domains share the same three-dimensional architecture and jellyroll topology observed in type I cohesin domains[24]. In particular, the hydrophobic character of most internal positions in type I cohesin domains is also conserved in type II domains. Since the corresponding dockerin domains borne by the catalytic components (type I) and by CipA (type II) are also closely similar to each other, it is expected that cohesin-dockerin complexes of both types display similar structural features, in spite of different binding specificities and different functional roles.

2.4 Cohesin-dockerin interactions

The crystal structures of the unliganded cohesin domains does not provide much information about the region of the protein involved in binding to complementary dockerin domains. However, the long strand-connecting loops at one end of the β-barrel (Figure 1a) and part of the adjacent four-stranded β-sheet involved in intermolecular crystal contacts appear as a likely candidate for the interacting region. A structural protrusion formed by two of these loops, which is found in all cohesin domains, has a well-defined conformation stabilised by several conserved hydrogen-bonding interactions that is also conserved in the two crystal structures[24,25]. Furthermore, type II domains have a long (15 amino acid residues) insertion in one of these loops with respect to type I domains that could account for the difference in binding specificities. On the other hand, a dockerin binding site involving part of the four-stranded β-sheet would also be consistent with the binding properties of polypeptides containing a pair of covalently-linked cohesin domains. Complexation of this polypeptide (containing two adjacent cohesin domains) to individual dockerin domains occurs with a strict 1:2 stoichiometry, but no complexes involving a single dockerin domain can be observed[26]. This result can be explained if the dockerin-binding site were masked by contacts between the two cohesin domains similar to those observed in the crystal structure.

The dockerin domain is a small segment of 60-70 residues that includes an EF-hand pair[28] similar to those found in a large number of proteins that bind a Ca^{2+} ion[29]. No high affinity Ca^{2+}-binding site was detected in the dockerin domain of endoglucanase CelD[30], nor in the dockerin domain of type II originating from the COOH terminus of CipA[31]. However, the stability of cohesin-dockerin complexes of type I[20] and type II[31] is enhanced in the presence of Ca^{2+}, the concentration dependence curve suggesting that two Ca^{2+} ions are involved in complex formation. Since the crystal structures of the cohesin domain have no obvious Ca^{2+}-binding site (crystals of CipA_7 were grown at high Ca^{2+} concentration[21]), calcium ions presumably bind to the EF-hand motif of the dockerin domain. The interactions with the cohesin domain may thus increase Ca^{2+} affinity by stabilising the structure of the EF-hand motifs and/or by contributing to the coordination of the Ca^{2+} ions. Further evidence supporting this hypothesis is provided by the electrophoretic behaviour of endoglucanase CelD. The enzyme carrying the uncomplexed dockerin domain has lower mobility than the CelD-cohesin complex on native gels in a polyacrylamide gradient[24], indicating that complexation results in a more compact, globular protein.

Taken together, these observations suggest a model for cohesin-dockerin interactions, according to which the EF-hand pair of the dockerin domain could be partially or totally unstructured in solution, and requires complexation with the cohesin partner for high affinity Ca^{2+}-binding and protein stability (Figure 2). If true, such a model would be in

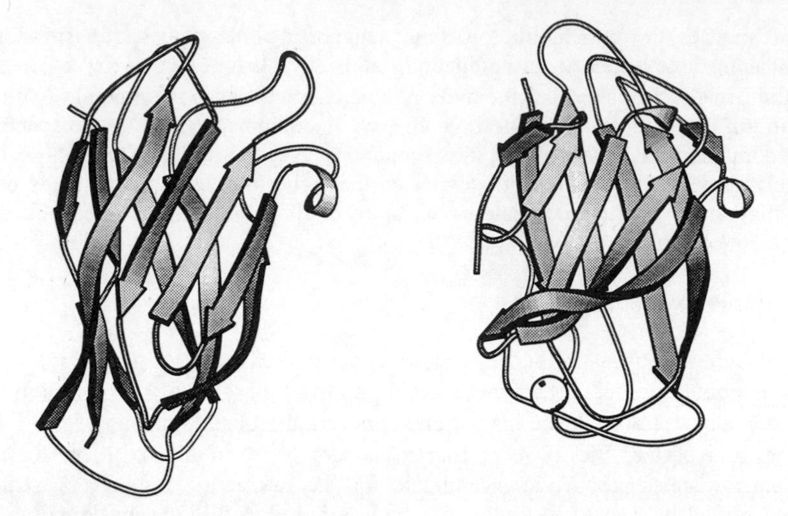

```
AVRIKVDTVNAKPGDTV-RIPVRFSGIPSKGIAN--CDFVYSYDPNVLEII--EIEPG
NLKVEFYNSNPSDTTNSINPQFKVTNTGSSAIDLSKLTLRYYYTVDGQKDQTFWCDHA

ELIVDPN----PTKSFDTAVYPD-------RKMIVFLFAEDSGTGAYAIT-EDGVFAT
AIIGSNGSYNGITSNVKGTFVKMSSSTNNADTYLEISFTG------GTLEPGAHVQIQ

IVAKVKSGAPNGLSVI--------------KFVEVGGFANNDLVEQKTQFFDGGVNVG
GRFAKNDWSNYTQSNDYSFKSASQFVEWDQVTAYLNGVLVWGKEPGRF
```

Figure 1 *The structures of the seventh cohesin domain (left, PDB code 1aoh) and cellulose-binding domain (right, PDB code 1nbc) of CipA are shown in a similar orientation. The structural alignment is shown below. Boxed residues indicate internal positions. Horizontal bars indicate the β-strands.*

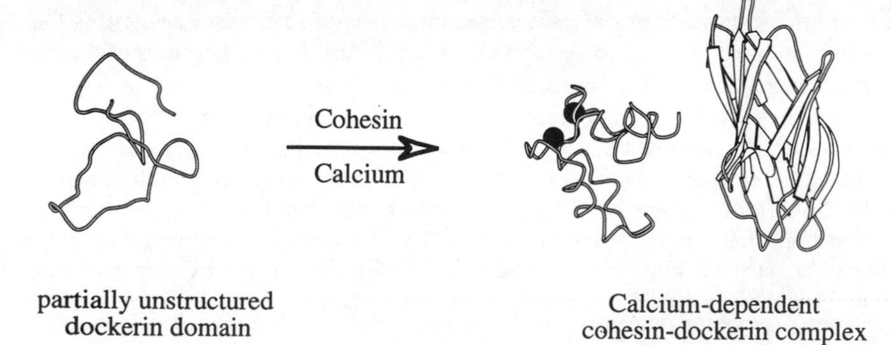

Figure 2 *Hypothetical model of cohesin-dockerin interactions.*

contrast to small cytosolic EF-hand proteins, where the EF-hand pair itself represents a stable unit capable of high affinity Ca^{2+}-binding[32].

3 STRUCTURAL COMPARISON OF COHESIN AND CELLULOSE-BINDING DOMAINS

The architecture and topology of the cohesin β-barrel is similar to that observed for two bacterial cellulose-binding domains (CBD), although no recognisable amino acid sequence similarity can be detected between these two families of proteins. The NMR structure of a family II CBD from *Cellulomonas fimi*[33] and the X-ray structure of the family III CBD from CipA[34] revealed that these bacterial domains are also organised in a nine-stranded β-barrel which display the same jellyroll topology as the cohesin domain. The structures, however, differ in the disposition of the C-terminal β-strand 9 (Figure 1), which belongs to a different β-sheet in each case.

The atomic models of the two CBDs and the cohesin domain are structurally similar. A total of 48 equivalent Cα positions between the family II and family III CBDs can be superimposed with an rms deviation of 1.66 Å[34], a value comparable to that reported[24] for a similar comparison between the cohesin domain and the family II CBD (1.57Å for 51 equivalent Cα positions). When the two subunits of CipA (the family III CBD and the seventh cohesin domain) are compared using an automatic alignment procedure[9], the atomic coordinates of 47 Cα positions can be superimposed with an rms deviation of 1.7 Å. The corresponding sequence alignment deduced from the structural comparison (Figure 1c) shows that only 11 amino acid residues (out of 125 aligned positions) are identical between the two CipA subunits. The major differences occur in the length and conformation of the strand-connecting loops at both ends of the barrel. The cohesin domain has an insertion in one of the loops at the top of the barrel (for the orientation shown in Figure 1), in a region that could be associated with dockerin binding, whereas the family III CBD has longer loops on the other end of the barrel (these loops form a Ca^{2+}-binding site in the CBD which is absent in cohesin domains). In spite of these differences, however, most internal positions involve hydrophobic residues in both CBDs and cohesin domains (boxed positions in Figure 1c), thus raising the possibility of a common (albeit distant) evolutionary origin for these two families of proteins.

4 CONCLUSION

Cellulosome assembly and function require not only a variety of glycosidases with distinct substrate specificities but also various non-catalytic protein modules that serve to integrate the enzymes within the complex (type I cohesin domains), to attach the cellulosome to the cell surface (type II cohesin domains) and to mediate enzyme-substrate interactions (cellulose-binding domains). It is worth to note that, whereas the hydrolytic components of the cellulosome display a striking diversity of molecular architectures, the two types of cohesin domains and the cellulose-binding domains represent functional variations of a single structural motif - the nine-stranded β-barrel - which evolved to acquire distinct ligand-binding specificities.

References

1. C.R. Felix and L.G. Ljungdahl, *Ann. Rev. Microbiol.*, 1993, **47**, 791.
2. J.H.D. Wu, 'Biocatalyst Design for Stability and Specificity', M.E. Himmel and G. Georgiu, (eds), American Chemical Society, Washington DC, 1993, Vol. 516, p, 251.
3. E.A. Bayer, E. Morag and R. Lamed, *Trends in Biotechnol.*, 1994, **12**, 379.
4. P. Béguin and M. Lemaire, *Crit. Rev. Biochem. Mol. Biol.*, 1996, 201.
5. R. Lamed, E. Setter, R. Kenig and E.A. Bayer, *Biotechnol. Bioeng. Symp.*, 1983, **13**, 163.
6. G.P. Hazlewood, M.P.M. Romaniec, K. Davidson, O. Grépinet, P. Béguin, J. Millet, O. Raynaud and J.-P. Aubert, *FEMS Microbiol. Lett.*, 1988, **51**, 231.
7. B. Henrissat and A. Bairoch, *Biochem. J.*, 1993,**293**, 781.
8. M. Juy, A.G. Amit, P.M. Alzari, R.J. Poljak, M. Claeyssens, P. Béguin and J.-P. Aubert, *Nature,* 1992, **357**, 89.
9. R. Dominguez, H. Souchon, S. Spinelli, Z. Dauter, K.S. Wilson, S. Chauvaux, P. Béguin and P.M. Alzari, *Nature Struct. Biol.*, 1995, **2**, 569.
10. R. Dominguez, H. Souchon, M.-B. Lascombe and P.M. Alzari, *J. Mol. Biol.*, 1996, **257**, 1042.
11. P.M. Alzari, H. Souchon and R. Dominguez, *Structure*, 1996, **4**, 265.
12. J. Jenkins, L. Lo Leggio, G. Harris and R. Pickersgill, *FEBS Lett.*, 1995, **362**, 281.
13. B. Henrissat, I. Callebaut, S. Fabrega, P. Lehn, J.-P. Mornon and G. Davies, *Proc. Natl. Acad. Sci. USA*, 1995, **92**, 7090.
14. V. Ducros, M. Czjzek, A. Belaich, C. Gaudin, H.-P. Fierobe, J.-P. Belaich, G. Davies and R. Haser, *Structure*, 1995, **3**, 939.
15. U.T. Gerngross, M.P.M. Romaniec, N.S. Huskisson and A.L. Demain, *Mol. Microbiol.*, 1993, **8**, 325.
16. T. Fujino, P. Béguin and J.-P. Aubert, *FEMS Microbiol. Lett.*, 1992, **94**, 165.
17. S. Salamitou, O. Raynaud, M. Lemaire, M. Coughlan, P. Béguin and J.-P. Aubert, *J. Bacteriol.*, 1994, **176**, 2822.
18. B. Lytle, C. Myers, K. Kruus and J.H.D. Wu, *J. Bacteriol.*, 1996, **178**, 1200.
19. E. Leibovitz and P. Béguin, *J. Bacteriol.*, 1996, **178**, 3077.
20. S. Yaron, E. Morag, E.A. Bayer, R. Lamed and Y. Shoham, *FEBS Lett.*, 1995, **360**, 121.
21. P. Béguin, O. Raynaud, M. Chaveroche, A. Dridi, A. and P.M. Alzari, *Prot. Sci.*, 1996, **5**, 1192.
22. O. Shoseyov, M. Takagi, M.A. Goldstein and R.H. Doi, *Proc. Natl. Acad. Sci. U.S.A.*, 1992, **89**, 3483.
23. S. Pagès, A. Belaich, C. Tardif, C. Reverbel-Leroy, C. Gaudinand J.-P. Belaich, *J. Bacteriol.*, 1996, **178**, 2279.
24. G.A. Tavares, P. Béguin and P.M. Alzari, *J. Mol. Biol.*, 1997, **273**, 701.
25. L.J.W. Shimon, E.A. Bayer, E. Morag, R. Lamed, S. Yaron, Y. Shoham and F. Frolow, *Structure*, 1997, **5**, 381.
26. I. Kataeva, G. Guglielmi and P. Béguin, *Biochem. J.,* 1997, **326**, 617.
27. T. Fujino, P. Béguin and J.-P. Aubert, *J. Bacteriol.*, 1993, **175**, 1891.
28. S. Chauvaux, P. Béguin, J.-P. Aubert, K.M. Bhat, L.A. Gow, T.M. Wood and A. Bairoch, *Biochem. J.*, 1990, **265**, 261.
29. H. Kawasaki and R.H. Kretsinger, *Protein profile*, 1994, **1**, 343.
30. K. Tokatlidis, P. Dhurjati and P. Béguin, *Protein Eng.*, 1993, **6**, 947.

31. E. Leibovitz, H. Ohayon, P. Gounon and P. Béguin, *J. Bacteriol.,* 1997, **179**, 2519.
32. N.C.J. Strynadka and M.N.G. James, *Ann. Rev. Biochem.*, 1989, **58**, 951.
33. G.-Y. Xu, E. Ong, N.R. Gilkes, D.G. Kilburn, D.R. Muhandiram, M. Harris-Brandts, J.P. Carver, L.E. Kay and T.S. Harvey, *Biochem.*, 1995, **34**, 6993.
34. J. Tormo, R. Lamed, A.J. Chirino, E. Morag, E.A. Bayer, Y. Shoham and Steitz, *EMBO J.*, 1996, **15**, 5739.
35. P.J. Kraulis, *J. Appl. Crystallogr.*, 1991, **24**, 946.

βα-BARREL GLYCOSIDASES AND β-HELIX PECTINASES

Richard Pickersgill, Gillian Harris, Leila Lo Leggio[1], Olga Mayans[2] and John Jenkins

Institute of Food Research, Reading Laboratory, Earley Gate, Whiteknights Road, Reading RG6 6BZ, UK

Current addresses: [1]Centre for Crystallographic Studies, Universitetsparken 5, DK-2100, Copenhagen, Denmark. [2]EMBL-Hamburg, Notkestrasse 85, 2000 Hamburg 52, Germany.

1 XYLANASES AND βα-BARREL GLYCOSIDASES

1.1 Xylanase biochemistry and structure

Xylan, the second most abundant plant cell wall polysaccharide, comprises a backbone of β1,4-linked xylose units which are substituted with acetyl groups and various sugars. The xylan backbone is hydrolysed by xylanases which can be classified as family 10 or family 11 glycosyl hydrolases on the basis of sequence similarities[1]. The previous nomenclature was family F and family G xylanases[2]. Hydrolysis of the glycosidic linkage by both family 10 and 11 xylanases occurs with retention of the anomeric configuration[3] and both families have catalytic glutamates at their active sites[4-5]. However, despite sharing a common mechanism these enzymes have clearly evolved independently since the catalytic core of the family 10 xylanases comprises an 8-fold βα-barrel[5-8] (Figure 1) while the family 11 xylanases are smaller and of predominantly β-architecture[4]. It is probably simple coincidence that both families of xylanases employ a retaining mechanism.

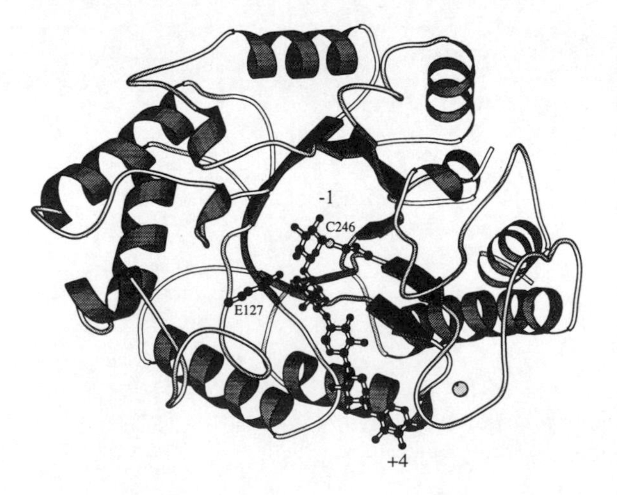

Figure 1 *The βα-barrel of the catalytic domain of* Pseudomonas fluorescens *xylanase A, showing the position of the active site glutamates at the ends of β4 (acid-base) and β7 (nucleophile, in the inactive enzyme shown here a cysteine). Xylopentaose is shown in ball and stick and calcium is represented by the small sphere. This is the architecture of the family 10 glycosyl-hydrolases (family F xylanase).*

1.2 Substrate binding

Xylopentaose was soaked into crystals of an inactive xylanase A in which the active site glutamate was replaced by cysteine and difference Fourier analysis was used to indicate the position of the five xylose residues[5]. The electron density corresponding to the substrate was at the carboxyl-end of the β-barrel across β-strands five and six, with the substrate occupying a cleft formed by the long loops that succeed β-strands four and seven. Ihese loops carry the two catalytic residues: glutamate 127 and glutamate 246 which is cysteine in this inactive enzyme. The quality of the electron density was improved by averaging the two non-crystallographic symmetry related molecules in the asymmetric unit[9]. The initial model of xylopentaose was constructed using the X-ray structure of xylobiose-hexaacetate as a building block, but with ϕ and ψ angles 175° and 135° consistent with the fibre diffraction models of xylan[10]. The final model after crystallographic refinement, shown in Figure 2, has a very similar conformation to the three-fold helix observed for xylan in fibre diffraction studies.

While the structural information alone does not completely resolve the ambiguity between the two possible orientations of the substrate, the one with the non-reducing end of the substrate closest to the active site fits the xylopentaose model best and is also the consistent with the identification of the Glu 127 as the acid-base and Glu 246 as the nucleophile. Subsites +4 to -1 are bound by the substrate. A sixth subsite, subsite -2, is indicated by the presence of three residues proximal to subsite -1, Glu 43, Asn 44 and Lys 47, which are absolutely conserved in the family 10 xylanases (Table 1). A crystal contact prevents the substrate accessing this subsite but the role of subsite -2 has been shown to include ensuring the productive binding of xylo-oligosaccharides[11]. The results of site specific mutagenesis studies are described in more detail by Prof. Harry Gilbert and colleagues elsewhere in these proceedings.

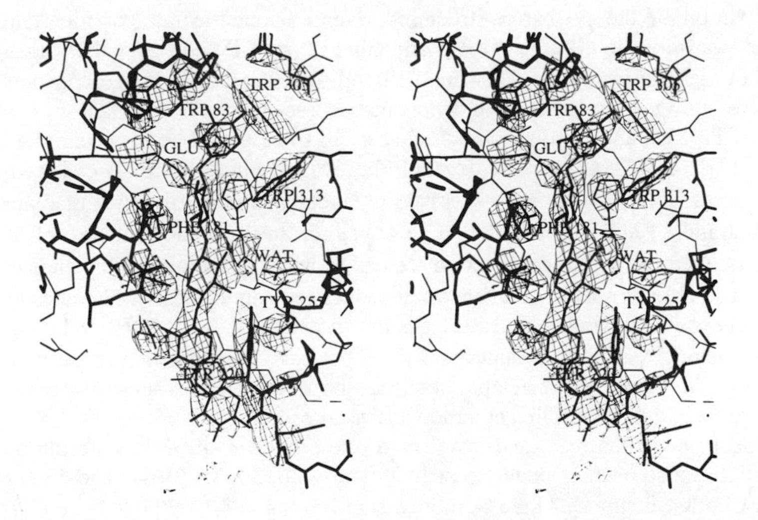

Figure 2 *Stereo-drawing showing xylopentaose binding to xylanase A. Electron density at 1.2 σ, resolution 3.2 Å, averaged map (two copies).*

Table 1 *Residues forming the substrate binding subsites in xylanase A. Absolutely conserved residues are in bold. Semi-conserved in italics. A dash represents a hydrogen bond.*

Subsite	*Residue*
+4	Tyr 220, Asn 253
+3	*Glu 185*, Tyr 255
+2	Arg 250, **Asn 182**-*Asp 131*
+1	Tyr 255, *Phe 181*, Trp 313, **His 215**
-1	**Glu 246, Glu 127, Trp 305, His 79, Lys 47, Trp 313, Trp 83, Gln 213-His 215**
-2	**Glu 43, Asn 44, Lys 47**

1.3 The 4/7-superfamily of βα-barrel glycosidases (clan GH-A)

Catalysis by many glycosylhydrolases involves two essential carboxylates, one acting as the nucleophile and the other as the proton donor[12]. Cleavage of the glycosidic bond results in either retention or inversion at the anomeric carbon. The typical separation of the acid-base and nucleophile is about 5Å in retaining enzymes, but in inverting enzymes a water molecule is accommodated between the nucleophile and the substrate and the separation of nucleophile and acid-base is around 10Å. Henrissat[1] has compared the available sequences of glycosylhydrolases (E.C. 3.2.1.x) using hydrophobic cluster analysis and classified these enzymes into families (currently 61 see web page http://expasy.hcuge.ch/cgi-bin/lists?glycosid.txt for updates). Each family comprises enzymes that can be seen to have evolved from a common ancestor on the basis of sequence analysis. The structures of two barley β-glucanases GHS and GHR (family 17)[13] looked similar in architecture and active site geometry to xylanase A[14]. Superimposition of the βα-barrel structures revealed that the nucleophile (Glu246) of xylanase A was close to Glu231 of GHS and Glu232 of GHR and gave the identity of the acid-base in the β-glucanases (Glu94 in GHS and Glu 93 in GHR). Among the residues which were shown to interact with xylopentaose (see above) in xylanase A, Gln213, Trp305 and Trp313, and form subsite -1 adjacent to the cleaved bond are similar residues in GHS and GHR (Asn166/Asn168, Phe274/Phe275 and Phe291/Trp291). The conserved βα-barrel architecture and conserved retaining mechanism involving glutamates at the ends of β-strands 4 and 7 led to the name 4/7-superfamily. The structure of *E. coli* β-galactosidase (family 2) though distorted seemed to have glutamates at the ends of β4 and β7 and the possible remnants of an ancestral sequence around these glutamates (Table 2). Further sequence analysis led to the suggestion that family 1, 2, 5, 10 and 17 glycosylhydrolases belong to the 4/7 superfamily of βα-barrels. The superfamily therefore includes enzymes with no detectable sequence homology and various activities which have nevertheless evolved from a common ancestral protein.

Subsequent structures have confirmed the validity of this superfamily. The superfamily has also been extended to include family 26, 30, 35, 39, 42 and 53 enzymes[15]. A number of other βα-barrels may be more distantly related[14].

Table 2 *Equivalent regions of sequences for families forming the 4/7 superfamily*

Family	*Activities*	*Acid-base*	*Nucleophile*
1	BGAL, BGLU, PBGLU, LPH	WLTFNEP	YITENGA
2	BGAL, BGLR	IWSLGNES	ILCEYAH
5	EG,EX	IYEIANEP	FVTEWGT
10	XYN, EG, CBH	WDVVNEA	KITELDV
17	LAM, GH, XLAM,LIC	YIAAGNEV	VVSESGW

2 PECTINASES AND PARALLEL β-HELIX PROTEINS

2.1 Introduction to pectins and pectin modifying enzymes

Pectins are complex polysaccharides located in the middle lamella and primary cell wall of plants. The linear α–1,4 linked D-galacturonate regions (homogalacturonan) are interspersed with highly branched regions containing alternating α–1,2 linked rhamnose and D-galacturonate (rhamno-galacturonate). The reader is referred to the contribution of Dr. J. Visser (elsewhere in these proceedings) for a more extensive description of this most complex carbohydrate and for some additional information concerning the enzymes. However it is important for the discussion that follows for the reader to understand the difference between high and low methylation pectins and these are illustrated in Figure 3. Also illustrated in Figure 3 are the bonds cleaved by pectin lyase, pectate lyase and polygalacturonase. Rhamnogalacturonase, which cleaves the rhamno-galacturonate backbone and polygalacturonase belong to the same family of inverting glycosidases, family 28[16].

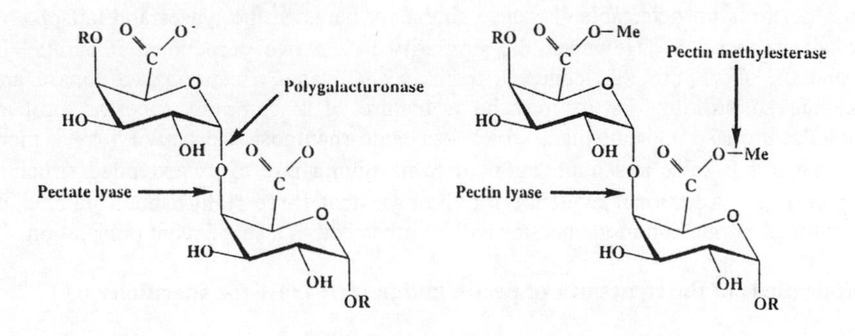

Figure 3 *Chemical structure of pectin and the site of action of the pectin modifying enzymes.*

2.2 Introducing the right-handed parallel β-helix proteins

The structures of pectate lyase PelC from *Erwinia*[17] and the alkaline protease from *Pseudomonas aeruginosa*[18] overthrew the general belief that purely parallel β-structures would be unstable. Within a year three further parallel β-helix structures were published, pectate lyases from *Bacillus subtilis*[19] and PelE from *Erwinia chysanthemi*[20] and a fragment of the P22 tailspike protein[21], as well the alkaline protease from *Seratia marcescens*[22]. These structures were reviewed[23-25] establishing the nomenclature for parallel β-helix for proteins similar to the pectate lyases and for the β-roll homologues and analogues of the alkaline proteases. The right-handed parallel β-helix consists of three β-sheets PB1, PB2 and PB3, sheets PB1 and PB2 form an antiparallel sandwich and PB3 is approximately perpendicular to PB2, connected by the T2 turn as discussed later (Figure 4). Two other turns T1 and T3 connect PB1 to PB2 and PB3 to PB1 of the next turn as the polypeptide chain coils round and round to form the overall architecture of the right-handed parallel β-helix (Figure 4). Subsequently, three left-handed parallel β-helix proteins have been solved[26-28]. These structures are distinct from the right-handed parallel β-helix proteins not only in hand but also in overall shape, viewed down the axis of the helix they are rather triangular in comparison to the 'L' shape of the right-handed parallel β-helices. Three new right-handed parallel β-helix structures have also been recently reported, P69 pertactin *from Bordetella pertussis*[29], rhamnogalacturonase A from *Aspergillus aculeatus*[30], and pectin lyase A from *Aspergillus niger*[31]. At the same time, the tailspike structure has been refined to higher resolution, characterised as an endorhamnosidase, and the binding of substrates was analysed crystallographically[32].

2.3 A superfamily of right-handed parallel β-helices

The most striking characteristics of the pectate lyase fold are the stacking of side chains and the residues in α_L-conformation that form the sharp turn from PB2 to PB3. *Aspergillus niger* pectin lyase A has 15% sequence identity after structural alignment with *Bacillus subtilis* pectate lyase; therefore they have clearly evolved from a common ancestor. There is no detectable sequence similarity between the lyases and tailspike or rhamnogalacturonase A. However, the similarity in the two perpendicular parallel β-sheets and the short turn that connects them in the lyases, rhamnogalacturonase and tailspike are so striking that it may be a feature of a common ancestral protein. Rhamnogalacturonase and tailspike, which has endo-rhamnosidase activity, have more turns of parallel β-helix and duplication of turns giving rise to an extended structure seems plausible. Additional evidence for the case that these right-handed parallel β-helices evolved from a common ancestor will be presented in a subsequent publication.

2.4 Implications of the structures of pectin and pectate lyase for specificity

Pectate lyases cleaves low-methoxy pectins through a calcium dependent β-elimination reaction. Pectin lyase cleaves the high-methoxy pectins by β-elimination but calcium is not essential for activity. The calcium binding site of *Bacillus subtilis* pectate lyase has been observed directly and the surrounding cleft identified as the substrate binding cleft. This substrate binding cleft is polar in nature, consistent with the binding of the highly charged low-methoxy pectins or polygalacturonase (Figure 5).

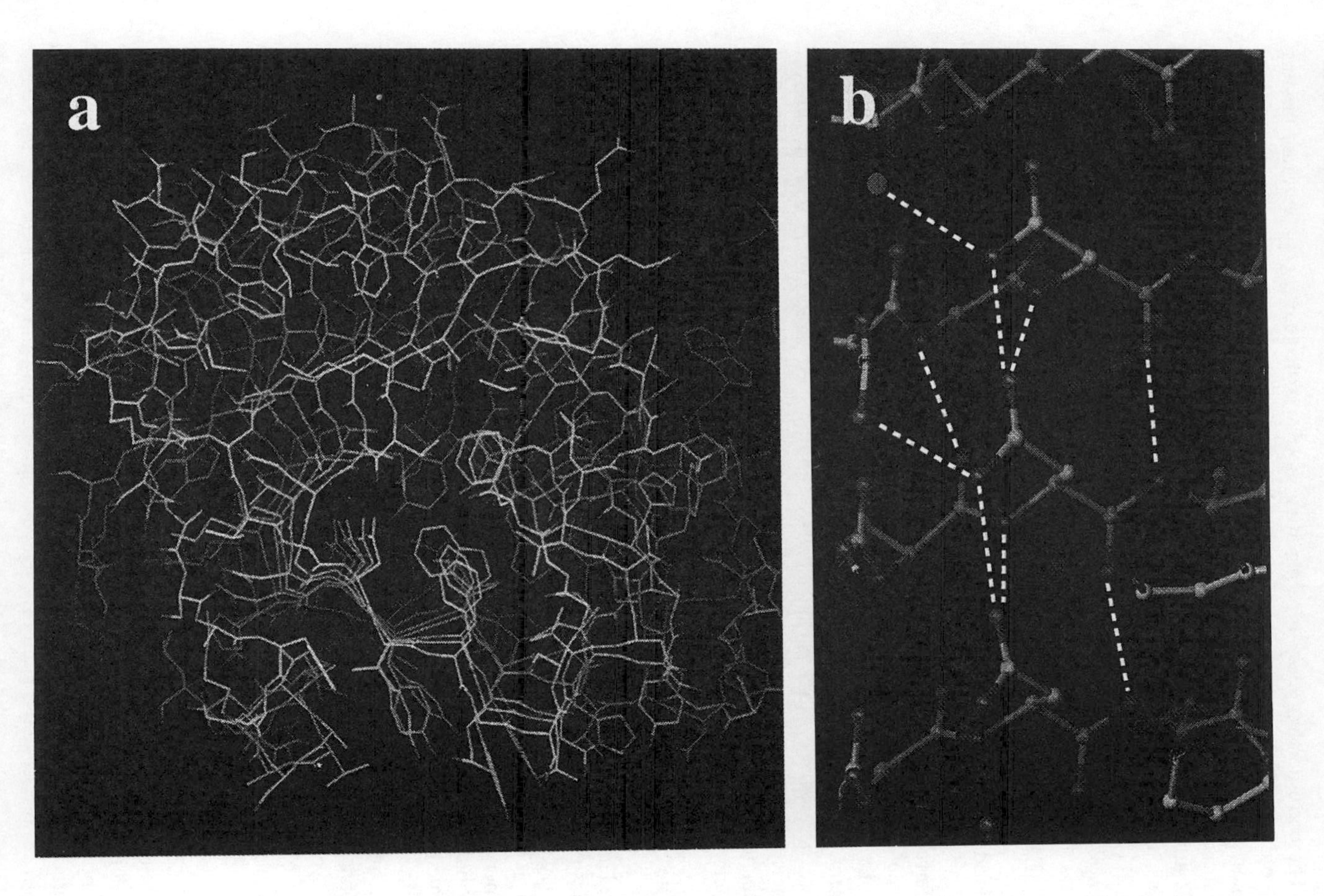

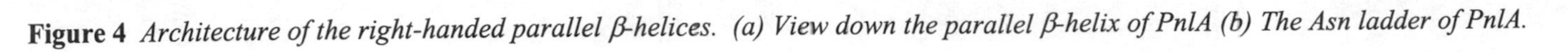

Figure 4 *Architecture of the right-handed parallel β-helices. (a) View down the parallel β-helix of PnlA (b) The Asn ladder of PnlA.*

The corresponding cleft in pectin lyase is also expected to constitute the binding site. Note that this is not the region identified as the active site from considerations of the sequence alone. The cleft of pectin lyase A is highly divergent from that of the pectate lyase structures available; this contrasts markedly with the notable structural similarity of the T2 turn region. Pectin lyase A exhibits a binding cleft dominated by aromatic residues, the charged amino acids being those for which a catalytic role is expected: D154, R176 and D236 (Figure 5). The aromatic residues are four tryptophans: 66, 81, 151 and 212; and three tyrosines 85, 211 and 215. They do not form stacks but instead the pairs W81-W151, W66-W212 and Y211-W212 are arranged in a edge-to-face fashion in which the positively polarised hydrogen atoms of one ring interact with the electron cloud of the second aromatic ring. The residues involved in this interaction pairs are conserved in the other pectin lyases produced by *Aspergillus niger*, while tyrosines Y85 and Y215, which do not form similar interactions, are not conserved. Therefore, it could be inferred that the three aromatic pairs described above contribute to maintain the architecture of the binding cleft. Fluorescence quenching on substrate binding by the highly homologous pectin lyase D[33] confirms that aromatics are involved in binding in solution. The structures allow identification of the aromatics involved (Figure 5). Further, the aromatic nature of the substrate binding cleft is consistent with the known specificity of pectin lyase A for the more hydrophobic high-methoxy pectins. The knowledge is beginning to fall into place to allow the rational manipulation of lyase specificity.

2.5 The lyase active site

Identification of the catalytic residues in the substrate binding clefts of the pectate lyases is no easy task since many residues in the substrate binding could be involved. The pectin lyase structure presents a simplified situation with only three reactive groups, D154, R176 and R236, within the substrate binding cleft. Although the catalytic machinery of these lyases have obviously diverged with calcium being required by pectate lyases but not by pectin lyases, it is expected that the key catalytic residues remain conserved and common basis for the mechanism are shared.

Prior to the structure determination of pectin lyase A, a reliable alignment of pectate and pectin lyases sequences at the binding cleft was not possible, given that this is formed by loops that constitute the most divergent part of the molecule. The structure of pectin lyase A confirms that D184 and R279 (BsPel numbering) are the only residues of the binding cleft absolutely conserved throughout the family. Three aspartates constitute the calcium binding site of BsPel: D184, D223 and D227. D184 and D227 are absolutely conserved among the pectate lyases of this family, while D223 is conservatively substituted for a glutamate. The structure of pectin lyases A reveals that D184 is in fact the only conserved carboxylate (D154 in PnlA) while D223 becomes R176 in PnlA and D227 is equivalent to V180 (Figure 5). Site directed mutagenesis of BsPel shows that both calcium binding and position as determined by alanine substitutions and arginine 279 are important for catalysis. Determination of the structure of an enzyme-substrate (analogue) complex will provide fresh insight.

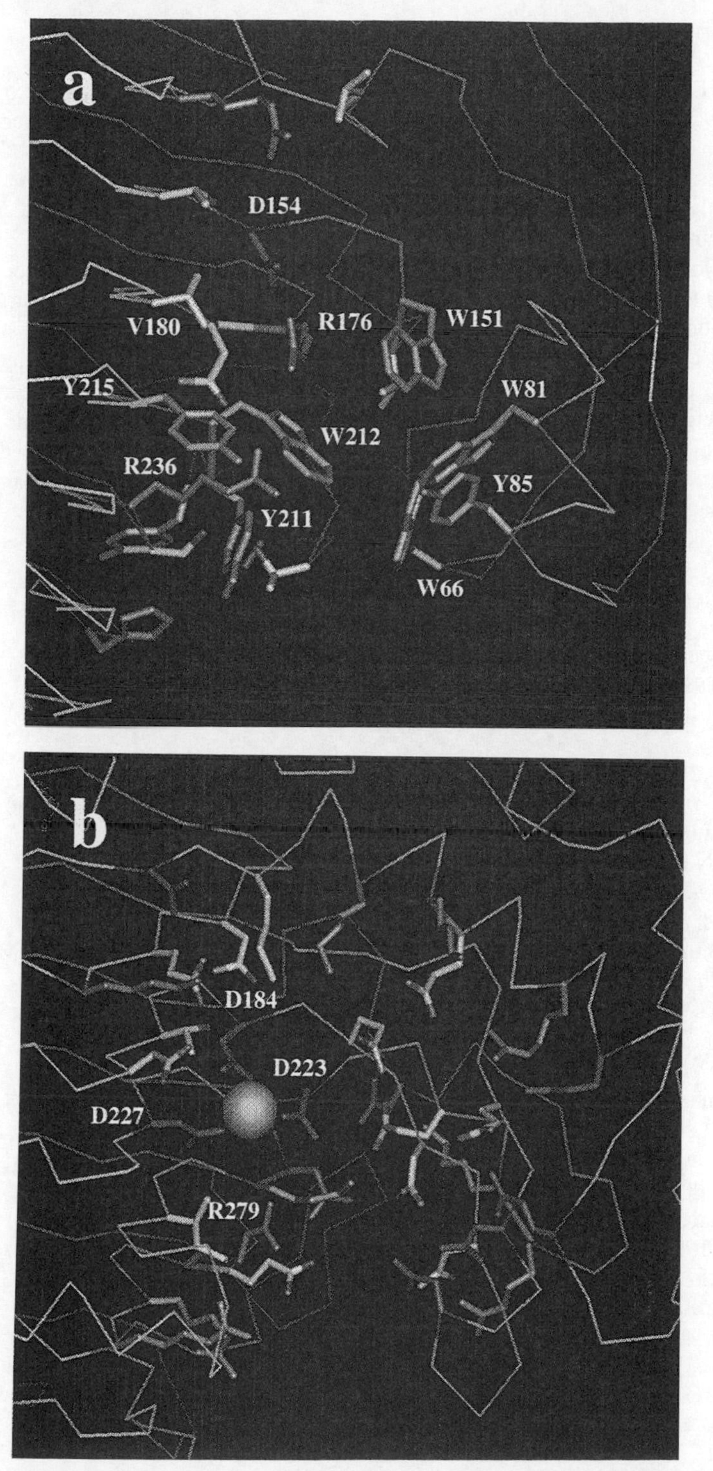

Figure 5 *Substrate binding clefts of (a) pectin lyase and (b) pectate lyase.*

2. 6 A pH-driven conformational change in pectin lyase

The pH optimum of catalysis arises as a combination of the dependency of the rate limiting step and the stability of the protein[33]. For the pectin lyases studied, a slow conformational change has been observed when increasing the pH of the media above neutral values. In all cases, it associates with a reversible decrease in activity.

Pectin lyase A from strain N400 was crystallised at pH 6.5, at which the enzyme is optimally active. The structure at this pH contains two aspartates inside the parallel β-helix, D186 in the fourth T1 loop and D221 in the consecutive turn. The distance between the OD2 atoms is 2.7Å. A hydrogen bond must be formed between these aspartates, one being deprotonated. The PnlA 4M-147 structure solved at pH 8.5 shows D221 still inside the core but D186 has sprung out, and is now exposed to the solvent. It appears that residue D221 is charged, permanently inside the β-helix, while the conformational change observed at high pH can be attributed to the deprotonation of residue D186. This provides a rational explanation for the puzzling biochemistry, explaining that what happens at high pH is the substrate binding cleft is affected by the conformational change accompanying the deprotonation of D186.

Acknowledgements

This work was funded by the BBSRC of the UK, MAFF and EU AIR2 projects CT94-1345 and CT92-1272.

References

1. Henrissat, B. (1991) *Biochem J.*, **280**, 309.
2. Gilkes, N.R., Henrissat, B., Kilburn, D.G., Miller, Jr., R.C. and Warren, A.J. (1991) *Microbiol. Rev.,* **55**, 303.
3. Gebler, J. *et al.* and Withers, S.G. (1992) *J. Biol. Chem.,* **267**, 12559.
4. Wakarchuk, W.W., Campbell, R.L., Sung, W.L., Davoodi, J. and Yaguchi, M. (1994) *Protein Sci.*, **3**, 467.
5. Harris, G.W., Jenkins, J.A., Connerton, I., Cummings, N., Lo Leggio, L., Scott, M., Hazlewood, G.P., Laurie, J.I., Gilbert, H.J. and Pickersgill, R.W. (1994) *Structure,* **2**, 1107.
6. Derewenda , U. *et al.* and Dewewenda, Z.S. (1994) *J. Biol. Chem.*, **269**, 20811.
7. White, A., Withers, S.G., Gilkes, N.R. and Rose, D.R. (1994) *Biochem.*, **33**, 12546.
8. Dominguez, R., Souchon, H., Spinelli, S., Dauter, Z., Wilson, K., Chauvaux, S., Beguin, P. and Alzari, P.M. (1995) *Nat. Struc. Biol.*, **2**, 569.
9. Lo Leggio, L. (1997) PhD Thesis. University of London. UK.
10. Atkins (1992) 'Xylans and Xylanases', Ed. J. Visser, Elsevier.
11. Charnock, S., Lakey, J., Virden, R., Hughes, N., Sinnott, M., Hazlewood, G.P., Pickersgill, R. and Gilbert, H.J. (1997) *J. Biol. Chem.,* **272**, 2942.
12. Sinnott, M.L. (1990) *Chem. Rev.,* **90**, 1171.
13. Varghese, J.N., Garrett, T.P.J., Colman, P.M., Chen, L., Høj, P.B. and Fincher, G.B. (1994) *Proc. Natl. Acad. Sci.,* **91**, 2785.
14. Jenkins, J., Lo Leggio, L., Harris, G. and Pickersgill, R. (1995) *FEBS Letters,* **362**, 281.

15. Henrissat, B., Callebaut, I., Fabrega, S., Lehn, P., Mornon, J.P. and Davies, G. (1995) *Proc. Natl. Acad. Sci.,* **92**, 7090.
16. Biely, P., Benen, J., Heinrichova, K., Kester, H.C.M. and Visser, J. (1996) *FEBS Letters*, **382**, 249.
17. Yoder, M.D., Keen, N.T. and Jurnak, F. (1993) *Science*, **260**, 1503.
18. Baumann, U., Wu, S., Flaherty, K.M. and McKay, D.B. (1993) *EMBO J.*, **12**, 3357.
19. Pickersgill, R., Jenkins, J., Harris, G., Nasser, W. and Robert-Baudouy, J. (1994) *Nat. Struct. Biol.*, **1**,717.
20. Lietzke, S.E., Yoder, M.D., Keen, N.T. and Jurnak, F. (1994) *Plant Physiol.*, **106**, 849.
21. Steinbacher, S., Seckler, R., Miller, S., Steipe, B., Huber, R. and Reinemer. (1994) *Science*, **265**, 383.
22. Baumann, U. (1994) *J. Mol. Biol.,* **242**, 244.
23. Yoder, M.D., Lietzke, S.E. and Jurnak, F. (1993) *Structure*, **1**, 241.
24. Jurnak, F., Yoder, M.D., Pickersgill, R. and Jenkins, J. (1994) *Curr. Opin. Struct. Biol.*, **4**, 802.
25. Yoder, M.D. and Jurnak, F. (1995) *FASEB J.*, **9**, 335.
26. Raetz, C.R.H. and Roderick, S.L. (1995) *Science*, **270**, 997.
27. Kisker, C., Schindelin, H., Alber, B.E., Ferry, J.G. and Rees, D.C. (1996) *EMBO J.*, **15**, 2323.
28. Beaman, T.W., Binder, D.A., Blanchard, J.S. and Roderick, S.L. (1997) *Biochem.,* **36**, 489.
29. Emsley, P., Charles, I.G., Fairweather, N.F. and Isaacs, N.W. (1995) *Nature*, **381**, 90.
30. Petersen, T.N., Kauppinen, S. and Larsen, S. (1997) Structure, **5**, 533.
31. Mayans, O., Scott, M., Connerton, I., Gravesen, T., Benen, J., Visser, J., Pickersgill, R. and Jenkins, J. (1997) *Structure,* **5**, 677.
32. Steinbacher, S., Baxa, U., Miller, S., Weintraub, A., Seckler, R. and Huber, R. (1996) *Proc. Natl. Acad Sci.,* **93**, 10584.
33. van Houdenhoven, F.E.A. (1975) Thesis. Agricultural University of Wageningen, The Netherlands.

Substrates and Industrial Applications

MONTE CARLO SIMULATION OF ENZYMATIC CELLULOSE DEGRADATION

Veljo Sild[‡,§,*], Anu Nutt,[‡] Göran Pettersson[‡] and Gunnar Johansson[‡]

‡ Department of Biochemistry, University of Uppsala, PO Box 576, S-75 123 Uppsala, Sweden

§ Institute of Molecular and Cell Biology, University of Tartu, Lai 40, EE2400, Tartu, Estonia

* Corresponding author. Telephone: +46-18-471 45 47. Fax: +46-18-55 21 39. E-mail: veljo.sild@biokem.uu.se

1 INTRODUCTION

In nature, fungi and bacteria degrade crystalline cellulose by means of several enzymes acting from the ends (exocellulases) or internally (endoglucanases) on the cellulose chains. This process is, for several reasons, impossible to describe by classical enzyme kinetics, firstly, because it takes place at a solid-liquid interface and, secondly, because the necessary adsorption of enzymes to the cellulose does not follow a standard Langmuir isotherm. The deviation from a single Langmuir isotherm has been interpreted as the existence of several different classes of binding sites, but the good numerical fit often obtained with such functions is not necessarily biologically relevant. Here we present a simulation process, which allows the basic events of association and dissociation to take place randomly, but with selected probabilities as a general approach to describe this binding process, also taking into account the overlapping of the binding sites.

It was believed for many years that cellobiohydrolases (exocellulases) attack the cellulose chain from the nonreducing end. Recent data indicate, however, that CBH I preferentially attacks the reducing end of cellooligosaccharides, whereas CBH II indeed prefers the nonreducing end (Barr, *et al.*, 1996; Biely, *et al.*, 1993).

End preference on insoluble cellulose may be influenced by the binding to the bulk cellulose, especially when the binding domain (Tomme, *et. al.*, 1988) is involved. This means that the mode of action on soluble substrates does not necessarily reflect the degradation pattern for insoluble cellulose, which undoubtedly is influenced by the structure of the cellulose, introducing separation and distortion of the cellulose chains as new kinetic parameters. Also, it is important to consider that cellulolytic enzymes can work processively without leaving the cellulose chain between the hydrolytic events (Rabinovich *et al.*, 1984).

The following questions now arise: What is the probability for hydrolytic attack at the reducing /nonreducing ends? How pronounced is the processivity in either direction? These questions are virtually impossible to answer by a conventional kinetic approach.

The simulation process suitable for binding processes can be extended also to degradation kinetics if additional parameters such as hydrolysis and migration on the surface are added.

In this study we also present a method for qualitative description of action of cellulases on crystalline cellulose using results obtained from experiments with *T. reesei*

CBH I, CBH II and the corresponding enzymes (CBH 58 and CBH 50) from *P. chrysosporium.* The method is based on a comparison between experimental and simulated kinetic curves of hydrolysis of radioactively end-labeled cellulose with respect to parameters such as end preference, enzyme concentration and processivity.

2 MATERIALS AND METHODS

2.1 Materials

2.1.1 Enzymes. CBH I and CBH II were purified from the culture filtrate of *Trichoderma reesei* QM 9414 as described in Bhikhabhai *et al.*, (1984) with a final purification step performed by affinity chromatography according to van Tilbeurgh *et al.* (1984). The corresponding enzymes from *P. chrysosporium* (CBH 58 and CBH 50) were purified according to Uzcategui *et al.*(1991).

2.1.2 Other chemicals. Sodium boro[^{3}H]hydride with a specific activity of 888 GBq/mmol was from Amersham International, England. *Acetobacter xylinum* BMCC was prepared from a commercially available product from Thep. Padung Porn Coconut Co. Ltd., Bangkok, Thailand. All other chemicals were of analytical grade.

2.2 Methods

2.2.1 Preparation of bacterial microcrystalline cellulose (BMCC). The cellulose pieces were washed with tap water for 2 days, followed by refluxing in 3 % NaOH for 2 days, changing the NaOH when it turned yellow. After washing with 1 % acetic acid and distilled water to neutral pH, the cellulose was then hydrolysed by refluxing in 2.5 M HCl for 1 day. Finally, the cellulose was washed repeatedly with distilled water and stored in 50 mM sodium acetate buffer (pH 5.0). The number average DP of the cellulose was calculated from the concentration of reducing end groups and total sugar determination as described below and found to be 240 glucose units/chain.

2.2.2 Labelling of the reducing end of BMCC with ^3H. Reduction of terminal aldehyde groups of cellulose was carried out as follows: to 500 mg BMCC in 100 ml 0.1 M NaOH was added 4.6 mmoles unlabeled $NaBH_4$ and 0.5 µmoles ^3H-$NaBH_4$ (0.7 mCi). The reaction mixture was boiled for 20 minutes. The reagent addition and boiling were repeated twice, after which the reaction mixture was acidified with HCl and washed to neutral pH. Under these conditions >80 % of the aldehyde groups became reduced. An alternative procedure with labelled reagent added first followed by the unlabeled reagent was not satisfactory.

2.2.3 Kinetic studies. BMCC (1 mg/ml) was incubated in 50 mM sodium acetate buffer (pH 5.0) at 0^oC with various enzyme concentrations under continuous magnetic stirring. The samples were withdrawn and the reaction was stopped by adding 1/9 sample volume of 1 M NaOH. After centrifugation the concentrations of total sugar, reducing end groups and radioactively labeled end groups in the supernatant were determined.

2.2.4 Determination of carbohydrate content. Total sugar was measured by the anthrone/H_2SO_4 method (Hörmann and Gollwitzer, 1962) using D-glucose as standard and absorbance measurements at 585 nm. The contribution from enzyme glycosylation was subtracted from the initial data.

2.2.5 Determination of reducing end groups. Determination of reducing end groups both in the supernatant and on the cellulose was performed according to the Somogyi-Nelson method (Somogyi, 1952; Nelson, 1944), using D-glucose as standard and absorbance measurements at 510 nm.

2.2.6 Protein concentration determination. Enzyme concentration was calculated from UV absorbance at 280 nm using the following molar extinction coefficients (M^{-1} cm^{-1}): CBH I, 78 800; CBH II, 92 000; CBH 50, 111 500; CBH 58, 71 100.

2.2.7 Modelling of enzyme binding and cellulose hydrolysis. The modeling was carried out using a recently developed simulation program (Sild *et al.*, 1996). The fiber surface to which the cellulase-shaped ligands were to bind consisted of 80 chains, each containing 120 cellobiose units (average length of the fibers in our experiment), giving 80x120 EBS (elementary binding sites) in the layer. The fiber used in the model was composed of four layers. The fiber surface was defined as completely anisotropic, which means nonequivalence in two dimensions, polarity in the vertical dimension and polarity in the horizontal dimension. We expect this type of surface to be relevant to the cellulose/cellulase system. The shape of the cellulases in the model was defined according to SAXS imaging (Abuja *et al.*, 1988a,b) (Figure 1).

The simulations were carried out to mimic the actual experimental situation where a fiber is introduced at time zero into a solution containing the cellulases. All of the elementary acts then proceed step by step. At every step a certain number of ligands 1) try to bind to the surface 2) will dissociate from the surface 3) try to hydrolyse the cellulose chain from the A (reducing)- end moving towards the B(non-reducing)-end. The number of ligands trying to perform a certain elementary act are calculated according to the mass action law:

$$L_{pot.ass} = k_1 \times L_{free} \times Nr_{EBS} \quad (1)$$

$$L_{diss} = k_2 \times L_{bound} \quad (2)$$

$$L_{pot.\ A \rightarrow B} = k_3 \times L_{prod.\ bound} \quad (3)$$

where

L_{diss} and $L_{pot.ass}$ are the number of ligands dissociating and trying to associate respectively;

$L_{pot.\ A \rightarrow B}$ is the number of ligands trying to hydrolyse processively the cellulose chain;

L_{free} is the concentration of free ligands and L_{bound} is the number of bound ligands;

$L_{prod.\ bound}$ is the number of productively bound ligands (*e.g.* exoenzymes bound at the "correct" end of the polymer chain)

Nr_{EBS} is the total number of EBS on the fiber surface;

k_1, k_2 and k_3 are the corresponding rate constants.

When moving the ligands, a bound ligand or potential surface binding site is chosen at random. Binding or movement up/down will occur only when the random site in the lattice is not occupied. The numbers of associating ligands L_{ass} and processing ligands $L_{A \to B}$ are always less than or equal to $L_{pot.ass}$ and $L_{pot.\ A \to B}$, respectively. The fate (release) of the initial cellulose chain ends was followed separately for comparison with the experiment with the end-labeled cellulose. The ratios of the constants k_3 / k_2 are defined here as processivity indices.

All simulations were repeated 2 times and the average value is shown in the figures. The standard deviation was in no case greater than 10 %.

Simulation of binding equilibrium was carried out by using only equations (1) and (2) for the process.

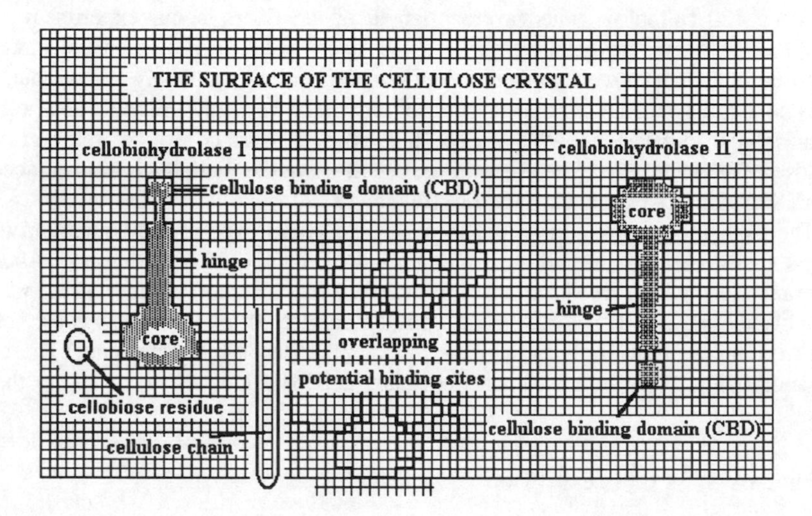

Figure 1 *The dimensions of* T. reesei *cellobiohydrolases in comparison with the parameters of cellulose.*

3 RESULTS

3.1 Simulation of ligand binding to a surface

The simulated time course of ligand binding to overlapping binding sites is shown in Figure 2. It is important to note that at higher ligand concentration an initial "apparent equilibrium" is attained at a stage where no more ligands can bind at the very moment despite the incomplete coverage of the surface. True equilibrium will require a very time consuming rearrangement of initially bound ligands. It is likely that the "apparent equilibrium" is representative for the results obtained in real experiments. It has been shown by us recently that there is a good fit between experimental data and the simulated binding of the CBH I core domain (Sild *et al.*, 1996).

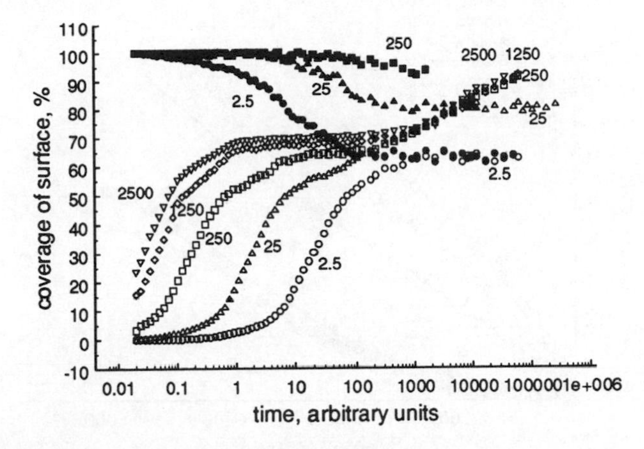

Figure 2 *Simulation of the kinetics for binding of square ligands with size 3x3 elementary binding sites to the inflnlte surface. Rate constant for association = 2.3, rate constant for dissociation = 0.256. Numbers at curves indicate the free ligand concentration in arbitrary units.*

3.2 Simulation of the hydrolysis of BMCC by cellulases

In this part of our work we simulated the kinetics of hydrolysis in order to deduce the typical pattern of the exo-hydrolysis from different ends at various enzyme concentrations.

The pattern of the release of the primary ends from the polymer crystal depends significantly on the end preference of the enzyme (Figure 3) and, also on the enzyme concentration. However, it should be pointed out that most informative here is the kinetics of the hydrolysis of the first two or of three layers of the polymer crystal, especially in the case of shorter fibers.

The "jump" in the release of the primary ends from the polymer crystals is dependent on the enzyme concentration if the enzyme can be described as follows: a) it is processive; b) it hydrolyses polymer mainly in the exo-mode from the labeled end; and c) it can bind nonproductively in the middle of the polymer chain.

3.3 Release of the end-groups of BMCC during hydrolysis

The ratio of released labeled ends to the released sugar residues (hereafter called end-release probability) showed systematic differences between the enzymes and changed during the course of the reaction (Figure 4). CBH I and CBH 58 showed virtually identical curves with a high end-release probability at the beginning of the hydrolysis. The end-release probability was notably lower and rather similar for CBH II and CBH 50.

The initial end-release probability shows a strong positive correlation with the concentration of CBH I (Figure 5), whereas no similar tendency was observed in the case of CBH II (data not shown).

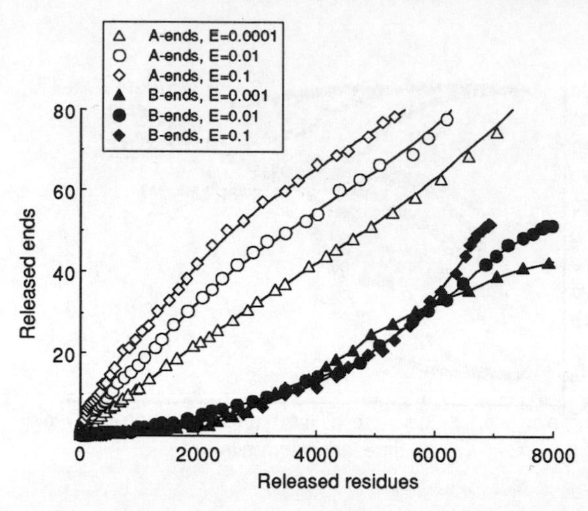

Figure 3 *Influence of the enzyme concentration on the simulated release of A- and B-end- release probability by a strict A-end enzyme with CBH I shape. The processivity of enzyme was 100. Open labels represent the release of A (reducing)-ends, closed labels the release of B (non-reducing)-ends during the simulated hydrolysis of cellulose at given concentration of enzyme.*

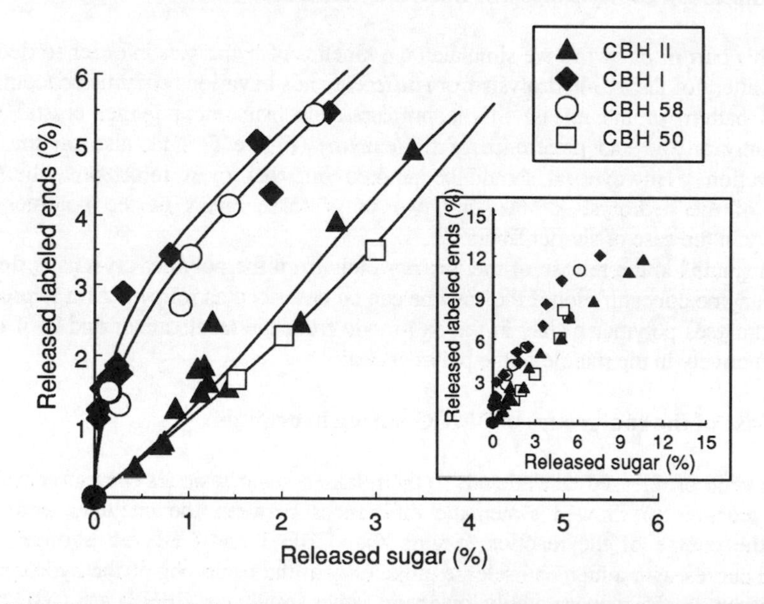

Figure 4 *Release of labeled end-groups as a function of the total released glucose residues during hydrolysis of BMCC (1 mg/ml) by 5 μM CBH I, CBH II, CBH 50, CBH 58. The hydrolysis experiments were carried out at 0°C in 50 mM NaAc buffer, pH=5.0. The inset shows the result after more extensive reaction.*

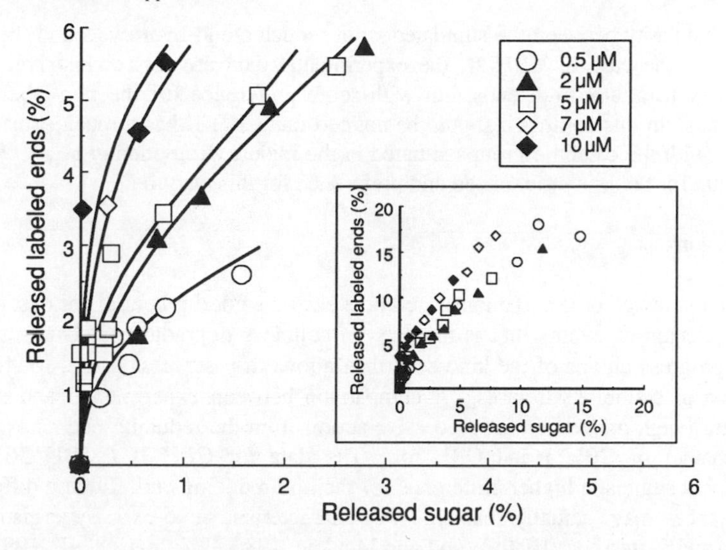

Figure 5 *Release of labeled end-groups as a function of the total released glucose residues during hydrolysis of BMCC (1 mg/ml) by CBH I. The hydrolysis experiments were carried out at 0°C in 50 mM NaAc buffer, pH=5.0. The inset shows the result after more extensive reaction.*

4 DISCUSSION

4.1 Simulation of binding

It is obvious that our simulation algorithm reproduces experimental binding data at least equally good as a two-site Langmuir function, but is free from the unjustified assumption about nonequivalent binding sites (Sild *et al.*, 1996). It should be remembered, however, that a simulation still is totally dependent on a correct definition of the basic events and the system components.

4.2 Simulation and measurement of degradation

Our study reveals systematic differences in the specific modes of attack of individual cellobiohydrolases on BMCC.

Recent studies of enzyme attack on oligomeric substrates revealed that CBH I attacks β-glucans in a manner which is distinctly different from that of CBH II (Biely *et al.*, 1993). Whereas the latter enzyme most frequently cleaves the second glucosidic bond from the nonreducing end of cellooligosaccharide, the CBH I hydrolyses the bonds at the reducing end, producing both glucose and cellobiose. Our results seem to be consistent with this differentiation. CBH I-type enzymes hydrolyse with considerably higher probability from the reducing end (in our experiment, the labeled end) than the CBH II - type (Figure 4).

The simulation data in Figure 3 clearly show that the end-release probability for the preferred end is enhanced by a high enzyme concentration. The comparison of the shapes of the experimental curves (Figure 5) with the simulated kinetics (Figure 3) reveals a remarkable similarity between the simulated strict reducing-end-hydrolysis and the kinetics of CBH I. In the case of CBH II, the experimental data are less conclusive, but may suggest ability to attack both ends, but with some preference for the non-labeled (non-reducing) ends. In this context it should be noticed that CBH II has a much shorter tunnel than CBH I with the catalytic groups situated in the middle (Rouvinen *et al.*, 1990). This could account for the less pronounced end preference for this enzyme.

4.3 Conclusions

The simulation algorithm demonstrated here shows a good potential for description of the complex chain of events that is involved in cellulose degradation. Our results also show that progress curves of the kind described above may serve as a tool for functional classification of cellobiohydrolases. A comparison between experimental and simulated data suggest a high preference for processive action from the reducing end as explanation for the behavior of CBH I and CBH 58. The data for CBH II / CBH 50 are less conclusive, but suggest a higher preference for the non-reducing end. Such a difference in end preference may actually explain the characteristic exo-exo synergistic effect (Fägerstam and Pettersson, 1980; Wood and McCrae, 1986, Uzcategui *et al.*, 1991).

5 ACKNOWLEDGMENTS

This work was supported by grants from the Swedish Research Council for Engineering Sciences (230 93-27), the Swedish Natural Science Research Council (K-AA/KU 02475-0306), the Sahlen foundation, the Swedish Pulp and Paper Research Foundation, the Swedish Royal Academy of Sciences, the Estonian Science Foundation and the Swedish Institute.

6 REFERENCES

Abuja, P. M., Pilz, I., Claeyssens, M. and Tomme, P. (1988a), *Biochem. Biophys. Res. Comm.*, **156**, 180.

Abuja, P., Schmuck, M., Pilz, I., Tomme, P., Claeyssens, M. and Esterbauer, H. (1988b), *Eur. Biophys. J.*, **15**, 339.

Bhikhabhai, R., Johansson, G. and Pettersson, G. (1984), *J. Appl. Biochem.*, **6**, 336.

Biely, P., Vršanská, M., and Claeyssens, M. (1993), *in* TRICEL 93 Symposium on Trichoderma reesei Cellulases and Other Hydrolases (Suominen, P. and Reinikainen, T., Eds.), pp 99-108, Foundation for Biotechnical and Industrial Fermentation Research, Espoo, Finland.

Fägerstam, L. and Pettersson, L. G. (1980), *FEBS Lett.*, **119**, 309.

Hörmann, H. and Gollwitzer, R. (1962), *Ann. Chem.*, **655**, 178.

Nelson, N. J. (1944), *J. Biol. Chem.*, **153**, 375.

Rabinovich, M. L., Klesov, A. A. and Berezin, I. V. (1984), *Dokl. Akad. Nauk. SSSR*, **274**, 758.

Rouvinen, J., Bergfors, T., Knowles, J. K. C. and Jones, T. A. (1990), *Science*, **249**, 380.

Sild, V., Ståhlberg, J., Pettersson, G. and Johansson, G. (1996), *FEBS Lett.*, **378**, 51.

Somogyi, M. (1952), *J. Biol. Chem.*, **195**, 19.

van Tilbeurgh, H., Bhikhabhai, R., Pettersson, G. and Claeyssens, M. (1984), *FEBS Lett.*, **169**, 215.

Tomme, P., van Tilbeurgh, H., Pettersson, G., van Damme, J., Vandekerckhove, J., Knowles, J., Teeri, T. and Claeyssens, M. (1988), *Eur. J. Biochem.*, **170**, 575.

Uzcategui, E., Ruiz, A., Montesino, R., Johansson, G. and Pettersson, G. (1991), *J. Biotechnol.*, **19**, 271.

Wood, T. M. and McCrae, S.I. (1984), *Biochem. J.*, **234**, 93.

ON THE STRUCTURE OF CRYSTALLINE CELLULOSE I

Andreas P. Heiner and Olle Teleman

VTT Biotechnology and Food Research, POB 1500, FIN-02044 VTT, Finland

1 ABSTRACT

The three-dimensional structure of monoclinic and triclinic cellulose I has been proposed by model fitting to two-dimensional fibre diffraction data [J. Sugiyama, R. Vuong, H. Chanzy, *Macromolecules*, 1991, **24**, 4168]. Nanosecond long molecular dynamics simulations were performed on the cellulose coordinates in order to refine them. The monoclinic form was more stable than the triclinic by 8.7 kJ·(mol cellobiose)$^{-1}$. The two molecules in the monoclinic unit cell were found not to be equivalent. Their different properties explain additional resonances in NMR spectra, for some of which it was possible to predict the chemical shift. Atomic Force Microscopy was used to record microfibrils from *Valonia macrophysa*. A resolution of 4 Å was achieved. Comparison in Fourier space with the proposed crystal surfaces enabled the identification of a monoclinic (1-10) surface. The AFM data corroborate the proposed cellulose structure, wherefore we submit that it is to be considered known. Computer simulations of the crystal surfaces showed that only the outermost molecular layer is affected by the solvent. The number of hydrogen bonds with the solvent indicate that the two "narrower" surfaces are more hydrophilic, consistent with hydrophobic effect assisted preference by the cellobiohydrolase I cellulose binding domain for a "wide" surface.

2 INTRODUCTION

In order to understand the mode of action of cellulases on their substrate, an in-depth knowledge of the structure of that substrate, crystalline cellulose, is as essential as knowing the structure of the enzyme itself. Cellulose is a polymer of β-(1→4) linked D-glucose, and packed in tiny fibres known as microfibrils. The microfibrils contain both crystalline and amorphous or less ordered domains.

Very little is known about the structure at a nanometer level of the less ordered domains. The three-dimensional structure of cellulose remained unclear until the eighties although it was the subject of high-resolution diffraction studies already very early[1]. The difference in crystal symmetry between cellulose produced by lower and higher organisms was particularly puzzling. Based on ^{13}C CP/MAS NMR spectra of natural cellulose of both lower and higher plants, Atalla and vanderHart proposed that natural cellulose I consists of a mixture of two phases, Iα (C4 resonance at δ = 90.4 ppm) and Iβ (C4 resonance at δ = 89.4 and 88.6 ppm).[2-4] Cellulose Iα is the dominant form in lower organisms, whereas cellulose Iβ preponderates in cellulose produced by higher organisms. Horii *et al.* showed, also using solid state NMR, that hydrothermal treatment (260°C, NaOH) completely converts cellulose Iα into cellulose Iβ, indicating that the latter form is thermodynamically more stable.[5] Evidence for the co-existence of cellulose Iα and Iβ in the same microfibril was obtained using Raman scattering and Fourier transform infrared spectroscopy. The unit cells of the two phases were finally determined by Sugiyama *et al.* using electron diffraction.[6] Atomic coordinates were then proposed by fitting a model to the two-dimensional diffraction data. The Iα phase has one chain per triclinic unit cell, the Iβ-phase contains two chains per monoclinic unit cell. This model accounts qualitatively for the fine structure of high-resolution ^{13}C CP/MAS NMR experiments.

The small microfibril diameter (a few to 20 nm) constitutes an experimental difficulty. This is especially true for the surface of cellulose, the part of cellulose interacting with cellulases. Molecular dynamics simulations do not have this limitation, and we have applied this method extensively over the past years. Likewise, Atomic Force Microscopy has proven capable of delivering direct non-destructive experimental data on the microfibril surface at high resolution.

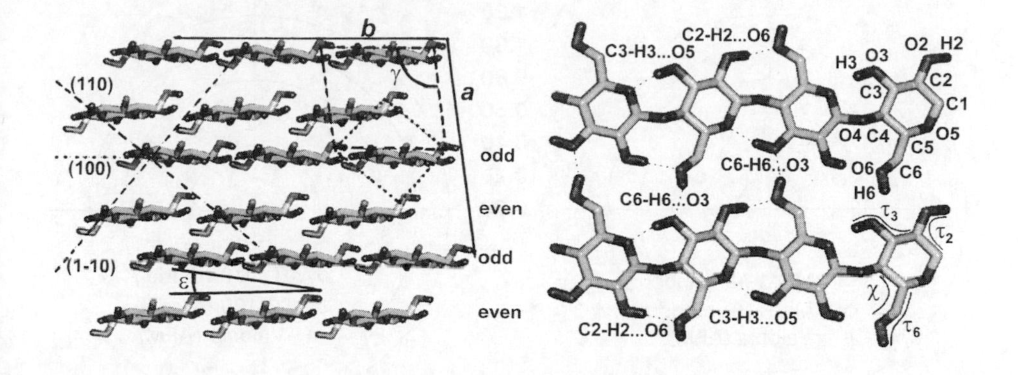

Figure 1 *a) Relative positions for the proposed structure of cellulose chains in the monoclinic form as viewed from the nonreducing end along the c-axis. The monoclinic a and b crystal axes are indicated by thick dashed lines, the projection of the triclinic a-b plane on the monoclinic a-b plane is indicated by thin dot-dashed lines. The crystal planes are indicated by dotted lines. Chains in the odd (100) planes are at a slight angle with respect to the even (100) planes. b) Nomenclature of the cellulose atoms and relevant dihedral angles. Intramolecular hydrogen bonds and hydrogen bonds between adjacent chains are depicted as dotted lines.*

3 THE CRYSTALLINE STATE OF NATURAL CELLULOSE

Figure 1 shows the crystal structure proposed for cellulose I (J. Sugiyama, personal communication). The molecular repeat of the crystal is the cellobiose unit. The main difference between the two phases is the staggering of chains in alternate sheets, where a sheet is a monomolecular layer in the monoclinic (100) (triclinic (110)) plane. In cellulose Iβ, chains are staggered along the c-axis according to a 0c, +1/4c, 0c, +1/4c pattern; in the triclinic phase the staggering is 0c, +1/4c, 2/4c, +3/4c, +4/4c=0c. The density of the phases differs slightly, 1.599 g·cm^{-3} for the Iβ phase and 1.582 g·cm^{-3} for the Iα phase. Chains in the sheets are kept together by strong hydrogen bonds. In the proposed monoclinic crystal structure, chains in alternate sheets are slightly off centre, and its glucose rings are at a small angle (≈7.4°) with respect to the plane.

We performed one-nanosecond simulations of cellulose Iα and Iβ using the GROMOS87 force field.[7] The simulations constitute a refinement of the atomic coordinates and explained several experimental observations.[8,9]

The enthalpy of cellulose Iβ was found to be 8.7 kJ·(mol cellobiose)$^{-1}$ lower than for Iα, consistent with the relative stability of the two allomorphs. The additional stability is mainly due to electrostatic interactions within (100) planes, especially within the even ones. The more parallel orientation may put strain between alternate chains (as indicated by a higher intra-planar van der Waals interaction). The net effect of van der Waals interactions (both inter- and intra sheet) to the relative stability is negligible.

The transition I$\alpha \rightarrow$Iβ at high temperatures was recently suggested to occur via sliding of alternate (100) planes.[10] This appears more viable than the suggestion by Hardy and Sarko, who suggest that the transition occurs via breaking and formation of hydrogen bonds, since the interaction between the stacked planes are mainly van der Waals forces.[11]

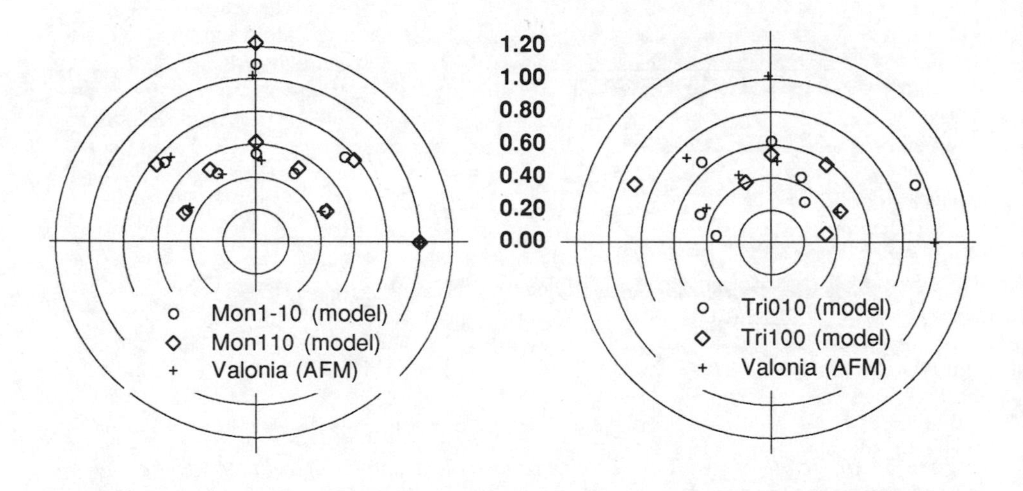

Figure 2 *Comparison of an experimental AFM micrograph to surfaces modelled from fibre diffraction experiments. The peaks are plotted in polar diagrams and represent periodicities in the Fourier transform of images. The radial scale is given in nm. The experimental image is represented by small plus signs and agrees closely with the monoclinic 1-10 surface. Two peaks from the model surface are not observed experimentally. We attribute their presence in the model images to the used Connolly surface, which is a very crude way to represent the interaction between surface and microscope tip.*

Molecules in alternating Iβ (100) planes behave quite differently compared to each other and to chains in cellulose Iα. The dihedral distributions of the three hydroxyl groups are quite different, as is the distribution of the exocyclic hydroxymethyl group. In diffraction studies the χ-dihedral (defined by C4-C5-C6-O6) defining the position of O6 is invariable in the tg-conformation (trans w.r.t. O5, gauche w.r.t. C4), thereby forming the hydrogen bond O6-H6···O3. In the Iβ/even sub-phase the tg-fraction is close to 1, but in the Iβ/odd sub-phase other states are also occupied (tg: 0.92, gg: 0.06, gt: 0.02), and even more so in the Iα phase (tg: 0.81, gg: 0.08, gt: 0.11).

The glucose ring conformation between Iα, Iβ/even and Iβ/odd differs slightly. The ring conformation is conveniently described by means of pucker parameters, which describe atomic position with respect to the ring plane.[12,13] The glucose ring always remains close to the $^4C_1$ (chair) conformation in all sub-phases. The differences between the sub-phases become clear when looking at the puckering azimuth Φ, which describes towards which form the ring deforms. The Iβ/odd subphase has a tendency to deform towards the ^2E conformation, whereas the even subphase does not have a clear preference. Glucose rings in the Iα phase have a preference for the ^OE, and a smaller one for the $^2H_3$ conformation.

Also the ring orientation with respect to the unit cell varies somewhat. We will describe this orientation by means of the pitch and camber angles. If the reader "sits" on a glucose ring looking along the polymer, the camber describes whether the glucose leans "sideways" and the pitch whether the ring is inclined "upwards" or "downwards" in the direction of looking.

The pitch is similar in all sub-phases (Iβ/even: 10.3°, Iβ/odd: 7.3°, Iα: 8.8°). In the triclinic crystal the difference between subsequent moieties is approximately 2°, indicating minor structural differences of the glycosidic bond angle and/or pucker parameters of these rings. In the monoclinic crystal the camber is rather distinct for the two sub-phases (Iβ/even: -7.2°, Iβ/odd: -1.4°, Figure 1a), but the chains possess a 2-fold screw axis. In the triclinic system the 2-fold screw axis is absent, as exemplified by the camber values (Iα/1st: 1.0°, Iα/2nd: -7.5°). The lack of two-fold screw symmetry has to be compensated along the chain, *e.g.* in the φ/ψ dihedrals of alternating groups or the ring shape, in order to maintain translational symmetry along the c-axis. Alternating rings in the Iα phase have camber values similar to those of the Iβ/odd and Iβ/even sub-phase, the origin of which one may speculate.

The hydrogen bonds (Figure 1b) proposed for the crystal structure remain in the simulations. The O3-H3···O5 is most frequently observed in the Iβ/odd sub-phase (Iβ/even: 0.73, Iβ/odd: 0.88, Iα:0.64), but the other two hydrogen bonds O2-H2···O6 and O6-H3···O3 are most frequently observed in the Iβ/even and least in the Iα phase (Iβ/even: 0.95, Iβ/odd: 0.80, Iα:0.64) and (Iβ/even: 0.87, Iβ/odd: 0.71, Iα:0.50). The intra-chain hydrogen bond O2-H2···O6 also occurs with donor-acceptor roles reversed. Taking that into account, the occupancy is similar in Iβ/even and Iβ/odd, and the occupancy in the Iα phase increases. The electrostatic interaction between alternate chains in a sheet is increased by hydrogen bonds between O6-H3···O3, O2-H2···O6 and O3-H3···O6, but these hydrogen bonds are considerably weaker than the one proposed for the crystal structure. Hydrogen bonds between (100) planes are rare and weak. The most frequent hydrogen bonds for each sub-phase (Iα, Iβ/even, Iβ/odd) give seven different adsorption frequencies for an Iα/Iβ mixture, in agreement with FT-IR experiments.[14-16]

Solid state NMR is a powerful tool to study cellulose, even though the relation between chemical shift and structural features is not well understood. The C6 chemical shift has been published for several rotamers.[17] To allow for dihedral periodicity we fitted these data to a sine function and used the fit to predict C6 shifts from the dihedral distributions obtained in the simulations. All three resonances are close 66 ppm, in close agreement with the experimental result.[9]

4 THE INTERFACE BETWEEN CRYSTALLINE CELLULOSE AND WATER

All biological and technical important processes involving cellulose take place at its surface. Solid state NMR spectroscopical data indicate that the surface of cellulose is different from the bulk and gives rise to further resonances.[18,19] Two peaks upfield of the C4 chemical shift were identified as surface peaks, and a very broad resonance was attributed to paracrystalline cellulose, cellulose with order somewhere in between crystalline and amorphous cellulose.

Another experimental method to study the surface is Atomic Force Microscopy (AFM), which produces an image of the sample by scanning a flexible cantilever mounted tip over it.[20,21] AFM has been used to image a wide variety of surfaces with molecular or even atomic resolution. Early AFM studies[22] of cellulose microfibrils provided an idea of the surface topology and we were later able to bring about a high resolution study.[23]

Valonia cellulose contains 60-70% of the Iα crystalline phase.[2] In *Valonia macrophysa* the cellulose microfibrils are highly organised and packed in parallel within the cell wall layers. From the cell wall it is easy to obtain low resolution AFM images, showing the arrangement of microfibrils. After correction for the convolution artefact, the typical microfibril diameter is estimated at 20 nm, such that it contains in the order of 1000 cellulose molecules. From the top of the microfibril surface it is possible to obtain high-resolution images [ref. 23, Figure 1] which contain clearly resolved molecular details.

We used the proposed crystal structures to calculate artificial AFM micrographs. The Connolly method was used to generate a z(x,y) map for each crystal face, *i.e.* the monoclinic 110 and 1-10 faces and the triclinic 100 and 010 faces. The amount of noise in the experimental images precludes identification of the observed surface by direct comparison to the model images. The Fourier transform of the image features several bright peaks, which indicate periodicities [ref. 23, Table 1]. These periodicities can be compared to the corresponding peaks from the model surfaces (Figure 2). The experimental peaks are drawn as small crosses and it is clear from the figure that the experimental image in this case agrees very well with the model representing the monoclinic 1-10 surface. These results corroborate the fibre diffraction structure of the monoclinic phase and show that AFM can be used to study cellulose surfaces with a lateral resolution of about 0.4 nm.

At present AFM provides information at the 0.5 nm level at best. Solid state NMR does not provide direct structural data, but probes chemical differences of the environment of atoms under investigation (mostly ^{13}C). To study the effect of water on the cellulose surface, we simulated the two monoclinic surfaces for 1.5 ns and the two triclinic surfaces for 1 ns[24,25].

A sketch of the cross section is given in Figure 3, and a schematic top-view of all the surfaces in Figure 4. The difference between the two monoclinic surfaces (and the two triclinic ones) is the larger distance between alternate molecules, being 0.6 nm for the monoclinic (110) and triclinic (010) surfaces, and 0.53 nm for the monoclinic (1-10) and triclinic (100) surfaces. We will refer to these surfaces as Mwide, Twide, Mnarrow and Tnarrow, respectively.

In all simulations, the distance between layers reduced slightly, but the relative orientation (camber, pitch) and intramolecular structure of the four central layers were not affected. The hydrogen bonding patterns were similar to those observed in the crystal simulation and the energy difference between Iα and Iβ remains.

The conformation of molecules in direct contact with the solvent was affected to some extent. The orientation of the glucose moieties in the monoclinic phase stay close to the crystal values, except for rings in the Mwide/odd phase with the hydroxymethyl group exposed to the solvent. These glucose rings rotate over $\approx +7°$, see Table 1. A similar glucose reorientation occurs at the Twide surface ($\approx +5°$). In contrast, the orientation of the glucose units at the narrow surfaces is similar to that in the respective core areas. The pitch of glucose moieties is not affected by the presence of solvent, and is $-9.6° \pm 1.2°$, averaged over all surfaces.

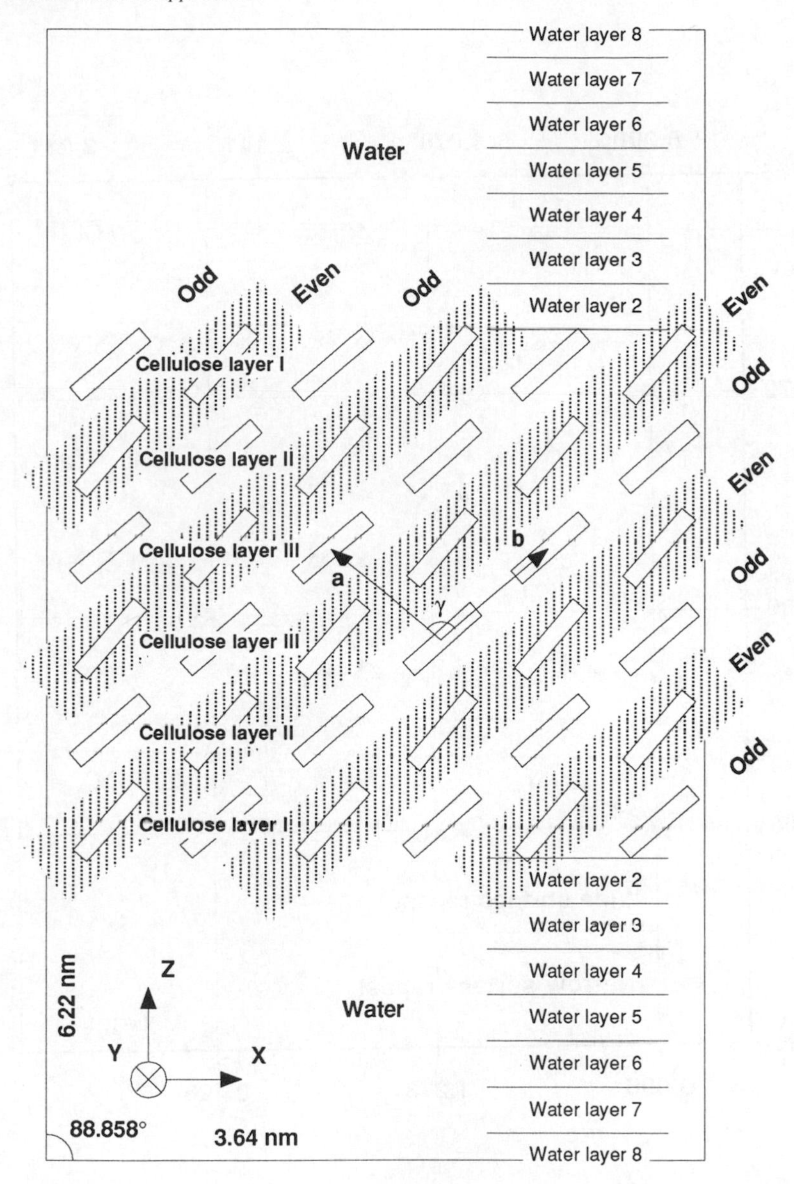

Figure 3 *Schematic placement of monoclinic crystalline cellulose in the periodic simulation box. Each cellulose layer has been numbered. In the triclinic systems, the upper surface (100 or 010) is the mirror image of the lower (-100 or 0-10), but this does not affect any geometric or energetic analysis, since water is not a chiral molecule. The numerical values in the figure are for the Mwide system.*

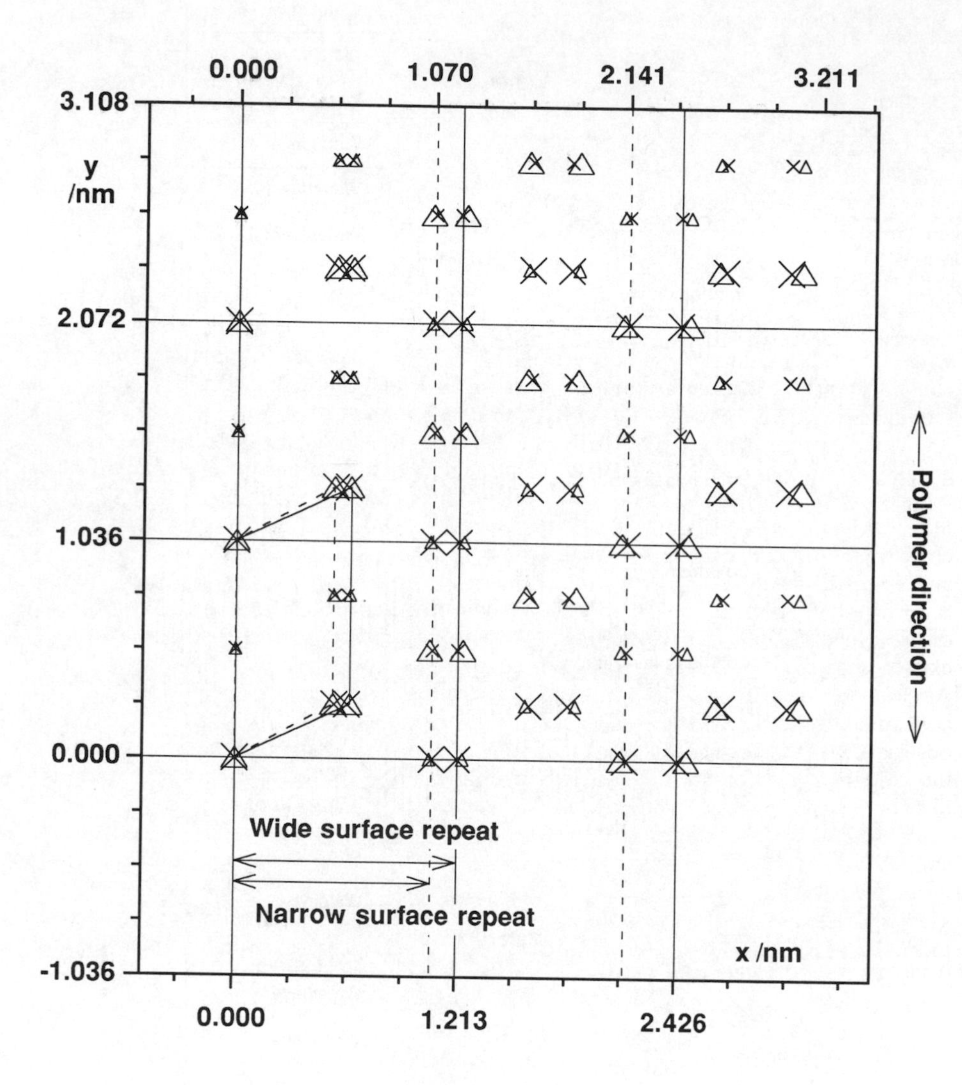

Figure 4 *Schematic representation of the surfaces* en face. *Crosses denote the two monoclinic surfaces' glucose positions; trangles denote the two triclinic surfaces' glucose positions. A larger mark denotes a C6-outward glucose and a smaller mark a C6-inward glucose. The cellulose polymers run from top to bottom in the picture, which also shows the narrow (1.070 nm) and the wide (1.213 nm) repeat distance. Two out of four polymers are out of phase regarding C6-outward or -inward between a monoclinic and triclinic surface, otherwise the similarity is striking.*

Table 1 *Camber angle of the glucose moieties.*

	Orientation	Mwide		Mnarrow		Twide	Tnarrow
		Even	Odd	Even	Odd		
Surface	C6-Inward	-6.9°	0.6°	-8.5°	-3.2°	2.0°	-2.7°
	C6-Outward	-8.9°	7.5°	-8.9°	-3.2°	-3.3°	-9.7°
Core	C6-Inward	-7.7°	1.0°	-6.9°	-3.2°	-0.7°	-2.0°
	C6-Outward	-7.3°	1.0°	-7.2°	-3.0°	-8.0°	-6.9°

The distribution of the pucker azimuth angle Φ at the narrow surfaces resembles the most to that of the core region. The distinction between glucose moieties belonging to the odd/even duplicity remains at the Mnarrow surface, but is lost at the Mwide surface, where the distribution of C6-outward pointing rings belonging to odd and even chains converges. The distribution of ring shapes at the triclinic surfaces is barely affected by the presence of the solvent.

As expected, the solvent affects the dihedral distributions of the solvent exposed groups. The population of the tg-conformation of the hydroxymethyl group (χ-dihedral) reduces mostly for the wide surfaces. At the Mwide surfaces the gt-conformation, and at the Twide the gg-conformation becomes dominant. At the narrow surfaces the tg-conformation remains marginally the dominant conformer. The difference between odd and even distributions is larger at the Mwide surface than at the Mnarrow surface. Surprisingly, alternate molecules at the Tnarrow surface display different dihedral distributions. The trans conformation of the $\tau 3$ dihedral (which correlates with the existence of the O3-H3···O5 hydrogen bond) reduces mostly for the narrow surfaces. Alternate molecules show different preferences for the gauche+ and gauche- conformations at the Tnarrow surface, but similar preferences at the Twide. This is also observed for the $\tau 2$ dihedral. The $\tau 2$ dihedral is mostly trans at the surface but gauche+ in the core to allow for the O2-H2···O6 hydrogen bond. The $\tau 6$-dihedral is uniformly distributed over all three rotamers.

At the narrow surfaces, the fraction of O2···O6 hydrogen bonds is considerably smaller than at the wide surfaces, with different fractions for alternate chains in both the monoclinic (odd/even) and triclinic surface. The fractions at the wide surface are very similar for alternate chains. The O3-H3···O5 hydrogen bond is found more often at the monoclinic surfaces (Mwide: 0.60, Mnarrow: 0.43) than at the triclinic surfaces (Twide: 0.36, Tnarrow: 0.27). For this property, the odd/even duplicity disappears for the Mnarrow surface, and alternate chains at the triclinic surface have similar fractions. The intermolecular O6···O3 hydrogen bond is absent from solvent exposed hydroxyl groups.

Hydrogen bonds to the solvent are abundant, especially for the O6-H6 and O2-H2 hydroxyl groups (averaged over all surfaces 1.30 and 1.26 respectively). Alternate molecules behave similarly at the wide surfaces, while significant differences in hydrogen bonding apply to the narrow surfaces. The difference in hydrogen bonding for alternate molecules, per cellobiose, is 0.12, 0.25, 0.0 and 0.32 for the Mwide, Mnarrow, Twide and Tnarrow surfaces. O3 is hydrogen bonded with the solvent 87% of the time. At the narrow surfaces it often serves as a donor, when the $\tau 3$ dihedral is gauche.

The fraction of hydrogen bonds per nm^2 is a measure for the relative hydrophilicity of the surfaces. The total number of hydrogen bonds per cellobiose is not that different among the four surfaces, but the different surface area per cellobiose make the narrow surfaces more hydrophilic (Mwide, 5.74, Mnarrow: 6.91, Twide: 5.66, Tnarrow: 7.20 hydrogen bonds per nm^2).

The assessment of the relative hydrophilicity is supported by cellulase binding data. The enzymic degradation of cellulose is initiated by the cellulase anchoring on a crystalline cellulose surface by means of a cellulose binding domain (CBD).[26] Electron microscopy data led Chanzy *et al.* to conclude that *Trichoderma reesei* cellobiohydrolase I preferentially adsorbs either to a wide surface or to its edge.[27] We later found that the binding capacity corresponds to a monomolecular layer on the cellulose surface, such that the enzyme primarily binds to the surface.[28] Site-directed mutagenesis has demonstrated the importance of certain aromatic residues on the CBD binding face but a hydrophobic effect also contributes to the affinity of the CBD to cellulose. The analysis presented here suggests that the wide surfaces are less hydrophilic than the narrow, which is consistent with a hydrophobic effect and the observed preferential binding to a wide surface.

5 CONCLUSIONS

The AFM and simulation results provide a consistent and adequate picture of crystalline cellulose. Native crystalline cellulose is found in two crystal forms, a triclinic (Iα) form with one cellobiose moiety per unit cell and a monoclinic (Iβ) form with two cellobioses per unit cell. The Iβ allomorph is the more stable of the two. The two chains in the Iβ form are non-equivalent owing to a slightly different orientation. This odd/even dichotomy is consistent with ^{13}C CP/MAS NMR or IR spectra, but can also be probed by calculating ring conformations (pucker azimuth Φ), dihedral angles ($\tau 2$, $\tau 3$, χ) or hydrogen bonding patterns. For monoclinic crystallites, the microfibril exposes (110) and (1-10) surfaces but no (100) surface. When a surface is exposed to solvent, only the outermost layer is affected. The change is larger for the glucose moieties where the hydroxymethyl group points into the solvent. The distinction between the odd and even chains disappears at the monoclinic wide surface (inter-chain distance 0.6 nm), but remains at the monoclinic narrow surface (inter-chain distance 0.53 nm). The behaviour of triclinic surfaces is similar to that of the monoclinic surfaces: alternate molecules differ at the narrow surface but not at the wide. That alternate molecules at the triclinic narrow surface show significant differences in structural properties constitutes the most remarkable observation from the simulations. Apparently the solvent and the tight packing of this surface favours slightly alternating conformations, although the triclinic unit cell ensures that all molecules are equivalent in the crystal interior.

References

1. K.H. Meyer and L. Misch, *Helv. Chim. Acta*, 1937, **11**, 534.
2. R.H. Atalla and D.L. vanderHart, *Science*, 1984, **223**, 283.
3. D.L. vanderHart and R.H. Atalla, *Macromolecules*, 1984, **17**, 1465.
4. P.S. Benton, S.F. Tanner, N. Cartier, H. Chanzy, *Macromolecules*, 1989, **22**, 1615.
5. F. Horii, H. Yamamoto, R. Kitamaru, M. Tanahashi and T. Higuchi, *Macromolecules*, 1987, **20,** 2946.
6. J. Sugiyama, R. Vuong and H. Chanzy, *Macromolecules*, 1991, **24**, 4168.
7. W.F. van Gunsteren and H.J.C. Berendsen, 1987, GROMOS Library manual, Biomos., Nijenborgh 4, Groningen, The Netherlands.
8. A.P. Heiner, J. Sugiyama and O. Teleman, *Carbohydr. Res.*, 1995, **273**, 207.
9. A.P. Heiner and O. Teleman, *Pure and Appl. Chem.*, 1996, **68**, 2187.
10. L.M.J. Kroon-Batenburg, B. Bouma, J. Kroon, *Macromolecules*, 1996, **29**, 5695.

11. B.J. Hardy and A. Sarko, *Polym. Prepr. (Am. Chem. Soc., Div. Polym. Chem.)*, 1995, **36**, 640.
12. D. Cremer and J.A. Pople, *J. Am. Chem. Soc.*, 1975, **97**, 1354.
13. M.K. Dowd, A.D. French and P.J. Reilly, *Carbohydr. Res.*, 1994, **264**, 1.
14. A.J. Michell, *Carbohydr. Res.*, 1988, **173**, 185.
15. J. Sugiyama, J. Persson, H. Chanzy, *Macromolecules*, 1991, **24**, 2461.
16. A.J. Michell, *Carbohydr. Res.*, 1993, **241**, 47.
17. F. Horii, A. Hirai, R. Kitamaru, *ACS Symp. Ser.*, 1987, **340**, 119.
18. R.H. Newman and J.A. Hemmingson, *Cellulose*, 1994, **2**, 95.
19. P.T. Larsson, K. Wickholm and T. Iversen, *Carbohydr. Res.*, 1997, **302**, 19.
20. G. Binnig, C.F. Quate and C. Gerber, *Phys. Rev. Lett.*, 1986, **56**, 930.
21. P.K. Hansma, V.B. Elings, O. Marti and C.E. Bracker, *Science*, 1988, **242**, 209.
22. S.J. Hanley, J. Giasson, J.-F. Revol and D.G. Gray, *Polymer*, 1992, **33**, 4639.
23. L. Kuutti, J. Peltonen, J. Pere and O. Teleman, *J. Microscopy*, 1995, **178**, 1.
24. A.P. Heiner and O. Teleman, *Langmuir*, 1997, **13**, 511.
25. A.P. Heiner, L. Kuutti and O. Teleman, *Carbohydr. Res.*, in press.
26. P. Tomme, R.A.J. Warren and N.R. Gilkes, *Adv. Microb. Physiol.*, 1995, **37**, 1.
27. H. Chanzy, B. Henrissat and R. Vuong. *FEBS Lett.*, 1984, **172**, 193.
28. T. Reinikainen, O. Teleman and T.T. Teeri, *Proteins*, 1995, **22**, 392.

THE POTENTIAL OF HYDROLYTIC ENZYMES TO MODIFY DOUGLAS-FIR DERIVED BLEACHED MECHANICAL PULPS

S. E. Ryan,[1,2] E. de Jong,[1,3] G. M. Gübitz,[1] M. Tuohy[2] and J. N. Saddler[1]

[1]Chair of Forest Products Biotechnology, Department of Wood Science, University of British Columbia, 270-2357 Main Mall, Vancouver, B.C. V6T 1Z4, Canada.
[2]Department of Biochemistry, University College Galway, Ireland.
[3]Current address: Agrotechnological Research Institute (ATO-DLO), P.O. Box 17, 6700 AA Wageningen, The Netherlands.

1 ABSTRACT

A range of commercial enzymes enriched in xylanase, mannanase and endoglucanase activities were assessed for their potential to improve the optical and physical properties of a bleached refiner mechanical pulp derived from Douglas-fir (*Pseudotsuga menziesii*). An endoglucanase preparation from *Gloeophyllum saepiarium* was also assessed to compare its efficiency against that of the commercial preparations. A synergistic effect was noted when a combination of xylanase and mannanase enzymes were used. A correlation was observed between the low molecular weight carbohydrates released after enzyme treatment and a corresponding reduction in interfiber bonding. Most enzymes increased pulp freeness while the addition of xylanase further enhanced action of the mannanase treatments. The ISO Brightness was not changed by any of the enzyme treatments.

2 INTRODUCTION

Currently kraft pulping is the predominant process used for paper making in the world.[1] However, the desire to more fully utilise the limited forest resource and meet the growing demands of the paper industry predict a greater utilisation of high yield mechanical pulping processes. To fulfill this market demand several wood species with less favorable characteristics, such as Douglas-fir, are being considered as potential wood furnish for mechanical pulping. The greater fiber wall thickness and coarseness of Douglas-fir reduce its potential as a mechanical pulp. The fibers tend to be rigid and inflexible resulting in poor fiber collapsability.[2,3] Therefore the resulting paper is lacking in strength as the fibers do not conform to each other, consequently influencing the bonding ability of the complete fiber mat.[4] The fibers also exhibit little external fibrillation and, as a result, the area available for bonding between individual fibers in the paper network is small. Due to the relatively long fiber length retained during refining, the resultant pulps generally exhibit a high tear strength but are brittle and lacking in burst strength. Mechanical pulps contain a high proportion of fines which may include as much as one third of the total pulp.[5] The importance of fines in mechanical pulps has been well characterised in relation to their contribution to pulp brightness and strength.[6]

In order to upgrade pulps such as those derived from Douglas-fir, various groups have evaluated the potential of hydrolytic enzymes, mainly hemicellulases and cellulases, to modify fiber characteristics and enhance the strength[7-9] and optical properties[10-12] of the resultant pulp. Although these enzyme preparations tend to be enriched in one particular component, such as cellulase or xylanase, in many cases these preparations include other enzyme activities. Although the presence of these contaminating side activities, such as hemicellulases in cellulase preparations, or vice versa, might be beneficial in increasing substrate accessibility, the "contaminating" activities can also mask the claimed effect by the enzyme of interest. Thus one of the objectives of this current work was to evaluate the synergistic effects of different ratios and loadings of xylanases, mannanases and cellulases.

To date most research has concentrated on kraft fibers primarily because they are considered to be more susceptible to enzyme attack than are mechanically derived pulp fibers.[12,13] Although a considerable amount of research has been directed towards xylanases/mannanases, and their ability to increase the final brightness of kraft pulps, there is still some debate as to the exact mechanism by which the enzyme pretreatments work.[14-18] Most past work has employed xylanases as a pretreatment stage prior to chemical bleaching. However, recent studies in our laboratory have indicated that a post bleaching xylanase treatment of peroxide bleached kraft Douglas-fir can result in a direct brightening effect.[19,20] Depending on the xylanase and the bleaching sequence used, a superior final brightness could be achieved when compared with a prebleaching treatment of a corresponding pulp. It has also been reported that the partially bleached kraft pulp was more susceptible to enzyme hydrolysis with more sugars being released compared to the prebleaching enzyme application, despite the fact that peroxide bleaching resulted in the dissolution of some of the carbohydrates present in the pulp.[19,21,22,23]

Unlike kraft pulps, mechanical pulps contain little if any redeposited xylan and hence it is the native xylan that is the target substrate during any possible xylanase mediated modifications. Although the hemicellulose has not undergone any specific chemical treatment, it may have been physically altered, depending on the conditions used for pulping. Studies to investigate the action of xylanases on mechanical pulps carried out by Jeffries and Lims[24] showed that a crude xylanase preparation from *Aureobasidium pullulans* containing negligible endoglucanase activity was not successful in releasing any significant amount of sugars from an Aspen derived Thermo Mechanical Pulp (TMP). However, dilute alkali treatments, which resulted in the partial removal of acetyl units, greatly increased the susceptibility of TMP to enzyme digestion. The strength properties were generally found to deteriorate after enzyme treatment, particularly the burst and tear strengths while the tensile strength remained essentially unchanged. Both alkali treated and untreated pulps showed an increase in brightness and opacity as a result of the enzyme treatment.

In related work, mechanical pulps derived from Douglas-fir wood chips were treated with a commercial cellulase preparation to assess the ability of the enzymes to modify fiber characteristics.[12] This cellulase preparation contained a relatively high xylanase activity and treatments of mechanical pulps, at low enzyme concentrations, resulted in a reduction in fiber coarseness and a slight improvement in handsheet roughness. However, these beneficial modifications were achieved at the expense of both the individual fiber strength and a reduction in interfiber bonding. Due to the high xylanase activity present in the cellulase mixture the demonstrated effects could not be solely attributed to the action of one group of enymes.

Overall, the use of hydrolytic enzymes to enhance mechanical pulp and fiber characteristics for paper making has met with limited success. In general, cellulase treatments are considered to be detrimental to pulp strength. However, treatments with cellobiohydrolases had little effect on the pulp, with no significant changes observed to the pulp brightness and strength.[25] Treatments with cellobiohydrolases have also resulted in substantial energy savings in the secondary refining of coarse mechanical pulp, as well as improvements in tensile strength.[26] To date the most significant impact of hydrolytic enzymes on TMP has been the application of acetyl esterases which have increased the pulp yield by as much as 1%. This probably occurred because of the adsorption of enzymatically deacetylated galactoglucomannans onto the mechanical pulp fibers.[27]

In the present work, we have investigated the potential of using hydrolytic enzymes to modify a peroxide bleached Douglas-fir mechanical pulp. We also assessed the modifications that occured in strength properties and determined if a direct brightening could be achieved by the action of the enzymes on the mechanical pulp.

3 MATERIALS AND METHODS

3.1 Pulp

Refiner Mechanical Pulp, provided by MacMillan Bloedel Research, was prepared by pulping steam pretreated Douglas-fir (*Pseudotsuga menziesii*) wood chips using a Sprout Waldron refiner under atmospheric pressure.

3.2 Chelation

Metal ions were removed from the pulp by complexing them with EDTA. The conditions for the chelation stage were: 3% (w/v) pulp consistency at pH 5.5 adjusted with HCl, with a loading of 1% EDTA at 50°C for 30 min. After the EDTA treatment the pulp was given a 1% water wash. The filtrate was passed through the pulp patty twice after each filtration step to collect the fines.

3.3 Peroxide Bleaching

Alkaline peroxide bleaching was carried out at a 10% pulp consistency for 20 min at 123°C in an autoclave.[28] The chemical dosages used were: 6% H_2O_2, 4% NaOH, 0.05% $MgSO_4$ and 4% Na_2SiO_3. The pulp was drained and 40 mls of the filtrate collected. The pulp was then given a 1% water wash in deionised water, with the pH adjusted to 5.5.

3.4 Enzyme Preparations and Assays

Five commercially available enzyme preparations were used: a xylanase, (Irgazyme 40S-4X) from *Trichoderma longibrachiatum,* (Genencor, CA, USA) and Xylanase I, Mannanase J, Cellulase E, Ecopulp T from *Trichoderma reesei* (Primalco Ltd. Biotech., Rajamaki, Finland). A partially purified endoglucanase from *Gloeophyllum saepiarium* was also included in the testing.[29] Mannanase and xylanase activities were determined according to Bailey et al,[30] 0.5% locust gum bean galactoglucomannan (Sigma) and 1% birchwood xylan (Sigma) were used as their respective substrates. Endoglucanase activity was measured according to Wood and Bhat[31] using 1% carboxymethylcellulose (Sigma) as the substrate. All assays were carried out at the optimum temperature and pH for each individual enzyme and expressed in nanokatals (where one nkat is the amount of enzyme that reduces 1 nmol of substrate per second).

3.5 Enzymatic Treatments

The enzymatic treatments were carried out at 10% pulp consistency for 1 h with agitation in polyethylene bags. The pH and temperature were dependent on the enzyme used. The xylanase loadings were based on the amount of D-xylose liberated from the pulp by enzyme treatment. Control pulps were treated in parallel with equivalent amounts of heat inactivated enzyme. After hydrolysis, pulp samples were boiled for 20 minutes to denature the enzyme. Filtrates were collected for sugar analysis.

3.6 Carbohydrate Analysis

Filtrates collected after peroxide bleaching and enzyme treatments were freeze dried prior to acid hydrolysis. Samples were resuspended in 28 ml of 4% H_2SO_4 and autoclaved (123°C) for 30 minutes according to TAPPI Test Method UM 250, (1991). Aliquots (1.5 ml) of each sample were filtered through 45 μm Millipore filters and the monosaccharides solubilised were detected using anion-exchange chromatography on a CarboPac PA-1 column and pulsed amperometic detection using a Dionex DX-500 HPLC system (Dionex, Sunnyvale CA, USA) controlled by Peaknet 4.3 software.[32] The carbohydrate composition of the bleached and unbleached pulps was determined as previously described.[20]

3.7 Handsheet Preparation

Optical handsheets were made after peroxide bleaching and after subsequent enzyme treatment according to Tappi Test Methods T272 om-92. Physical handsheets were prepared after enzyme treatment (Tappi Test Methods T205 om-88), with the fines recycled to minimise yield loss. The handsheet properties and pulp freeness were determined after enzyme treatment using standard TAPPI procedures.

4 RESULTS AND DISCUSSION

As previous work indicated that both the individual fractions as well as mixtures of xylanase, mannanase and endoglucanase could affect pulp and paper properties in different ways, we first assayed a range of enzymes to determine their specificities and activities (Table 1). The activities were determined at the pH and temperature optima for each individual enzyme. It was apparent that all of the enzyme fractions that were enriched in a specific activity also contained various levels of "contaminating" side activities.

In earlier work[12,13] it was found that mechanical pulps demonstrated a low susceptibility to enzyme hydrolysis. For this reason relatively high enzyme concentrations could be used before any substantial yield or strength loss might occur. A range of individual and or combinations of different enzyme loadings and mixtures were prepared to assess the possibility of enhancing the pulp fiber characteristics (Table 2). The potential of the Cellulase E preparation (which contains a mixture of endoglucanase and xylanase activities) was assayed, as previous work had indicated that beneficial changes could be obtained when this type of enzyme mixture was added to unbleached mechanical pulp.[12] To determine the effects that might occur due to the action of endoglucanase alone, purified endoglucanase from *G. saepiarium* was included. As the endoglucanase enzymes within these two preparations differed substantially in their activities on model substrates it was anticipated that they would also exhibit different effects on the pulp.

The individual xylanase loadings that were used with the Ecopulp T, Xylanase I and Irgazyme samples were based on the liberation of D-xylose from the substrate. The Mannanase J loading was established in a similar fashion. As it was expected that the hemicellulase enzymes would be less degradative than the cellulases, substantially higher concentrations of the individual and "cocktail" mixtures of xylananses and mannanases were used (Table 2).

Table 1 *The enzyme activities of various enzyme preparations assayed at their optimum temperature and pH*

Enzyme	Temp (°C)	pH	Endo-glucanase (nkat/ml)	Xylanase (nkat/ml)	Mannanase (nkat/ml)	Ratio (endo:xyl:man)
Mannanase J	50	7.0	160	18,000	112,000	1: 112 :700
Xylanase I	50	7.0	135	250,000	500	1: 1,850 :3.7
Irgazyme	50	7.0	20	285,000	400	1: 14,250 :20
Ecopulp T	70	7.0	20	286,000	47,000	1: 14,300 :2,35
Cellulase E	45	4.8	33,000	25,000	4,000	1: 0.76 :0.12
Endoglucanase	45	4.8	2,700	43	36	1: 0.02 :0.01

Table 2 *The enzyme loading and combinations used for the subsequent treatment of unbleached mechanical pulp*

Enzyme Loading	Endoglucanase (nkat/g pulp)	Xylanase (nkat/g pulp)	Mannanase (nkat/g pulp)
Cellulase E	2100.00	1600.00	250.00
Endoglucanase	1000.00	16.00	13.30
Ecopulp T	0.20	2800.00	450.00
Irgazyme	0.05	800.00	1.00
X_I 9000*	5.20	9450.00	17.40
X_I 1000	0.70	1260.00	2.30
M_J 8000*	12.00	1350.00	8250.00
M_J 4000	6.00	675.00	4125.00
M_J 1000	1.60	180.00	1100.00
M_J 8000+X_I 1000	12.70	2610.00	8252.00
M_J 4000+ X_I 500	6.30	1305.00	4126.00
M_J 1000+ X_I 1000	2.30	1440.00	1102.00
M_J 500+ X_I 500	1.10	720.00	551.00

*M_J = Mannanase J and X_I = Xylanase I

Various enzyme loadings and combinations of different enzymes were added to bleached mechanical pulp to determine the effects on the optical and physical properties of the pulp. Several studies have assessed the potential of enzymes to modify either bleached or unbleached kraft pulps and in some cases the bleached samples showed a lower response to enzyme treatment.[33] This was thought to be due to the fact that the bleaching process resulted in the dissolution of carbohydrates, thus reducing the amount of substrate available to the enzyme. However, several other reports have stated[12,13] that the bleaching stage enhanced the accessibility of the enzyme to the substrate and that a direct brightening effect could be demonstrated after a subsequent enzyme treatment.[19,20] To determine if the bleaching stage increased enzyme accessibility, we peroxide-bleached the pulp, and the carbohydrate composition of the substrate was analysed before and after bleaching to determine if there were any changes in the sugars released (Table 3). We next determined what effect the various enzyme additions had on the strength properties of the pulp and individual fibers. Previously we had established that it was desirable to maintain high fiber strength after the enzyme treatment while tolerating yield losses of less than 4%.[12] When xylanases were applied to the pulp, little or no loss in zero span breaking length (Figure 1A) was observed with all the readings falling within the standard deviation of the control, regardless of the loading that was used. This agreed with earlier reports which showed removal of xylan did not significantly decrease the zero span breaking length.[34] Previous work carried out on unbleached Douglas-fir kraft pulp suggested that the partial removal of xylan slightly reduced the intrinsic fiber strength, although the effect was diminished by subsequent bleaching.[7]

Table 3 *Carbohydrate composition of Douglas-fir mechanical pulp (RMP) before and after bleaching*

Substrate	Carbohydrate Composition %				
	Mannose	Xylose	Glucose	Galactose	Arabinose
Unbleached Pulp	13.7	4.8	47.0	3.2	1.4
Bleached Pulp	14.5	5.0	51.0	2.9	1.5

There has been little previous work carried out on the potential of mannanases to modify mechanical pulp properties. However, earlier work on kraft pulp indicated that a mannanase prebleaching step resulted in a reduced loading of bleaching chemicals and was accompanied by no significant loss in handsheet strength properties.[35] Our work indicated that the addition of Mannanase J resulted in a greater reduction in fiber strength than did the xylanases, with substantial reductions occurring at the higher enzyme loadings. An almost linear decrease from 8.4 km to 7.8 km was observed with loadings from 1000 nkat to 8000 nkat. It is possible that the fiber weakening observed was due to the action of contaminating endoglucanase that becomes more apparent at the higher enzyme loadings.

Previous reports indicated that cellulase treatments, particularly the action of endoglucanases, could detrimentally decrease pulp strength.[36] Enzyme treatments with a complete cellulase mixture caused severe fiber strength reduction in kraft pulps,[12] while other reports indicate that cellobiohydrolases had no effect on pulp strength.[37] In our work it was apparent that the addition of Cellulase E, a CBH minus preparation, caused a substantial decrease in zero span breaking length of the mechanical pulp (Figure 1).

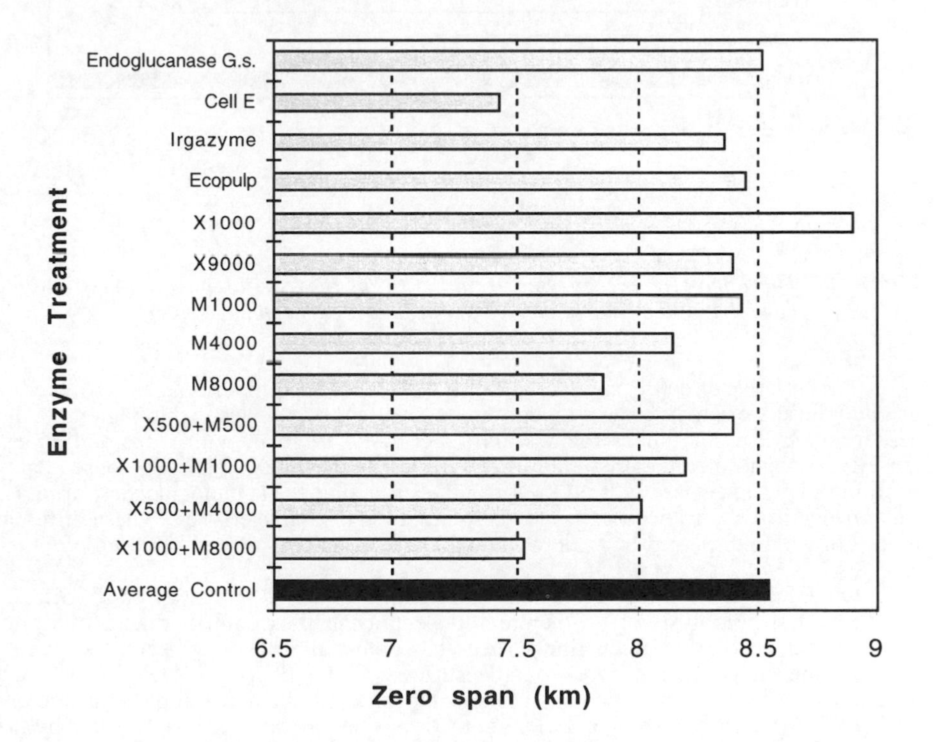

Figure 1 *The effects of the enzyme treatments on the zero span breaking length of bleached mechanical pulp (the standard deviation of the control is +/-0.44)*

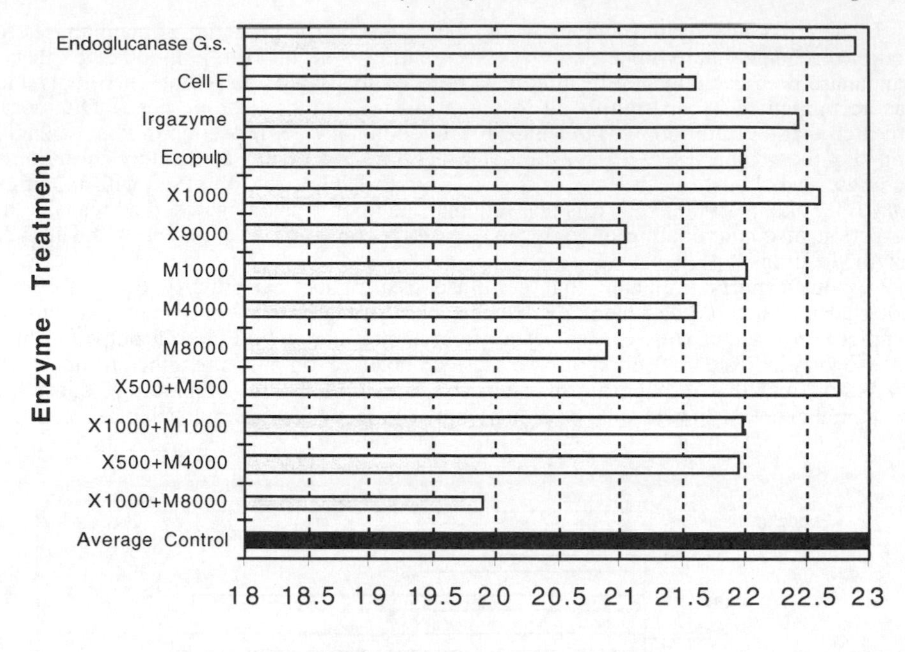

Figure 2 *The effects of the enzyme treatments on the tensile index (the standard deviation of the control is +/-1.67) of bleached mechanical pulp*

The negative effect resulting from some cellulases has primarily been attributed to its endoglucanase activity as it is thought to hydrolyse the more accessible regions of the cellulose. As the amorphous regions are structurally irregular, with kinks and nodes, these are thought to be preferentially hydrolysed by some endoglucanases, consequently resulting in detrimental effects on the fibers.[38] Surprisingly, the endoglucanase from *G. saepiarium*, had shown no changes to the pulp strength characteristics. These different modes of action displayed by various endoglucanases have also been reported by other workers.[32,33,36,37]

One of the goals of the various enzyme treatments was to increase fiber fibrillation by the partial hydrolysis of the hemicellulose portion thus contributing to the area available for bonding between fibers. Another potentially positive mechanism would occur if the enzyme could work on the surface of the fibers by "peeling off" the accessible hemicellulose components. It was apparent that the removal of hemicellulose was generally accompanied by a decline in the interfiber bonding as demonstrated by the change in the tensile breaking length (Figure 2) and the burst index (Figure 3). In all cases the tensile and burst values decreased with increasing enzyme loading. High loadings of Xylanase I 9000 nkat/g, Mannanase J 8000 nkat/g and the combined treatments of Xylanase I 1000 nkat/g and Mannanase J 8000 nkat/g resulted in the highest removal of hemicellulose (Figure 4) consequently both the tensile and burst values were significantly decreased. Several groups have indicated that the removal of hemicellulose from the pulp was detrimental to the interfiber bonding.[12,34]

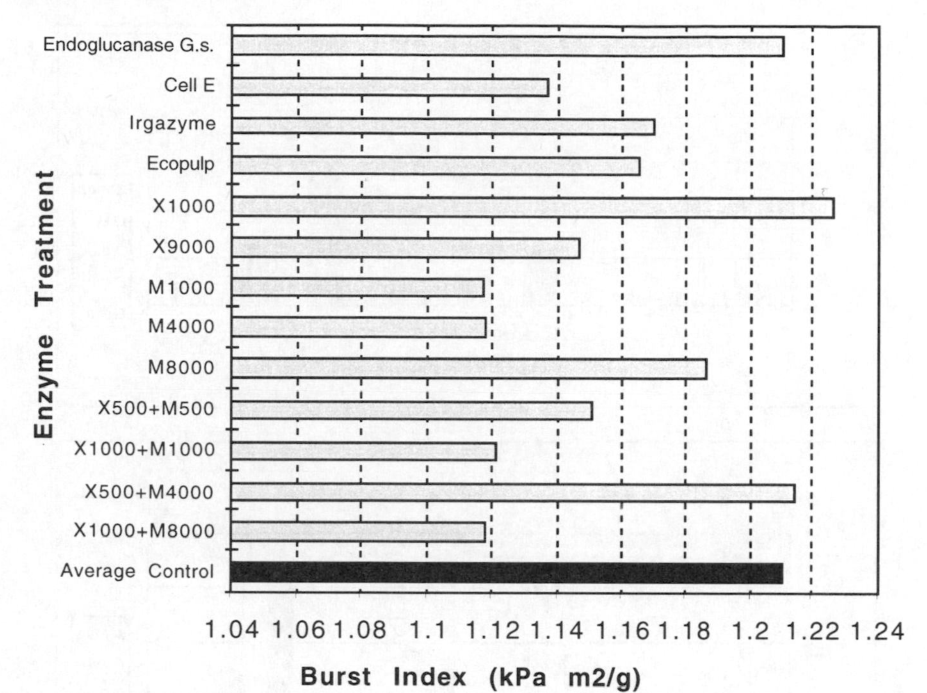

Figure 3 *The effects of the enzyme treatments on the burst index (the standard deviation of the control is +/-0.074) of bleached mechanical pulp*

Degradation of both the cellulose and hemicellulose was also evident after the Cellulase E treatment (Figure 4C) and the observed decrease in interfiber bonding was in the same range as that suffered after the combined xylanase and mannanase treatment at the higher loadings. Again the addition of endoglucanase from *G. saepiarium* did not result in any changes to the interfiber bonding.

To assess if there was a correlation between losses in the strength properties and the amount of sugars liberated, the neutral wood sugars released were analysed by HPLC, after acid hydrolysis of the soluble sugars (Figure 4). A mannanase treatment of 1000 nkat/g of pulp hydrolysed about 0.4% of the available carbohydrate, similar to the values reported by other workers.[34] This low level of hydrolysis is surprising as galactoglucomannan comprises a significant part of the hemicellulose in softwood. However, as was found with the xylanases and endoglucanases, the origin of the mannanase appears to be one of the factors that determines its ability to hydrolyse pulp mannan, with different species incorporating a different mode of action.[39] It is also recognised that the origin of the pulp and the pulping procedure used will also influence the ability of the enzymes to solubilise softwood mannan.[40-42] It appears that the mannan is largely inaccessible to the enzyme, possibly shielded by the xylan which is more readily hydrolysed by the contaminating xylanase (Table 2) which resulted in the liberation of twice as much xylose as mannose. As the galactoglucomannans appeared to be the primary polysaccharides liberated as a result of peroxide bleaching of the RMP (Table 3), it was possible that less of the substrate was available to the mannanase.[43] Higher enzyme loadings were accompanied by an increase in xylan hydrolysis. Individual loadings of mannanase at 1000 nkat/g and 4000 nkat/g resulted in xylose being liberated as the main hydrolysis product (Figure 4A).

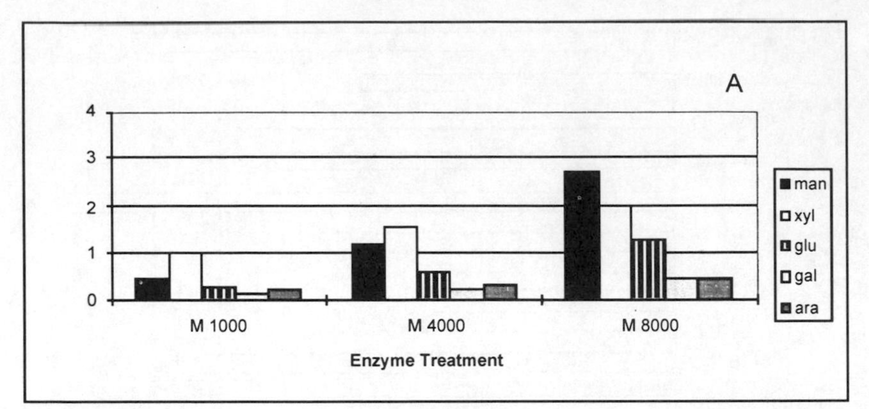

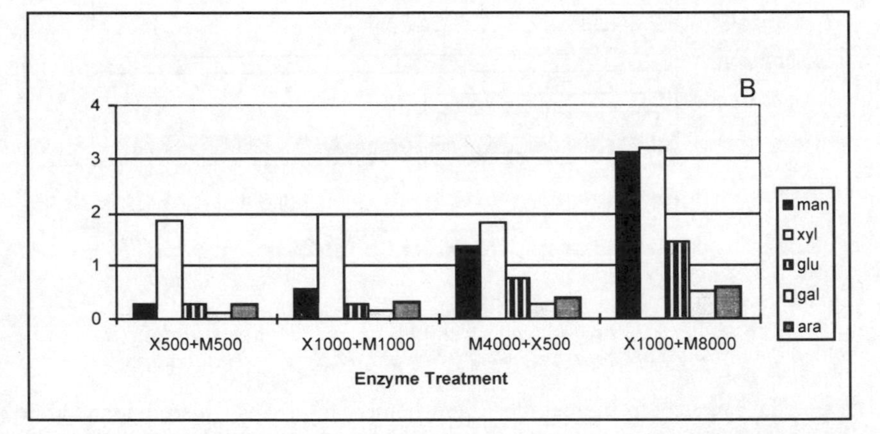

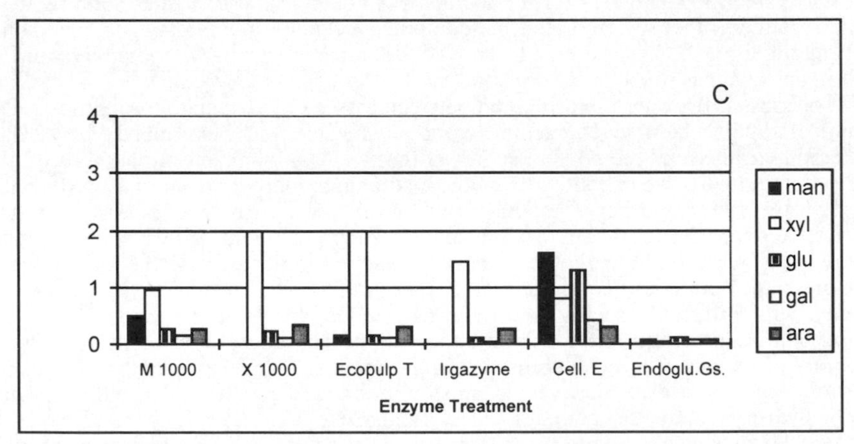

Figure 4 *The carbohydrates solubilised during the enzyme treatment*

The highest mannanase loading resulted in the solubilisation of 0.9% of the pulp. The ratio of the solubilised sugars, mannose: glucose: galactose was approximately, 2: 1: 0.4, with this ratio remaining constant with increasing mannanase loadings.

Supplementing the mannanase with additional xylanase (Xylanase I) enhanced the release of mannose and a synergistic effect was demonstrated (Figure 4B). The most pronounced effect was observed after the combined treatment of mannanase 8000 nkat/g with the xylanase at a loading of 1000 nkat/g. It appears that the combined treatment with the xylanase further increased the accessibility of the galactoglucomannan to the mannanase, implying that the xylan and galactoglucomannan are closely associated with each other on the surface of the fiber. The additional xylanase activity promoted galactoglucomannan hydrolysis and resulted in a 50% increase in mannose released over the action of the individual enzymes. This synergistic trend was also obvious in the lower, combined enzyme loadings.

The profile of sugars released (Table 2) indicated that Irgazyme and Xylanase I (1000 nkat/g) only acted on the pulp xylan with 0.2% and 0.3% respectively of the original xylan hydrolysed. Surprisingly, Ecopulp T also demonstrated a similar hydrolysis profile with relatively little mannose released despite having a higher mannanase side activity (Table 2). All three xylanases approximately released the same ratio of xylose to arabinose (1:0.125). It has been reported[24] that unbleached TMP treated with a xylanase (6000 nkat/g of pulp), from *Aureobasidum pullans* resulted in a 1% hydrolysis of the pulp. Of all the enzymes tested, the endoglucanase from *G. saepiarium* was the least effective and caused a negligible change to the carbohydrate content of the pulp. In contrast, the commercial endoglucanase, Cellulase E, hydrolysed 0.45% of the pulp, with a surprisingly high levels of mannose (0.16%) released.

The important contribution of fines in determining both the optical and physical properties of mechanical pulps has been shown by several groups.[4,44] Freeness is essentially a measure of the drainage of the pulp and has been associated with the accumulation of fines. Typically mechanical pulps contain up to 30% fines and these fines generally contain a high proportion of lignin. After enzyme treatments both positive and negative changes in pulp freeness were observed (Figure 5). As cellulases are known to act preferentially on the fines due to their relative high surface area[7] the 10% increase in freeness resulting from treatment with Cellulase E was expected (Figure 5). Similarly as hemicelluloses are known to be hydrophilic in nature, their partial solubilisation should lead to an increase in the rate of water removal.[33] Thus, the significant increases in freeness were obtained after the combined action of the mannanases (8000 nkat/g) and xylanase (1000 nkat/g). Surprisingly, the *G. saepiarium* endoglucanase which had previously produced no changes on the pulp increased the freeness to the same order of magnitude, as was observed with the other enzymes. The various xylanase treatments, although loaded at similar concentrations, produced quite different changes to the pulp freeness. Addition of Ecopulp T increased the freeness by about 6%. The Irgazyme treatment showed no significant change over the control, while Xylanase I treatments resulted in decreased freeness irrespective of the loading. This was the result of fibrillation of the fibers or partial hydrolysis that contributes to the fines content.

The overall effect of the various enzyme treatments on pulp brightness was negligible, with no direct brightening obtained. The brightness achieved directly after peroxide bleaching varied between 56.4% and 61.4% ISO (Table 4). Although a subsequent treatment with most of the enzyme resulted in a slight decrease in brightness, Cellulase E reduced the ISO brightness by about 1%. When reference bleached pulps were treated with heat inactivated enzyme, the resultant brightnesses were also lower and were similar to the brightness values obtained after the treatment with active enzymes. This reduction in brightness resulting from treatments with both the active and deactivated enzyme may be due to the rather dark colour of the enzyme formulation. The Cellulase E application was particularly dark and it was also loaded at the higher volume.

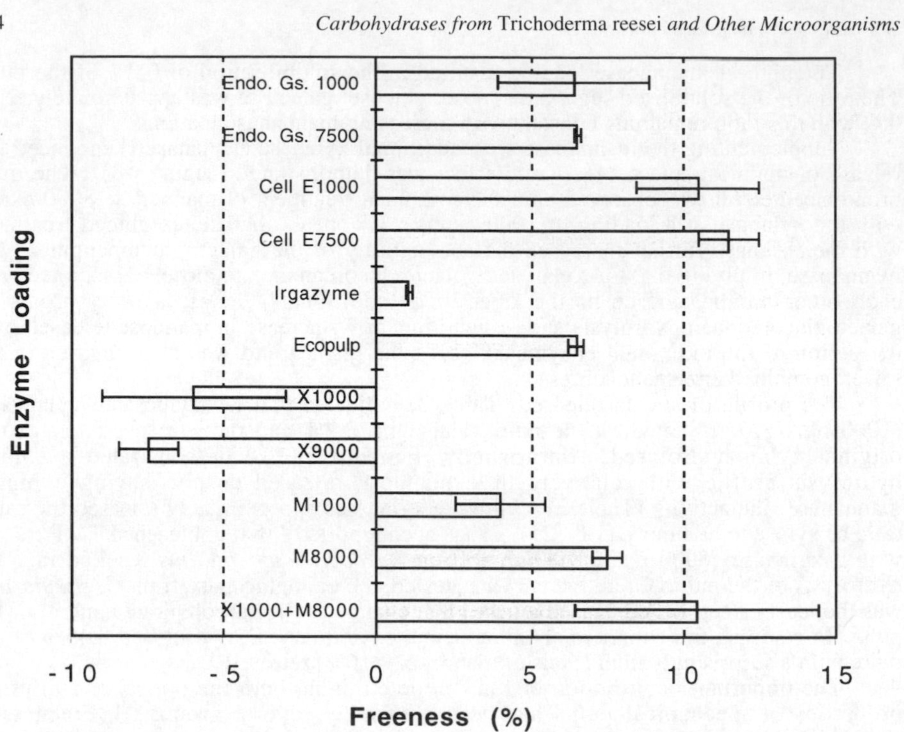

Figure 5 *The percentage of change in freeness after the various enzyme treatments*

Table 4 *The ISO Brightness of peroxide bleached pulp and the corresponding enzyme treated pulp*

Enzymes[a] Treatment	Pulp Brightness (% ISO)			
	Bleached Pulp Control[b]	Denatured Enzyme[b]	Bleached Pulp Control[c]	Enzyme Treated[c]
M_J 8000+ X_I 1000	61.4	59.4	59.0	58.7
M_J 8000	N/A	N/A	60.2	59.7
X_I 1000	N/A	N/A	60.4	60.6
Irgazyme	58.6	58.2	58.9	58.9
Ecopulp T	57.1	56.8	56.4	56.4
Cellulase E	61.1	58.5	59.0	58.0

a . M_J = Mannanase J and X_I = Xylanase I, refer to Table II for enzyme loadings.

b. Control and corresponding denatured enzyme treated pulp brightnesses.

c. Control and corresponding enzyme treated pulp brightnesses.

N/A. Samples not assessed

Previous work[19,20] indicated that, when peroxide bleached Douglas-fir kraft pulps were treated with a xylanase (Irgazyme), a direct brightness gain of 1.2% ISO was achieved. Further studies on fractionated Douglas-fir kraft pulp revealed that, following an Irgazyme treatment, the smallest fibers showed the highest response after a subsequent bleaching stage.[7]

However, this present study indicated that both the Irgazyme and the other enzymes resulted in no significant change to the pulp brightness of the mechanical pulp. This was probably due to the different target substances in the bleached mechanical pulp. Previously a slight increase in brightness was observed after a xylanase treatment of an unbleached TMP. This was reported to occur without any significant change in the xylan content of the pulp.[24] Although both the xylanase and mannanase treatment of the unbleached mechanical pulps increased the bleachability of the pulp a direct brightening effect was not observed.[21]

5 CONCLUSIONS

The commercial enzyme preparations resulted in some fiber modification although the degree of modification was dependent on the type of enzyme and dosage used. A synergistic effect was observed when a combination of xylanase and mannanase was applied to the bleached pulp. The xylan appeared to be the most accessible hemicellulose component available to the enzymes. The endoglucanase from *G. saepiarium* had little effect on the pulp while the Cellulase E caused a substantial decrease in pulp strength. The fact that Cellulase E solubilised a significant amount of hemicellulose suggested that the other enzyme activities within the preparation greatly enhanced pulp hydrolysis when the patterns of sugars released after treatments with the purified endoglucanase from *G. saepiarium* were compared.

6 ACKNOWLEDGMENTS

We would like to thank the Pulp and Paper Research Institute of Canada (Vancouver) for the use of their handsheet making and testing facilities. We extend our gratitude to the National Diagnostic Research Center of Ireland and Coford Ireland for the financial support of this project and also to Shawn Mansfield for his assistance in editing this paper.

References

1. G.A. Smook, 'Handbook for Pulp and Paper Technologists'. 2nd Edition, Angus Wilde Press, Vancouver BC., 1992, p. 35.
2. U.-B. Mohlin, *Tappi Proc. Int. Mech. Pulp. Conf.*, 1995, 71.
3. T.H. Quick, M.A. Siebers, D.M. Hanes, *Tappi J.*, 1991, **74**(10), 107.
4. O.L. Forgac, *Pulp Pap. Can.*, 1963, **C**, T89.
5. G. de Silveira, X. Zhang, R. Berry, J.R. Wood, *J. Pulp Pap. Sci.*, 1996, **22**, J315.
6. U-B. Mohlin, *Svensk Papperstidn.*, 1977, **3**, 84.
7. S.D. Mansfield, K.K.Y. Wong, E. de Jong, J.N. Saddler, *Appl. Microbiol. Biotechnol.*, 1996, **46**, 319.
8. B. Saake, T. Clark, J. Puls, *Holzforschung*, 1995, **49**, 60.
9. R.P. Kibblewhite, T.A. Clark, *Appita J.*, 1996, **49**, 390.
10. L. Viikari, M. Ranua, A. Kantelinen, J. Sundquist, M. Linko, *Proc. 3rd Int. Conf. Biotechnol. Pulp Pap. Ind.*; Stockholm, 1986, 67.
11. G.M. Gübitz, W. Schnitzhofer, H. Balakrishnan, W. Steiner, *J. Biotechnol.*, 1996, **50**, 181.
12. S.D. Mansfield, K.K.Y. Wong, E. de Jong, J.N. Saddler, *Tappi J.*, 1996, **79**(8), 125.

13. T.W. Jeffries, 'Materials and Chemicals from Biomass', 1992, p. 313.
14. A. Suurnakkii, A. Heijneson, J. Buchert, U. Westermark, L. Viikari, *J. Pulp Pap. Sci.*, 1996, **22**, J91.
15. A. Kantelinen, B. Hortling, J. Sundquist, M. Linko, L. Viikari, *Holzforschung*, 1993, **47**, 318.
16. J.L. Yang, K.-E.L. Eriksson, *Holzforschung*, 1992, **46**, 481.
17. T.A. Clarke, D. Steward, M.E. Bruce, A.G. McDonald, A.P. Singh, D.J. Senior, *Appita J.*, 1991, **44**, 389.
18. K.K.Y. Wong, P. Clarke, S.L. Nelson, *ACS Symp. Ser.*, 1995, **618**, 352.
19. K.K.Y. Wong, L. Martin, F.M. Gama, J.N. Saddler, E. de Jong, *Biotechnol. Bioeng.* 1997, (in press).
20. E. de Jong, K.K.Y. Wong, L. Martin, S.D. Mansfield, F.M. Gama, J.N. Saddler, *ACS Symp. Ser.*, 1996, 655, 44.
21. T.A. Clark, R.P. Kibblewhite, R.W. Allison, *Appita Annu. Gen. Conf.*, Auckland, 1996, 47.
22. D.J. Senior, P.R. Mayers, D. Miller, R. Sutcliffe, U.L. Tan, J.N. Saddler, *Biotechnol. Letters*, 1988, **10**, 907.
23. S. Nelson, K.K.Y. Wong, J.N. Saddler, R.P. Beatson, *Pulp Pap. Can.*, 1995, **96**(7), 42.
24. T.W. Jeffries, C.W. Lins, *Proc. 4th Int. Conf. Biotechnol. Pulp Pap. Ind.*, T. Kirk, H.-M. Chang, Eds.; Butterworth-Heinemann, Boston, 1990, 191.
25. J. Pere, M. Siika-aho, J. Buchert, L. Viikari, *Tappi J.* , 1995, **78**(6), 71.
26. J. Pere, S. Liukkonen, M. Siika-aho, J. Gullichsen, L. Viikari, *Proc. Tappi Pulp. Conf. Nashville TN* , 1996, 693.
27. J. Thornton, M. Tenkanen, R. Ekman, B. Holmbom, L. Viikari, *Holzforschung*, 1994, **48**, 436.
28. P. Kyriacou, E. de Jong, C.L. Johansson, R.P. Chandra, J.N. Saddler, (submitted).
29. G.M. Gübitz, S.D. Mansfield, J.N. Saddler, (submitted).
30. M.J. Bailey, P. Biely, K. Poutanen, *J. Biotechnol.*, 1992, **23**, 257.
31. T.M. Wood, G.R. Bhat, *Methods Enzymol.*, 1988, **160**, 160.
32. E. de Jong, K.K.Y. Wong, J.N. Saddler, *Holzforschung*, 1997, **51**, 19.
33. R.P. Kibblewhite, T.A. Clarke, *Appita J.*, 1996, **49**, 390.
34. J.C. Roberts, A.J. McCarthy, N.J. Flynn, P. Broda, *Enzyme Microb.Technol.*, 1990, **12**, 210.
35. T.A. Clarke, A.G. Mc Donald, D.J. Senior, P.R. Mayers, *Proc. 4th Int. Conf. Biotechnol. Pulp Pap. Ind.*; T.K. Kirk, H.-M. Chang, Eds., Butterworth-Heinemann, Boston, 1990, 153.
36. J. Buchert, M. Ranua, M. Siika-aho, J. Pere, L. Viikari, *Appl. Microbiol. Biotechnol.*, 1994, **40**, 941.
37. J. Pere, M. Siika-aho, J. Buchert, L. Viikari, *Tappi J.*, 1995, **78**(6), 71.
38. N. Gurnagul, D.H. Page, M.G. Paice, *Nord. Pulp Pap. Res. J.*, 1992, **7**, 152.
39. G.M. Gübitz, D. Haltrich, B. Latal, W. Steiner, *Appl. Microbiol. Biotechnol.* (in press).
40. J. Buchert, A. Kantelinen, M. Rättö, M. Siika-aho, M. Ranua, L. Viikari, *Proc. 5th Int. Conf. Biotechnol. Pulp Pap. Ind.*, M. Kuwahara, M. Shimada, Eds., Uni Publishers Co. Ltd.: Tokyo, Japan, 1992, 139.
41. A. Suurnakkii, T.A. Clarke, R.W. Allison, L. Viikari, J. Buchert, *Tappi J.*, 1996, **79**(7), 111.
42. J. Buchert, J. Salminen, M. Siika-aho, M. Ranua, L. Viikari, *Holzforschung*, 1993, **47**, 473.
43. J. Thorton, R. Ekman, B. Holmbom, F. Örsa, *J. Wood Chem. Technol.*, 1994, **14**, 159.
44. E. Retulainen, K. Nieminen, *Paperi Ja Puu*, 1996, **78**(5), 305.

IDENTIFICATION, PURIFICATION AND CHARACTERISATION OF THE PREDOMINANT ENDOGLUCANASES FROM TWO BROWN-ROT FUNGAL STRAINS OF *GLOEOPHYLLUM*

S.D. Mansfield[1], B. Saake[2], G. Gübitz[1], E. de Jong[3], J. Puls[2] and J.N. Saddler[1]

[1]Forest Products Biotechnology, University of British Columbia, Vancouver, BC, Canada; [2]Bundesforschungsanstalt für Forst- und Holzwirtschaft, Leuschnerstr. 91, 21027 Hamburg, Germany; [3]Agrotechnological Research Institute, ATO-DLO, 6700AA Wageningen, The Netherlands

1 INTRODUCTION

Our understanding of cellulolytic enzymes is primarily based on the study of relatively few bacteria, such as *Clostridium* or *Cellulomonas*, and fungi such as *Trichoderma* and *Humicola*. However, there is little information available on the mechanism of cellulose hydrolysis by brown-rot fungi. The brown-rot fungi are among the most damaging group of microorganisms to the wood products industry through their ability to attack the carbohydrate components and modify the lignin in the plant cell wall. However, the enzymatic machinery involved in this extensive degradation is not fully understood. These fungi are unique in that they process the cellulose and hemicellulose components of lignocellulose substrates without significant lignin removal. During the initial stages of decay they rapidly reduce the degree of polymerisation of cellulose in wood from about 10,000 to residual fragments with a DP of about 300, without significant weight loss.[1-3] In contrast to the substantial carbohydrate degradation, the lignin moieties have been shown to be relatively undigested by brown-rot attack, undergoing modifications such as hydroxylation,[4,5] demethoxylation[6] and the accumulation of oxidised polymeric compounds.[7,8] The degradation/modification mechanisms for lignocellulosic constituents by brown-rotters is still unclear, however, it has been suggested that the observed alterations are the result of both oxidative and hydrolytic pathways.[9,10]

It has been shown[11] that white- and brown-rot fungi bring about degradation of wood in very different ways. While the white-rot fungi appear to degrade both the crystalline and amorphous regions of cellulose with no rapid depolymerisation of the cellulose occurring, the brown-rot fungi have been shown to depolymerise the cellulose very rapidly with little degradation of the fibres being evident by microscopic examination.[1,12] To date, only a few cellulolytic enzymes have been isolated and purified from brown-rot fungi, including endoglucanases from *Tyromyces palustris*,[13] *Polyporus schweinzii*,[14] *Gloeophyllum* (*Lenzites*) *trabea*[15] and *Gloeophyllum* (*Lenzites*) *sepiarium*.[16] Most brown-rot fungi do not possess the complete cellulolytic system needed for degradation of crystalline cellulose, with strains such as *G. trabeum* showing only incomplete hydrolysis.[17] However, some of the brown-rot fungi such as *Coniophora puteana* have been shown to produce cellobiohydrolases.[18,19]

Most brown-rot fungi produce endoglucanases, which are thought to initiate the process of degradation by cleaving internal glycosidic bonds[20] contributing to the observed reduction in the degree of polymerisation of the cellulose. Studies on isolated cellulose-depolymerising endoglucanases from cultures of the brown-rot fungus *Merulipora incrassata* suggest that they predominantly cleave within the amorphous regions of cotton cellulose microfibrils.[21] The endoglucanases, which have been isolated from different brown-rot fungi, were found to have rather small molecular weights, ranging from 16 to 45 kDa. These acidic cellulases generally appear to be constitutively expressed and are not readily catabolically repressed by glucose.[11,20,21]

However, there is indirect evidence that the biochemical agents which cause the initial depolymerisation are small diffusible molecules, as enzymes have been shown to be too large to penetrate wood and reach the internal cellulose.[1,10,12,22-30] To date, attempts to identify the "cellulose depolymerising agent" produced by the brown-rot fungi have been unsuccessful. However, putative agents include the generation of Fenton's type reagents (H_2O_2 and ferrous salts)[17,31-33] and oxalic acid.[34,35] Another possible mechanism of depolymerisation is by production of cellobiose oxidising enzymes. A brown-rot fungus (*Coniophora puteana*) has recently has been reported to produce a cellobiose dehydrogenase,[36] and this enzyme has recently been shown to aid in the depolymerisation of cellulose.[37]

In this current investigation, five common strains of brown-rot fungi were screened for their production of both hydrolytic and oxidative enzymes, in an attempt to further elucidate the mechanism of lignocellulose degradation. During the screening process it became evident that two different strains of *Gloeophyllum* produced significantly higher amounts of endoglucanases than did the other brown-rot strains. These two distinct endoglucanases were subsequently purified and characterised.

2 MATERIALS AND METHODS

2.1 Fungal growth and sampling

Five fungal strains of brown-rot fungi (*Gloeophyllum trabeum* MAD617, *Gloeophyllum sepiarium* MB 135, *Coniophora puteana* UBC 01, *Lentinus lepideus* MAD534, and *Postia placenta* MAD698) were maintained on 2 % malt agar plates at 28°C. Three 8 mm plugs from 5 day old cultures were inoculated into a basal salts medium[38] supplemented with 5 % (w/v) glucose and 5 % (w/v) carboxymethylcellulose. Cultures were maintained at 28°C without shaking for 30 days. Each culture was sampled every 5 days and assayed for different enzyme activities.

2.2 Protein determination

The protein content of the fungal growth filtrates and purified steps was quantified using the bicinchoninic acid kit from Pierce (Rockford, USA) according to the manufacturer's suggested instructions.

2.3 Hydrolytic enzyme assays

The hydrolytic activities of the culture filtrates were determined using carboxymethylcellulose (1 % CMC, Sigma, St. Louis, USA), filter paper (No.1 Whatman), xylan (1 % Birchwood xylan, Sigma), and mannan (1 % locust bean gum and 1 % guar gum, Sigma) in sodium citrate buffer (pH 4.5) and methods described previously.[39-41] For all of the hydrolytic assays, reducing sugars were determined using the dinitrosalicylic acid (DNS) method. The β-glucosidase assay was performed using 1 mM *p*-nitrophenyl β-D-glucoside in 50 mM sodium acetate buffer pH 4.5 as a substrate. A 400 μL aliquot of culture filtrate was added to 100 μL substrate, 100 μL buffer, and 400 μL H_2O, and mixed. After 5 minutes, the reaction was stopped with 200 μL 1 M sodium carbonate and the adsorption was read on a spectrophotometer at 400 nm using a blank of 400 μL basal salts media, 100 μL substrate, 100 μL buffer, and 400 μL H_2O.[39]

2.4 Oxidative enzyme assays

2.4.1 Laccase (Polyphenol oxidase). The oxidation of 2,6-dimethoxyphenol (DMP) to an orange/brown dimer was used to measure laccase activity. The reaction mixture contained up to 600 μl culture filtrate in 50 mM sodium malonate buffer (pH 4.5), 1 mM manganese (II) sulphate and 1 mM DMP in a total volume of 900 μl. The reaction was followed spectrophotometrically for a minimum of 2 minutes at 468 nm.[42]

2.4.2 Manganese Peroxidase (MnP). The H_2O_2-dependent formation of an orange/brown product from DMP was used to assay for MnP activity. The reaction mixture contained 50 mM sodium malonate buffer (pH 4.5), 1 mM manganese (II) sulphate and 1 mM DMP and up to 600 μl culture filtrate in a total volume of 1 mL. The reaction was started by the addition of 0.4 mM H_2O_2 and corrected for any laccase activity that was present.[42]

2.4.3 Oxidase Activity. Aryl (veratryl) alcohol oxidase, glucose oxidase and glyoxal oxidase activity were monitored by the oxidation of veratryl alcohol, glucose and glyoxal, respectively. The reaction mixture contained 50 mM sodium 2,2-dimethylsuccinate (DMS) buffer (pH 4.5), 10 mM horseradish peroxidase (type II), 0.01 % (w/v) phenol red, and 10 mM aryl alcohol (3,4-dimethoxybenzyl alcohol) or glucose or glyoxal and up to 600 μl culture filtrate in a total volume of 950 μl. The reaction was allowed to proceed for 1 hour and then stopped by the addition of 50 μl of 2N NaOH and monitored spectrophotometrically at 610 nm.[42]

2.4.4 Oxidoreductase Activity. Cellobiose dehydrogenase (CDH) activity was assayed by the reduction of cytochrome *c* at 550 nm (ε=28 $mM^{-1}\cdot cm^{-1}$). The assay mixture contained 3 mM cellobiose, 20 mM succinate, pH 4.5, 12.5 μM cytochrome *c* and varying amounts of culture filtrate making a total of 1 mL. 2,6-Dichlorophenolindophenol (DCPIP) was used to measure the combined CDH and CBQ activity at 515 nm (ε=6.8 $mM^{-1}\cdot cm^{-1}$). The DCPIP assay mixture contained 3 mM cellobiose, 20 mM succinate, pH 4.5, 7.5 μM DCPIP and varying amounts of culture filtrate making a total of 1 mL. All assays were performed at 23°C.[37]

2.5 Protein purification

After 30 days of growth the proteins were harvested from the fungal culture. The mycelium was separated from the culture fluid by vacuum filtration. Proteins in the solution were precipitated on ice by slowly adding $(NH_4)_2SO_4$ until 90 % saturation. Following precipitation the proteins were collected by centrifugation at 8000 rpm for 2 hours at 4 °C.

The precipitated protein was dissolved in 10 mM ammonium acetate-buffer (pH 4.5). This solution was brought to 400 mM $(NH_4)_2SO_4$ and subjected to hydrophobic interaction chromatography on a Phenyl-Sepharose CL-4B column, which was pre-equilibrated with 10 mM ammonium acetate-buffer (pH 4.5). Elution of the protein was carried out by a linear gradient of 400 mM to 0 mM $(NH_4)_2SO_4$ in 10 mM ammonium acetate-buffer (pH 4.5). The samples were collected with a fraction collector and assayed for cellulase activities. All fractions exhibiting activity were pooled and concentrated in an Amicon ultrafiltration unit equipped with a YM10 (10,000 MW cut-off) membrane. The concentrated protein was then washed thoroughly with 10 mM 1-methylpiperazine (pH 4.8).

This concentrated protein solution was further purified by anion-exchange chromatography using a MonoQ HR5/5 (Pharmacia Biotech) column by FPLC. The column was pre-equilibrated with 10 mM 1-methylpiperazine (pH 4.8) before protein loading. Elution of the protein from the column was achieved by a linear gradient of 0 to 250 mM NaCl in 10 mM 1-methylpiperazine (pH 4.8). Samples were collected, assayed for activity, pooled and concentrated by Amicon ultrafiltration.

The final stage of purification involved gel filtration chromatography using a Superose 6 HR 10/30 column (Pharmacia Biotech) using 10 mM 1-methylpiperazine/100 mM NaCl (pH 4.8) at a flow rate of 0.2 mL/min.

2.6 Isoelectric focusing (IEF) and SDS-PAGE

The pI values were determined by isoelectric focusing (IEF). Standard marker proteins with pI values ranging from 2.5 to 6.5 were run simultaneously (Pharmacia). Proteins were separated by SDS-PAGE under denaturing conditions and the bands were compared to those of molecular weight standards (range 14.4 kDa to 97 kDa, Biorad, Richmond, USA).

2.7 N-terminal protein sequencing

Proteins were transblotted from SDS-PAGE gels onto PVDF membranes. Electroblotting was carried out in 10 mM CAPS-buffer at pH 11 using the BioRad Trans-Blot Electrophoretic Transfer Cell. N-terminal protein sequencing was performed on a Perkin-Elmer Applied Biosystems Model 476A automated protein sequencer using standard gas-phase Edman chemistry. The sequencer was equipped with an on-line reverse phase HPLC and utilised the 610A data analysis system.

2.8 Transglycosylation

Transglycosylation activity was monitored by thin layer chromatography using *p*-nitrophenyl ß-cellobioside (Sigma) as a substrate. The purified endoglucanases were incubated with 50 mM *p*-nitrophenyl ß-cellobioside in sodium citrate buffer, pH 4.2 at

60°C. Aliquots of the transglycosylation solution were removed at various times, blotted onto Whatman AL SIL G/UV Silica gel (250 μm layer) thin layer chromatography plates and eluted with an 7:2:1 ethylacetate:methanol:water solution. A charring solution of 9:1 methanol:H_2SO_4 was used to visualise cellooligosaccharides.

2.9 Temperature and pH profiles

The pH profiles of the purified endoglucanases were established by determining maximal endoglucanase activity on solutions of 1 % carboxymethylcellulose (CMC) in 50 mM McIlvaine buffer (pH 2.6 - 7.6) at 50 °C for 30 min. Following the determination of the pH optima, the temperature optima were determined on 1 % carboxymethylcellulose at temperatures ranging from 20 °C to 70 °C at the individual substrate pH optima of both endoglucanases for 30 min.

2.10 Substrate activity

Hydrolysis experiments were conducted on a number of substrates (10 g/L) at pH 4.2 (50 mM sodium citrate) and 30°C using 1 nkatal/mL and 10 nkatals/mL of the purified enzyme. These included carboxymethylcellulose (CMC, Sigma), hydroxyethylcellulose (HEC, Sigma), methylcellulose (MC, Sigma), microcrystalline cellulose (Avicel, Fluka), filter paper (Whatman), phosphoric acid swollen cellulose (PASC) prepared from Avicel, bacterial microcrystalline cellulose (BMCC, kindly donated by Dr. Tony Warren), ivory nut mannan (Megazyme, Sydney, Australia), locust bean galactomannan (Sigma), carob galactomannan (Megazyme, Sydney, Australia), konjac glucomannan (Megazyme, Sydney, Australia), *Tsuga canadiensis* galactoglucomannans (purified at BFH, by Dr. J. Puls), birchwood xylan (Sigma), larchwood xylan (Sigma), oat spelts xylan (Sigma), xyloglucan (Megazyme, Sydney, Australia), laminarin (Sigma), lichenan (Megazyme, Sydney, Australia) and pachyaman (Megazyme, Sydney, Australia). Released reducing sugars were determined using the DNS assay.

2.11 Cellooligosaccharide degradation

Ten mM soluble cellooligosaccharides (cellobiose to cellohexose, Seikagaku, Rockville, USA) were incubated with 2 nkatals endoglucanase for 30 minutes and the degradation products were quantified by high performance anion-exchange chromatography.

The cellooligosaccharides were separated on a CarboPac PA-1 column using a Dionex DX-500 High Performance Liquid Chromatography (HPLC) system (Dionex, Sunnyvale, CA, USA) controlled by Peaknet 4.30 software. The column was equilibrated with 150 mM NaOH and 50 mM sodium acetate and regenerated with 300 mM NaOH. After injecting 20 μL of sample using a SpectraSYSTEM AS3500 autoinjector (Spectra-Physics, Fremont, CA, USA), the oligosaccharides were eluted using a 50-200 mM gradient of sodium acetate (over 20 min.) at a flow rate of 1 mL/min. The oligosaccharides were monitored using a Dionex ED40 electrochemical detector (gold electrode), with parameters set for pulsed amperometric detection of sugars using deionised water at a flow rate of 1 mL/min.

2.12 Viscosity measurement

1.5 % (w/v) CMC-solutions in 50 mM sodium citrate buffer (pH 4.2) were incubated with 2 nkatals/mL of the purified endoglucanases at 30°C. The decrease in viscosity was monitored using a Rheoscan 115 system with a MS0-115 cell (Contraves, Zürich, Switzerland) at a constant shear rate of 594 s^{-1}. The released reducing sugars were determined using the DNS method.

2.13 Molecular weight distribution by size exclusion chromatography (SEC)

A high protein concentration of 22 nmol per mg substrate was applied for both enzymes for 3 days incubation to obtain a complete degradation pattern on the substrates. SEC analysis of cellulose ethers was performed using a sample concentration of 0.2 % and an injection volume of 100 µl. The samples were separated using TSK columns (G5000PW_{XL}, G4000PW_{XL}, G3000PW_{XL}, 300 x 7.8 mm each and a precolumn G2500PW_{XL}) with 0.1 M $NaNO_3$ used as eluent at 40°C. Detection was performed using an RI-detector (Shodex RI-71), a two angle laser light scattering detector (Precision Detectors PD 2000) and a viscosimetric detector (Viscotek H502). For molar mass determination by viscosimetry a universal calibration curve using dextran standards was applied. Collection and calculation of the data was performed using the WINGPC 4.0 software of PSS (Mainz, Germany), including light scattering and viscosity modules.

Molar mass determination for most samples was calculated using the viscosity, especially for enzyme treated CMCs, because viscosity detection is more sensitive to low molar mass degradation products while these components are underestimated by light scattering. For enzymatically degraded MC, the fragmentation was very intense and molar mass determination was not possible by viscosimetry. Therefore the MC samples were evaluated using a conventional dextran calibration curve. For the CMC substrate blanks, light scattering was the method of choice for molecular weight determination. These samples developed high viscosities (700 mL/g CMC 0.6 and 730 mL/g CMC 1.6) and correspondingly a high hydrodynamic volume. For this reason most of the elution curve was out of the range of universal calibration.

3 RESULTS AND DISCUSSION

3.1 Hydrolytic enzymes

Initially we evaluated the production of both extracellular hydrolytic and oxidative enzymes of five commonly isolated brown-rot fungi over a 30 day growth period using a number of different carbon sources. Previously it had been shown that no or only trace amounts of extracellular hydrolytic activity were detected in culture filtrates of brown-rot fungi when supplemented with simple sugars or non-cellulosic polysaccharides.[38] Thus, we initially evaluated the supplementation of a basal salts media with 5 % (w/v) of either Avicel (microcrystalline cellulose), cotton, carboxymethylcellulose or Douglas-fir kraft pulp. The supplementation of the basal salts medium with the 5 % (w/v) carboxymethylcellulose resulted in the largest production of protein and enzymes (Figure 1 and 2), when compared to the other carbon sources (data not shown). This medium was therefore used to monitor the extracellular enzyme production over a prolonged incubation period (Figure 1 and 2). In all cases, with the exception of *L. lepideus*, both endoglucanase and ß-glucosidase activities increased with days of incubation, with *G. trabeum* and *G. sepiarium* demonstrating significantly higher endoglucanase activities than did any of the other fungal strains. The ß-glucosidase activity produced by *L. lepideus* increased within days of incubations, as was found with the other fungal strains. However, endoglucanase activity was maximal early in growth and decreased to almost no cellulolytic activity during later stages of growth (Figure 1).

The filter paper assay is a common method used for measuring total cellulolytic activity. Generally, for filter paper degradation to occur, both endoglucanases and exoglucanases are required in the culture filtrate, and therefore this measurement can act as indirect indicator of exoglucanase activity. At no time during the sampling period did any of these brown-rot fungal cultures exhibit any filter paper activity, indicating the absence of any substantial exoglucanase activity. This is contrary to the previously reported existence of exoglucanase activity found in strains of *C. puteana.*[18]
During the evaluation of xylanase and mannanase activities (Figure 2) it was apparent that, generally, these strains of brown-rot fungi progressively produced greater amounts of hemicellulolytic enzymes with increasing duration of incubation. Again the exception was *L. lepideus* which initially demonstrated low levels of xylanase activity. However, as growth proceeded this activity quickly disappeared (Figure 2). Despite the reported initial high removal rates of hemicellulose during lignocellulosic degradation by this group of fungi,[28,43] the xylanase and mannanase activities produced by these fungal strains was not as significant as the endoglucanase values (Figure 1).

3.2 Oxidative enzymes

As mentioned earlier, brown-rot degradation is generally characterised by the rapid depolymerisation of the carbohydrate moieties of lignocellulosics.[1,20,44] However, it is recognised that even the smallest cellulases are too large to penetrate the pores of wood to cause this type of substrate modification.[12] To date, the production of oxidative enzymes by brown-rot fungi has been limited. The few reported cases include the production of a laccase and peroxidase by *Gloeophyllum trabeum,*[45] a lignin peroxidase by *Polyporus ostreiformis,*[46] a manganese peroxidase by *Piptoporus betulinus,*[47,48] and cellobiose dehydrogenase by *C. puteana.*[36] However, in general, brown-rot fungi do not produce polyphenol oxidases. Therefore, it has been proposed that the oxidative agent(s) involved in the depolymerisation is nonenzymatic.[49,50] It has been suggested that the putative mechanism involves the production of extracellular hydrogen peroxide[29] which is a small molecule capable of penetrating into the wood cell structures,[24,51,52] and the presence of transition metals, such as iron, or a metal chelate[53,54] which can participate in (a) Fenton's type reaction(s) generating hydroxyl radicals.[31,33,55] The hydroxyl radicals subsequently cleave glycosidic bonds between chains and cause substantial depolymerisation of polysaccharides.[31,37,56]

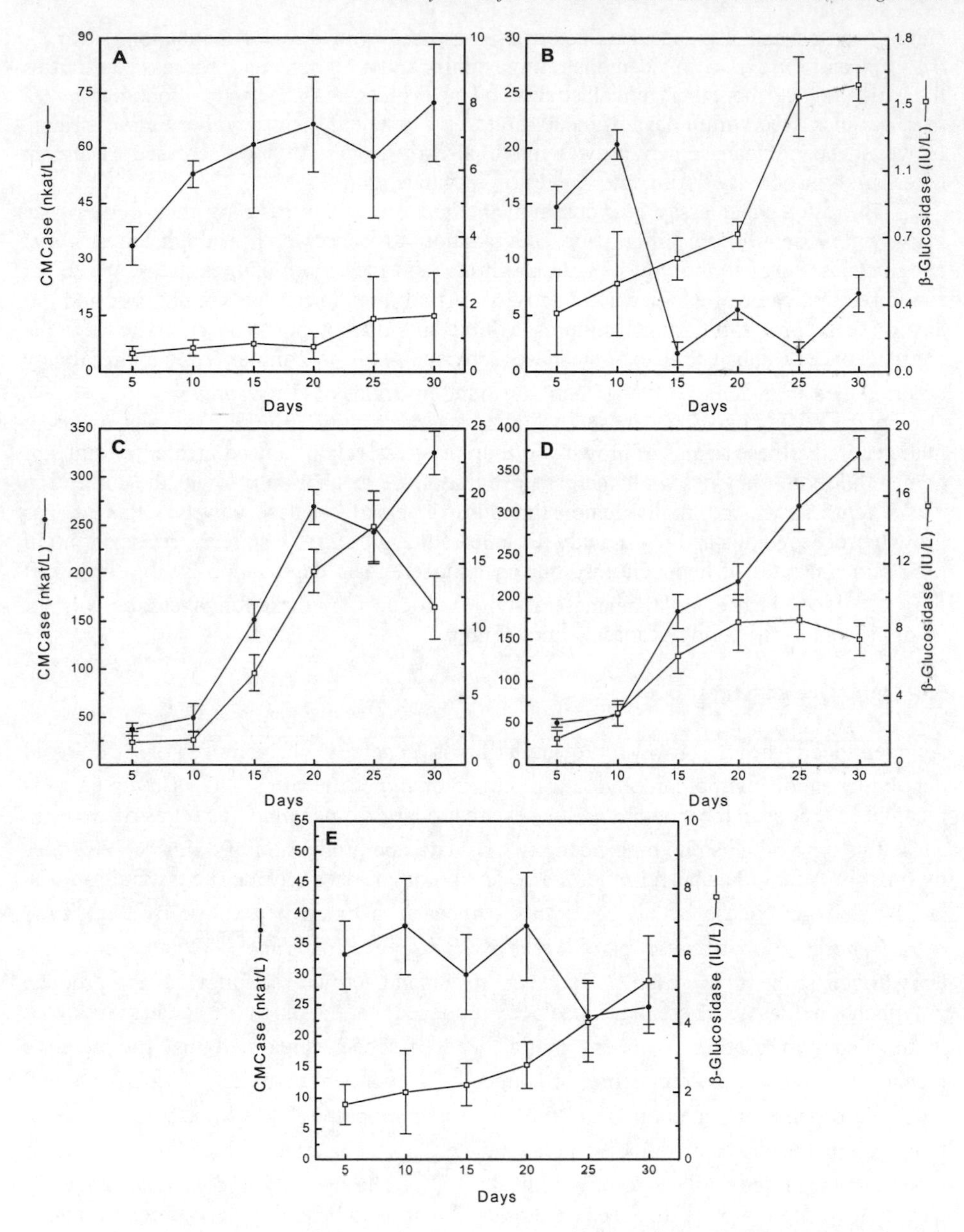

Figure 1 Endoglucanase (CMC) and ß-glucosidase activities produced by (A) *P. placenta*, (B) *L. lepideus*, (C) *G. trabeum*, (D) *G. sepiarium* and (E) *C. puteana* over 30 days incubation.

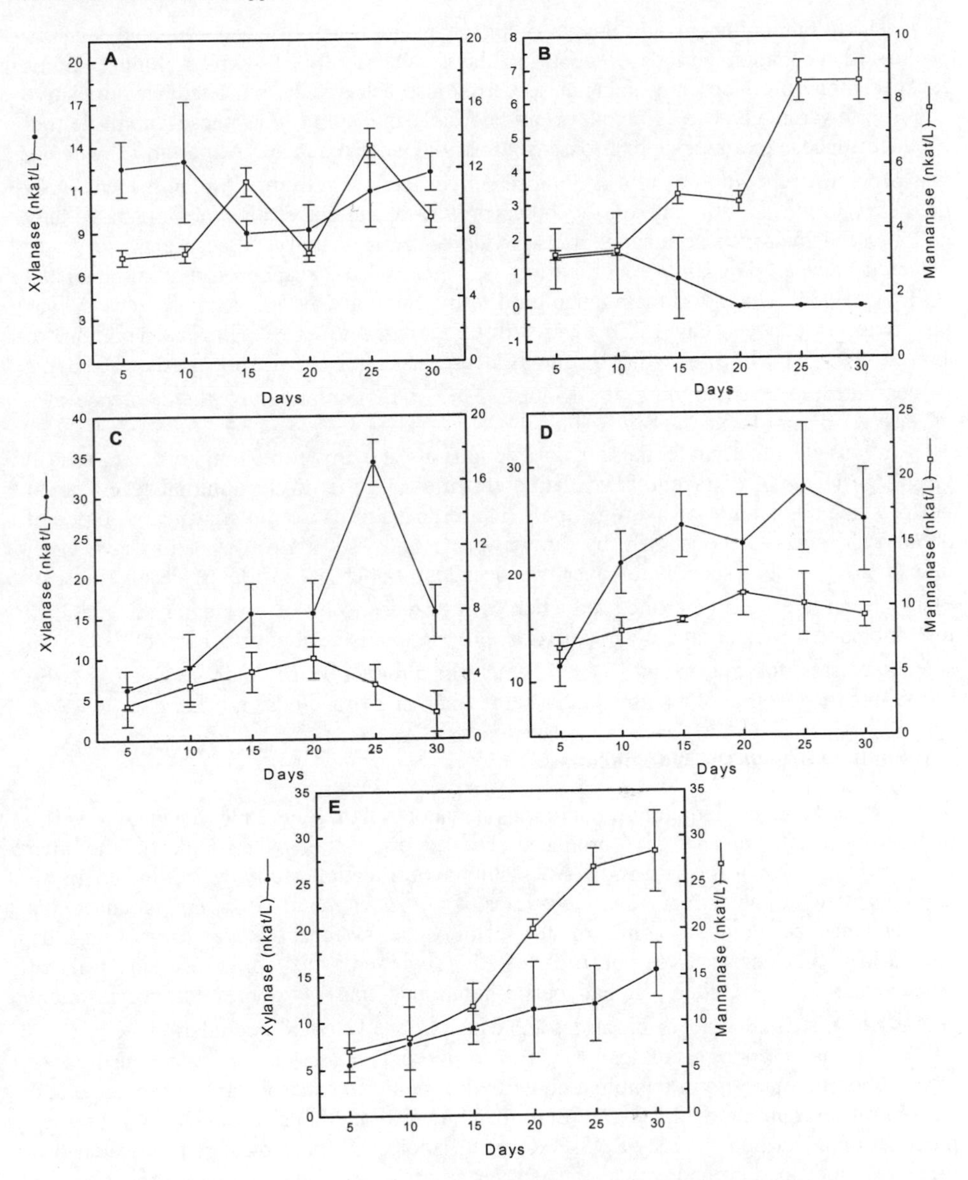

Figure 2 *Xylanase and mannanase activities produced by (A)* P. placenta, *(B)* L. lepideus, *(C)* G. trabeum, *(D)* G. sepiarium *and (E)* C. puteana *over 30 days incubation.*

Using our culture conditions, very little oxidative activity was detected, even at prolonged incubation periods of up to 50 days. Of the five brown-rot fungal strains assayed, none produced any manganese peroxidase activity or exhibited any oxidative activity towards glucose, glyoxal or arylalcohol, indicating that these fungi do not produce glucose oxidase, glyoxal oxidase or arylalcohol oxidase. Although *C. puteana* was previously shown to exhibit cellobiose dehydrogenase activity,[36] in the present study neither *C. puteana* nor any of the other brown-rot fungal strains demonstrated any cellobiose dehydrogenase nor cellobiose:oxidoreductase activity. Laccase activity was detected between day 20-25 of growth of *G. trabeum*, and reached a maximum of 0.9 IU/L at day 35, which was maintained until termination at day 50. Laccase activity was also detected between day 15-20 of growth of *P. placenta* and reached a maximum of 1.77 IU/L at day 40 after which it slowly decreased until termination at day 50. This agrees with the recent finding of D'Souza *et al.*[45] who demonstrated the existence of a laccase gene homologue in *P. placenta*.

It was clear that, under these culture conditions, the brown-rot fungi did not produce very high levels of extracellular oxidative enzymes and that these minimal levels could not produce the amount of hydrogen peroxide required to cause the significant degree of depolymerisation demonstrated by brown-rot fungi. Brown-rot systems have been investigated without success for over two decades in attempts to identify the vehicle of extracellular hydrogen peroxide production.[29] It remains to be seen whether this inability to conclusively detect hydrogen peroxide production is a function of the lack of a selective assay for this reagent[44] or whether the production of hydrogen peroxide by brown-rot fungi is that of an intracellular rather than an extracellular mechanism.

3.3 Purified brown-rot endoglucanases

Under the specified culture conditions it was clear that the endoglucanase was the predominant enzyme activity produced by all the strains, particularly the two *Gloeophyllum* brown-rot strains. As a result, a purification study was initiated in an attempt to isolate the endoglucanase enzymes of *G. trabeum* and *G. sepiarium* and to try and enhance our understanding of the cellulolytic system of brown-rot fungi by comparing the degradative potential of the two enzymes on different isolated polysaccharides. Previously it had been shown that these two basidiomycetes could actively degrade native polysaccharides and cause extensive wood degradation.[57]

The major endoglucanases of *G. trabeum* and *G. sepiarium* were purified to electrophoretic purity by a combination of hydrophobic interaction, anion-exchange and gel filtration chromatography (as described in the methods section). Following purification, electrophoresis by SDS-PAGE and isoelectric focusing (IEF) indicated that these two independent endoglucanase had very similar molecular masses and pI values (Table 1). The *G. trabeum* and *G. sepiarium* (40.5 and 45.1 kDa, respectively) endoglucanases had molecular masses that were in a similar range to those previously reported for brown-rot fungi such as *Tyromyces palustris* (39.5 kDa)[13] and *Polyporus schweinzii* (45 kDa).[14] However, they differ from the previously isolated endoglucanases of *Gloeophyllum* (*Lenzites*) *sepiarium*[16] and *Gloeophyllum* (*Lenzites*) *trabeum*[15] which had molecular masses of 85 kDa and 29 kDa, respectively. Both of the *Gloeophyllum* endoglucanases had acidic pI values, 3.1 (*G. trabeum*) and 3.8 (*G. sepiarium*) and a pH optima in the region of pH 3-5, which is very common for endoglucanases isolated from brown-rot fungi (Table 1). The pH and temperature profiles of these two purified endoglucanases are depicted in figure 3. Both enzymes had similar profiles, with the *G. trabeum* endoglucanase showing a slightly broader pH tolerance, as well as being slightly more thermotolerant (Figure 3, Table 1).

N-terminal amino acid sequence analysis revealed that both of the *Gloeophyllum* endoglucanases have an identical sequence of their first 21 amino acids (Table 1), suggesting that these two enzymes are closely related. A computer search (BLASTp database) showed that part of the sequences (12aa) from *Gloeophyllum* endoglucanases had homology with endoglucanases from *Macrophomina phaseolina* (EG2), *Trichoderma reesei* (EG3) and *Penicillium janthinellum* (EG2). However, these endoglucanases belong to different cellulase families.[58,59]

Table 1 *Properties of purified endoglucanases from* G. sepiarium *and* G. trabeum

	G. Sepiarium	*G. Trabeum*
Molecular mass [kDa]	45.1	40.5
Isoelectric point [pI]	3.8	3.1
N-terminal Sequence	VTGPAPLKFAGVNIAGFDFGC	VTGPAPLKFAGVNIAGFDFGC
Transglycosylation	Yes	Yes
pH - Optimum*	4.1	4.2
Temp. - Optimum* [°C]	59	62
Specific activity [µkat/mg]*	1.9	2.3

*measured on carboxymethylcellulose (D.S. 1.6).

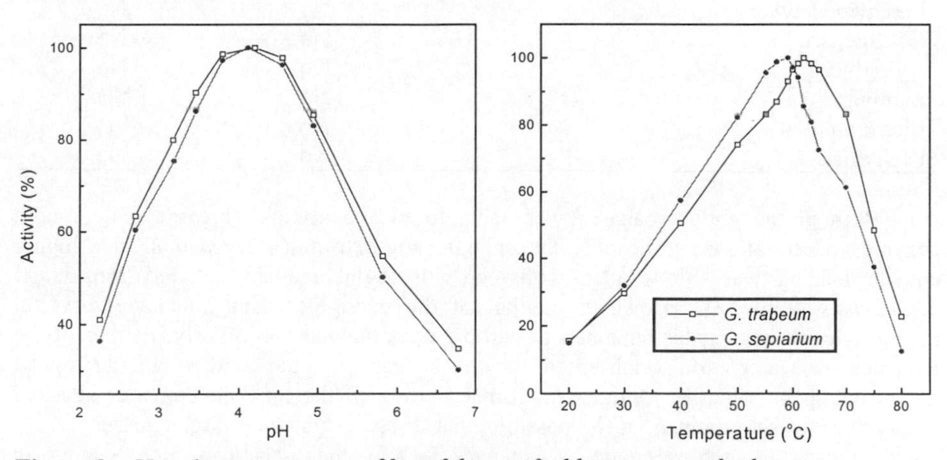

Figure 3 *pH and temperature profiles of the purified brown-rot endoglucanases measured on carboxymethylcellulose (DS 1.6).*

3.4 Substrate specificities of the endoglucanase

The purified brown-rot endoglucanases were added to a number of different polysaccharide substrates at both a high (10 nkatal/mL) and low (1 nkatal/mL) enzyme loading (Table 2). Both the *G. trabeum* and *G. sepiarium* endoglucanase hydrolysed carboxymethylcellulose (CMC), hydroxyethylcellulose (HEC), methylcellulose (MC) and phosphoric acid swollen cellulose (PASC), all of which are "amorphous" in nature.

However, they were unable to hydrolyse any of the more "crystalline" cellulosic substrates such as Avicel (microcrystalline cellulose), filter paper or bacterial microcrystalline cellulose (BMCC), even at the higher enzyme loadings. Similar cellulolytic behaviour has also been reported for other fungal endoglucanases, such as those purified from *M. phaseolina*,[60] *A. niger*[61] and *T. palustris*[13] (Table 2).

Table 2 *Substrate specificities of purified endoglucanases from* G. sepiarium *and* G. trabeum

	G. sepiarium	*G. trabeum*
Carboxymethylcellulose	Yes	Yes
Hydroxyethylcellulose	Yes	Yes
Methylcellulose	Yes	Yes
Avicel (microcrystalline cellulose)	No	No
Filter Paper	No	No
Bacterial microcrystalline cellulose (BMCC)	No	No
Phosphoric acid swollen cellulose (PASC)	Yes	Yes
Ivory nut mannan	No	No
Locust bean galactomannan	No	No
Carob galactomannan	No	No
Konjac glucomannan	Yes	Yes
Tsuga canadiensis galactoglucomannan	Yes	No
Birchwood xylan	No	No
Larchwood xylan	No	No
Oat Spelts xylan	No	No
Xyloglucan	No	No
Laminarin	No	Yes
Lichenan	No	Yes
Pachyman	No	No

Both of the endoglucanases were able to cleave konjak glucomannan at both enzyme concentrations, while only the endoglucanase from *G. sepiarium*, at a higher enzyme loading, was able to attack softwood galactoglucomannan isolated from *Tsuga canadiensis* (Table 2). However, neither of the endoglucanases attacked ivory nut mannan, locust bean galactomannan or carob galactomannan. Similarly, neither of the enzymes were active on xylan-based substrates such as birchwood xylan, larchwood xylan, oat spelts xylan or xyloglucan. Since neither of the purified enzymes attacked mannan or galactomannan, it is possible that these isolated endoglucanases cleave glucomannans at glucose-mannose linkages, while the galactose side chains in galactoglucomannan restrict the enzyme action. This is in agreement with the results of Tomme *et al.*[62] who showed that the *Cellulomonas fimi* endoglucanases CenB and CenC were able to cleave glucomannan but not galactomannan. However, other enzymes such as EGIII from *Trichoderma reesei* also act on galactomannan.[63]

The substrate activity of these purified brown-rot endoglucanases was also assessed on 1,3-1,4-ß-D-glucans (laminarin and lichenan) and 1,3-ß-D-glucan (pachyman). Although the *G. trabeum* endoglucanase showed activity on both the 1,3-1,4-ß-D-glucan substrates, neither of the endoglucanases were able to cleave the 1,3-ß-D-glucan polysaccharide (Table 2). This suggested that glycosidic cleavage occurs at the 1,4-ß-glycosidic linkages of the substrate. The ability of endoglucanases to cleave 1,3-1,4-ß-D-glucans has previously been observed with the recombinant *C. fimi* CenA, CenB, CenC and CenD endoglucanases.[62]

3.5 Activity on soluble cellooligosaccharides

Both of the purified endoglucanases showed similar degradation patterns when they were incubated with soluble cellooligosaccharides (cellobiose to cellohexose). Neither of the endoglucanases were able to cleave cellooligosaccharides smaller than cellotetraose and cellotetraose was cleaved only to cellobiose. Cellopentaose and cellohexose were cleaved to glucose, cellobiose, and cellotriose (Figure 4), with cellobiose being the major product (> 80 %). Furthermore, both of these purified brown-rot endoglucanases demonstrated transglucosidation activity (Table 1), suggesting that these enzymes cleave ß(1→4)-glucosidic bonds of the cellulosic substrates by a retaining type mechanism.[62]

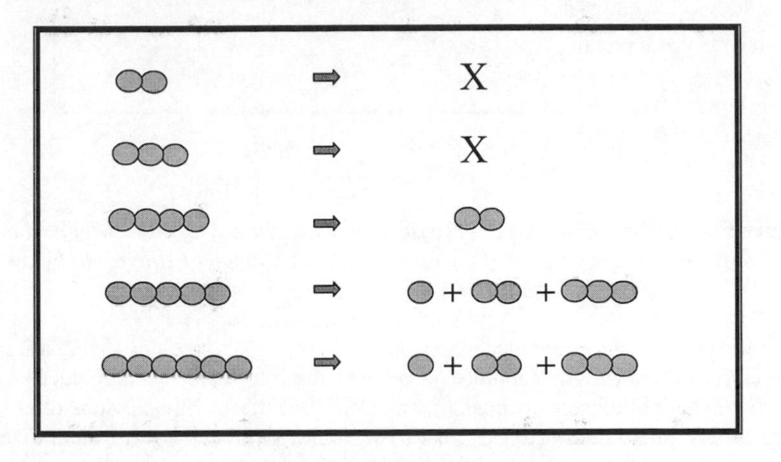

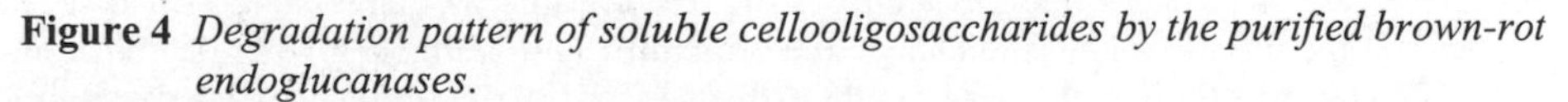

Figure 4 *Degradation pattern of soluble cellooligosaccharides by the purified brown-rot endoglucanases.*

3.6 Degradation of substituted soluble cellulose

We next compared the two brown-rot endoglucanases for their endo-character by following the increase in the specific fluidity of a carboxymethylcellulose (CMC) solution versus the liberation of reducing sugars (Figure 5). The action of both endoglucanases resulted in a rapid decrease in viscosity. When the specific fluidity (η^{-1}) of CMC versus the reducing sugars released was plotted, straight lines were obtained, with a steeper slope for the endoglucanase from *G. sepiarium*. This suggested a more random type of degradation of the polysaccharides during the early stages of hydrolysis.[64]

To determine the degradation products of latter stages of hydrolysis, the purified endoglucanases were incubated with carboxymethylcellulose and methylcellulose for extended periods (3 days) prior to analysis by gel permeation chromatography. This comparison was initiated to investigate whether the substituent side-groups on the cellulosic backbone influenced the degradation mechanism. Concurrently, the purified endoglucanases were incubated with carboxymethyl- and methylcellulosic substrates having different degrees of substitution, in order to examine if steric hindrances influence the degree of degradation between the two endoglucanases.

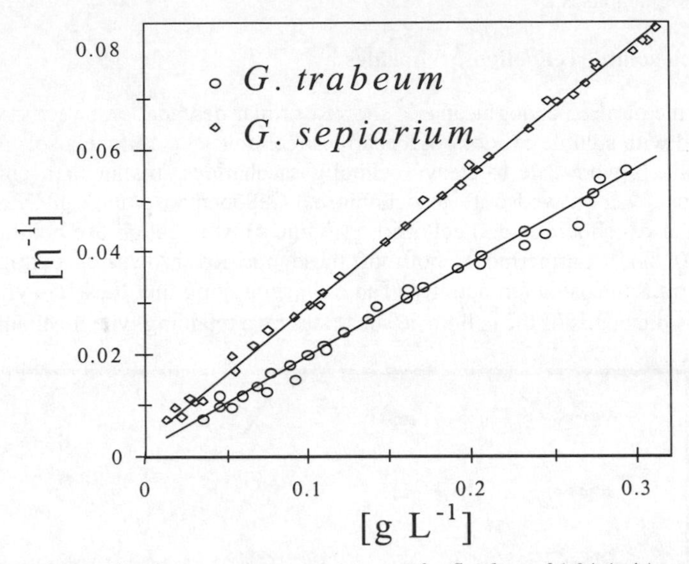

Figure 5 *Relationship between increase in specific fluidity of 1 % (w/v) carboxymethylcellulose solution and the amount of reducing sugar liberated by two purified brown-rot endoglucanases.*

In general, both endoglucanases showed similar degrees of degradation (Table 3). Although the methylcellulose were more readily degraded than the carboxymethylcellulose substrates, it was clear that that cellulosic substrates which were more highly substituted (*i.e.* DS 1.6/1.5) were not degraded to the same extent as those that were less substituted (*i.e.* DS 0.6/0.5) (Figure 6, Table 3). This suggested that an increase in the number of substituents on the substrates limited the degree of degradation, probably due to steric hindrances. While the substrates did influence the degree of degradation, both enzymes degraded the substrates to relatively the same extent.

When the DS 1.6 carboxymethylcellulose was used as a substrate, the endoglucanase purified from *G. sepiarium* resulted in slightly more degradation of the substrate than did the *G. trabeum* endoglucanase.

Table 3 Weight average *molecular weight* [g/mol] *of carboxymethylcellulose (CMC) and methylcellulose (MC) after incubation with purified brown-rot endoglucanases.*

Sample	Control		*G. trabeum*		*G. sepiarium*	
CMC (DS 1.6)	344,000	1)	235,000	2)	217,000	2)
CMC (DS 0.6)	289,000	1)	12,200	2)	13,100	2)
MC (DS 1.5)	72,400	2)	1,860	3)	2,180	3)
MC (DS 0.5)	--		640	3)	770	3)

DS indicates degree of substitution

Determination of MW by 1) light scattering; 2) viscosimetry; 3) dextran calibration

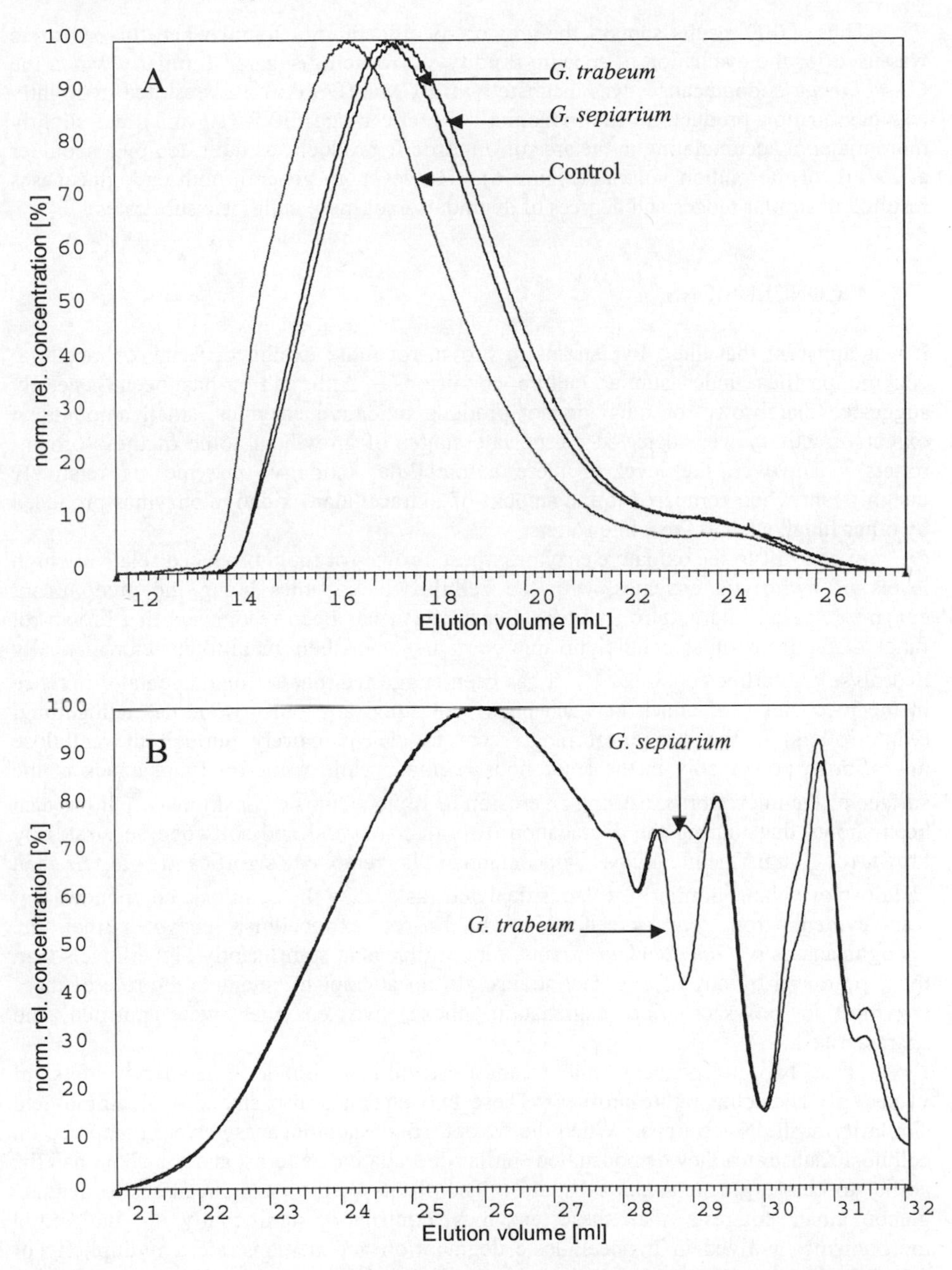

Figure 6 *Gel permeation chromatography (GPC) of carboxymethylcellulose (CMC) with a degree of substitution of (A) 1.6 and (B) 0.6 treated with purified brown-rot endoglucanases.*

These GPC results support the greater specific fluidity found when this substrate was used for the evaluation of specific fluidity vs. reducing sugars. Similarly, when the *G. sepiarium* endoglucanase was incubated with CMC (DS 0.6), this resulted in slightly less degradation products in the monomeric/dimeric range (30.5-31.5 mL) and slightly more material accumulating in the area of oligomeric products, as indicated by a shoulder at 29 mL of the elution volume (Figure 6). However, in general, both endoglucanases resulted in similar modes and degrees of degradation of these cellulosic substrates.

4 CONCLUSIONS

It was apparent that these five strains of brown-rot fungi exhibit differing extracellular enzyme profiles under similar culture conditions. Although it has been generally suggested that brown-rot fungi do not produce oxidative enzymes, small amounts of oxidative activity were detected during late stages of growth of some of these brown-rotters. However, the level of these extracellular oxidative enzymes is relatively insignificant when compared to the amount of extracellular oxidative enzymes produced by other fungi, such as the white-rotters.

In contrast to the oxidative enzymes, these brown-rot fungi produced relatively high levels of hydrolytic enzymes, with the cellulolytic enzymes being the predominant enzyme released into culture filtrates. Generally, it has been recognised that brown-rot fungi differ from other cellulolytic microorganisms in their inability to enzymatically hydrolyse crystalline cellulose.[38,65] It has been suggested that the fundamental difference in the mechanism of attack between brown-rot fungi and other wood-degrading fungi (white-rotters) is that brown-rot fungi seem to cleave entirely through the cellulose microfibrils, presumably in the amorphous regions, while white-rot fungi attack at the surface of the microfibrils, causing an erosion of lignocellulosic constituents.[66] It has also been shown that, during the degradation of both hardwood and softwood substrates by brown-rot fungi, hemicellulose glucomannan is removed significantly faster than cellulose or xylan. Similarly, xylan is depleted faster than the cellulosic component.[43] It was evident from our screening of brown-rot extracellular enzymes that the endoglucanases of *Gloeophyllum* strains were produced at significantly higher levels than those produced by any of the other strains. In an attempt to elucidate the role of these enzymes in polysaccharide degradation, these two enzymes were purified and characterised.

These two independent endoglucanases exhibited similar molecular weights, pI values, pH and temperature profiles. These two enzymes also shared a 21 amino acid similarity at the N-terminus. When the two purified endoglucanases were compared on cellulosic substrates they demonstrated similar degradative patterns, and they both had the ability to cleave glucomannan substrates. The ability of the endoglucanases to degrade glucomannan suggests that these enzymes contribute significantly to the initial mechanisms involved in lignocellulosic degradation and demonstrate a multiplicity of function, namely cellulose depolymerisation and glucomannan removal.

5 ACKNOWLEDGEMENT

We would like to thank BC Science Council and Weyerhaeuser Inc. for a scholarship held by S. Mansfield. Part of this work was funded from a Canadian–German Scientific Cooperation grant in Agricultural Research (project No. FOR 1/95). We would also like to thank Heinz Seiler for his skilful technical assistance.

References

1. E.B. Cowling. (U.S. Department of Agriculture, Technical Bulletin No. 1258, Washington, DC, 1961, pp. 79.
2. T.K. Kirk and E.B. Cowling, *in* 'Chemistry of solid wood', ed. R.M. Rowell, Advances in Chemistry Series, 1984, **207**, 455.
3. T.K. Kirk, R. Ibach, M.D. Mozuch, A.H. Conner and L.T. Highley, *Holzforschung*, 1991, **45**, 239.
4. T.K. Kirk and E. Adler, *Acta Chem. Scand.*, 1969, **23**, 705.
5. T.K. Kirk, S. Larsson and G.E. Miksche, *Acta Chemic. Scand.*, 1970, **24**, 1470.
6. T.K. Kirk and E. Adler, *Acta Chem. Scand.*, 1970, **24**, 3379.
7. L. Jin, T.P. Schultz and D.D. Nicolas, *Holzforschung*, 1990, **44**, 133.
8. L. Jin, T. Sellers, T.P. Schultz and D.D. Nicolas, *Holzforschung*, 1990, **44**, 207.
9. T.L. Highley, *Mat. Org.*, 1977, **12**, 25.
10. A. Enoki, H. Tanaka and G. Fuse, *Wood Sci. Technol.*, 1989, **23**, 1.
11. T. Nilsson, *Mat. Organ.*, 1974, **9**, 173.
12. E.B. Cowling and W. Brown, *in* 'Cellulases and Their Applications', eds. G.J. Hajny and E.T. Reese, 1969, 152.
13. M. Ishihara and K. Shimizu, *Mokuzai Gakkaishi*, 1984, **30**, 79.
14. G. Keilich, P.J. Bailey, E.G. Afting and W. Liese, *Biochim. Biophys. Acta*, 1969, **185**, 392.
15. D. Herr, F. Baumer and H. Delweg, *Arch. Microbiol.*, 1978, **117**, 287.
16. B. Bhattacharjee, A. Toy and A.L. Majumber, *Biochem. Molecul. Biol. Int.*, 1993, **30**, 1143.
17. A.C. Ritschkoff, J. Buchert and L. Viikari, *Mat. Org.*, 1992, **27**, 19.
18. D.R. Schmidhalter and G. Canevascini, *Applied Microbiol. Biotechnol.*, 1992, **37**, 431.
19. D.R. Schmidhalter and G. Canevascini, *Arch. Biochem. Biophys.*, 1993, **300**, 551.
20. T.L. Highley, *Holzforschung*, 1988, **42**, 211.
21. K.M. Kleman-Leyer and T.K. Kirk, *Appl. Environ. Microbiol.*, 1994, **60**, 2839.
22. T.L. Highley, *Mat. Org.*, 1983, **18**, 161.
23. K. Ruel, K. Ambert and J.-P. Joseleau, *FEMS Microbiol. Rev.*, 1994, **13**, 241.
24. T.L. Highley, *Mat. Org.*, 1982, **17**, 205.
25. T.L. Highley and L.L. Murmanis, *in* 'International Research Group on Wood Preservation', (IRG/WP 1256), 1985, 1.
26. T.L. Highley and L.L. Murmanis, *Mat. Org.*, 1986, **20**, 241.
27. T.L. Highley, *FEMS Microbiol. Lett.*, 1987, **48**, 373.
28. T. L. Highley, *in* 'International Research Group on Wood Preservation', (IRG/WP 1319), 1987, 1.
29. T. L. Highley and D. S. Flournoy, *Recent Advances in Biodeterioration and Biodegradation*, 1994, **2**, 191.
30. A. Enoki, S. Yoshioka, H. Tanaka and G. Fuse, *in* 'International Research Group on Wood Preservation', (IRG/WP 1445), 1990, 1.
31. G. Halliwell, *Biochem. J.*, 1965, **95**, 35.
32. A. C. Ritschkoff, M. Rättö, J. Buchert and L. Viikari, *J. Biotechnol.*, 1995, **40**, 179.

33. J. K. Koenigs, *Wood Fiber*, 1974, **6**, 66.
34. S. Takao, *Appl. Microbiol.*, 1965, **13**, 732.
35. C.J. Schmidt, B.K. Whitten and D.D. Nicholas, *American Wood-Preservers' Association*, 1981, 157.
36. D.R. Schmidhalter and G. Canevascini, *Arch. Biochem Biophys.*, 1993, **300**, 559.
37. S.D. Mansfield, E. de Jong and J.N. Saddler, *Appl. Environ. Microbiol.*, 1997, **63**, 3804.
38. T.L. Highley, *Wood Fiber*, 1973, **5**, 50.
39. T.M. Wood and M.K. Bhat, *Methods Enzym.*, 1988, **160**, 87.
40. M.J. Bailey, P. Biely and K. Poutanen, *J. Biotechnol.*, 1992, **23**, 257.
41. G.M. Gübitz, M. Hayn, G. Urbanz and W. Steiner, *J. Biotechnol.*, 1996, **45**, 165.
42. E. de Jong, F.P. de Vries, J.A. Field, R.P. van der Zwan and J.A.M. de Bont, *Myc. Res.*, 1992, **96**, 1098.
43. T.L. Highley, *Mat. Org.*, 1987, **21**, 39.
44. F. Green III and T.L. Highley, *in* 'International Research Group on Wood Preservation', (IRG/WP 95-10101), 1995, 1.
45. T.M. D'Souza, K. Boominathan and C.A. Reddy, *Appl. Environ. Micro.*, 1996, **62**, 3739.
46. S. Dey, T.K. Maiti and B.C. Bhattacharyya, *J. Ferment. Bioeng.*, 1991, **72**, 402.
47. G.D. Szklarz, R.K. Antibus, R.L. Sinsabaugh and A.E. Linkins, *Mycologia*, 1989, **8**, 234.
48. M. Freitag and J.J. Morrell, *Can. J. Microbiol.*, 1992, **38**, 811.
49. M.M. Chang, T.Y.C. Chou and G.T. Tsao, *in* 'Advances in Biochemical Engineering Bioenergy 15', ed. A. Fieshter, Springer-Verlag, Berlin, 1981.
50. B. Phillip, D.C. Dan and H.P. Fink, *in* '1981 Int. Symp.Wood Pulping Chem.', TAPPI PRESS, Atlanta, 79.
51. J.W. Koenigs, *Phytopathology*, 1972, **62**, 100.
52. J.W. Koenigs, *Mat. Org.*, 1972, **7**, 133.
53. J. Jellison, B. Goodell, F. Fekete and V. Chandhoke, *in* 'International Research Group on Wood Preservation', (IRG/WP 1442), 1990, 1.
54. J. Jellison, V. Chandhoke, B. Goodell and F.A. Fekete, *Appl. Microbiol. Biotechnol.*, 1991, **35**, 805.
55. J.W. Koenigs, *Biotechnol. Bioeng. Sym. No. 5*, 1975, **5**, 151.
56. B.C. Gilbert, D.M. King and C.B. Thomas, *Carbohyd. Res.*, 1984, **125**, 217.
57. C. Job-Cei, J. Keller and D. Job, *Mat. Org.*, 1996, **30**, 105.
58. G. Mernitz, A. Koch, B. Henrissat and G. Schulz, *Curr. Genet.*, 1996, **29**, 490.
59. U. Arunachalam and J.T. Kellis, *Biochemistry*, 1996, **35**, 11379.
60. H.Y. Wang and R.H. Jones, *Gene*, 1995, **158**, 125.
61. S. Akiba, Y. Kimura, K. Yamamoto and H. Kumagai, *J. Ferm. Bioeng.*, 1995, **79**, 125.
62. P. Tomme, E. Kwan, N.R. Gilkes, D.G. Kilburn and R.A.J. Warren, *J. Bacteriol.*, 1996, **178**, 4216.
63. R. Macarrón, C. Acebal, M.P. Castillón and M. Claeyssens, *Biotechnol. Lett.*, 1996, **18**, 599.
64. H. Esterbauer, J. Schurz and A. Wirtl, *Cell. Chem. Technol.*, 1985, **19**, 341.
65. T.L. Highley, *Wood Fiber*, 1975, **6**, 275.
66. K. Kleman-Leyer, E. Agosin, A.H. Conner and T.K. Kirk, *Appl. Environ. Microbiol.*, 1992, **58**, 1266.

USE OF CELLULASES IN PULP AND PAPER APPLICATIONS

Liisa Viikari, Jaakko Pere, Anna Suurnäkki, Tarja Oksanen and Johanna Buchert
VTT Biotechnology and Food Research
P.O. BOX 1501
FIN-02044 VTT, Finland

1 ABSTRACT

Individual cellulases can be used in mechanical or chemical pulp production for specific modifications of cellulose in the fibre structure. When applying cellulases, it is important to minimise the detrimental effect of cellulases on the yield and strength properties of fibres. Depending on the type of substrates and cellulases used, different technical targets can be achieved. In mechanical pulping, enzymatic modification of the coarse fibres leads to higher outer fibrillation and consequent energy savings. The enzymatic treatment is carried out after the primary refining on coarse fibres prior to the secondary refining. In this application, the cellobiohydrolase I of *Trichoderma reesei* was shown to be superior to the other cellulases tested. In laboratory scale experiments CBH I resulted in energy savings of 20 - 40%, depending on the type of refiner used. In chemical pulps, the effect of cellulases is dependent on the stage of cellulase addition. Thus, when cellulases are used for the modification of refined pulp, improved drainage and runnability of the paper machine is obtained. A cellulase treatment prior to the refining leads to improved beatability of the pulp. In both of these applications, endoglucanases have been found to be the superior group of cellulases. In this paper, the mode of action and effects of the two types of cellulases, cellobiohydrolases and endoglucanases, are reviewed in the light of present applications.

2 INTRODUCTION

During recent years, new applications of cellulases have been developed especially in the textile industries. In the pulp and paper industry, the strength properties of the fibres, described as tensile, burst and tear indexes, are of primary importance and are sensible for impairment by cellulases. Some applications are, however, based on the action of certain types of cellulases alone or in combination with hemicellulases. These include attempts to decrease the energy consumption in mechanical refining, to improve the drainage of especially recycled pulps (removal of water adsorbed to the fine material) in order to increase the runnability of the paper machine (Jokinen *et al.* 1991, Pommier *et al* 1989, Pommier *et al.* 1990), to enhance the beatability of chemical pulps (*i.e.* fibrillation and inter-fibre bonding), (Noe *et al.* 1986), to increase the solubilisation of dissolving pulps (Rahkamo *et al.* 1996) or to improve the deinking processes for recycled papers. In spite of extensive research, the underlying mechanisms are not yet clear.

The presence of cellulases in enzyme preparations used for processing of pulps has traditionally been considered detrimental due to their negative effects on pulp strength properties. *Trichoderma reesei* produces two exoglucanases, cellobiohydrolases I and II, which are known to be especially efficient in hydrolysing highly crystalline substrates, and five endoglucanases (EG I, EG II, EG III, EG IV, EG V) acting on amorphous regions of cellulose. The impact of the four major *T. reesei* cellulases *ie.* EG I, EG II, CBH I and CBH II on the fibre properties of unbleached softwood kraft and dissolving pulps have been investigated by *e.g.* Pere *et al.* (1995) and Rahkamo *et al.* (1996). The two cellobiohydrolases and endoglucanases studied exhibited significant differences in their mode of action on the pulps. The cellobiohydrolases (CBH) had only a very modest effect on pulp viscosity. This is in agreement with the results of Kleman-Leyer *et al.* (1996). On the other hand, the endoglucanases (EG) dramatically decreased viscosity, even at low enzyme dosages. Similar results have also been obtained when unbleached kraft pulp were treated with monocomponent *T. reesei* cellulases prior to bleaching, and therefore, the enzyme preparations used for increasing the bleachability should not contain cellulase activities (Buchert *et al.* 1994). Of the endoglucanases, EG II was shown to decrease the viscosity most drastically, suggesting that EG II attacks cellulose at sites where even low levels of hydrolysis result in large decreases in viscosity and, consequently, a dramatic deterioration of the tensile index (Pere *et al.* 1995). In comparison, the CBH I treatment had no effect on the handsheet properties of chemical pulps even after a refining stage, indicating that CBH I caused no structural damage to the fibres. The negative effect of cellulases on pulps is even enhanced when cellulases are acting synergistically. The degree of synergy obtained, however, depends on the nature of the substrate and the concentration and ratios of the enzymes used. The highest degree of synergy is observed between pairs of cellulases; *e.g.* EG I has been reported to pretreat cellulose more efficiently for CBH I than for CBH II (Nidetzky *et al.* 1994). Thus, the negative effects of cellulase synergism should be minimised in cellulase preparations used in the pulp and paper industry.

In the applications studied, several types of fibres have been used as substrates. The action of enzymes in pulp is generally affected by the accessibility of substrates in the fibre matrix. The main factors limiting the access of enzymes in woody materials are the specific surface area and the porosity, *i.e.* the median pore size of fibres (Stone *et al.* 1969, Grethlein 1985). Also the molecular organisation of the fibre components and the surface chemistry; *i.e.* relative amounts of lignin and carbohydrates may have a significant impact on the action of enzymes on the fibre bound substrates. In general, the median size of pores in woody fibres, corresponding to mechanical pulps, is about 1 nm, whereas in chemical pulps the value is about 5 nm (Stone and Scallan 1968). Thus, the mechanical fibres are much less accessible to enzymes than chemical pulps. The presence of lignin in mechanical pulps can also be expected to limit the access of enzymes.

Although the mode of action of individual cellulases on pure substrates is well understood today, their mechanistical differences on practical substrates have not been fully solved and moreover, these characteristics are not exploited in various processes. In this work, we summarise the results obtained by purified cellulases, used in different applications on mechanical and chemical pulps.

3 THE ACTION OF CELLULASES ON DIFFERENT TYPES OF PULPS

The pattern of the cellulolytic enzymes produced by *Trichoderma reesei* has been exploited in a number of application studies in the pulp and paper industry. However, in fairly few publications the effect of pure enzymes has been investigated. The enzymes used in the present experiments were purified from culture filtrates of *T. reesei* as described previously (Pere *et al.* 1995). The ability of the individual cellulases to hydrolyse different types of pulp fibres are compared in Table 1. The fibres originated from mechanical pulping (thermomechanical pulping, TMP) and chemical pulping (kraft sulphite) processes (Pere *et al.* 1995, 1997, Rahkamo *et al.* 1996). The differences in the composition and structure of fibre substrates, due to the different pulping processes and their modifications, is reflected in the degree of hydrolysis obtained. In mechanical pulps, the cellulose content is approximately 50%, whereas in chemical pulps it is usually about 75%.

Table 1 *Hydrolysis of different pulps by* T. reesei *cellulases. The enzyme dosage was 0.5 mg/g and hydrolysis time with chemical pulps 2h and with mechanical pulps 6h. CSF: Canadian Standard Freeness.*

Pulp	Enzyme	% of pulp d.w. hydrolysed
Mechanical pulp,	CBH I	0.17
CSF 560 ml	CBH II	0.25
	EG I	0.85
	cellulase mixture	1.20
Unbleached softwood	CBH I	0.20
kraft pulp	CBH II	0.10
	EG I	0.80
	EG II	0.50
Dissolving pulp	CBH I	0.15
	CBH II	0.21
	EG I	0.70
	EG II	0.65

The hydrolysis experiments of coarse spruce fibres revealed that using the same amount of individual cellulases (CBH I and II, EG I and II), the endoglucanase released by far the highest amount of carbohydrates, whereas the two cellobiohydrolases hydrolysed the coarse pulps at the same efficiency (Table 1). Using a cellulase mixture, a clear synergistic action was demonstrated. When unbleached softwood kraft pulp was treated with the same enzymes, generally similar degrees of hydrolysis were obtained, however, within shorter hydrolysis times. Also in this case, the endoglucanases hydrolysed the pulp more efficiently. The hydrolysis degree by EGI was higher due to the simultaneous action on xylan and cellulose (Pere *et al.* 1995). When dissolving pulp with very low hemicellulose content was used as substrate, no such difference was observed (Table 1).

4 CELLULASES IN MECHANICAL PULPING

In mechanical pulping, wood logs or chips are fiberised by mechanical means. Refining is comprised of a series of compressions and decompressions which lead to the separation and fibrillation of individual fibres with simultaneous generation of fines. The yield in mechanical pulping processes is up to 95%. The main disadvantage of mechanical pulping is its high energy consumption, especially in the case of thermomechanical pulping (TMP). One way to reduce the energy consumption in thermomechanical pulping is to modify the raw material by biotechnical means prior to refining. So far, the main focus has been in the pre-treatment of wood chips with ligninolytic fungi. By biomechanical pulping with selected white-rot fungi, substantial energy savings and improvements in handsheet strength properties have been obtained (Leatham *et al.* 1990, Akhtar 1994). Encouraging results have been obtained in laboratory and pilot scales. From the chemical point of view, no correlation between lignin degradation and energy savings has been found, and evidently slight modification of both lignin and carbohydrates is needed for the observed positive effects.

The refining process consists of several steps, during which the fibres are separated from each others. After the primary refining stage, the freeness value (Canadian Standard Freeness, CSF, describing the extent of refining) is typically 400 - 500 ml. In the enzyme-aided refining concept, the enzymatic treatment is carried out at this stage. A remarkable part of the total energy demand is consumed at the secondary refining step, producing pulp with freeness values of about 50 - 100 ml. The aim of the enzymatic treatment has been to reduce the specific energy consumption in the secondary refining stages, after an intermittent enzymatic step, as visualised in Figure 1. In the enzymatic modification of mechanical pulps, the target hydrolysis level has generally been adjusted to correspond a degree of hydrolysis lower than 1% of the pulp, by choosing both the enzyme dosage and treatment time. The treatment time has typically been two hours.

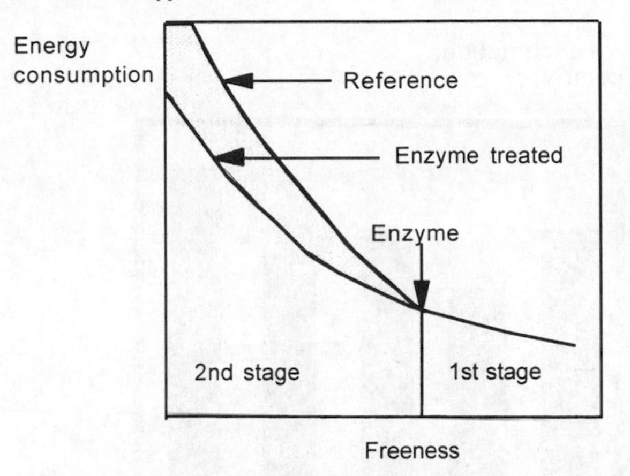

Figure 1 *Aim of the enzyme-aided refining process*

Preliminary experiments with different enzymes demonstrated that a slight modification of cellulose by CBH I gave rise to an energy saving of 20% in a laboratory-scale disk refiner. Treatment with EG I decreased the energy consumption slightly but only at the expense of pulp quality, by sensitising fibres to breaking. Interestingly, no positive effect on energy consumption was detected with the CBH II, neither with a cellulase mixture (Figure 2). When the refining was performed with a low-intensity refiner (wingdefibrator), the positive effect of CBH I in reducing the energy consumption was further enhanced. At a CSF level of 100 ml the energy consumption in the secondary refining was reduced by 40% with CBH I, as compared with the untreated reference (Pere *et al.* 1996). The laboratory refinings with different types of refiners indicated that in order to obtain a maximal benefit of the enzymatic treatment, the secondary refining should be performed with a rather low intensity.

The results obtained with the laboratory refiners were further verified in a pilot-scale experiment, where 900 kg of TMP reject was treated with CBH I prior to the secondary refining. In a two-stage secondary refining an energy saving of 10-15% with CBH I was obtained (Pere *et al.* 1996). These results clearly confirmed the previous results on smaller scale, although no optimisation of the process parameters in the pilot experiments has yet been carried out. The cellulase treatment with CBH I did not have any detrimental effects on pulp quality. In fact, the tensile index was even higher for the CBH I treated pulp than for the reference. The increase in tensile index could be explained by the intensive fibrillation induced by the CBH I treatment. The good optical properties were also maintained after the CBH I treatment (Pere *et al.* 1996).

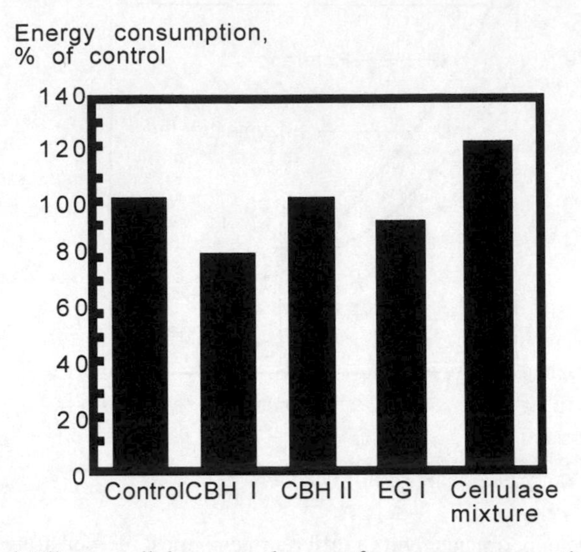

Figure 2 *Effect of different cellulases on the specific energy consumption in mechanical pulping.*

Most of the energy in refining is consumed in the loosening and partial unravelling of the layered and highly ordered cell wall structure of wood fibres. Therefore, it can be anticipated that the beneficial effect of CBH I treatment on the energy consumption is due to the limited hydrolysis of the surface cellulose chains in the microfibrils leading to enhanced delamination or loosening of the fibre structure. Neither the cellulase mixture nor CBH I was found to induce evident morphological modifications in the coarse and rigid TMP fibres during a short incubation. However, significantly more intensive fibrillation was observed after the secondary refining in the CBH I treated pulp than in that treated with the cellulase mixture or in the untreated control. Furthermore, a fibrillation index of 60 % was attained with CBH I at a 30 % lower energy input than in the case of the untreated control, which could be an indirect implication of the decreased interfibrillar cohesion inside the fibre wall due to the action of CBH I. Thus, it is possible that the CBH I acts according to the mechanism suggested by Karnis (1994), in which fibre development during refining is suggested to proceed through delamination and peeling-off reactions in the P- and S-layers of the cell wall as a function of energy input.

5 CELLULASE TREATMENTS OF CHEMICAL PULPS

In chemical pulping the yield is usually less than 50% because of the solubilisation of lignin and hemicellulose during the cooking procedure. Due to a variety of raw materials (softwood or hardwood species) and the pulping parameters, the chemical composition of the fibres varies considerably. After pulping and bleaching, the cellulose content is, however, generally about 75%. Hemicellulases of *T. reesei* are used in order to improve the bleachability of kraft pulps. The role of different cellulases in bleach boosting of chemical pulps has also been tested but only EG I, hydrolysing both xylan and cellulose, was found to improve the bleachability (Buchert *et al.* 1996).

After bleaching, beating, *i.e.* mechanical refining, is used to obtain the desired paper-making properties in chemical pulp fibres. Beating leads to improved comformability, flexibility and inter-fibre bonding of fibres through a process where both internal delamination and external fibrillation of the fibre cell wall take place. However, generation of fines during beating is accompanied by increased water sorption and subsequent decreased dewatering and drainage properties of fibres in the paper machine. When cellulases are used for the modification of the fibre properties, the enzyme stage can be carried out either prior to or after refining of the pulps. By carrying out the enzymatic stage prior to the refining process the aim has been to improve the beating response or other fibre properties. The principal challenge in using enzymes to enhance fibre bonding is to increase fibrillation without reducing pulp viscosity (Kirk *et al.* 1996). However, in the enzymatic treatment of refined or recycled pulps, the main focus is in the improvement of the dewatering, *i.e.* drainage properties of the pulps, affecting the speed of paper machine operation. The different approaches are schematically represented in Figure 3.

The effect of purified *Trichoderma* cellulases and hemicellulases on the beatability and paper technical properties of bleached kraft pulps has been investigated by Kamaya (1996) and Oksanen *et al.* (1997). Pretreatment of the pulp with CBH I or CBH II had practically no effect on the development of pulp properties, whereas endoglucanases, especially EG II, were found to improve the beatability of the pulp as measured by SR (Schopper-Riegler) value, sheet density and Gurley air resistance (Kamaya 1996, Oksanen *et al.* 1997).

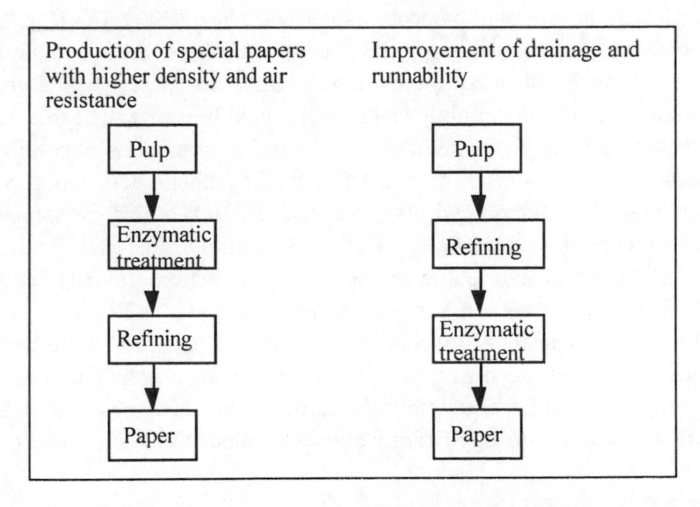

Figure 3 *Process schemes for modification of chemical pulps with cellulases.*

Drainage rate is one of the key parameters controlling the runnability of a paper machine. Several authors have described the use of enzymes in drainage control of pulps with different origin and furnish (Fuentes and Robert 1986, Pommier *et al.* 1989, Pommier *et al.* 1990, Stork *et al.* 1995, Kamaya 1996, Oksanen *et al.* 1996). Endoglucanase activity has been shown to be essential for drainage improvement of recycled pulps. According to Oksanen *et al.* (1996), EG I and EG II were equally efficient in decreasing the Schopper-Riegler value of recycled softwood kraft pulp although the amount of solubilised carbohydrates were lower with EG II. Obviously, the limited hydrolysis of hydrophilic amorphous cellulose by the EGs was essential, because the purified *T. reesei* xylanases and mannanases had only a slight positive impact on drainage of recycled fibres.

6 THE ROLE OF CBD IN CELLULASE ACTION IN PULP

The endoglucanases EG I, EG II, EG IV and EG V and the cellobiohydrolases CBH I and CBH II from *T. reesei* are composed of two main regions, *i.e.* active core domain and cellulose binding domain (CBD), which are bound together by a flexible linker. Today, using genetic engineering, both domains, *i.e.* BCDs and cores, can be produced separately. In addition, core proteins are found in traditionally produced cellulase mixtures. The removal of CBD from CBH II is reported to have no influence on the activity of the enzyme when soluble substrates are used, but using crystalline substrates both its binding and activity were clearly impaired (Skrisoduk *et al.,* 1993). The disrupture or swelling of cellulose structure by the binding domain has been claimed to be a reason for the higher activity of the intact enzyme towards crystalline cellulose in the case of CBHs. The CBD has been claimed to be less important in the action of endoglucanases as they are considered to act mainly on noncrystalline cellulose.

In order to understand the importance of CBD in the action of *T. reesei* cellulases in chemical pulp fibres, the effects of monocomponent cellulases (EG I, EG II, CBH I, CBH II) and their core proteins on pulp properties has recently been studied (Suurnäkki *et al.,* 1997). All the enzymes and core proteins used in the work were purified from culture liquors produced by genetically modified strains. It could be anticipated that the removal of the CBD would affect the site of cellulase action in fibres and thus hopefully minimise their detrimental effects in pulp. The role of CBD in cellulase action in the pulp was visualised mainly in the cellulose hydrolysis levels. The intact enzyme solubilised more cellulose than the core protein lacking the CBD. In addition, the CBD of EG I, active against both cellulose and xylan, directed the enzyme action towards hydrolysis of cellulose. Irrespective of the differences in the hydrolysis efficiency, at the same hydrolysis level the detrimental effect of intact enzyme and the corresponding core protein on pulp properties, such as viscosity of the unbeaten pulp and strength of the beaten pulp was, however, approximately the same. Thus, the presence of CBD in *T. reesei* cellulases appears not to significantly alter the action of enzyme on pulp substrate.

7 CONCLUSIONS

In spite of the vast amount of information on the genetics, biochemistry and structure of *T. reesei* cellulases, their mode of action on practical cellulosic substrates is still poorly understood. Cellulases are, however, potential tools for modification of fibres for various purposes. In these applications, the treatment usually has an optimum between the positive effects and the negative effects as a function of the treatment efficiency. Therefore, the negative effects should be minimised by carefully controlling the composition of the enzyme preparation, the enzyme dosage and the treatment time.

The new concept of using *T. reesei* CBH I before the secondary refining of coarse mechanical pulps has been shown to be successful in laboratory and pilot scale. The applicability of this method will depend on further success in optimising the enzymatic treatment and integrating the enzymatic step to present refining systems. Endoglucanases have been shown to be efficient in fibre modification for improved beatability or drainage. One commercial enzyme preparation is already available for this purpose. The use of endoglucanases, in spite of their positive effects, may however be hampered by their negative effect on pulp viscosity already with low dosages. In the future, more targeted commercial cellulase preparations can be expected to be available.

References

Akhtar M. (1994), *Holzforschung*, **48**, 199.

Béguin P. and Aubert J.-P. (1994), *FEMS Microbiol. Rev.*, **13**, 25.

Buchert, J., Ranua, M., Kantelinen, A. and Viikari, L. (1994), *Appl. Microbiol. Biotechnol.*, **40**, 941.

Chanzy H. and Henrissat B. (1985), *FEBS Lett.*, **184**, 285.

Chanzy H., Henrissat B. and Vuong R. (1984), *FEBS Lett.*, **172**, 193.

Chanzy H., Henrissat B., Vuong R. and Schülein M. (1983), *FEBS Lett.*, **153**, 113.

Grethlein H. (1985), *Bio/technology*, Feb., 155.

Fuentes, J.L. and Robert, M. (1986), *French Pat.* 2604198.

Hoshino E., Nomura M., Takai M., Okazaki M., Nisizawa K. and Kanda T. (1994), *J. Ferment. Bioeng.*, **77**, 496.

Hoshino E., Sasaki Y., Mori K., Okazaki M., Nisizawa K. and Kanda T. (1993), *J. Biochem.*, **114**, 236.

Henrissat B., Driguez H., Viet C. and Schulein, M. (1985), *Bio/Technology*, **3**, 772.

Jokinen, O., Kettunen, J., Lepo, J., Niemi, T., and Laine, J.E. (1991), *United States Pat.* 5,068,009.

Kamaya, Y. (1996), *J. Ferm. Bioeng.*, **82(6)**, 549.

Karnis A. (1994), *J. Pulp and Paper Research* **20(10)**, 280.

Kirk, T.K. and Jeffries, T.W. (1996), 'Roles of microbial enzymes in pulp and paper processing', *in* Enzymes in Pulp and Paper Processing, ACS Symp. Ser. 655, eds. Jeffries, T.W. and Viikari, L., American Chemical Society, Washington, USA, pp. 2-14.

Kleman-Leyer, K.M., Siika-aho, M., Teeri, T. and Kira, T.K. (1996), *Appl. Environ. Microb.*, **62**, 2883.

Leatham G., Myers G. and Wegner T. (1990), *Tappi J.*, **73**, 197.

Nidetzky B., Steiner W., Hayn M. and Claeyssens M. (1994), *Biochem. J.*, **298**, 705.

Noé, P., Chevalier, J., Mora, F. and Contat, J. (1986), *J. Wood Chem. Technol.*, **6**, 167.

Oksanen, T. Buchert, J., Pere, J. and Viikari, L. (1996), 'Treatment of recycled kraft pulps with hemicellulases and cellulases', *in* Biotechnology in the Pulp and Paper Industry,

Proc. 6th Int. Conf. Biotechnology in the Pulp and Paper Industry, eds. Srebotnik, E. and Messner, K., Facultas-Universitätsverlag, Vienna, pp. 177-180.
Oksanen, T., Pere, J., Buchert, J. and Viikari, L. (1997), *Cellulose*, **4**,1.
Pere, J., Siika-aho, M., Buchert, J. and Viikari, L. (1995), *Tappi J.*, **78(6)**, 71.
Pere J., Liukkonen S., Siika-aho M., Gullichsen J. and Viikari L. (1996), *Tappi Pulping Conference*, October 27-31, Nashville, TN.
Pommier, J.C., Fuentes, J.L. and Goma, G. (1989), *Tappi J.*, **72**, 187.
Pommier, J.C., Goma, G., Fuentes, J.L., Rousset, C. and Jokinen, O. (1990), *Tappi J.*, **73(12)**, 1.
Rahkamo, L., Siika-aho, M., Vehviläinen, M., Dolk, M., Viikari, L., Nousiainen, P. and Buchert, J. (1996), *Cellulose*, **3**, 153.
Skrisoduk, M., Reinikainen, T., Teeri, T. (1993), *J. Biol. Chem.*, **268**, 20756.
Sprey B. and Bochem H. (1991), *FEMS Microbiology Letters*, **78**, 183.
Sprey B. and Bochem H.-P. (1992), *FEMS Microbiology Letters*, **97**, 113.
Stone J.E. and Scallan A.M. (1968), *Pulp Paper Mag. Can.*, **69(6)**, T288.
Stone J.E., Scallan A.M., Donefer E. and Ahlgren E. (1969), *Adv. Chem. Ser.*, **96**, 219.
Stork, G., Pereira, H., Wood, T., Dusterhöft, E.M., Toft, A. and Pulps, J. (1995), *Tappi J.*, **78(2)**, 79.
Sundström L., Brolin A. and Hartler N. (1993), *Nordic Pulp and Paper Research Journal*, **8**, 379.
Suurnäkki, A., Siika-aho, M., Tenkanen, M., Buchert, J., Viikari, L. (1997), 'Effects of *Trichoderma* cellulases and their structural domains on pulp properties', Proc. 1997 TAPPI Biological Sciences Symp., San Francisco, pp. 283-286.
Wood T.M. and Garcia-Campayo V. (1990), *Biodegradation*, **1**, 147.
Woodward, J. (1991), *Bioresource Technology*, **1**, 67.
Woodward J., Affholter K.A., Noles K.K., Troy N.T. and Gaslightwala S.F. (1992), *Enzyme Microb. Technol.*, **14**, 625.

THERMOSTABLE XYLANASES PRODUCED BY RECOMBINANT *TRICHODERMA REESEI* FOR PULP BLEACHING

Marja Paloheimo*, Arja Mäntylä, Jari Vehmaanperä, Satu Hakola, Raija Lantto, Tarja Lahtinen, Elke Parkkinen, Richard Fagerström and Pirkko Suominen*

Roal Oy* and Röhm Enzyme Finland Oy
Tykkimäentie 15
FIN-05200 Rajamäki, Finland

1 INTRODUCTION

1.1 The principle of enzymatic kraft pulp bleaching

In kraft pulp bleaching the aim is to increase the brightness of pulp through removal of the lignin that is left after cooking. This is traditionally done by using chlorine-containing chemicals but because of their toxicity, new and more environmentally friendly technologies have been developed. One alternative approach, already commercially available, is enzyme-aided bleaching, a hemicellulase pretreatment that is done for kraft pulp after cooking, before the actual bleaching stages (for a review, see reference 1). After partial hydrolysis of the hemicellulose component(s) more efficient removal of lignin in the following bleaching stages can be achieved. In addition to a reduced amount of organochlorine compounds in effluents, chemical cost reduction and/or increased pulp brightness are achieved.

1.2 Towards better performance of xylanase products

The idea of using xylanases as bleach-boosters was first reported by Viikari *et al.* in 1986.[2] The first mill trials were run in 1989 - 1990 in Finland and several mills started regular use in North America and in Finland in the beginning of the 90's. Most commercial xylanases available to date are not very thermotolerant, especially when neutral or alkaline pH conditions are used. Most of the commercial products are native enzymes produced by *Trichoderma reesei* and *Aspergillus niger* and these fungal xylanases act most optimally at pH 5 - 6 and at temperatures from 50 ^{O}C to 60 ^{O}C.[3] They are devoid of harmful side activities, *i.e.* cellulases and are produced in high yields as a result of the development by gene technology and fermentation optimisation.

Parallel to commercial xylanase development, bleaching technology has progressed and many of the mills today use conditions in which the native fungal xylanases are inefficient or inactive. This is because oxygen delignification which is nowadays often used as pretreatment increases the overall process temperature. Also, the use of chlorine dioxide instead of chlorine demands higher temperatures. The commercial enzymes should be able to act at 45 - 85 ^{O}C and at pH 5.0 - 8.5. A high pH optimum is desirable because it reduces the need to acidify the pulp before the enzyme stage and helps to avoid the corrosion problem.

We have answered the need for more thermostable xylanase products by producing new thermostable xylanases in *T. reesei* at industrially feasible levels.

1.3 Production of thermostable xylanases at industrially feasible levels

For industrial production of thermostable xylanases, two prerequisites have to be met: first, the enzyme must have the desired properties, and second, the enzyme of interest has to be produced at industrially feasible levels. From the literature it is evident that there are substantially more bacterial xylanases with good thermostability than fungal xylanases and thus the thermostability condition may be harder to fill when fungi are used as the source for the enzyme. Protein engineering for higher thermostability of the first generation mesophilic xylanases may also prove difficult.

Filamentous fungi are known to be good producers of industrial enzymes. Genetically modified strains produce both homologous and heterologous fungal enzymes at industrially feasible levels in *T. reesei.*[4-7] Efficient and cost-effective industrial production of bacterial thermostable xylanases is a challenge because they mainly originate from relatively unstudied bacteria which often produce only minimal amounts of xylanase. In addition, there may be little or no experience of cultivating these microbes in a fermentor or these microbes may otherwise be unsuitable for industrial scale production. The best reported yields of bacterial enzymes from filamentous fungi have been at the level of 20 mg/l[8] and many of the enzymes produced have been detected only intracellularly (for reviews, see references 9 and 10).

To circumvent possible problems in properties and production, we used two approaches to develop a new thermostable xylanase product. First, fungal genes, including mutagenised *T. reesei* xylanase 2 (*xln2*[11]) and genes coding for heterologous fungal xylanases having high thermostability were expressed in *T. reesei*. Second, genes coding for thermostable bacterial xylanases were expressed in *T. reesei*.

2 PRODUCTION OF THERMOSTABLE FUNGAL XYLANASES

2.1 Constructions used for expressing fungal xylanases in *Trichoderma reesei*

The fungal xylanase gene, either homologous mutated *xln2* or a heterologous gene, was expressed in *T. reesei* under the control of the strong cellobiohydrolase 1 (*cbh1*) promoter. This is equal to what we have previously done in the case of native xylanase expression to obtain high production levels.[4] The cassettes for fungal xylanase gene expression in *T. reesei* also contained a marker gene for screening the transformants and flanking region(s) for targeting the cassette to a specific locus. When heterologous fungal xylanase genes were expressed, *cbh1* terminator sequences were inserted downstream of the xylanase gene.

2.2 Targeted mutagenesis of *Trichoderma reesei xylanase 2* gene

T. reesei xln2 cDNA (coding from Ala23, with 11 amino acids of the prosequence) was inserted after *Bacillus amyloliquefaciens* -amylase promoter, RBS and signal sequence in low-copy plasmids pWSK29 and pWSK129.[12] The *E. coli* strains harbouring the recombinant plasmids produced a clearing zone on RBB-xylan plates, and produced a xylanase with a slightly larger molecular weight than *T. reesei*, presumably due to the remaining part of the prosequence. Virtually all of the xylanase was released into the growth medium. *T. reesei xln2* cDNA was modified by site-directed mutagenesis (Megaprimer[13]). Thermostability of the mutated xylanases was tested in *E. coli* culture supernatants. The chosen mutations were reconstructed in a *Trichoderma* vector and transformed into a suitable production strain.

N38 was chosen as a primary target for site-directed mutagenesis because in the homologous *Bacillus pumilus* xylanase mutations in the corresponding amino acid G (G38D and G38S) resulted in higher thermostability of the enzyme.[14] Also, analysis of the *T. reesei* XYLII 3-D structure suggested that N38 has a higher temperature (B-) factor than other amino acids on the same surface, and reduction of the factor might enhance the thermostability (A. Törrönen and J. Rouvinen, University of Joensuu, personal communication). Eight mutations were introduced at N38 (shown below). N38 is a putative N-glycosylation site, but XYLII does not appear to be glycosylated in *T. reesei*.

1) N38D N38S: known to enhance thermostability in *B. pumilus* xylanase
2) N38T N38Y N38E: amino acids similar to D or S, the codons of which were not possible to evolve from the *B. pumilus* sequence (GGC) by a single-base substitution
3) N38R N38Q: amino acids present in other homologous xylanases at this location
4) N38G: the amino acid in native *B. pumilus* xylanase at this location

Other sites elsewhere in the *xln2* were also mutagenised, but these did not result in higher thermostability (T. Laakso, J. Rouvinen and J. Vehmaanperä, unpublished).

Five of the mutations, N38 T, Y, E, R and Q, made the xylanase more thermostable than the native enzyme (the N38S mutation could not be established). The thermostability was estimated in *E. coli* by incubating the culture supernatants at 57 and 60 °C, and measuring the residual activity at 42 °C (Figure 1). The time, t_{50}, after which the xylanase still retained 50 % of its activity, was 1.3 - 1.6 times longer in the two best mutant xylanases N38E and N38Y compared to the native enzyme.

The mutations N38E and N38Y were introduced into the genomic copy of *xln2* gene, placed under the *cbh1* promoter and transformed into *T. reesei*.

The production levels of the mutated xylanases from *T. reesei* were high as expected, *i.e.* grams/litre levels, comparable to the levels obtained when homologous genes are expressed under the control of the *cbh1* promoter. The mutant xylanases (Figure 1) retained their thermocharacteristics when produced in *Trichoderma*. The temperature (T_{op50}) at which the enzyme retained 50 % of its maximal measured activity was increased in the mutant xylanases about 1.6 °C (from 62 to 63.6 °C). The specific activity appeared to be about the same and the mutant xylanases had expected bleach-boosting effect.

In conclusion, the improvement after mutagenesis in one site was not high enough for a novel thermostable product but rather the improved products will perform more consistently in pulp mills which use, in their enzyme stage, temperatures suggested for mesophilic enzymes.

2.3 New xylanases with improved properties from thermophilic fungi

About 150 thermophilic fungal isolates were screened and the most promising enzyme preparations were further characterised in more detail by determining relative enzyme activities at 50 - 80 °C at different pH-values using 60 min incubation time for the enzyme reaction. Several strains were found to produce xylanase activity with good thermostability, giving higher activities at 70 °C than at 60 °C (Figure 2).

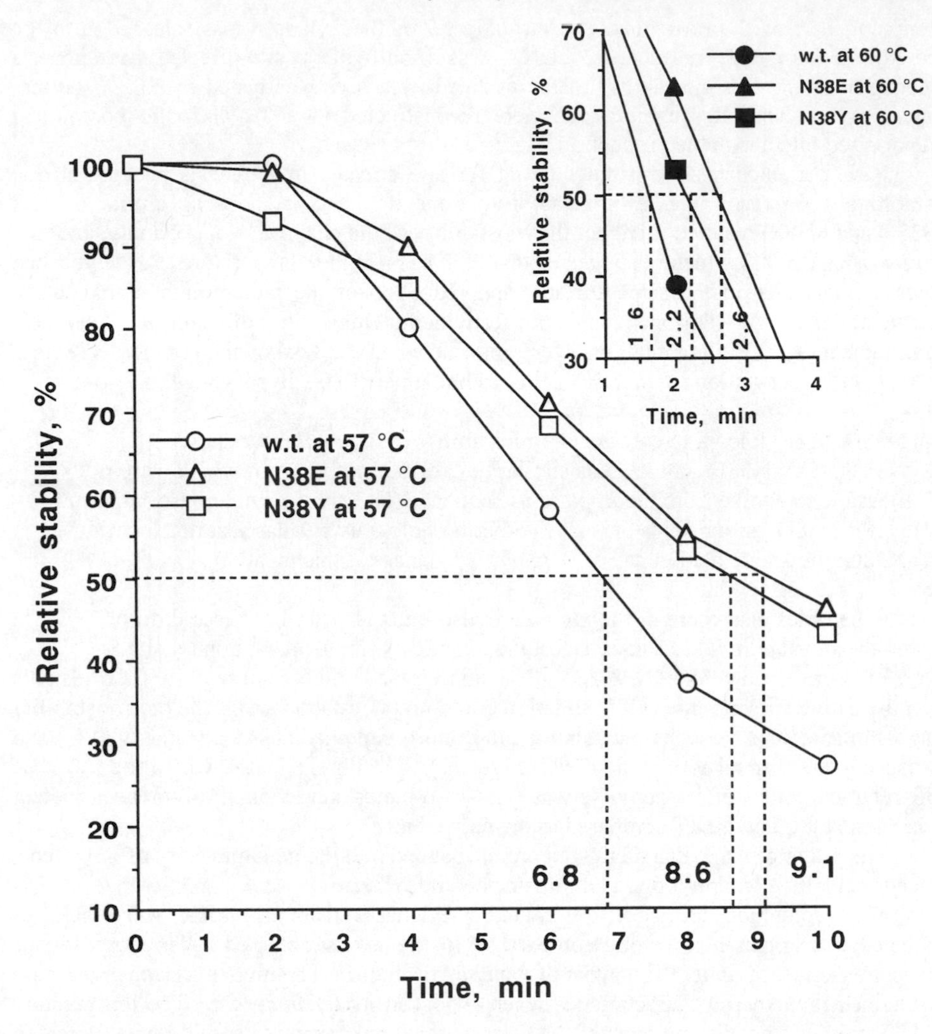

Figure 1 *Relative enzyme stabilities of the wild type and mutant xylanases at 57 and 60 °C. The residual activity was measured after various time points (5 min reaction at 42 °C, pH 5.3). The t_{50} times, after which xylanases still retain 50 % of their activity, are shown.*

The enzyme mixtures secreted into the culture supernatant were also tested for their bleach-boosting efficiency at 70 °C, pH 7.

Several fungal xylanases were purified from the above mentioned enzyme mixtures and several genes were cloned. Considerable variation between the thermostabilities of enzymes isolated from a single strain was observed.

Xylanase genes were expressed in *T. reesei* by using the *cbh1* promoter. Expression similar to this of native *T. reesei* xylanases was obtained.

The properties and function in bleaching of the recombinant fungal xylanases produced by *T. reesei* seemed to be very similar to those produced in the original host.

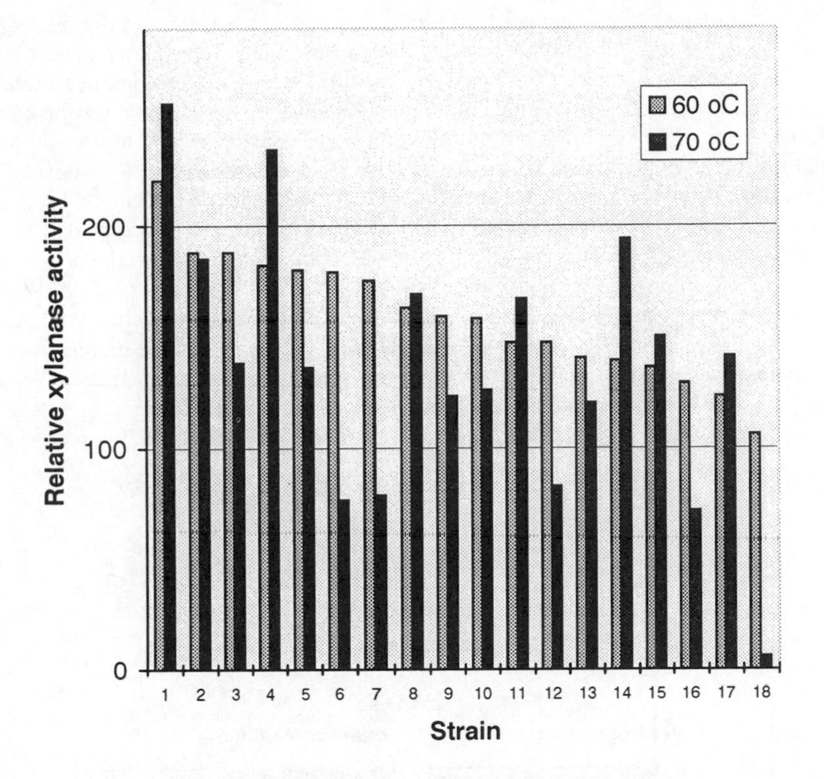

Figure 2 *Relative xylanase activities from culture supernatants of selected thermophilic fungi. The enzyme reaction was done at 50, 60 or 70 ^{o}C, pH 7 with a 60 min reaction time. The relative activities at 50 ^{o}C are set to 100.*

3 PRODUCTION OF THERMOSTABLE BACTERIAL XYLANASES

3.1 Screening and cloning of bacterial xylanase genes

Several bacterial strains were screened for thermostable xylanase activities in the same way as described in the case of fungal xylanases (see paragraph 2). An actinomycete strain, *Actinomadura flexuosa* ALKO3685, was one of the chosen strains for gene cloning.

The *A. flexuosa* genomic library was screened on RBB-xylan plates, and six different clones with xylanolytic activity were isolated. Expression of *am35* gene coding for a 35 kDa xylanase in *T. reesei* is described below.

3.2 Expression of *Actinomadura flexuosa* xylanase gene in *Trichoderma reesei*

Fusion strategy was chosen for production of AM35 in *T. reesei*. Gene fusions have been previously used in *Trichoderma* and *Aspergillus* for expressing heterologous mammalian proteins and from 5 to 1000 fold increases in yields, compared to nonfused constructions, have been obtained (resulting in protein levels from 5 mg/l to 250 mg/l; for a review, see reference 15).

The constructions that were used for producing AM35 are shown in Figure 3. Gene expression is under the *cbh1* promoter in all the constructions. The *man1* gene's core + hinge region was chosen as a carrier because it has the same kind of domain structure as cellulases[16] and is not detrimental but advantageous in bleaching application at temperatures around 50 °C.[1] It is also known that both mannanase I (MANI)[7] and MANI core (Paloheimo *et al.*, unpublished) can be produced in high amounts in *T. reesei*.

The expression cassette pALK1118 is a control where the *am35* gene is fused to the *cbh1* promoter and *T. reesei* mannanase 1 (*man1*) signal sequence (amino acids 1 - 19).[16] Expression cassettes pALK945, 948, 1021 and 1022 contain, as a carrier, the *man1* core + hinge region encoding amino acids M1 - G406 (pALK945 and 948) or M1 - G410 (pALK1021 and 1022). The additional R in pALK945 and a synthetic KEX2-like protease processing signal (RDKR) in pALK948 and 1022 precedes the mature AM35 to ensure proteolytic cleavage of the xylanase from the carrier. The expression cassette pALK1021 controls whether an unspecific protease can process the fusion protein.

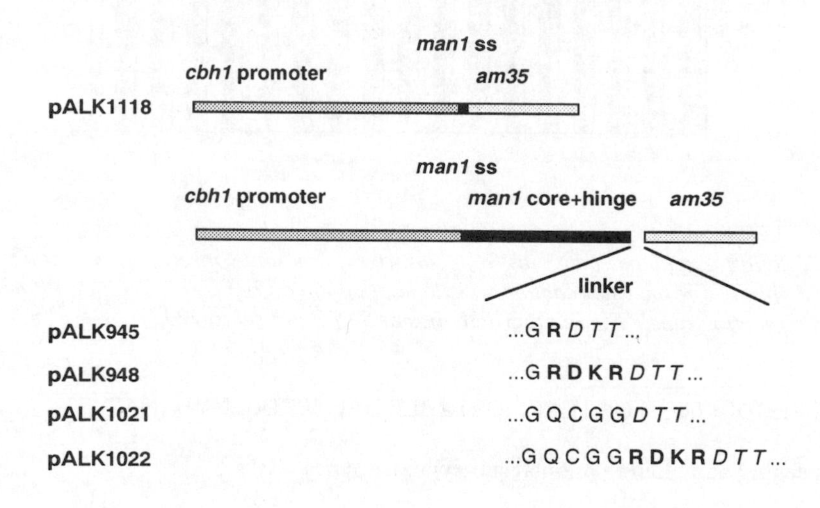

Figure 3 *Expression cassettes for* Actinomadura flexuosa *xylanase, AM35, production in* T. reesei. *The* am35 *gene was fused to the* man1 *signal sequence in pALK1118 or to the* man1 *core + hinge carrier (described in text, 3.2). The amino acids of the* man1 *sequence are in regular font, those of* am35 *are in italics and the synthetic amino acids for proteolytic cleavage are in bold. All the expression cassettes also contain the* cbh1 *promoter, a marker gene and the* cbh1 *3'-flanking region for targeting the expression cassettes into the* cbh1 *locus.*

3.3 Production of AM35 in *Trichoderma reesei*

The relative xylanase activities obtained in shake flask cultivations of the recombinant *T. reesei* strains producing AM35 are shown in Table 1. The lowest activity was obtained in transformants in which xylanase was produced without a carrier protein (pALK1118). Even in this case the level of xylanase activity was increased 70 fold compared to the gene donor strain. A fusion to the carrier further increased the production level and the strains having pALK945, 948 and 1022 expression cassettes produced about equal amounts of xylanase activity, more than 200-fold the amount of the bacterial donor strain and over three times as much as the construction without the fusion. In the fusion strains the lowest activity was obtained from the transformant containing the pALK1021 fusion construct which did not have a synthetic protease processing site. This result is explained (see 3.4) by the fact that in this strain the fusion is only partially cleaved and the fusion protein does not have xylanase activity.

Table 1 *Production of AM35 in* T. reesei. *Relative xylanase activities obtained from shake flask cultivations of one-copy transformants in which the corresponding expression cassette has been targeted into the* cbh1 *locus are shown. The enzyme assay was done at pH 7, 70 °C using 60 min reaction time.*

Transformed Fragment	Relative Xylanase Activity
Actinomadura flexuosa (gene donor)	1
T. reesei host	0.5
pALK1118 *man1* ss + *am35*	69
pALK945 *man1* core+hinge / G + R + *am35*	224
pALK948 *man1* core+hinge / G + RDKR + *am35*	235
pALK1021 *man1* core+hinge/GQCGG + *am35*	181
pALK1022 *man1* core+hinge / GQCGG + RDKR + *am35*	246

3.4 Properties of the recombinant AM35

The culture supernatants of strains harbouring the expression cassettes with gene fusion showed one or two major bands on Western blots (Figure 4). The strains having expression cassettes pALK945 and pALK948 had a major protein band migrating at the same speed as the purified 35 kDa protein of *A. flexuosa*. This recombinant AM35 has the native AM35 N-terminus showing that the cleavage of the fusion protein has been as desired. The pALK1022 strain has in addition a major band with about 2 - 3 kDa higher MW. This form of AM35, also faintly seen in the supernatants of the other transformants, is probably cut at a different position of the fusion. The strain having the pALK1021 expression cassette produced as major proteins both the 35 kDa form and also a protein of about 70 - 80 kDa which correlates with the molecular weight of the unprocessed fusion.

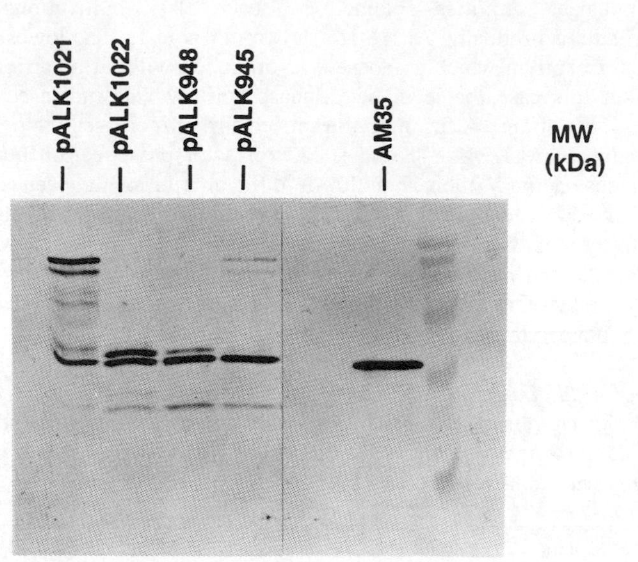

Figure 4 *Production of the AM35 xylanase in* Trichoderma reesei *by gene fusions. A rabbit polyclonal AM35 antibody was used to detect the AM35 protein in Western blot. Purified AM35 protein from* A. flexuosa *was run beside the supernatants of strains harbouring the expression cassettes.*

The unprocessed form could also faintly be seen in the culture supernatants of the other strains. It seems that the fusion is cut more efficiently in pALK948 and pALK1022 transformants having the synthetic KEX2 processing site than in the pALK945 transformant because substantially less fusion protein is visible. In all the culture supernatants also a smaller fragment of about 30 kDa was visible. This form of AM35 has been cut near its C-terminus. In addition to the major proteins the pALK1021 strain also has several faint bands of molecular weights between 35 and 80 kDa possibly representing proteolytically cleaved forms.

The fusion protein has mannanase but not xylanase activity which explains the lower xylanase activities measured from the culture supernatant of the pALK1021 strain as compared to the others (Table 1).

The thermal properties of the recombinant xylanases seem to be the same compared to the authentic xylanase except the pALK1021 product (fusion protein) which is clearly less stable (Figure 5).

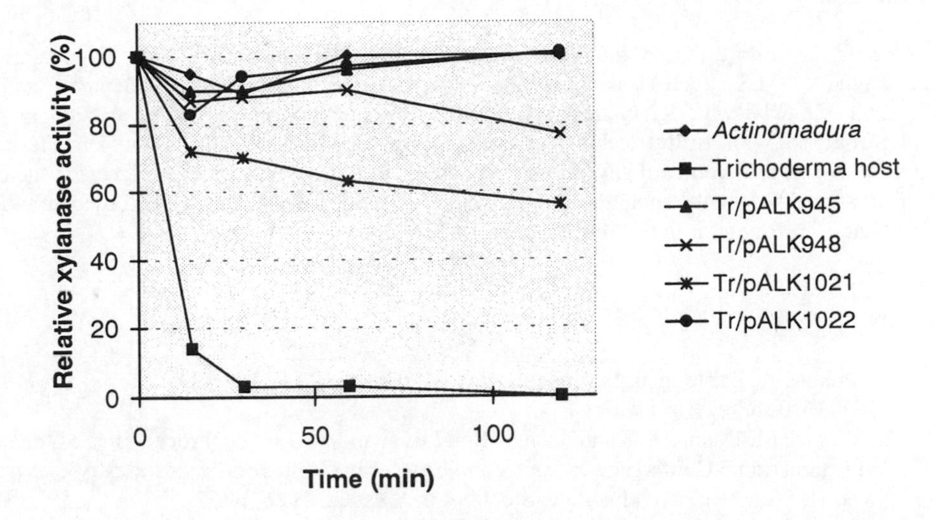

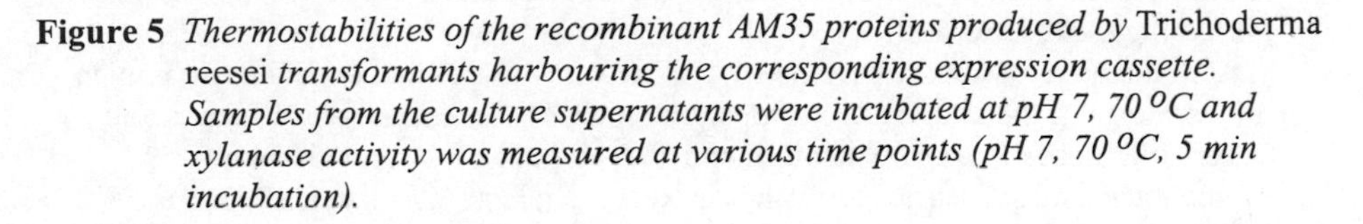
Figure 5 *Thermostabilities of the recombinant AM35 proteins produced by* Trichoderma reesei *transformants harbouring the corresponding expression cassette. Samples from the culture supernatants were incubated at pH 7, 70 ^{o}C and xylanase activity was measured at various time points (pH 7, 70 ^{o}C, 5 min incubation).*

4 ECOPULP-TX ENZYMES IN MILL SCALE

The novel thermophilic xylanase products have been tested and are in use in mill scale. 15 - 20 % reductions in consumption of chlorine chemicals have been obtained.

Thermostable xylanases are essential for enzyme-aided bleaching technology. As the process conditions in mills have changed to demand more thermo- and alkali-tolerant enzymes, thermostable products can serve over 90 % of kraft pulp mills whereas only about 30 % of the mills have conditions suitable for conventional fungal xylanase products.

5 SUMMARY

We have produced in *T. reesei*, by using the *cbh1* promoter, heterologous thermophilic xylanases active at temperatures up to 80 ^{o}C and at pH 4 - 9. Industrially feasible levels, grams/litre, of both heterologous fungal and bacterial xylanases can be achieved by fermentations. High yields of bacterial xylanases could be produced by using a fusion to a carrier protein. The properties of the products are very similar to those of the authentic enzyme. *T. reesei* mutant xylanases with higher thermostability, compared to the native enzyme, were also constructed by site-directed mutagenesis and produced in *T. reesei*.

Acknowledgements

We thank Dr. John Londesborough for critical reading of the manuscript and for revising the language. M.Sc. Heli Haakana is thanked for sequencing and Merja Helanterä, Outi Könönen, Outi Nikkilä, Kirsti Leskinen, Riitta Tarkiainen, Sirpa Holm, Leena Oinonen and Sirpa Okko for skilful technical assistance. Prof. George Szakacs and Dr. Maria Szakacs-Dobozi, Technical University of Budapest, Hungary, are acknowledged for their help in screening the thermophilic fungal strains. We also thank Dr. Juha Rouvinen and Dr. Anneli Törrönen for useful discussions.

References

1. L. Viikari, A. Kantelinen, J. Sundquist and M. Linko, *FEMS Microbiol. Reviews*, 1994, **13**, 335.
2. L. Viikari, M. Ranna, A. Kantelinen, J. Sundquist and M. Linko, 'Proceedings of the 3rd International Conference on Biotechnology for the Pulp and Paper Industry', Swedish Wood Research Institute, Stockholm, Sweden, 1986, p. 67.
3. D. Haltrich, B. Nidetzky, K.D. Kulbe, W. Steiner and Silvia Zupancic, *Bioresource Technology*, 1996, **58**, 137.
4. M. Paloheimo, A. Miettinen-Oinonen, T. Torkkeli, H. Nevalainen and P. Suominen, 'Proceedings of the 2nd Tricel Symposium on *Trichoderma reesei* Cellulases and Other Hydrolases', Eds. P. Suominen and T. Reinikainen, Foundation for Biotechnical and Industrial Fermentation Research, Espoo, Finland, 1993, Vol. 8, p. 229.
5. J.V. Joutsjoki, T.K. Torkkeli, and K.M.H. Nevalainen, *Curr. Genet.*, 1993, **24**, 223.
6. A. Miettinen-Oinonen, T. Torkkeli, M. Paloheimo and H. Nevalainen, *J. Biotechnol.*, 1997, in press.
7. A. Mäntylä, M. Paloheimo and P. Suominen, '*Trichoderma* and *Gliocladium*', Eds. C.P. Kubicek and G.E. Harman, in press.
8. D.I. Gwynne, F.P. Buxton, S.A. Williams, S. Garven and R.W. Davies, *Bio/Technology*, 1987, **5**, 713.
9. C.A.M.J.J. van den Hondel, P.J. Punt and R.F.M. van Gorcom, 'More Gene Manipulations in Fungi', Eds. J.W. Bennet and L.L. Lasure, Academic Press, Inc., San Diego, U.S., 1991, Chapter 18, p. 396.
10. D.J. Jeenes, D.A. MacKenzie, I.N. Roberts and D.B. Archer, 'Biotechnology and Genetic Engineering Reviews', Intercept Ltd., Andover, UK, 1991, Vol. 9, Chapter 9, p. 327.
11. R. Saarelainen, M. Paloheimo, R. Fagerström, P.L. Suominen and K.M.H. Nevalainen, *Mol. Gen. Genet.*, 1993, **241**, 497.
12. R.F. Wang and S.R. Kushner, *Gene*, 1991, **100**, 195.
13. G. Sarkar and S.S. Sommer, *BioTechniques*, 1990, **8**, 404.
14. A. Arase, T. Yomo, I .Urabe, Y. Hata, Y. Katsube and H. Okada, *FEBS Letters*, 1993, **316**, 123.
15. R.J. Gouka, P.J. Punt and C.A.M.J.J. van den Hondel, *Appl. Microbiol. Biotechnol.*, 1997, **47**, 1.
16. H. Stålbrand, A. Saloheimo, J. Vehmaanperä, B. Henrissat and M. Penttilä, *Appl. Environ. Microbiol.*, 1995, **61**, 1090.

Gene Regulation and Expression

REGULATORY MECHANISMS INVOLVED IN EXPRESSION OF EXTRACELLULAR HYDROLYTIC ENZYMES OF *TRICHODERMA REESEI*

A. Saloheimo, M. Ilmén, N. Aro, E. Margolles-Clark and M. Penttilä

VTT Biotechnology and Food Research
P.O.Box 1500
FIN-02044 VTT, Finland

1 INTRODUCTION

Trichoderma reesei is known and used as a producer of large amounts of extracellular hydrolytic enzymes. These provide the fungus with nutrients and energy when grown on complex plant material. In the presence of an easily metabolisable carbon and energy source, production of these enzymes would be an extra load to the organism. Thus, regulation of gene expression is of great importance to the organism. From the biotechnical point of view, continuous improvements in enzyme production are very important. Understanding of the regulation of the production promoters provides tools to enhance or modulate production of industrial enzymes or enzyme mixtures in a controlled way.

T. reesei produces two cellobiohydrolases, at least five endoglucanases and a β-glucosidase capable of hydrolysing in synergy crystalline cellulose into glucose. Furthermore, *T. reesei* produces a group of enzymes needed for efficient hydrolysis of hemicelluloses. Two endoxylanases and an endomannanase hydrolyse the backbones of xylans and mannans, respectively. Different sets of side-group cleaving activities are additionally needed for the total hydrolysis of differently substituted hemicelluloses. These include acetylxylan esterase, α-arabinofuranosidase, α-galactosidase and α-glucuronidase activities. β-Xylosidase and β-mannosidase are needed for the final steps of the hydrolyses into simple sugars.

Genes encoding most of these hydrolytic enzymes have been isolated[1,2]. Thus it is possible to study their regulation at the molecular level.

2 REGULATION OF CELLULASE EXPRESSION

In order to study the regulation of cellulase gene expression in *T. reesei* QM9414, a set of cultivations was performed on different carbon sources and the steady-state mRNA levels of the cellulase genes were analysed. No cellulase mRNAs could be detected in media containing glucose, while very high levels were obtained from Solka floc cellulose-containing cultures. Thus the regulation of cellulase expression operates at the transcriptional level. Sorbitol and glycerol were shown to be neutral carbon sources in respect to cellulase expression. They did not themselves promote cellulase expression but addition of small amounts (1-2 mM) of the known inducer of *T. reesei* cellulase expression, sophorose, caused accumulation of very high mRNA levels[3]. When sophorose was added to similar glucose cultures, or when glucose was added to cellulose cultures, no cellulase mRNAs could be detected[3]. These results clearly show that both induction and glucose repression are operating in the regulation of cellulase expression.

In order to compare the relative efficiencies of different carbon sources in induction, a slot blot experiment of RNA samples from different cultivations was carried out. The parallel filters were hybridised with cellulase gene probes of similar length and specific activity. The relative proportions of different cellulase mRNAs were found out to be the same in all inducing conditions analysed, the amount of *cbh1* mRNA always being the highest. The induction levels caused by cellulose or sophorose were very high. No signals could be detected from glucose or sorbitol cultures, even when overloading the analyses, whereas very small amounts of sophorose or cellulose-induced RNAs gave signals, the difference being at least a thousand fold between fully induced and uninduced state[3]. Moderate cellulase mRNA levels could be detected three hours after addition of sophorose to sorbitol cultures. Cellobiose and lactose induced all the cellulase genes to a much lower level. The cellulase genes appear to be coordinately regulated. However, this does not necessarily hold true for the β-glucosidase gene, which is expressed at a very low level even in the presence of sophorose or cellulose[2].

Constitutively produced cellulase enzymes could be a means for the fungus to initially attack cellulose whenever there are no simple sugars present. No cellulase mRNAs could be detected when glucose was present in our cultivations. However, when the fungus was let to consume all the glucose, clear cellulase signals were detected by northern analysis after depletion of glucose without the addition of an inducer. This level was about 10% of the expression level of each of the genes in sophorose-induced conditions. Carbon or nitrogen starvation were not as such responsible for this derepression since transfer of the mycelia to fresh media lacking either carbon or nitrogen did not lead to derepression[3]. It is possible that sophorose or some other inducing compound is formed from the glucose or released from the cell walls during the growth and can induce only when glucose repression is relieved. In nature, this could be a way of avoiding the constitutive expression of the hydrolase genes and still allow rapid induction as soon as the enzymes are needed.

3 REGULATION OF HEMICELLULASE GENES

The regulation of the hemicellulase genes encoding the two β-xylanases, *xyn1* and *xyn2*, the β-xylosidase, *bxl1*, the α-arabinofuranosidase, *abf1*, the α-glucuronidase, *glr1*, the acetylxylanesterase, *axe1*, the β-mannanase, *man1,* the three α-galactosidases, *agl1, agl2* and *agl3,* and, as a reference, the cellulase gene *cbh1* was studied at the transcriptional level[2] (Table 1). Additional cultivations were performed using as carbon sources different hemicelluloses as well as mono- or disaccharides generated during the hydrolysis of hemicelluloses. Also the effects of the sugar alcohols xylitol and arabitol were tested.

A common feature of all the genes was the lack of expression on glucose-containing media. When glucose was, however, exhausted from the medium induction of all the genes except *xyn1* and *agl3* to rather high levels could be observed. In the glucose-derepressed strain Rut-C30 most of the genes were expressed on glucose, with the exception of *xyn2* and *man1*. These results support the view that most, if not all of the hemicellulase and cellulase genes are subject to glucose repression in *T. reesei*.

Table 1 *Expression levels of hemicellulase genes in* T. reesei *strain QM9414 grown for 3 days on different carbon sources. The data are summarised from visual judgement of signal intensities from repeated northern hybridisations and various exposures of the films. Signal intensities from strong to non-detectable:* ++++, +++, ++, +(+), +, (+), -.

Carbon source[a]	Probe										
	cbh1	*xyn1*	*xyn2*	*bxl1*	*abf1*	*glr1*	*axe1*	*man1*	*agl1*	*agl2*	*agl3*
Glucose	-	-	-	-	-	-	-	-	-	-	-
Glucose depl	++++	-	++	++	+++	+	++	+	+++	+++	-
Cellulose	++++[b]	++[b]	++[b]	++[b]	-	+[b]	+++[b]	++[b]	+[b]	+[b]	+
Xylan	(+)	++	-	++	-	+	-	-	-	-	-
MeGlc-Xylan	+	+	++	+(+)	-	(+)	++	-	-	-	-
OS Xylan	+	+++	+	++	+++	+	(+)	-	+	+	-
Xylose	-	-	-	-	-	-	-	-	-	-	-
Xylitol	(+)	-	-	-	+	-	-	-	+	+	-
Arabinose	-	-	-	+	+	-	-	-	-	-	-
Arabitol	++	++	++	+++	++++	+	+	-	++	+++	-
Sorbitol	-	-	(+)	-	-	-	-	-	-	+	-
Sorb/Soph	++++	+	+(+)	++	+	-	+	-	+	+(+)	(+)
Sorb/Xylob	+	-	(+)	++	+	-	-	-	+	+	-
Glycerol	-	-	-	-	-	-	-	-	+	-	-
Glyc/Mannob	(+)	-	-	-	(+)	-	-	-	+	+	-
Glyc/Xylob	+	++	+++	++++	-	++	+	-	+	+(+)	-
Mannose	-	-	-	-	-	-	-	-	+	-	-
Galactose	-	-	-	-	-	-	++	-	+++	++++	-
Lactose	++	-	-	+	-	-	-	-	-	-	-

[a] Glucose depl: glucose depleted, Xylan: unsubstituted birch wood xylan, MeGlc-Xylan: 4-O-Methylglucuronoxylan, OS Xylan: oat spelt arabinoxylan, Sorb/Soph: sorbitol+ sophorose, Sorb/Xylob: sorbitol+xylobiose, Glyc/Mannob: glycerol+mannobiose, Glyc/Xylob: glycerol+xylobiose.

[b] Amount of RNA loaded was 0.5 μg. Other samples contained 5 μg RNA.

Induction patterns of the hemicellulase genes were variable. Sophorose and cellulose, the inducers of the cellulase genes, induced most but not all of the genes. The cellulose preparation used, Solka floc cellulose, is known to contain a significant portion of xylan possibly enhancing induction of xylan inducible genes. The expression patterns of *xyn1* and *xyn2* were somewhat different suggesting differences in their regulation. *bxl1* was expressed always when one of the xylanases was. Different types of polymeric xylans, xylobiose and sophorose were among the best inducing compounds for these genes.

The genes coding for side group cleaving activities were in general more efficiently induced by substituted xylan containing side groups than by unsubstituted xylan. The hydrolysis products which are liberated from the polymers into the growth medium are likely to be the critical factors modulating the expression patterns observed. Interestingly, arabitol which has been reported to induce arabinofuranosidases of *Aspergilli* promoted significant expression of *cbh1* and all hemicellulases with the exception of *man1* and *agl3* which were in general poorly expressed.

The monosaccharides galactose, mannose and xylose given to the fungus in unlimited amounts were poor inducers for practically all genes analysed with the exception of the strong and specific galactose induction of *agl1, agl2* and *axe1*. None of these genes was induced by lactose, a disaccharide of glucose and galactose. The poor inducing capacity of mannose and mannobiose is in accordance with the observations that *T. reesei* was unable to grow on mannan and that *man1* transcription was undetectable except on cellulose-containing medium.

In conclusion, the regulation of hemicellulase genes appears to be more complex than that of the cellulase genes. Both general and specific induction mechanisms may play a role in regulating their expression.

4 FUNCTIONAL ANALYSIS OF THE *CBH1* PROMOTER

Since the expression of the *cbh1* gene was clearly the highest of all studied hydrolase genes of *T. reesei,* and the strong *cbh1* promoter is used for industrial protein production, we decided to study it more closely. No previous knowledge existed on the regulatory sequences or on the regulatory proteins mediating carbon source dependent regulation of cellulase gene expression. The *cbh1* promoter (2.2 kb) was fused to the *E. coli lacZ* reporter gene. A series of deletions and site specific mutations were introduced in the *cbh1* promoter and the effects of the mutations on *lacZ* expression were analysed by X-gal plate assays and northern analyses of the *T. reesei* transformants *in vivo*[4]. The main findings are illustrated in Figure 1.

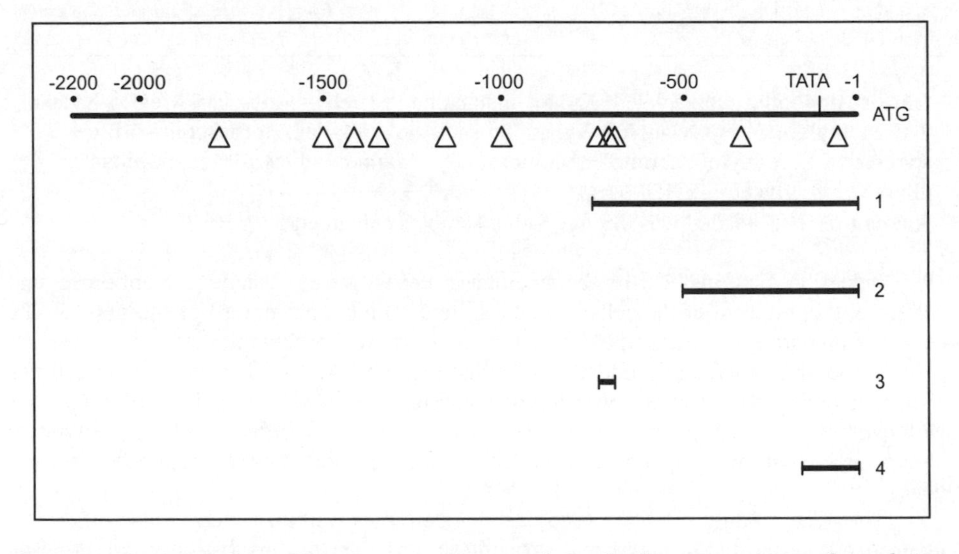

Figure 1 *Regions of the cbh1 promoter of* T. reesei *involved in transcriptional regulation. 1, glucose repressible promoter fragment; 2, derepressible fragment; 3, CREI binding sites mediating glucose repression; 4, sophorose inducible promoter fragment. Consensus binding sites for the* A. nidulans *CREA (SYGGRG) are marked with triangles.*

The analysis of the deletion series indicated that a very short promoter fragment, covering sequences from 30 bp upstream of the TATA-box to ATG, was inducible by sophorose. Thus a regulatory protein may bind close to the TATA-box and activate transcription. Another deletion removing sequences upstream of -500 from the initiator ATG remarkably relieved the *cbh1* promoter from glucose repression. Examination of the nucleotide sequence revealed that the deleted region contains a number of 6-bp elements similar to those recognised by the glucose repressor proteins MIG1 of *S. cerevisiae* and CREA of *A. nidulans*. Site directed mutagenesis of the putative glucose repressor binding sites was subsequently carried out. Transformants having a single copy of the *cbh1-lacZ* expression cassette integrated in the *cbh1* locus were used in the study. Comparison of *lacZ* transcript levels produced by transformants grown in the presence of glucose indicated that mutations within the 40 bp region around -700, which contains three clustered CREA-binding sites, caused derepressed *lacZ* expression. In contrast, mutagenesis of the putative repressor binding sites at -1500 or at -1000 caused no or only very weak derepression, respectively, indicating that these more upstream sites are less important for repression[3,4]. These results strongly suggested that glucose repression of *cbh1* transcription is mediated by a protein having a similar DNA-binding specificity as the previously identified fungal glucose repressors.

5 THE GLUCOSE REPRESSOR CRE I

In order to further investigate molecular mechanisms of glucose repression the glucose repressor gene *cre1* was isolated from *T. reesei*. PCR primers were designed based on conserved regions of the *A. nidulans* and *A. niger* CREA. The CREA/CREI proteins contain a highly conserved DNA-binding domain with two zinc fingers of the Cys_2His_2 type, and another highly conserved proline-serine rich region of unknown function[5].

The discovery that the well known hypercellulolytic *T. reesei* strain Rut-C30 has a mutation in the *cre1* gene made it possible to study the function of *cre1*. This mutant form of the gene, named *cre1-1*, is predicted to encode a peptide of 95 amino acids containing only the N-terminal zinc finger. It can be expected that this mutation is likely to cause defects in glucose repression. In accordance with this it was detected that the Rut-C30 strain, unlike the QM9414 strain, produces cellulase and also hemicellulase transcripts when grown in the presence of glucose suggesting that glucose repression is relieved in this strain due to the *cre1-1* mutation. This was proved to be the case, since introducing the full length copy of the *cre1* gene into the Rut-C30 strain restored glucose repression of *cbh1* expression in the transformants[5]. In addition of mediating glucose repression of cellulase expression, the *cre1* gene also controls expression of many genes encoding hemicellulases. At least *xyn1*, *bxl1*, *agl1*, *agl2*, *agl3* and *axe1* are under transcriptional control mediated by *cre1*, as shown by repression of their expression following the introduction of the *cre1* gene into the Rut-C30 strain.

6 ACTIVATORS OF CELLULASE EXPRESSION

The finding that cellulase genes are coordinately regulated and need an inducer led us to the attempt to isolate gene(s) encoding activator(s) responsible for the induction. The isolation was not possible using traditional methods because of the total lack of data on the corresponding protein(s). Sequence comparison of the cellulase promoter regions gave little hint of possible binding sites for activator proteins useful in the One-Hybrid cloning system. Furthermore, the cellulase-negative mutants of *T. reesei* produced earlier at VTT[6] seemed not to be regulatory mutants suitable for complementation cloning. Therefore we set up a method to isolate positively-acting regulatory genes by genetic selection in yeast which did not rely on any previous protein or binding site data.

The 1.1 kb promoter fragment upstream of the TATA box of the major cellulase gene *cbh1* was linked to the *S. cerevisiae HIS3* gene used as a reporter gene. The *HIS3* gene fragment contains a minimal promoter with the TATA box to be recognised by the basic transcription machinery of yeast. An auxotrophic yeast strain transformed with the reporter construct could not grow without added histidine showing that the *Trichoderma* promoter, or the minimal yeast promoter, could not drive the expression of the reporter gene.

An expression cDNA library prepared from *T. reesei* grown in hydrolase-inducing conditions was constructed into a yeast multicopy expression vector. The strong and constitutive PGK promoter was used. The library was transformed into the reporter yeast strain and screened for growth on medium containing no histidine. Only yeast cells containing the *Trichoderma his3* gene, or a cDNA encoding an activator protein capable of binding its binding sites in the *cbh1* promoter fragment and activating *HIS3* transcription could survive.

The growth of the putative activator-containing yeast clones in the absence of histidine was confirmed to be dependent on the reporter plasmid with the *cbh1* promoter. The cDNA inserts were sequenced and found out to encode two different proteins named ACE I and ACE II (Activator of Cellulase Expression). In the ACE I protein there is a bipartite nuclear targeting signal followed by three zinc fingers of the Cys_2His_2 type (Figure 2). The function of the second finger as a DNA-binding element is uncertain since there are extra amino acids in the predicted DNA recognition loop compared to the consensus sequence of zinc fingers. In the ACE II protein, there is an N-terminal Zn_2Cys_6 binuclear cluster domain unique to the DNA-binding proteins of fungal origin (Figure 2).

When the sequences of the ACE I and ACE II proteins were scanned against the protein data bases, no significant similarities were detected outside the putative DNA-binding regions, not even when the whole yeast genome was analysed. This suggests that the proteins are not components of the general transcription machinery well conserved between species. Some short stretches of similarity to many different DNA-binding regulatory proteins were, however, detected, such as areas rich in glutamine and/or proline residues possibly involved in the activation function, and in ACE II a region rich in histidines.

7 CONCLUSIONS

The regulation of the cellulase and hemicellulase genes was studied at the transcriptional level. The cellulase genes were coordinately expressed while the regulation of the hemicellulase genes was more variable and probably subject to both general and specific induction mechanisms. Most of the genes are subject to glucose repression mediated by the CREI repressor protein.

The promoter region of the major cellulase gene *cbh1* was studied by deletion analysis. Separate areas important for the glucose repression and the induction were identified.

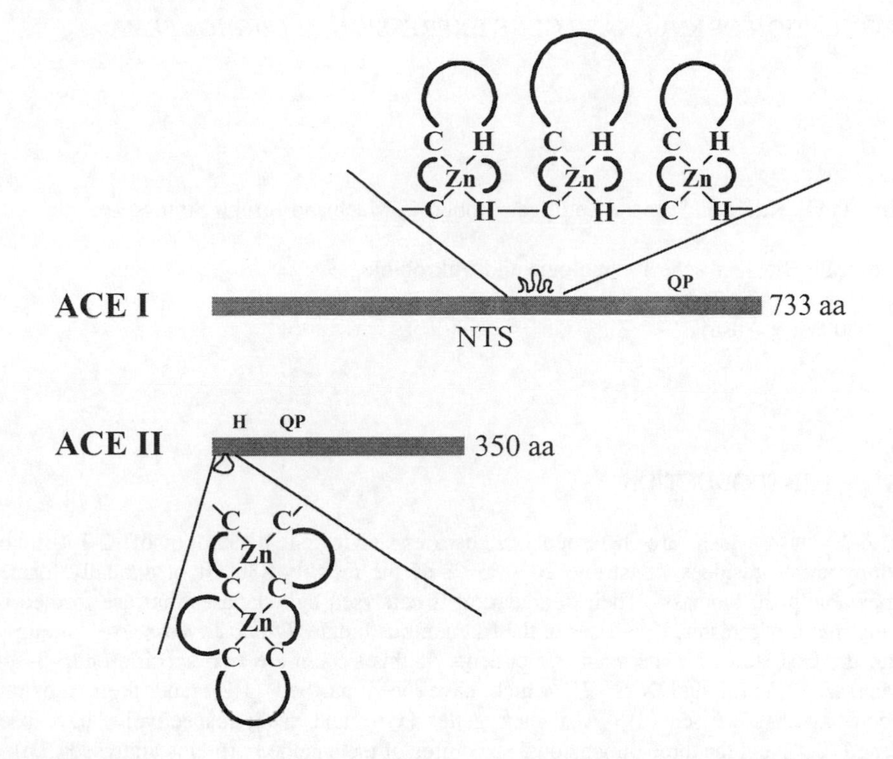

Figure 2 *Schematic representations of the activator protein sequences and the putative DNA-binding elements. The nuclear targeting signal (NTS) and areas with high local concentration of certain amino acids (H, P, Q) are marked.*

A method was developed for screening of transcriptional activator proteins based on no previous data on the regulatory proteins or their binding sites in the promoter. In contrast to the yeast One-Hybrid System, the binding itself is not sufficient for selection in our system, but the regulatory protein has to possess the activation function as well. Thus genes having relevance to the specific activation of the promoter *in vivo* are more likely to appear instead of basic transcription factors. Two putative activator genes were isolated.

References

1. H. Nevalainen and M. Penttilä, "The Mycota II", Kück (Ed.), Springer-Verlag, Berlin/Heidelberg, 1995, Chapter 18, p. 303.
2. E. Margolles-Clark, M. Ilmén and M. Penttilä, *J. Biotechnol.,* 1997, **57**, 167.
3. M. Ilmén, A. Saloheimo, M.-L. Onnela and M.E. Penttilä, *Appl. Environ. Microbiol.*, 1997, **63**, 1298.
4. M. Ilmén, M.-L. Onnela, S. Klemsdal, S. Keränen and M. Penttilä, *Mol. Gen. Genet.*,1996, **253**, 303.
5. M. Ilmén, C. Thrane and M. Penttilä, *Mol. Gen. Genet.*, 1996, **251**, 451.
6. K.M.H. Nevalainen and E.T. Palva, *Appl. Environ. Microbiol.,* 1978, **35**, 11.

REGULATION OF XYLANASE GENE EXPRESSION IN *TRICHODERMA*

Christian P. Kubicek, Susanne Zeilinger, Robert L. Mach and Joseph Strauss

Institut für Biochemische Technologie und Mikrobiologie
TU Wien
A-1060 Wien, Austria

1 INTRODUCTION

ß-1,4-Xylans, which are heteropolysaccharides having a backbone of ß-1,4-linked xylopyranosyl residues, constitute 20 - 35 % of the roughly 830 Gt of annually formed renewable plant biomass. Their degradation is catalysed by xylanases that are formed by many micro-organisms. Xylanases of the filamentous fungus *Trichoderma reesei* are up to date the best studied. The main components of this system are two specific endo-ß-1,4-xylanases, XYN I and XYN II, which have been purified (1,2), and their substrate specificity characterised (3). Also their genes (*xyn1* and *xyn2*; respectively) have been cloned (2,4), and the three dimensional structures of the encoded proteins analysed (5,6).

The regulation of their formation has only received incomplete attention. Hrmová *et al.* (7), using isoelectric focusing gels coupled with staining of xylanase and cellulase activities showed that part of this xylanase activity in *T. reesei* is due to EG I activity. Consequently, most of previous studies, which made use of enzyme activity assays alone, did not provide reliable information on the regulation of formation of the two specific xylanases. The same authors also observed that upon induction by sophorose, only one of both xylanases and EG I were formed whereas induction by xylobiose led to the formation of both xylanases but not of endoglucanase I, thereby further complicating the picture.

Having the genes encoding the two xylanases of *T. reesei* in hand, we have initiated an investigation towards identifying the regulatory mechanisms governing *xyn1* and *xyn2* gene expression. Here we will briefly review the present state of the information obtained.

2 REGULATION OF *XYN1* GENE EXPRESSION

Transcriptional studies, using both Northern blotting as well as slot-blotting techniques, indicated that the *xyn1* gene product is formed during growth on xylan, but not on glucose. In washed mycelia, only xylose - but not xylobiose or xylitol - was able to induce the formation of the *xyn1* transcript. These preliminary studies suggested that *xyn1* gene expression may be regulated by carbon catabolite repression and/or induction.

In order to discriminate between these possibilities, we first investigated whether *xyn1* is indeed subject to carbon catabolite repression (8). To this end, we first carried out a promoter deletion analysis. A 534-bp fragment of the *xyn1* promoter, fused to the *E. coli hph* structural gene, conferred essentially the same transcriptional regulation as of native *xynl*, and thus contains all the essential information for *xynl* regulation. When these 534-bp were subjected to promoter deletion analysis, the removal of a 213-bp area between -534 and -321 simultaneously abolished inducibility by xylose and repression by glucose. Instead, the truncated promoter was expressed at a low constitutive level, indicating that *xynl* is indeed subject to glucose repression.

Sequence analysis of the relevant 213-bp of the promoter revealed the presence of three binding sites for the catabolite repressor protein Cre1 (9-11), two of them organised as a tandem repeat. All of them were able to bind a Cre1:GST (glutathione S-transferase) fusion protein *in vitro*. Since studies on other promoters in *Aspergillus nidulans* provided evidence that only doublets of the CreA (*A. nidulans* homologue of Cre1) binding site are physiologically active targets (12,13), we disrupted the two sites (3 and 4, which were arranged as doublets) *in vivo*, and analysed the expression of *xynl* in several of the transformants. It is evident that the mutated promoter of *xynl* was no more subject to glucose repression but expressed at a basal constitutive level. This level correlated with the level observed on lactose, a carbon catabolite derepressing carbon source. In contrast, the strain carrying the mutated promoter expressed the same amounts of transcript from the *xynl* promoter on glucose and on lactose. These findings prove the involvement of the inverted-repeat of the Cre1-target sequence in carbon catabolite repression of *xynl*.

In order to further proof that Cre1 is involved in this repression, we analysed the transcriptional regulation of *xynl* in the mutant strain *T. reesei* RUT C-30. This mutant has been obtained by selection for cellulase production on glycerol plus 2-deoxyglucose, and was shown by Ilmen *et al.* (11) to contain an extremely truncated copy of Cre1, and therefore to be unable to repress the formation of *cbhl* on glucose. In accordance with the results obtained with the mutant promoter (see above), a low basal level of *xynl* transcript was observed in RUT C-30 during growth on glucose and lactose, respectively. This proved that Cre1 is involved in carbon catabolite repression of *xynl*.

Interestingly, both in strain RUT C-30 as well as in the transformants with the mutated promoter, addition of xylose induced the formation of the *xynl* transcript clearly over the constitutive level, and this was also observed in the simultaneous presence of xylose and glucose. Hence, *xynl* appears to be regulated by both carbon catabolite repression and induction, but the former only acts on the basal level of expression, whereas induction is not affected by glucose. A scheme of this model is depicted in Figure 1.

To identify nucleotide targets confering this induction by xylose, we have used the 213-bp promoter fragment (see above) as DNA-probe in gel retardation assays with cell-free extracts of *T. reesei* mycelia cultivated on glucose or induced by xylose. Two different, specific DNA-protein complexes were obtained, which were representative for the negative conditions: upon growth on glucose, a complex of slow mobility (indicative of a high molecular weight) was observed. Formation of this complex could be competed by the addition of an excess of an unlabelled oligonucleotide containing a functional Cre1-binding site, but not with oligonucleotides containing mutated sites. The putative involvement of Cre1 in the formation of this complex was finally proven by demonstrating its presence by Western blotting.

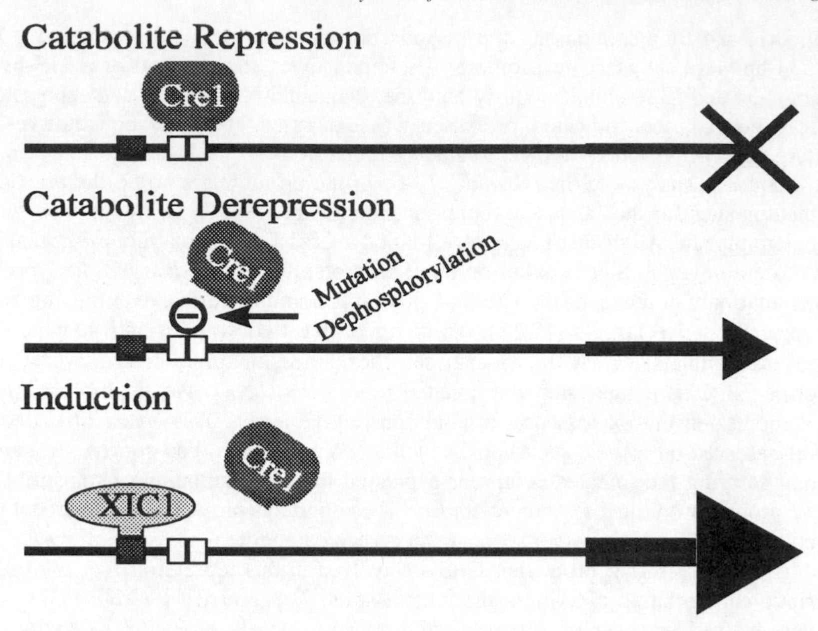

Figure 1 *Model for the regulation of xyn1 gene transcription in* T. reesei. Abbreviations: XIC_1, xylan-dependent transcription inducing complex

Upon induction of *xyn1* gene transcription by xylose, EMSAs with cell-free extracts led to a faster-migrating DNA-protein complex (indicating lower molecular weight), which could no more be competed with Cre1-binding oligonucleotides. Instead, specific competition could be obtained by an excess of an oligonucleotide containing a CCAAT-box, but no additional homologous areas on it. Mutation of this element to CCCTT abolished competition, thereby indicating the involvement of the two dA's of the putative CCAAT element in protein-DNA binding. Studies are now in progress to characterise the involved components in more detail.

3 REGULATION OF *XYN2* GENE EXPRESSION

Xylanase II is usually secreted during growth on xylan, but the enzyme cannot be detected by activity assays or immunological methods during growth on glucose. Consistent with this finding, no *xyn2*-mRNA is detected in Northern blots from glucose cultures. However, using the more sensitive slot-blot technique, a low basal level of *xyn2*-transcript on glucose could be shown (14). These findings were also supported by the use of the *hph*-gene as a reporter, which led to slow growth on glucose/hygromycin.

Growth on xylan, as well as addition of xylobiose and sophorose to washed mycelia, induced transcription roughly tenfold over the basal level. Interestingly, addition of glucose inhibited this induction, and resulted only in the basal level of transcription. The mechanism of this "repression" by glucose is not known, but as it does not occur in the *cre1*⁻ mutant *T. reesei* RUT C-30, it apparently involves Cre1. It should be noted that growth on lactose, which has been shown to relieve *xyn1* transcription from carbon catabolite repression (see above), did not lead to higher transcript levels of *xyn2*. This can be interpreted in such a way as that the derepression of the transactivator alone is insufficient for *xyn2* gene transcription, and that the presence of an inducer is essential.

In order to identify the promoter area responsible for induction by xylan, xylobiose and sophorose, promoter deletions analogous to those with *xyn1* (see above) were carried out. They delimitated the active region to 55-bp's 40-nt's upstream of the TATA box. Further deletions resulted in a complete loss of transcription. In contrast to *xyn1*, and in support of the assumption of absence of carbon catabolite repression directly acting on *xyn2*, no deletion was found which led to increased transcription of *xyn2* on glucose.

No significant similarity with nt-motifs conferring transcriptional regulation in other promoters was detected in the 55-bp of the *xyn2* promoter, with the exception of an inverted CCAAT-motif, and a single Cre1-binding site.

The presence of the single Cre1 binding site was puzzling as it clearly bound Cre1 *in vitro*, whereas *xyn2* did not seem to be subject to Cre1 regulation. Since the active Cre1-binding site in the *xyn1* promoter is organized as an inverted repeat, we tested whether the presence of such a site would cause Cre1-dependent repression of *xyn2*. Hence, this site and one further upstream were mutated to yield inverted and/or tandem Cre1-binding motifs, and introduced into *T. reesei*, using the *hph* reporter system. Although several transformants were investigated, none of them showed any gain of glucose repression. These findings are intriguing, since of both constructs one showed the same distance to the transcription start as the *in vivo* functional Cre1-binding site in *xyn1*, and the other the same distance to the inducing CCAAT-motif.

To identify potential nucleotide motifs in the *xyn2* promoter involved in xylan induction, cell-free extracts were prepared from induced (xylan) and non-induced (glucose) mycelia, and used in EMSA's with the 55-bp fragment (14). They showed the presence of a fast-migrating (low molecular weight) DNA-protein complex with cell-free extracts from glucose-grown mycelia, which was shifted into a slow migrating (high molecular weight) complex in mycelia induced by xylan. Using an excess of an unlabelled oligonucleotide exhibiting no homology to the 55-bp fragment apart from the inverted CCAAT-motif, both the constitutive as well as the induced DNA-protein could be competed.

Mutation of the CCAAT-motif to CCTTT in the competing oligonucleotide did not show competition for the inducing complex. These data suggest that regulation of *xyn2*-gene transcription occurs by an element binding to the CCAAT-like motif or another sequence motif overlapping with this box. The validity of this interpretation has more recently been supported by *in vivo* footprinting analyses, which clearly demonstrated the occupation of the two dA's of the CCAAT element during induction by sophorose (Zeilinger *et al.*, ms in preparation).

Based on these data, a model of the regulation of *xyn2* is proposed in Figure 2, which postulates the presence of a constitutive transcription complex binding to the CCAAT-like element, which allows basal transcription.

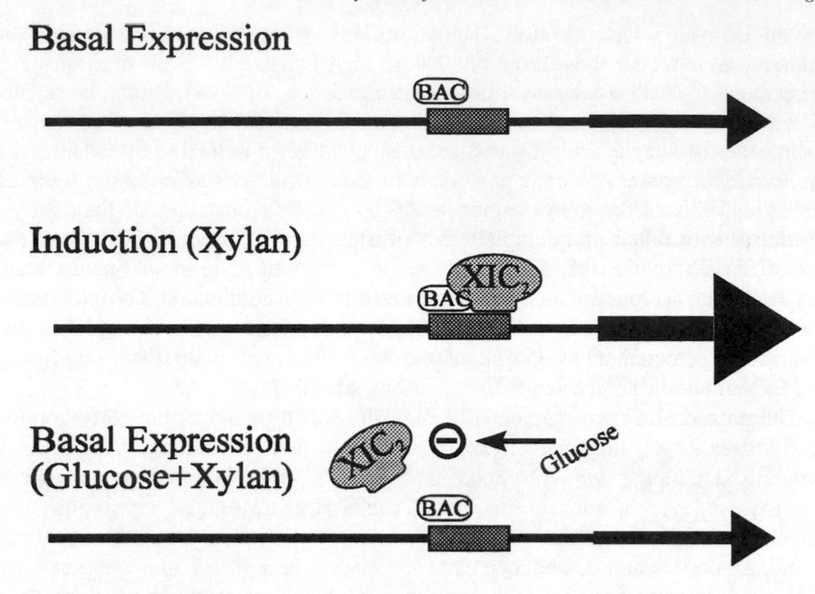

Figure 2 *Model for the regulation of xyn2 gene transcription in* T. reesei. Abbreviations: BAC, basal transcriptional activation complex; XIC_2, xylan-specific transcription-inducing complex

Upon growth on xylan/cellulose or induction by xylan/cellulose-derived inducers, (an) additional transactivating factor(s) bind(s) to the DNA and this complex, and increase(s) transcription. Another possibility, equally consistent with the data, would be the replacement of the basal CCAAT-binding complex by another, induced CCAAT-binding complex. While it remains to be established which of these two models is correct, it is evident that the induced transactivating factor or binding complex is subject to repression by the *cre1*-gene product.

4 ACKNOWLEDGEMENTS

Work reported in this paper has been supported by grants from the Austrian Science Foundation (P8979-CHE; P10793-GEN).

References

1. M. Tenkanen, J. Puls and K. Poutanen, *Enzyme Microb. Technol.*, 1992, **14**, 566.
2. A. Törrönen, R.L. Mach, R. Messner, R. Gonzalez, A.M. Harkki, N. Kalkinnen and C.P. Kubicek, *Bio/Technol.*, 1992, **10**, 1461.
3. P. Biely, M. Vršanská, L. Kremnicky, M. Tenkanen, K. Poutanen and M. Hayn, "*Trichoderma reesei* cellulases and other hydrolases" (P. Suominen and T. Reinkainen, eds.), Foundation of Biotechnical and Industrial Fermentation Research, 1993, p. 125.

4. M. Saarelainen, M. Paloheimo, R. Fagerström, P.L. Suominen and K.M.H. Nevalainen, *Mol. Gen. Genet.*, 1993, **241**, 497.
5. A. Törrönen, A. M. Harkki and J. Rouvinen, *EMBO J.*, 1994, **13**, 2493.
6. A. Törrönen and J. Rouvinen, J., *Biochemistry* 1995, **34**, 847.
7. M. Hrmová, P. Biely and M. Vršanská, *Arch. Microbiol.*, 1986, **144**, 307.
8. R.L. Mach, J. Strauss, S. Zeilinger, M. Schindler and C.P. Kubicek, *Mol. Microbiol.*, 1996, **21,** 1273.
9. J. Strauss, R.L. Mach, S. Zeilinger, G. Stöffler, M. Wolschek, G. Hartler and C.P. Kubicek, *FEBS Letts.*, 1995, **37,** 103.
10. H. Takashima, A. Nakamura, H. Iikura, H. Masaki and T. Uozumi, *Biosci. Biotechnol. Biochem.,* 1995, **60**, 173.
11. M. Ilmén, C. Thrane and M. Penttilä, *Mol. Gen. Genet.*, 1996, **251**, 451.
12. M. Mathieu and B. Felenbok, *EMBO J.*, 1994, **13**, 4022.
13. B. Cubero and C. Scazzochio, *EMBO J.*, 1994, **13**, 407.
14. S. Zeilinger, M. Schindler, P. Herzog, R.L. Mach and C.P. Kubicek, *J. Biol. Chem.*, 1996, **271**, 25624.

CLONING OF HYDROLASES FROM *TRICHODERMA REESEI* AND OTHER MICROORGANISMS IN YEAST

W. H. van Zyl[1], D. C. la Grange[1], J. M. Crous[1], M. Luttig[1], P. van Rensburg[2] and I. S. Pretorius[2]

[1] Department of Microbiology and [2] Institute for Wine Biotechnology
University of Stellenbosch, Stellenbosch 7600, South Africa

1 INTRODUCTION

Cellulose and hemicellulose, together with lignin, are the major polymeric constituents of plant cell walls and form the largest reservoir of fixed carbon in nature (Figure 1). The hydrolysis of these polymers to allow the resulting sugars to be assimilated, remains a major technological and economic challenge. The degradation of polysaccharides can be carried out by acid or alkaline treatment, but this is expensive, time consuming and generates toxic by-products and waste that are expensive to treat. The enzymatic hydrolysis of polymeric substances by extracellular enzymes, such as cellulases and hemicellulases, is therefore preferred to chemical depolymerisation. The yeast *Saccharomyces cerevisiae* has long been associated with the alcoholic beverage and baking industries, and it has also been pivotal in the production of fuel ethanol and probiotic animal feed (single cell protein). The main limitation of this versatile organism of GRAS status (Generally Regarded As Safe), is its inability to utilise a wide range of carbohydrates, including cellulose and hemicellulose.

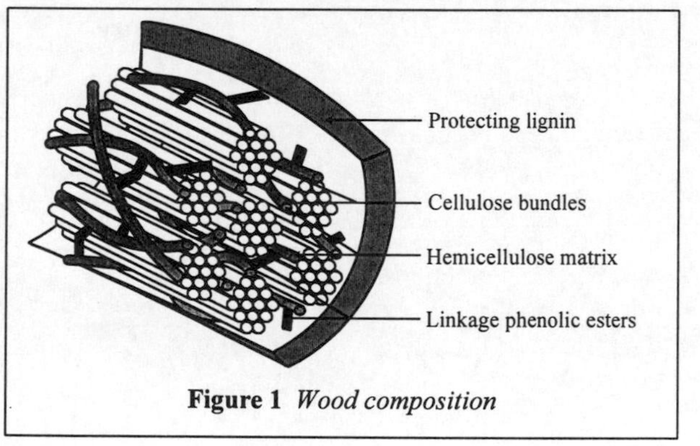

Figure 1 *Wood composition*

2 YEAST AS HETEROLOGOUS HOST

S. cerevisiae has been successfully used as expression host for the production of numerous industrial enzymes and pharmaceuticals and is considered the most popular model for fundamental molecular genetic studies[1]. *S. cerevisiae* thus provides an ideal microorganism for both basic research and biotechnological exploitation. The aim of our research is the cloning and expression of carbohydrase (hemicellulase and cellulase) genes in *S. cerevisiae* and the characterisation of the enzymes produced in yeast. We also aim at expanding the genetic potential of *S. cerevisiae* to enable it to utilise polysaccharide-containing agricultural crops and biowaste as substrates, or to convert them to commercial commodities, such as fuel extenders and nutritionally fortified animal feeds.

Genes encoding carbohydrases were obtained from fungal and bacterial donor organisms, manipulated through standard cloning procedures and joined to appropriate yeast expression signals (promoter, secretion and terminator sequences). These constructs (called expression cassettes) were subsequently introduced into *S. cerevisiae* yeast cells with the aid of yeast plasmid vectors to allow stable maintenance and expression of the appropriate genes in yeast (Figure 2).

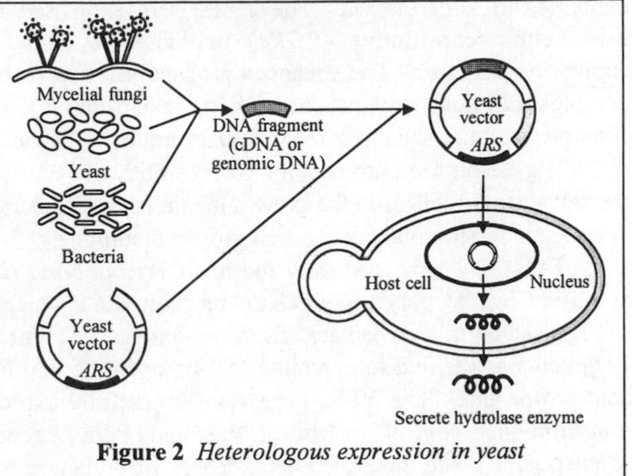

Figure 2 *Heterologous expression in yeast*

3 PRODUCTION OF HEMICELLULASES IN YEAST

3.1 Hemicellulose structure

Hemicellulose is second only to cellulose in natural abundance. Unlike cellulose, hemicellulose is comprised of complex polymers of which xylan is the major component. Xylan consists of a β-D-1,4-linked polyxylose backbone substituted with acetyl, arabinofuranosyl and glucuropyranosyl side chains. Hydrolysis of the xylan backbone is catalysed by endo-β-1,4-D-xylanases and β-D-xylosidases. Endo-β-D-xylanases act on xylans and xylooligosaccharides, producing mainly mixtures of xylooligosaccharides. β-Xylosidases hydrolyse xylooligosaccharides to D-xylose. The removal of the xylan side-chains requires side-chain cleaving enzymes, such as α-D-glucuronidase, α-L-arabinofuranosidase, acetylxylan esterase and phenolic esterases (feruloyl and *p*-coumaroyl esterase) (Figure 3).

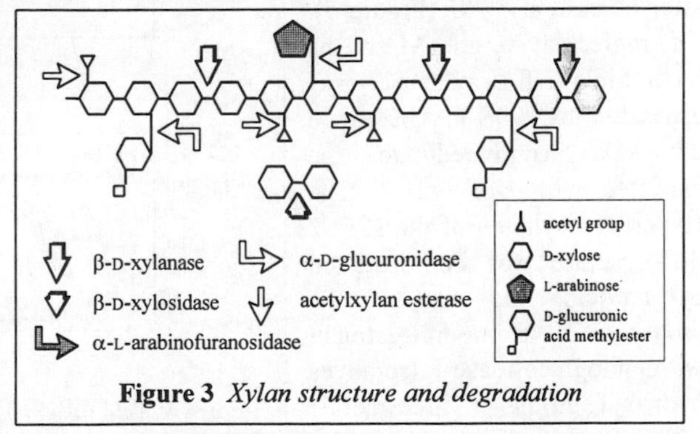

Figure 3 *Xylan structure and degradation*

3.2 Production of hemicellulases in yeast

Genes encoding four different fungal β-D-xylanases, a bacterial β-D-xylosidase, and two different fungal α-L-arabinofuranosidases have been cloned and successfully expressed in *S. cerevisiae.* The genes were expressed from yeast multicopy plasmids using either constitutive (*PGK1*) or inducible (*ADH2*) yeast promoter / terminator expression cassettes. The enhanced production of recombinant enzymes in non-selective complex medium, without the risk of losing the episomal vector, was obtained by constructing auto-selective recombinant strains. The uracil phosphoribosyl-transferase *(FUR1)* gene on the chromosome of recombinant yeast was disrupted, which prevented the salvage of uracil from the growth medium. This ensured auto-selection for the *URA3*-bearing expression plasmids in rich growth medium[2].

The *XYN2* gene encoding the main *Trichoderma reesei* QM 6a β-D-xylanase was amplified by the polymerase chain reaction (PCR) technique from first strand cDNA synthesised on mRNA isolated from the fungus[3]. The nucleotide sequence of the cDNA fragment was verified to contain a 699-bp open reading frame that encodes a 223-amino acid pro-peptide. The *XYN2* gene was successfully expressed in *S. cerevisiae* under the transcriptional control of the *ADH2* and *PGK1* gene promoters and terminators, respectively. The heterologous XYN2 β-D-xylanase was functional and efficiently secreted from the yeast cells. In non-selective complex medium *S. cerevisiae* strains produced 1,200 and 160 nkat/ml of β-D-xylanase activity under the control of the *ADH2* and *PGK1* promoters, respectively (Figure 4A&B). The recombinant enzyme showed highest activity at pH 6 and 60°C and retained more than 90% of its activity after 60 min at 50°C (Figure 5). The molecular weight (M_r) of the recombinant β-D-xylanase was estimated by SDS-PAGE to be 27 kDa, compared to the expected M_r of 21 kDa. Hyperglycosylation of the XYN2 β-D-xylanase was demonstrated by reducing the size of the protein to 21 kDa upon treatment with endoglucosidase-F (removes hydroxyl groups). Three acidic β-D-xylanases from *Aspergillus* species were also cloned by PCR and successfully expressed in *S. cerevisiae*[4,5]. The XYN3 β-D-xylanase of *Aspergillus kawachii*

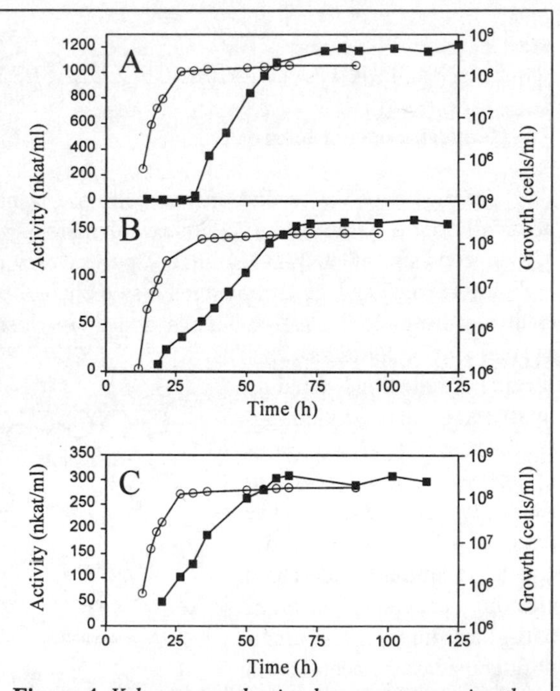

Figure 4 *Xylanase production by yeast expression the* XYN2 *gene under transcriptional control of the* ADH2 *(**A**) or* PGK1*(**B**) promoters, and the* XYN3 *gene under control of the* PGK1 *promoter (**C**). Xylanase activity is depicted by (■) and the growth by (○)*

IFO4308 yielded the highest activity in non-selective complex medium (300 nkat/ml) with the *XYN3* gene under transcriptional control of the *PGK1* promoter sequences (Figure 4C). The XYN3 β-D-xylanase has a pH and temperature optimum of <pH 3 and 60°C, respectively (Figure 5).

Two β-D-xylanase-encoding genes, amplified from *Aspergillus niger* ATCC 90196 mRNA, were inserted between the yeast *ADH2* promoter and terminator sequences and designated *XYN4* and *XYN5*, respectively. The β-D-xylanase activities of 91 nkat/ml and 56 nkat/ml for XYN4 and XYN5, respectively, were considerably lower than those observed for XYN2 and XYN3. Both the XYN4 and XYN5 enzymes have pH and temperature optima of pH 4 and 60°C, respectively (Figure 5). The nucleotide sequences of the *XYN4* and *XYN5* genes revealed that both genes encode 211-amino acid proteins that are 92% identical to each other. Surprisingly, *XYN4* proved to be 96% identical to *XYN3* (also 211 amino acids long) with only two amino acid differences in the mature protein portion (glycine to alanine at position 42 and threonine to serine at position 80). These conserved amino acid substitutions can be directly responsible for the difference in pH optimum. They can also improve the stability of the enzyme produced in yeast, particularly at low pH and indirectly be responsible for the difference in pH optimum.

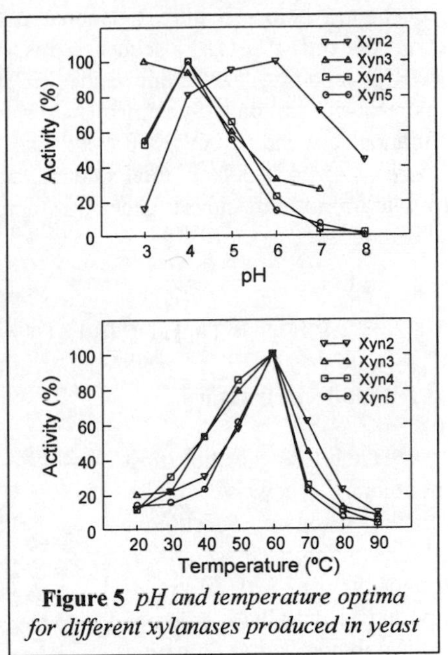

Figure 5 *pH and temperature optima for different xylanases produced in yeast*

We also cloned the β-D-xylosidase gene (*xynB*) of the bacterium *Bacillus pumilus* from a genomic DNA library[6]. The *xynB* gene was fused in-frame to the secretion signal sequence of the yeast mating pheromone α-factor ($MF\alpha 1_S$) and expressed under control of the *ADH2* promoter and terminator sequences (designated *XLO1*). The total β-D-xylosidase activity produced in *S. cerevisiae* remained cell associated with a mere activity of 1.40 nkat/ml obtained when the recombinant *S. cerevisiae* strain was grown in non-selective complex medium. Co-expression of the *XYN2* gene of *T. reesei* and *XLO1* gene of *B. pumilus* surprisingly yielded a total β-D-xylanase activity of 3,500 nkat/ml compared to the 1,200 nkat/ml for *XYN2* expression only. Analysis of enzyme products released from birch wood xylan using recombinant XYN2 and XYN2/XLO1 preparations showed that for both enzyme preparations xylobiose and xylotriose are the major hydrolysis products with no detectable D-xylose released. Subsequent β-D-xylosidase assays confirmed that XLO1 exhibits aryl-activity against *p*-nitrophenyl-β-D-xylopyranoside but very little activity against β-xylobiose. It is therefore tempting to speculate that the enhanced β-D-xylanase activity of the recombinant *XYN2/XLO1* *S. cerevisiae* strain is not due to synergism between the two enzymes, but rather enhanced secretion of the XYN2 β-D-xylanase in the presence of XLO1 β-D-xylosidase that remains cell associated.

The α-L-arabinofuranosidase gene (*abf*B) of *A. niger* MRC11624 was amplified by PCR from mRNA and the cDNA fragment was inserted between the yeast *PGK1* promoter and terminator sequences (designated *ABF2*)[7]. The *ABF2* gene was expressed successfully in *S. cerevisiae* but the α-L-arabinofuranosidase protein (ABF2) remained cell-bound before mature functional α-L-arabinofuranosidase was eventually released from the cell. The *ABF2* sequence was verified to contain a 1497-bp open reading frame (ORF) encoding 499 amino acids. ABF2 was characterised and also shown to have hyperglycosylation. Synergism was shown to occur between the ABF2 α-L-arabino-furanosidase and a XYN3 β-D-xylanase of *A. kawachii*. This interaction led to an increase of up to 32% in total xylanase activity on oat spelt xylan as substrate. A second α-L-ara-binofuranosidase gene was amplified from *A. kawachii*, but has not been characterised.

4 PRODUCTION OF CELLULASES IN YEAST

4.1 Cellulose structure

Cellulose is the most abundant renewable biological resource. Cellulose chains comprised of linear polymers of β-1,4-glucopyranosyl units with a degree of polymerization ranging from 30 to 15,000 units. Cellulose is arranged in insoluble fibres known as microcrystalline cellulose, interrupted with short amorphous regions. Extensive hydrolysis of crystalline cellulose requires synergistic action of three different types of cellulases: endo-β-1,4-D-glucanase, cellobiohydrolase, and cellobiase (β-D-glucosidases) (Figure 6). Cellodextrinases form a fourth class of an exo-acting cellulase which specifically hydrolyses cellotetraose, cellopentaose and cellohexaose and releases predominantly cellobiose. Different cellulolytic bacteria and fungi produce various arrays of these enzymes, many of which have different degrees of specificity.

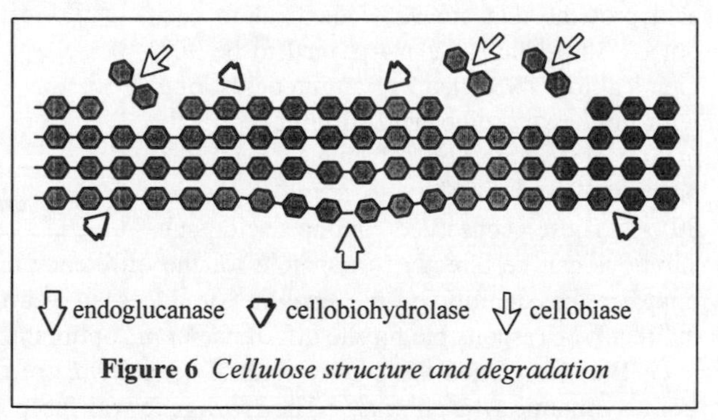

Figure 6 *Cellulose structure and degradation*

4.2 Production of cellulases in yeast

Several cellulases and glucanases from both bacterial and fungal sources were cloned and successfully expressed in *S. cerevisiae*. The endo-β-1,4-D-glucanase gene (*end1*) of *Butyrivibrio fibrisolvens*[8], cellodextrinase gene (*celA*) of *Ruminococcus flavefaciens*[9] and the endo-β-1,3-1,4-D-glucanase gene (*beg1*) of *Bacillus subtilis*[10] were fused to the yeast secretion signal sequence ($MF\alpha 1_S$) and expressed in *S. cerevisiae*. In addition, the cellobiohydrolase gene (*cbh1-4*) of *Phanerochaete chrysosporium*[11] and the cellobiase gene (*BGL1*) of *Endomyces fibuliger*[12] were expressed in *S. cerevisiae,* as well as the exo-β-1,3-D-glucanase gene (*EXG1*) of *S. cerevisiae*[12], which was over-expressed in the same yeast. The *EXG1* and *BGL1* genes were expressed under the control of their native promoters. All the enzymes were successfully secreted from the yeast. Both CBH1 and BGL1 were secreted using their authentic leader peptides. Simultaneous expression of the *EXG1*, *begl* and *endl* in auto-selective yeast strains allowed for the effective degradation of glucan (barley-β-glucan)[10]. When co-expressing the *endl* gene of *B. fibrisolvens* with the *Erwinia chrysanthemi* pectate lyase gene (*pelE*) and *E. carotovora* polygalacturonase gene (*pehl*), a recombinant *S. cerevisiae* strain that could degrade carboxymethylcellulose and polypectate was constructed[8]. The final construction of an engineered yeast expressing four cellulolytic enzymes (*endl*, *cbhl-4*, *begl*, *BGL1*) yielded a strain that could degrade different cellulose / glucan substrates, such as carboxymethylcellulose, hydroxyethylcellulose, laminarin and barley-β-glucan. However, no significant activity could be demonstrated against amorphous cellulose[12] (Table 1). A breakthrough was the construction of a *S. cerevisiae* yeast that now not only degrade cellobiose (end product of cellulose degradation) but can grow on this substrate[12] (Figure 7).

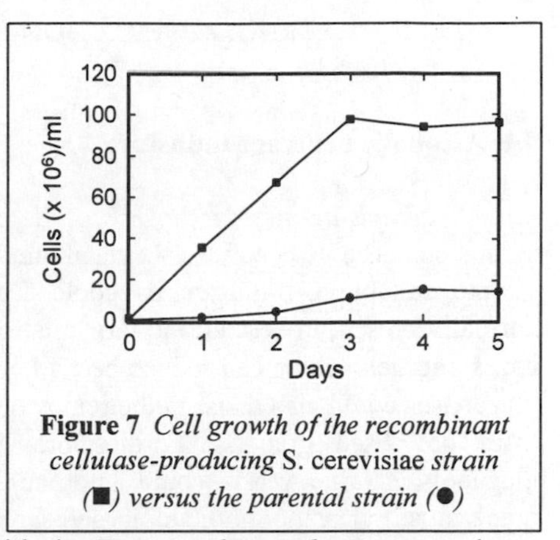

Figure 7 *Cell growth of the recombinant cellulase-producing* S. cerevisiae *strain* (■) *versus the parental strain* (●)

Table 1 *Enzymatic hydrolysis of cellulosic substrate by recombinant S. cerevisiae Y294 strains*

Substrate	*Reducing sugars released**	
	Y294[YEp352]	Y294[pECCB]
Carboxymethylcellulose (CMC)	28·2	300·1
Hydroxyethylcellulose (HEC)	20	55
Laminarin	63	257
Amorphous cellulose	88·3	101·3
Barley-β-glucan	32·6	409
Polygalacturonic acid (polypectate)	3	48·5
Birch wood glucuronoxylan	800	984
Cellobiose	0·5	9 340

* One enzyme unit is defined as 1 nMole reducing sugars . ml^{-1} . 60 min^{-1} . 10^6 $cells^{-1}$

5. INDUSTRIAL APPLICATION OF POLYSACCHARIDE-DEGRADING YEASTS

5.1 Alcoholic beverage industry

S. cerevisiae strains capable of degrading polysaccharides such as glucans and xylans could be of benefit to the alcoholic beverage industry. Brewing yeast strains that secrete suitable β-D-glucanases could facilitate the degradation of barley β-glucans normally present in beer during fermentation and thus prevent the development of glucan hazes and gels which can reduce beer filterability[13]. Wine yeast strains that secrete β-D-glucanases could also assist in the removal of haze-forming glucans found in wines made from botrytised grapes. Furthermore, it has been suggested that pectolytic and glucanolytic wine yeasts could improve liquefaction, clarification and filterability of grape must, releasing more colour and flavour compounds entrapped in suspended solids and grape skins, thereby making a positive contribution to the wine bouquet[14].

5.2 Baking industry

Apart from the basic ingredients (flour from cereals, water, yeast and salt) of bread and other baked foods, it is common practice to add ingredients that have a positive effect on the processing of the dough and/or the quality of the baked products. It has been shown that the addition of β-D-xylanases is an important contributor to improve bread volume, especially when lower quality flours with low gluten content are being used[15].

5.3 Animal feed supplements

Scientists from around the world have reported significant improvements in meat production when using certain strains of *S. cerevisiae* in both ruminant and non-ruminant diets. Some of the effects of yeast culture supplements on ruminants that have been reported are increased feed intake, increased milk production, increased fat and protein content in milk, improved weight gain and improved animal health[16]. A polysaccharolytic *S. cerevisiae* strain could thus be used as a probiotic protein supplement, and to increase the digestibility of animal feed.

6. REFERENCES

1. C. Hadfield, K. K. Raina, K. Shashi-Menon and R. C. Mount, *Mycol. Res.*, 1993, **97**, 897.
2. G. Loison, M. Nguyen-Juilleret, S. Alouani and M. Marquet, *Bio/Technology*, 1986, **4**, 433.
3. D. C. la Grange, I. S. Pretorius and W. H. van Zyl, *Appl. Environ. Microbiol.*, 1996, **62**, 1036.
4. J. M. Crous, I. S. Pretorius and W. H. van Zyl, *Curr. Genet.,* 1995, **28,** 467.
5. M. Luttig, I. S. Pretorius and W. H. van Zyl, *Biotechnol. Lett.* 1997, **19**, 411.

6. D. C. la Grange, I. S. Pretorius and W. H. van Zyl, *Appl. Microbiol. Biotechnol.* 1997, **47**, 262.

7. J. M. Crous, I. S. Pretorius and W. H. van Zyl, *Appl. Microbiol. Biotechnol.*, 1996, **46**, 256.

8. P. van Rensburg, W. H. van Zyl and I. S. Pretorius, *Curr. Genet.* 1994, **27**, 17.

9. P. van Rensburg, W. H. van Zyl and I. S. Pretorius, *Biotechnol. Lett.*, 1995, **17**, 481.

10. P. van Rensburg, W. H. van Zyl and I. S. Pretorius, *J. Biotechnol.*, 1997, **55**, 43.

11. P. van Rensburg, I. S. Pretorius and W. H. van Zyl, *Curr. Genet.*, 1996, **30**, 246.

12. P. van Rensburg, W. H. van Zyl and I. S. Pretorius, *Yeast,* 1998, (In Press).

13. M. Penttilä, M.-L. Suihko, U. Lehtinen, M. Nikkola and J.K.C. Knowles, *Curr. Genet.*, 1987, **12**, 413.

14. I. S. Pretorius and T. J. van der Westhuizen, *SA J. Enol. Vitic.*, 1991, **12**, 3.

15. J. Maat, M. Roza, J. Verbakel, H. Stam, M. J. Santos da Silva, M. Bosse, M. R. Egmond, M. L. D. Hagemans, R. F. M. von Gorcom, J. G. M. Hessing, C. A. M. J. J. von der Hondel and C. von Rotterdam, 'Xylans and Xylanases', Elsevier Science Publishers B.V., Amsterdam, 1992, p. 349.

16. G. Annison, 'Biotechnology in the Feed Industry', Nottingham University Press, Nottingham, 1997, p. 115.

CLONING OF GLUCANASES AND CHITINASES OF *TRICHODERMA HARZIANUM* AND FUNCTIONAL ANALYSES IN HOMOLOGOUS AND HETEROLOGOUS SYSTEMS

B. Cubero[1], I. García[1], J. Delgado-Jarana[2], M.C. Limón[2], T. Benítez[2] and J.A. Pintor-Toro[1]

[1]Instituto de Recursos Naturales y Agrobiología, CSIC, Apdo. 1052, Sevilla 41080, Spain;
[2]Departamento de Genética, Universidad de Sevilla, Apdo. 1095 Sevilla 41080 Spain.

1 SUMMARY

Trichoderma hydrolases have been attributed a significant role in antagonism against phytopathogenic fungi. We have isolated and characterised several hydrolytic enzymes secreted by *Trichoderma harzianum* and have cloned the corresponding genes by peptide microsequencing, PCR and screening of cDNA libraries. The *chit*42 and *chit*33 genes encode two endo-chitinases, *bgn*13.1 encodes an endo-β(1→3) glucanase and *bgn*16.2 encodes an endo-β(1→6) glucanase. The expression of all these genes is strongly repressed by glucose, but whereas glucose derepressing conditions are enough to achieve maximal expression of *chit*33 and *bgn*16.2, the gene *chit*42 requires additional specific induction by chitin or fungal cell walls and the expression of *bgn*13.1 is strictly dependent on substrate induction.

One of the main objectives of this work was to obtain new *T. harzianum* strains with higher antagonistic abilities. We have constructed vectors carrying the ORFs of *chit*33 and *bgn*16.2 under the control of the constitutive promoter *pki* and the cellobiohydrolase II terminator of *Trichoderma reesei*, and have obtained stable multicopy transformants *chit*33 and *bgn*16.2. Transformants *bgn*16.2 show a positive correlation between the number of copies integrated in the genome and mRNA and protein accumulation, as well as extracellular specific activity, in glucose-derepressed, chitin-induced conditions. Other conditions tested (glucose repression with or without induction) failed to produce detectable extracellular specific activity, due probably to proteolysis and/or defective protein secretion. Experiments of dual cultures show that transformants *bgn*16.2 kill and/or inhibit fungal growth faster than the wild type strain in all conditions tested. Transformants *chit*33 are able to process correctly the Chit33 protein, although it seems that a gene dosage effect in the expression of this gene does not exist. Transformants *chit*33 show ten-fold higher extracellular chitinase activities than the wild type in glucose-repressed conditions and, as for transformants *bgn*16.2, experiments of dual cultures show that some of the *chit*33 transformants are significantly better control agents than the *Trichoderma* wild type strain in all conditions tested.

The effect of chitinase overexpression in plants has also been tested. We have transformed *Nicotiana tabacum* with the *chit*42 gene and with the chimaeric *pschit*42 gene, which carries an apoplastic plant protein signal peptide instead of the corresponding fungal signal peptide. In both types of transgenic plants the presence of active Chit42 protein is associated with a significant increase of resistance against phytopathogenic fungi.

2 INTRODUCTION

In recent years, biological control has appeared as an alternative to the use of pesticides in the prevention of plant diseases. Soilborne pathogens are the main cause of crop losses, and amongst them fungi are by far the most aggressive.[1,2] Several species of *Trichoderma* have been described as antagonists of phytopathogenic fungi,[3] and much effort is currently directed towards the obtaining improved strains by genetic engineering.

As antagonists, the properties of *Trichoderma* strains are the result of a combination of different mechanisms which may act coordinately or sequentially and include the competition for limiting nutrients in the soil, antibiosis and mycoparasitism. Some species of *Trichoderma* produce antibiotics such as gliotoxin, gliovirin, trichodermin, alameticin and suzukamicin[3,4] and some of these have been associated to processes of biological control in soil.[4,5] Mycoparasite strains of *Trichoderma* are necrotrophic, and obtain nutrients from their hosts in a process which involves several steps: recognition of the host, coiling, developing of specialised penetration structures, such as appresoria, and digestion of the host's cell walls and cytoplasm.[3,6,7]

The role of lytic enzymes in the antagonic process, either as a part of the mycoparasitic process, or as a means to enhance *Trichoderma* saprophytic abilities is well established.[8-10] Among the hydrolases that have shown antifungal properties are chitinases,[11-14] glucanases,[14-17] proteases,[18,19] mannanases and other hydrolases.[20] Transformants of *Trichoderma, Rhizobium* and *Pseudomonas* expressing a bacterial chitinase show an increased efficiency as biological control agents.[12,21] The overexpression of a protease from *Trichoderma harzianum* also improves the biocontrol ability of *T. harzianum* transformants.[22] Regulation analyses of some of the *Trichoderma* chitinases and glucanases have shown that in most cases the control of enzyme biosynthesis occurs at the transcriptional level, and that, in general, maximal expression is achieved when fungal cell walls are used as the sole carbon source.[23]

Plant chitinases and glucanases are associated with the pathogenic response[27,28] and their defensive role has been demonstrated by their antifungal activity *in vitro*, alone[27-30] or in combination with other pathogenesis related proteins.[31] As it is the case for fungal hydrolases, synergistic effects on the antifungal activity of plant chitinases when combined with glucanases have also been observed.[27,32] Some transgenic plants overexpressing these enzymes show enhanced resistance against fungal pathogens,[33,34] and constitutive co-expression of chitinases and glucanases has provided even higher levels of resistance to certain plant diseases.[35,36] The expression of a *Serratia* chitinase in bean and tobacco is also associated to a higher resistance to different pathogens.[37] Even if data on the effect of hydrolases overexpression in plants are still scarce, it would seem that the use of heterologous genes in transformation experiments gives positive results more often than those observed in the case of homologous transformation. A possible explanation would be the synergistic effect of a foreign hydrolase with the endogenous ones. Also, the introduction of heterologous genes prevents the occurrence of the phenomenon of cosupression, not rare in plants, which takes place when several copies of homologous genes coexist in the same genome.[38]

The interest of *Trichoderma* extracellular hydrolases in relation to their use as a biocontrol agent is then based on the possibility of obtaining new, more aggressive genetic variants, with higher levels of enzyme production and on the improvement of the mechanisms of plant defence against phytopathogenic fungi, by expressing these hydrolases in plants.

We have obtained constitutive overexpression of a chitinase, Chit33, and a glucanase, Bgn16.2, in *Trichoderma harzianum* by transformation, and have determined the effects of this overexpression in the antagonistic properties of different multicopy transformants. Another *T. harzianum* chitinase, Chit42, has been expressed in transgenic tobacco plants and the response to phytopathogens has been evaluated. All these results will be presented and discussed.

3 RESULTS AND DISCUSSION

3.1 Overexpression of the endochitinase Chit33 in *Trichoderma harzianum*

Chitinases as purified enzymes, or as a product expressed and secreted by engineered microorganisms, have been shown to be promising as a means to control fungal pathogens. *Trichoderma harzianum* produces at least three extracellular chitinases with an endolytic mode of action[11] which are glucose-repressed and induced by chitin, fungal cell walls and starvation conditions. The chitinase-encoding gene *chit*33 was placed under the control of the constitutive promoter *pki*, and stable *chit*33 transformants of *T. harzianum* CECT 2413 were obtained.[39] Northern blots showed constitutive expression of *chit*33 in the transformants, but no correlation between the levels of mRNA accumulation and the number of *chit*33 integrated copies for each of the transformants was detected. Western blot assays showed no significant increases on the level of secreted protein in the transformants in non-repressing conditions. However, in repressing conditions, where there is no wild type expression of the chitinase, the transformants exhibit constitutive expression of a correctly processed protein.

Chitinase activity was also determined in the wild type strain and in the *chit*33 transformants in repressing (glucose) and non-repressing (chitin) conditions (Table 1). Whereas the transformants grown in repressing conditions displayed higher levels of chitinase activity as compared with the wild type, no significant difference with the wild type activity was observed in chitin-containing medium. The unexpectedly low levels of extracellular chitinase activity of the transformants in these conditions could be due to limitations in the secretory pathway, or could be attributed to the comparatively small contribution of Chit33 to the total extracellular chitinolytic activity of *Trichoderma*.

Table 1 *Chitinase activity of the wild type (wt) and different* chit*33 transformants*

		Chitinase activity ($mU.mg^{-1}.10^{-3}$)	
Strain	Number of copies	Glucose	Chitin
wt	1	6.24	571
T1	1	4.08	516
T2	5-6	10.75	546
T6	8-12	41.56	538
T30	8-12	51.56	554
T51	3-4	19.48	506
T52	3-4	13.09	587
T65	5-6	21.19	592

3.1.1 Effects of Chit33 overexpression on antagonism. Transformants of *chit*33 were cocultivated with the phytopathogenic fungus *Rhizoctonia solani* in liquid medium, in repressing (glucose) and non-repressing (chitin) conditions, and their antagonistic effect was measured by placing samples of the cultures in *Trichoderma*-suppressive medium. In both conditions the transformants behaved similarly, producing a substantially faster decrease in the relative growth of *Rhizoctonia* than did the wild type (Figure 1). Replacement experiments, in which *Rhizoctonia solani* mycelium was placed in solid medium where *chit*33 transformants had previously been grown, demonstrated that, at least in the conditions tested, the antagonistic effect of *Trichoderma* did not require physical contact with its host, and that its enhancement in the transformants was due to the constitutive synthesis of Chit33 (Figure 2).

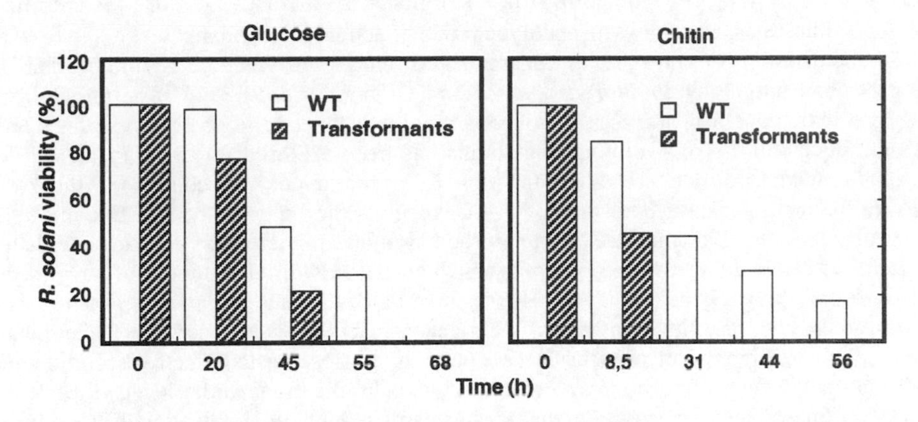

Figure 1 *Antagonistic effect of wild type (wt) and* chit*33* Trichoderma harzianum *transformants on* Rhizoctonia solani. *Glucose, repressing conditions; chitin, non-repressing conditions*

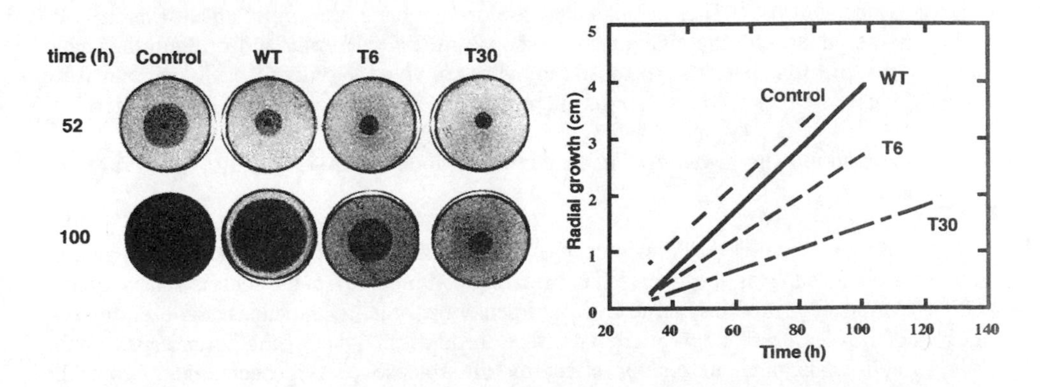

Figure 2 *Replacement tests.* Rhizoctonia solani *(5-mm diameter) inocula were placed in glucose containing medium on which* Trichoderma *wild type (wt) or* chit*33 transformants (T6 and T30) conidia had previously been placed onto cellophane disks and removed after 4 days of growth. The diameter of the* Rhizoctonia *colony was measured periodically after inoculation*

All these results confirm the important role of the chitinolytic system of *Trichoderma* strains during antagonism.[40] The increased chitinase levels of the transformants in repressing conditions, and the accelerated chitinase production both in glucose and in chitin-containing media, result in improved *in vitro* antagonistic abilities. These results are very encouraging with regards to the use of these strains as biocontrol agents in the field.

3.2 Expression of the β(1→6) glucanase Bgn16.2 in *Trichoderma harzianum*

Trichoderma harzianum produces at least two extracellular β(1→6) glucanases, of 51 and 43 kDa., which are induced by starvation conditions and -to a lesser extent- by chitin, and repressed by glucose. The purified β(1→6) glucanase II (that of 43 kDa.) is specific for β(1→6) linkages, and shows an endolytic mode of action on pustulan.

When combined with other glucanases and/or chitinases, Bgn16.2 is able to inhibit the growth of fungi such as *Botrytis cinerea* and *Gibberella fujikuroi*.[16] The possibility that, as it is the case for chitinases, the overexpression of this enzyme be of importance on the antifungal activity of overproducing strains has been explored. Transformants of *T. harzianum* with multiple copies of the *bgn*16.2 gene under the control of the *pki* constitutive promoter have been obtained. As expected, they showed constitutive mRNA accumulation, as well as increased protein production and higher enzymatic activity than the wild type in non-repressing conditions, which correlated with the number of integrated copies of the *bgn*16.2 gene (Figure 3). However, in glucose-containing medium no protein was detected, and very little increase of the glucanase activity was observed. This effect could be due to a rapid proteolytic action mediated by glucose, or to defective secretion of the constitutively expressed protein. Preliminary data on the intracellular levels of protein accumulation and specific activity in the transformants point to a combination of these two causes to explain this apparent lack of concordance between the constitutive expression of the *bgn*16.2 gene and the repressed production of the secreted enzyme.

In confrontation experiments with the phytopathogenic fungus *Rhizoctonia solani*, *bgn*16.2 transformants showed higher antagonistic capacity, both in repressing and non-repressing conditions.[41] These results confirm, on one hand, the importance of hydrolases other than chitinases in the antagonistic response of *Trichoderma*, and on the other hand suggest that the mycoparasitic response can bypass *in vivo* the glucose-mediated repressing processes that are observed *in vitro* for the production and correct processing of Bgn16.2.

3.3 Expression of the *Trichoderma harzianum* endochitinase-encoding gene *chit*42 in plants

Plants have developed a wide range of mechanisms to protect themselves against fungal attacks, such as the expression of a large number of pathogenesis-related genes (PR). Among PR proteins, some hydrolytic enzymes, mainly glucanases and chitinases, have a dual role: a direct, antimicrobial effect, through the lysis of the hyphal apex of the fungal cell walls, and an elicitor effect by the release of oligosaccharides from the pathogen cell wall that could induce the general mechanisms of defence response.

The antifungal activity of many plant chitinases have been demonstrated *in vitro*.[28] Furthermore, the overexpression of chitinases in plants often results in a best resistance to the fungal attack.[33,34,36] In other cases, genetic silencing of endogenous chitinase by the homologous expression of enzymes has lead to resistance levels similar to untransformed control plants.[42]

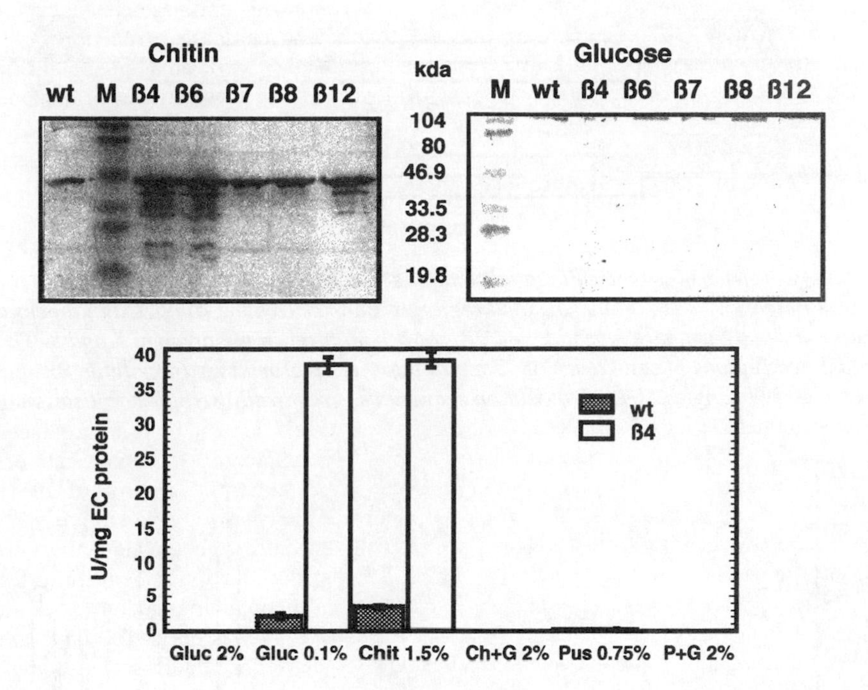

Figure 3 *Upper panels: Western blot analyses of extracellular proteins of* Trichoderma *wild type (wt) and different* bgn*16.2 transformants in repressing (glucose) and non-repressing (chitin) conditions. M, molecular weight marker. 15 μg of extracellular protein were separated by SDS-PAGE and Bgn16.2 was detected by polyclonal rabbit anti-Bgn16.2 antibodies. Lower panel: Extracellular β(1→6) glucanase activity of* Trichoderma *wild type and transformant β4 in different media*

In order to improve their antifungal capacity, tobacco plants have been transformed with a *T. harzianum* chitinase cDNA, CHIT42.[24] The heterologous expression of a *chit*42 avoids genetic silencing because it has no homologous DNA sequences in tobacco genome, and the biochemical characteristics of this enzyme are different to those of vegetal chitinases isolated until now.[28,43] We have constructed two kinds of tobacco transgenic plants expressing an active Chit42 protein (Figure 4). The fungal chitinase is secreted in both kinds of transgenic plants to the apoplast, showing the ability of this fungal signal peptide to promote the secretion of proteins in tobacco. Nevertheless, levels of Chit42 expression in *chit*42 plants are 10-fold lower than in *pschit*42 plants (Figure 5). Our results suggest that this difference is due to a differential mRNA *chit*42 accumulation in both types of transgenic plants. Some stabilisation and destabilisation mRNA sequences, whose presence in *chit*42 and *pschit*42 constructs has been observed (Figure 4), have been described in plants.[44,45]

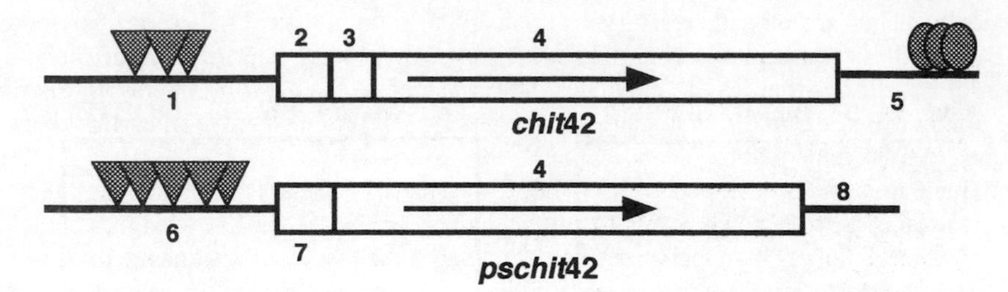

Figure 4 *Constructs introduced in tobacco plants* via Agrobacterium tumefaciens *infection of leaf disks. 1, 5' untranslated region of* chit42 *cDNA; 2, Chit42 signal peptide; 3, putative Chit42 activation peptide; 4, ORF of the mature Chit42 protein; 5 and 8, 3' untranslated regions of* chit42 *cDNA; 7, signal peptide of a tomato extracellular PR protein. Grey triangles, RNA stabilisation sequences; grey ovals, RNA destabilisation sequences*

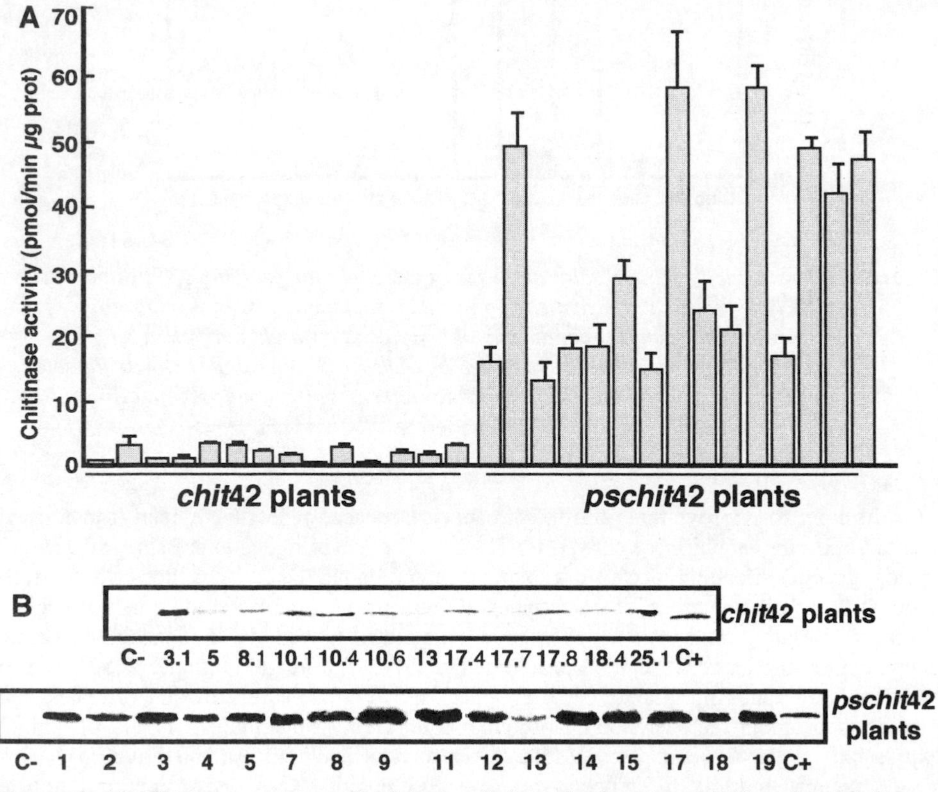

Figure 5 *(A) Chitinase activity of leaf tissue of plants harbouring* chit42 *and* pschit42 *constructs. Each bar corresponds to the mean of three independent assays. Untransformed plants (C-) did not show any activity on 4-methylumbelliferyl-ß-D-N,N',N''-triacetyl-chitotriose. (B) Detection of Chit42 in protein leaf extracts of plants. For each plant there is a positive correlation between chitinase activity and estimated amounts of Chit42. C+ corresponds to 20 ng of Chit42 isolated from* T. harzianum

Plate and soil resistance tests to *R. solani* have shown that both *chit*42 and *pschit*42 transgenic plants are able to resist the attack by this fungus better than untransformed plants, regardless of the levels of chitinase activity they show.[46] Our interpretation of this lack of correlation is that Chit42 enhances the resistance to *R. solani* by liberating elicitors from fungal cell walls in the first stages of infection, therefore triggering an earlier defence response in transgenic plants.

There are several examples of how the expression of chitinases in plants has lead to a better resistance of transgenic plants to fungi. However, and to our knowledge, this is the first time that a fungal mycoparasite has been used as a source of chitinases to improve plant resistance. The obtained results are very promising with regards to the biotechnological application of biocontrol organisms. Our effort is currently directed towards the study of the induction of defence responses in non-transformed, *chit*42 and *pschit*42 plants. This may help to understand the mechanisms through which Chit42 expression confers resistance to transgenic plants.

References

1. G.N. Agrios, 'Plant Pathology', Academic Press, London, 1988.
2. D. Hornby, 'Biological Control of Soil-Borne Plant Pathogens', CAB International, Oxford, 1990.
3. G.C. Papavizas, *Ann. Rev. Phytopathol.*, 1985, **23**, 23.
4. R.D. Lumsden, S.T. Locke, J.F. Adkins and C.J. Ridout, *Phytopathology*, 1992, **82**, 230.
5. S.E. Wilhite, R.D. Lumsden and D.C. Straney, *Phytopathology*, 1994, **84**, 816.
6. U. Vancura and F. Kunc, 'Interrelationships between Microorganisms and Plants in Soil', Elsevier, Amsterdam, 1989.
7. J.P. Nakas and C. Hagedorn, 'Biotechnology of Plant-Microbe Interactions', McGraw-Hill Publishing Co., London, 1990.
8. M. Cherif and N. Benhamou, *Phytopathology*, 1990, **80**, 1406.
9. P. Jeffries and T.W.K. Young, 'Interfungal Parasitic Relationships', CAB International, Wallingford, U.K, 1994.
10. N. Benhamou and Y. Chet, *Phytopathology*, 1996, **86**, 405.
11. J. De la Cruz, A. Hidalgo-Gallego, J.M. Lora, T. Benítez, J.A. Pintor-Toro and A. Llobell, *Eur. J. Biochem.*, 1992, **206**, 859.
12. J.R. Kinghorn and G. Turner, 'Applied Molecular Genetics of Filamentous Fungi', Blackie Academic and Professional, London, 1992.
13. Y. Sitrit, Z. Barak, Y. Kapulnik, A.B. Oppenheim and Y. Chet, *Mol. Plant-Microbe Interact.*, 1993, **6**, 293.
14. M. Lorito, G. Harman, C.K. Hayes, R.M. Broadway, A. Tronsmo, S.L. Woo and A. Di Pietro, *Phytopathology*, 1993, **83**, 302.
15. C.J. Ridout, J.R. Coley-Smith and J. M.Lynch, *Enzyme Microb. Technol.*, 1988, **10**, 180.
16. J. De la Cruz, J.A. Pintor-Toro, T. Benítez and A. Llobell, *J. Bacteriol.*, 1995, **177**, 1864.
17. J. De la Cruz, J.A. Pintor-Toro, T. Benítez, A. Llobell and L.C. Romero, *J. Bacteriol.*, 1995, **177**, 6937.
18. M. Schirmbock, M. Lorito, Y. Wang, C. K. Hayes, I. Arisan-atac, F. Scala, G. Harman and C. Kubicek, *Appl. Environ. Microbiol*, 1994, **60**, 4364.

19. R.A. Geremia, G.H. Goldman, D. Jacobs, W. Ardiles, S.B. Vila, M. van Montagu and A. Herrera-Estrella, *Mol. Microbiol.*, 1993, **8**, 603.
20. I. Labudova and L. Gogorova, *FEMS Microbiol. Lett.*, 1988, **52**, 193.
21. E.C. Tjamos, G.C. Papavizas and R.J. Cook, 'Biological Control of Plant Diseases. Progress and Challenges for the Future', Plenum Press, New York, 1992.
22. A. Flores, I. Chet and A. Herrera-Estrella, *Curr. Genet.*, 1997, **31**, 30.
23. J. De la Cruz, M. Rey, J.M. Lora, A. Hidalgo-Gallego, J.A. Pintor-Toro, A. Llobell and T. Benítez, *Arch. Microbiol.*, 1993, **159**, 316.
24. I. García, J.M. Lora, J. de la Cruz, T. Benítez, A. Llobell and J.A. Pintor-Toro, *Curr. Genet.*, 1994, **27**, 83.
25. M.C. Limón, J.M. Lora, I. García, J. de la Cruz, A. Llobell, T. Benítez and J.A. Pintor-Toro, *Curr. Genet.*, 1995, **28**, 478.
26. J.M. Lora, J. de la Cruz, T. Benítez, A. Llobell and J.A. Pintor-Toro, *Mol. Gen. Genet.*, 1994, **242**, 461.
27. M.B. Sela-Buurlage, A.S. Ponstein, S.A. Bres-Vloeman, L.S. Melchers, P.J.M. Van den Elzen and B.J.C. Cornelissen, *Plant Physiol.*, 1993, **101**, 857.
28. L. Graham and M. Sticklen, *Can. J. Bot.*, 1994, **72**, 1057.
29. E. Kombrink, M. Schröder and K. Hahlbrock, *Proc. Natl. Acad. Sci. USA*, 1988, **85**, 782.
30. F. Mauch, B. Mauch-Mani and T. Boller, *Plant Physiol.*, 1988, **88**, 936.
31. R. Leah, H. Tommerup, Y. Svendsen and J. Mundy, *J. Biol. Chem.*, 1991, **266**, 1564.
32. L.S. Melchers, M. Apotheker-de-Groot, J.A. van der Knaap, A.N. Ponstein, M.B. Sela-Buurlage, J.F. Bol, B.J.C. Cornelissen, P.J.M. van den Elzen and H.J.M. Linthost, *Plant J.*, 1994, **5**, 469.
33. K.E. Broglie, I. Chet, M. Holliday, R. Cressman, P. Biddle, C. Knowlton, C.J. Mauvais and R. Broglie, *Science*, 1991, **254**, 1194.
34. W. Lin, C.S. Anuratha, K. Datta, I. Potrykus, S. Muthukrishnan and K. Datta, *Bio/Technology*, 1995, **13**, 686.
35. Q. Zhu, E.A. Maher, S. Masoud, R.A. Dixon and C.J. Lamb, *Bio/Technology*, 1994, **12**, 807.
36. G. Jach, B. Görnhardt, J. Mundy, J. Logemann, E. Pindsdorf, R. Leah, J. Schelland and C. Maas, *Plant J.*, 1995, **8**, 91.
37. A.B. Oppenheim and I. Chet, *Tibtech*, 1992, **10**, 392.
38. C. Kunz, H. Schöb, M. Stam, J.M. Kooter and F. Jr. Meins, *Plant J.*, 1996, **10**, 437.
39. M.C. Limón, J.A. Pintor-Toro and T. Benítez, *Phytopathology*, 1997, submitted.
40. S. Haran, H. Schickler, A. Oppenheim and I. Chet, *Phytopathology*, 1996, **86**, 980.
41. J. Delgado-Jarana, unpublished data.
42. J.M. Neuhaus, P. Ahl-Goy, U. Hinz, S. Flores and F. Meins, *Plant Mol. Biol.*, 1991, **16**, 141.
43. D.B. Collinge, K.M. Kragh, J.D. Mikkelsen, K.K. Nielsen, U. Rasmussen and K. Vad, *Plant J.*, 1993, **3**, 31.
44. D.R. Gallie, *Ann. Rev. Plant Phys and Plant Mol. Biol.*, 1993, **44**, 77.
45. M.L. Abler and P.J. Green., *Plant Mol. Biol.*, 1996, **32**, 63.
46. I. García, unpublished data.

RHAMNOGALACTURONASES AND POLYGALACTURONASES OF *ASPERGILLUS NIGER*

Jacques A.E. Benen, Harry C.M. Kester, Lucie Parenicova and Jaap Visser

Section Molecular Genetics of Industrial Microorganisms
Wageningen Agricultural University
Dreijenlaan 2, 6703 HA Wageningen, The Netherlands.

1 INTRODUCTION

Pectin as one of the major cell wall polysaccharides has adverse effects on industrial processes like maceration, liquefaction and clarification. To overcome these problems various ill-defined cocktails of pectinases, mostly originating from *Aspergilli*, have found widespread use in these processes. One of our *Aspergillus* research programs is directed at the elucidation of the full pectinase spectrum of this industrially important fungus. Our aim is not only to clone all genes encoding pectinolytic enzymes but also to investigate the biochemical properties of the respective enzymes. This will lead to a better understanding of the individual roles played by these enzymes in pectinolysis and will allow the design of defined cocktails to be used in the industrial processes mentioned above. Furthermore, the individual enzymes or combinations are assessed for their applicability in pectin manufacturing. An overview of the research carried out for the *A. niger* rhamnogalacturonases (RHGs) and polygalacturonases (PGs) is presented here.

1.1 Pectin Structure

Among plant cell wall polysaccharides such as cellulose, hemicelluloses and pectin, the latter is the most complex carbohydrate. While this complexity was recognised already quite early, only in the last decade significant progress has been made in elucidating its fine structure. This is primarily due to the availability of individually purified and characterised pectinases allowing dissection of the pectin molecule and to the further development of strong analytical techniques like MALDI-TOF spectrometry, NMR and HPLC.

The structure of the pectin molecule is nowadays fairly well understood and is schematically presented in Figure 1. Panel A shows the homogalacturonan part which consists of α1-4 linked D-galacturonate residues which are usually methylesterified. The degree of methylation is generally up to 70 %. Acetylation is also observed at C2 and/or C3 whereas also to some degree at these positions xylose residues can be attached. Panel B shows the more complex rhamnogalacturonan I[1] (RGI), also called the 'hairy region'[2] part of the pectin molecule. This part of the molecule is also referred to as the ramified part of the pectin molecule.

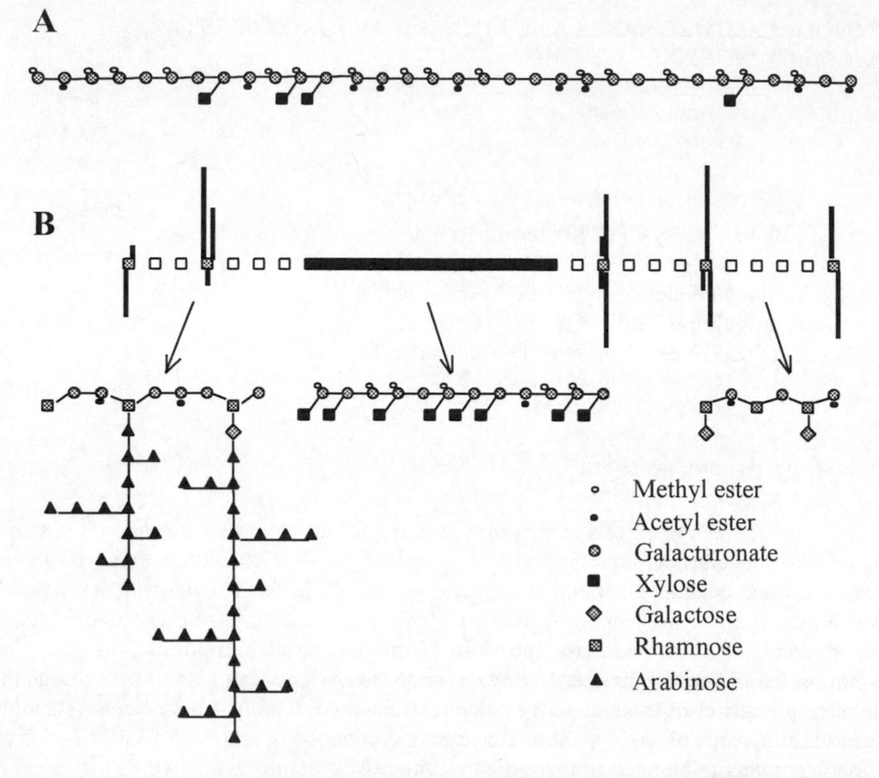

Figure 1 *Schematic representation of the pectin structure. Panel A, Homogalacturonan; Panel B, Rhamnogalacturonan I / hairy regions (After: H. Schols, 1995)*[3].

Alternating D-galacturonate residues α1-2 linked to rhamnose residues are characteristic for RGI. The stretches in which α-D-galacturonate and α-L-rhamnose are alternating may in some tissues be interrupted by stretches of galacturonate residues with a high degree of xylose attached at C2 and/or C3. The rhamnose residues often carry side-chains of galactans and arabinans. An even more complex pectin structure is found in the RGII (not shown). This fairly small molecule has a rather defined structure and consists of a galacturonan backbone with linked to it a multitude of different carbohydrates like glucose, glucuronate, fucose, 2-methyl-fucose, apiose, mannose, aceric acid, 2-keto-3-deoxy-D-manno-octulosonate and 3-deoxy-D-lyxo-2-heptulosanate[4].

1.2 Enzymatic degradation of pectin

The complexity of the pectin molecule dictates that for its effective breakdown by microorganisms such as *A. niger*, a vast array of enzymes acting consecutively and/or enzymes acting in concert is necessary. Table I lists pectinolytic enzymes which have been identified in several *Aspergilli*. From *A. niger* we have cloned a family of six pectine lyase (*pel*) genes[5], a family of seven endoPG (*pga*) genes[6], two RHG (*rhg*) genes[7], a pectin methylesterase (*pme*)[8], an exoPG (*pga*X)[9], a rhamnogalacturonan lyase (*rgl*A)[9], a rhamno galacturonan acetyl esterase (*rgae*A)[9] and a pectate lyase (*ply*A)[9].

Table I

Aspergillus *enzymes involved in pectin degradation*
endo-polygalacturonase
exo-polygalacturonase
pectin lyase
endo-pectate lyase
xylogalacturonanhydrolase
rhamnogalacturonan rhamnohydrolase
rhamnogalacturonan galacturonohydrolase
rhamnogalacturonan hydrolase
rhamnogalacturonan lyase
rhamnogalacturonan acetylesterase
pectin methyl esterase
pectin acetyl esterase
feruloyl esterase
coumaryl esterase
endo-arabinase
α-L-arabinofuranosidase
β-galactosidase

Almost all of these genes have been fused to a strong, constitutive promoter which allows overexpression of the enzymes. The individual pectinases have been characterised biochemically using first the simplest substrates possible. For the study of PGs we used polygalacturonate and (reduced) oligogalacturonates of defined chain length (DP 2 to 8) and saponified hairy regions to investigate RHGA and B.

2 HYDROLASES BELONGING TO FAMILY 28 ARE INVERTING ENZYMES

Based on their sequence identity endo- and exoPGs as well as RHGs group together in family 28 of hydrolases[10]. Within each family the mechanism of hydrolysis, either via a single or a double displacement mechanism, is conserved[11]. In a single displacement mechanism the anomeric configuration of the newly formed reducing end will be inverted while a double displacement leads to a retained configuration. Knowledge of the mechanism is important for correct interpretation of kinetic data. Retaining, in contrast with inverting enzymes, are likely to catalyse condensation and transglycosylation reactions.

In collaboration with Dr. P. Biely (Bratislava University, Slovakia) we addressed the study of the mechanisms of endoPGI and II and of exoPG by NMR following the hydrolysis of reduced pentagalacturonate (rG5) for endoPGI and II and rG3 for exoPG[12]. For all three enzymes it was clear that hydrolysis of the α-glycosidic bond resulted in the immediate formation of the β-anomer at the newly formed reducing end which gradually mutarotated into the α and β mixture (not shown). This clearly established that family 28 hydrolases are inverting enzymes and that kinetic analyses are not complicated by side-reactions.

3 DIFFERENCES BETWEEN RHGA AND RHGB

Comparison of the amino acid sequences of both RHGA and RHGB reveals that apart from the fact that both enzymes have a high degree of sequence identity, RHGB also has a C-terminal extension of 112 amino acids[7]. This extension is particularly rich in serines, prolines and alanines. Furthermore, a fourfold repeated sequence gly-glu-gln is present at the extreme C-terminus. In *A. aculeatus* only the RHGA counterpart has been cloned and sequenced[13]. While the function of the tail region of RHGB is as yet unknown, it is tempting to speculate that it interacts with the side chains of the substrate.

Analysis of the reaction products after hydrolysis of saponified hairy regions revealed that both *A. niger* and *A. aculaetus* RHGA yield similar profiles[7] (see Figure 2). However, RHGB clearly results in formation of a putative decamer, a product not observed with RHGA and a larger proportion of the octamer as well as smaller amounts of the lower compounds. This suggests that RHGB is responsible for formation of the higher oligomers.

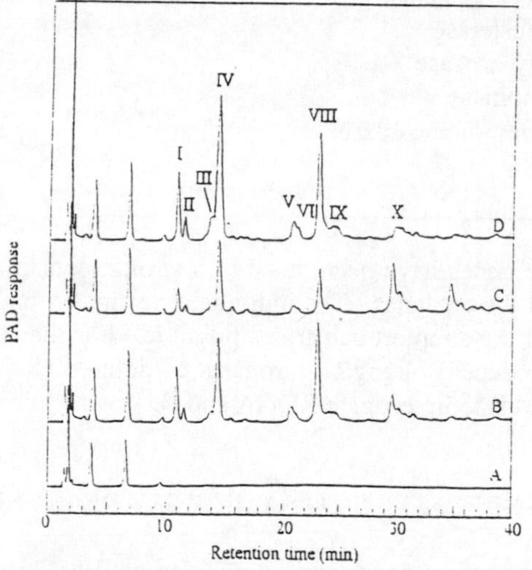

Figure 2 *Mode of action of rhamnogalacturonases as published in ref. 7. HPAEC-PAD analysis of RHG hydrolysis products of saponified hairy regions. 2 % saponified hairy regions in McIlvaine buffer pH 3.5, 4.1 and 4.5 for* A. niger *RHGA, RHGB and* A. aculeatus *RHGA, respectively, were incubated for 1 h at 40 °C with 1 μg RHG.* ***A****, blank;* ***B****,* A. niger *RHGA;* ***C****,* A. niger *RHGB;* ***D****,* A. aculeatus *RHGA. Peaks were identified as defined by Schols[3]. For compound X and the putative decamer the position of galactose residues is not known.*

I
II
III
IV
V
VI
VIIIA
VIIIB
IX
X

4 EXOPOLYGALACTURONASE

Although exoPG is formally classified to belong to hydrolase family 28, which also comprises the endo acting RHGs and PGs, the phylogenetic tree analysis clearly shows that exoPG is only distantly related to fungal RHGs and endoPGs[14]. In fact, the fungal exoPG is closely related to plant enzymes classified as polygalacturonases. These latter enzymes play an important role during plant development.

ExoPG of *A. tubingensis* is a true exo-acting enzyme as was shown by both TLC and HPAEC-PAD analysis: only monomers were detected during early and later stages of hydrolysis[14]. Reduced oligogalacturonates of defined chain length were used to establish which end of the substrate is attacked by exoPG. While exoPG is able to hydrolyse digalacturonate (G2), reduced G2 (rG2) appeared not to be a substrate but rather an inhibitor (Ki = 0.4 mM). Hydrolysis of rG3 resulted in the formation of G1 and rG2 which demonstrates that exoPG attacks the substrate from the non-reducing end. This finding was confirmed using rG4 up to rG6.

According to the methodology developed by Hiromi[15], the kinetic data (hydrolysis rates) obtained for cleavage of G2 up to G6 and rG3 up to rG6 were used to calculate subsite energies as shown in Table II. The energy distribution is similar to that found in other exo-glycanases: a low or negative affinity at subsite –1 and a very high affinity at subsite +1. Subsites +2 and +3 seem to contribute to the binding as well. The fact that at subsite +1 a high affinity was calculated may account for the observation that as in the case of rG2, G1 is an effective inhibitor (Ki = 0.3 mM). Furthermore, whereas k_{int} was calculated to be 716 s^{-1}, the maximal turnover number at the plateau (DP = 4), was only 204 s^{-1}, strongly suggesting that the product stays bound to the enzyme in a non-productive mode due to the high affinity at subsite +1 [14].

Table II *Subsite affinities of exopolygalacturonase from* A. tubingensis. *Data are taken from Kester* et al. *(1996)*[14]. *The cleavage takes place between subsites –1 and +1.*

Subsite	-1	+1	+2	+3	+4	+5
Affinity (kJ/mole)	-1.6	24.5	1.5	1.1	0.6	0.4

5 ENDOPOLYGALACTURONASES

5.1 The endopolygalacturonase gene family

As already mentioned before, a family of endoPG genes is present in *A. niger* comprising seven members, *pga*I, *pga*II, *pga*A, *pga*B, *pga*C, *pga*D and *pga*E[6]. The genes *pga*I and *pga*II, encoding PGI and PGII, were cloned via reverse genetics[16,17]. PGI and PGII appeared to be the most abundant endoPG activities present in a commercial pectinase preparation (Rapidase)[18]. The other genes were obtained via screening of an *A. niger* chromosomal library under non-stringent conditions. All genes have been sequenced.

Thus far four enzymes were characterised biochemically: PGI, PGII, PGC and PGE[19]. All four genes have a strong structural resemblance. The number of introns varies from 1 (*pga*II) to 3 (*pga*C and *pga*E) while *pga*I has two introns. The introns appear to be conserved with respect to their location in the coding sequence. All four enzymes are synthesised as pre-pro proteins. The maturation takes place in two steps: after cleavage by the signal peptidase following von Heynes[20] rules, processing further takes place via a monobasic (in case of PGII) or dibasic peptidase, the latter in analogy to the kex2 peptidase[21]. N-terminal sequencing demonstrated correct processing for PGI, PGII and PGE. The primary structure of mature PGC was deduced from sequence similarity. The mature enzymes are almost equal in length (335 to 343 amino acids) and have all, except for PGI, one conserved potential glycosylation site. In PGI an additional potential glycosylation site is present, located towards the C-terminus. Amino acid identity ranges from as low as 54.3 % for PGII and PGE to as high as 77.6 % for PGC and PGE.

5.2 Kinetic properties of individual endopolygalacturonases on polymeric substrate

In order to establish whether the endoPGs under investigation are true endo-acting enzymes we studied the initial product distribution after limited polygalacturonate hydrolysis. In Figure 3 the product distribution is presented by the rate of formation of a particular oligomer during initial stages of hydrolysis of polygalacturonate. In contrast to exoPG, which only showed the formation of G1, the action of all four endoPGs led to the formation of oligomers up to G12-15. Only G1 to G8 are presented here. These data demonstrate that all four endoPGs studied are indeed endo-acting enzymes. Upon prolonged incubation the rates of formation of individual oligomers changes depending on the enzyme used. Eventually, all enzymes yield G1 to G3 in different ratio's (not shown). Comparison of the initial rates for the individual enzymes shows that especially PGI, and to a lesser extent PGC, have a strong preference to form G1 and G2 during the early stages of hydrolysis. For PGII and PGE a much more even distribution of different oligomers is observed. This will be discussed in more detail in section 3.3.

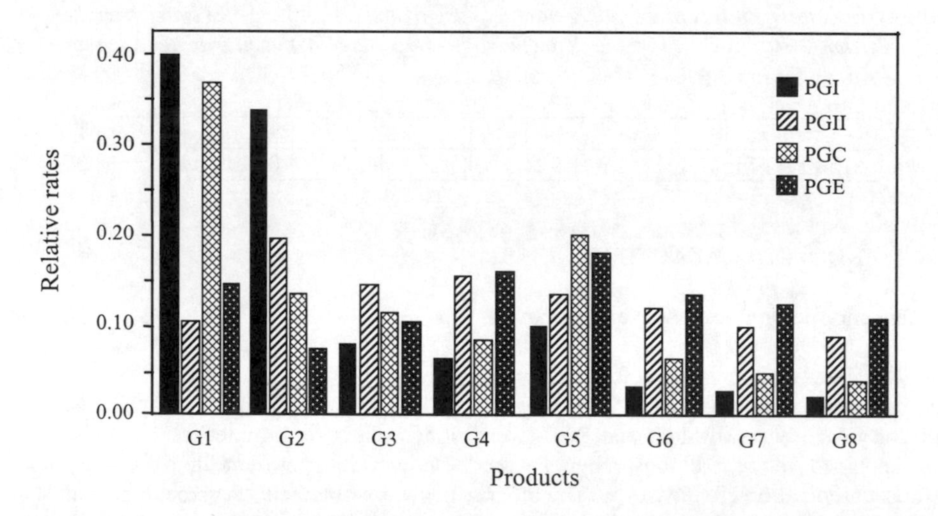

Figure 3 *Initial rate of product formation of* A. niger *endoPGI, II, C and E on polygalacturonate. 0.5 ml of polygalacturonate (0.25 % mass/vol.) in 50 mM Na-acetate, pH 4.2, 30 °C, was incubated with endoPGs at 4 U/ml. Rates were determined from the increase in [Gn] from t = 0 min to t = 10 min. Products were analysed by HPAEC-PAD. Quantitation was done using calibration mixtures of G1 to G8.*

Although the product distribution already indicated differences between the endoPGs when acting on polygalacturonate, striking dissimilarities were revealed upon comparison of the specific activities. In 50 mM Na-acetate, pH 4.2, for 0.25 % (mass/vol.) polygalacturonate at 30 °C the specific activities were 600 U/mg, 2000 U/mg, 38 U/mg and 24 U/mg for PGI, PGII, PGC and PGE, respectively. Thus, while polygalacturonate is a good substrate for PGI and PGII, this seems not to be the case for PGC and PGE, suggesting that the latter two enzymes prefer another, as yet unidentified, substrate.

5.3 Oligogalacturonate hydrolysis by endopolygalacturonases

In order to gain insight into the bond cleavage frequencies, and hence in the product distribution after polygalacturonate hydrolysis, we studied the cleavage patterns of oligogalacturonates of defined chain length by the individual endoPGs. The smallest oligogalacturonate hydrolysable by the endoPGs was G3. All enzymes cleaved G4 into G1 and G3 while endoPGI also produced a small amount of G2. Products arising from G5 were mainly G1 and G4 for PGC and PGE, while PGI and PGII resulted in the formation of G1/G4 and G2/G3 in almost equal ratio's. Using rG4 as a substrate resulted in formation of exclusively rG2 and G2 by PGI, PGII and PGC, while in addition to rG2 and G2 a very small amount of rG1 and G3 was observed for PGE. Hydrolysis of rG5 resulted also in a shift of the preferred bond of cleavage to formation of almost exclusively rG2 and G3 by all four enzymes. These results demonstrate that the enzymes hydrolyse the substrate from the reducing end. This is in contrast to the true exoPG where hydrolysis takes place from the non-reducing end. Thus, the endoPGs seem to have an exolytic activity on smaller oligogalacturonates. However, in case of G6 and G7 with PGII, PGC and PGE, increasing cleavage of the third glycosidic bond counted from the reducing end is observed, demonstrating the true endolytic character of the enzymes. For PGI using G6 and higher oligogalacturonates, the stoichiometry of product pairs (G_n is converted into $G_{n-x} + G_x$) was no longer observed. In these cases high amounts of G1 were observed and disproportional amounts of G4. From product progression analysis in relation to substrate consumption it was concluded that PGI shows extreme processivity or multiple attack on a single chain when DP > 5. The processivity ends when the oligogalacturonate is hydrolysed to DP 5 or 4. For PGC and to a lesser extent PGII and PGE some processivity is observed for G8 hydrolysis. In view of these results the data presented in Figure 3 can easily be explained *i.e.* the high rate of formation of G1 for PGI and PGC is due to processivity (this is also reflected in the low rate of G6-G8 formation).

6 ENDOPOLYGALACTURONASE II AS A MODEL FOR GLYCOSYLATION

In order to study the effect of glycosylation on enzyme activity and specificity we have addressed the glycosylation analysis of endoPGII in collaboration with Dr. C. Bergmann (The Complex Carbohydrate Research Centre, Athens, GA, USA). PGII was chosen as a model since the enzyme has a high specific activity, only one glycosylation site and the recombinant enzyme appeared to be homogeneous on IEF. Despite its homogeneous character on IEF, MALDI-TOF analysis of the untreated tryptic fragment carrying the glycan and the fragment treated with α1-2 mannosidase and *N*-glycanase, showed that PGII is heterogeneously glycosylated[22]. The glycan was shown to be of the high mannose type consisting of a consecutive series of seven glycoforms ranging from mannose-5 to mannose-11. The exact configuration of the glycan structures awaits further NMR analyses. The kinetic properties of the glycosylated enzyme will be compared to those of the deglycosylated enzyme.

References

1. A.G. Darvill, M. McNeil and P. Albersheim, *Plant Physiol.*, 1978, **62**, 418-422.
2. J.A. de Vries, F.M. Rombouts, A.G.J. Voragen and W. Pilnik, *Carbohydr. Polym.*, 1982, **2**, 25-33.
3. H. Schols, PhD Thesis, Agricultural University, Wageningen, The Netherlands, 1995.
4. A.J. Whitcombe, M.A. O'Neil, W. Steffan, P. Albersheim and A.G. Darvill, *Carbohydr. Res.*, 1995, **271**, 15-29.
5. J.A.M. Harmsen, M.A. Kusters-van Someren and J. Visser, *Curr. Genet.*, 1990, **18**, 161-166.
6. H.J.D. Bussink, J.P.T.W. van den Homberg, P.R.L.A. van den IJssel and J. Visser, *Appl. Microbiol. Biotechnol.*, 1992, **37**, 324-329.
7. M.E.G. Suykerbuyk, H.C.M. Kester, P.J. Schaap, H. Stam, W. Musters and J. Visser, *Appl. Environm. Microbiol.*, 1997, **63**, 2507-2515.
8. J. Visser, M.E.G. Suykerbuyk, M.A. Kusters-van Someren, R. Samson and P.J. Schaap, In: Fungal Identification Techniques (L. Rossen, V. Rubio, M.T. Dawson, and J. Frisvad, eds.) EUR 16510 EN, 1996, 194-201.
9. J.A.E. Benen and J. Visser, unpublished.
10. B. Henrissat, *Biochem. J.*, 1991, **280**, 309-316.
11. M.L. Sinnott, *Chem. Rev.*, 1990, **90**, 1171-1202.
12. P. Biely, J.A.E. Benen, K. Heinrichová, H.C.M. Kester and J. Visser, *FEBS Lett.*, 1996, **382**, 249-255.
13. L.V. Kofod, S. Kauppinen, S. Christgau, L.N. Andersen, H.P. Heldt-Hansen, K. Dörreich, and H. Dalbøge, *J. Biol. Chem.*, 1994, **269**, 29182-29189.
14. H.C.M. Kester, M.A. Kusters-van Someren, Y. Müller and J. Visser, *Eur. J. Biochem.,* 1996, **240**, 738-746.
15. K. Hiromi, Y. Nitta, C. Numata and S. Ono, *Biochim. Biophys. Acta*, 1973, **302**, 362-375.
16. H.J.D. Bussink, H.C.M. Kester and J. Visser , *FEBS Lett.*, 1990, **273**, 127-130.
17. H.J.D. Bussink, K.B. Brouwer, L.H. De Graaff, H.C.M. Kester and J. Visser , *Curr. Genet.*, 1991, **20**, 301-307.
18. H.C.M. Kester and J. Visser, *Biotechn. Appl. Biochem.*, 1990, **12**, 150-160.
19. L. Parenicová, J.A.E Benen, H.C.M. Kester and J. Visser, *Eur. J. Biochem.*, 1997, accepted for publication.
20. G. J. von Heyne, *Mol. Biol.*, 1985, **184**, 99-105.
21. T. Achstetter and D.H. Wolf, *EMBO J.*, 1985, **4**, 173-177.
22. Y. Yang, C. Bergmann, J. Benen and R. Orlando, *Rapid Commun. Mass Spectrometry*, 1997, **11**, 1257-1262.

INDUCTION AND REGULATION OF THE XYLANOLYTIC ENZYME SYSTEM OF *ASPERGILLUS NIGER*

N.N.M.E. van Peij, J. Visser and L.H. de Graaff

Section Molecular Genetics of Industrial Micro-organisms
Wageningen Agricultural University
Dreijenlaan 2, NL-6703 HA Wageningen, The Netherlands

1 INTRODUCTION

Degradation of polysaccharides is an important process in recycling carbohydrates from plant cell walls. Micro-organisms play a dominant role in this recycling process, due to their capacity to secrete a vast array of polysaccharide degrading enzymes. Filamentous fungi, such as *Trichoderma reesei* and *Aspergillus niger*, are for example capable of secreting cellulolytic, xylanolytic and pectinolytic enzymes.

In general, the expression of polysaccharidases in *A. niger* is controlled by carbon catabolite repression *via* CreA and by induction *via* a specific transcriptional activator. The formation of a low molecular weight inducer is a prerequisite for induction, since the polysaccharide itself, due to its size, can not enter the organism. Therefore, the micro-organism has the need for a mechanism by which it is able to generate such a low molecular weight inducer, in order to exploit its capacity to synthesise these extracellular enzymes and to liberate carbohydrates from polysaccharides, which are then used as a carbon source.

The presence of an inducer activates an induction pathway, which results in the synthesis of the enzyme system. Basically, this pathway consists of an uptake system for the inducer which may or may not be constitutive, and of a transcriptional activator since the expression of polysaccharide degrading enzymes is regulated at the level of transcription. The transcriptional activator drives the transcription of the polysaccharidase encoding structural genes. Whether the formation or activation of this transcriptional activator is the result of the direct interaction with the inducer, or whether the signal is transmitted *via* one or a number of intermediate steps, is not known; this may also vary with the particular system.

In this contribution we will describe and discuss the results of our studies on the induction and regulation of the xylanolytic enzyme system of *Aspergillus niger*. These studies are focussed on two steps of the induction process: the mechanism of inducer formation and the role of the transcriptional activator of the system.

2 THE XYLANOLYTIC ENZYME SYSTEM OF *ASPERGILLUS NIGER*

Xylan, a heteropolymer consisting of a backbone of β-1,4-linked D-xylose residues, can be decorated by various substituents. 1,2-Linked α-D-glucuronic acid or 4-O-methyl-α-D-glucuronic acid residues can be present, as well as 1,2- and 1,3-linked α-L-arabinose residues. In some plant cell walls, these latter residues may esterified with ferulic and *p*-coumaric acid, which enables cross-linking to the lignin matrix. In addition, and depending on the source of xylan, the D-xylose residues can be modified by acetylation at the C-2 or C-3 position.[1,2]

Due to this heterogeneous composition, a mixture of enzymes is necessary for complete degradation of the xylan polymer. *A. niger* is able to degrade xylan completely. It produces endoxylanases (E.C. 3.2.1.8) which cleave the main chain β-1,4-glycosidic bond, and a β-xylosidase which is involved in the conversion of the xylo-oligomers (resulting from the endoxylanase action) into D-xylose. In addition, a number of enzymes involved in hydrolysis of the substituents are being produced, *e.g.* acetyl(xylan)esterase (E.C.3.1.1.6), α-L-arabinofuranosidase B (E.C. 3.2.1.55), arabinoxylanhydrolase, α-glucuronidase (E.C.3.2.1.139), feruloyl esterase and coumaryl esterase.

The biochemistry and mode of action of the xylanolytic enzymes, involved in degradation of the backbone, has been studied extensively and is relatively well understood. Although a large number of genes encoding xylanolytic enzymes has been cloned from *A. niger* and the closely related *Aspergillus tubingensis*, the mechanism governing the induction and control of the genes is poorly understood.

3 THE ROLE OF β-XYLOSIDASE IN INDUCTION OF THE XYLANOLYTIC ENZYME SYSTEM OF *A. NIGER*

Due to the strong transglycosylation activity, β-xylosidase can play an important role in the formation of signal molecules, which leads to the expression of xylanolytic genes.[3,4] The molecular cloning of the gene encoding the β-xylosidase in *A. niger* provides the possibility to investigate the role and importance of this enzyme in induction of xylanase expression.

3.1 The *A. niger* β-xylosidase encoding *xln*D gene

The gene encoding β-xylosidase in *A. niger* was cloned in the following way. The enzyme was purified from the culture broth, after growth of the fungus on a liquid medium containing 3% xylan as a sole carbon source. The purified protein was used to characterise the enzyme, but it was also used to raise antibodies against the protein in mice. These antibodies were then used to screen a xylan-induced cDNA expression library in *Escherichia coli*, for the expression of the protein. From this screening, four cDNA clones were obtained, from which one was used to clone the β-xylosidase encoding *xln*D gene from *A. niger*.[5] The resulting clone was then used to determine the primary structure of the gene and to study its role in induction of xylanolytic gene expression.

The *A. niger xln*D gene encodes a protein of 804 amino acids, which contains a putative signal peptide of 26 amino acids. The derived molecular mass of the protein is 85 kDa, which is markedly lower than the 110 kDa of the purified protein . The difference in molecular mass results from N-glycosylation of the enzyme, since deglycosylation of the enzyme results in a molecular mass corresponding to the calculated molecular mass, as determined by SDS-PAGE. More surprisingly however, the encoded β-xylosidase shows a clear homology to specific regions of family 3 type β-glucosidases, specifically in the active site region. This homology explains the low activity of the β-xylosidase towards *p*-nitrophenyl β-D-glucopyranoside (0.2 U/mg), which is about 3% of the activity towards *p*-nitrophenyl β-D-xylopyranoside. The structural homology of β-xylosidase and β-glucosidases and the low β-glucosidase activity of the β-xylosidase enzyme suggest an evolutionary relationship between both classes of enzymes.

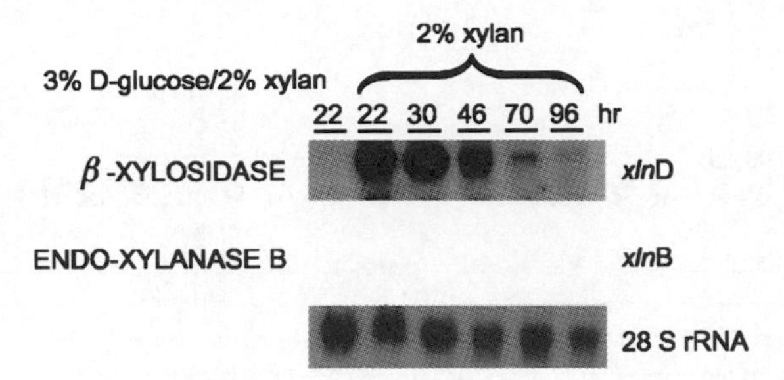

Figure 1 *Expression of β-xylosidase in* A. niger. *Northern analysis of the time course of induction of* xln*B encoding endoxylanaseB and* xln*D encoding β-xylosidase in* A. niger *grown on xylan as a carbon source. The 28 S rRNA was used as a loading control.*

3.2 β-Xylosidase gene expression and its role in induction of xylanolytic genes

The level of transcription of the *A. niger xln*D gene upon growth on xylan as a carbon source, was investigated at different time intervals. The time course of transcription was compared to the time course of transcription of the endoxylanase B encoding *xln*B gene (Figure 1). Relative to the *xln*B transcription, the transcription of the *xln*D gene occurs in the initial phase of growth.

While the *xln*B transcription is initially low, it increases and reaches its maximum at about 46 h. This is in contrast to the *xln*D transcription level, which is high in the early phase of growth and decreases after 30 h. This high level of transcription of the *xln*D gene in the early stage of induction indicates a potential for β-xylosidase in generating the inducer of the xylanolytic enzyme system.

To study this more precisely, the gene encoding the enzyme was inactivated by gene disruption. The resulting strain, which lacks extracellular β-xylosidase activity, was then used to study the transcription and expression patterns of other xylanolytic genes upon growth on the monomeric inducer D-xylose and on xylan.

During culturing on 1% oat spelts xylan as a carbon source, the degree of hydrolysis of the xylan was followed by HPLC analysis of the culture filtrate (Figure 2). In the culture filtrate of the wild type strain, hydrolysis products accumulated in the 9 h and 28 h samples. These were mainly D-xylose and xylobiose, while minor amounts of L-arabinose and xylotriose were no longer detected. After 28 h, due to the increase in biomass, the hydrolysis of the xylan becomes rate limiting and none of the sugars or oligomers can be detected. In the β-xylosidase disrupted strain however, xylobiose and to a lesser extent xylotriose accumulate over the whole culture period. The β-xylosidase disruption strain is unable to hydrolyse xylobiose and xylotriose.

A Northern analysis was performed to analyse the effect of the lack of β-xylosidase acitivity, in D-xylose and xylan grown cultures, on the expression of the xylanolytic enzyme spectrum (Figure 3). For the xylan induced cultures, a significant increase in endoxylanase B expression was found in the β-xylosidase disruption strain, while in strains having multiple copies of the *xln*D gene the effect is less clear. Some of the strains have slightly decreased *xln*B transcription levels, while in other *xln*D multiple copy strains a significant decrease in *xln*B transcription is initially found, which increases to wild type levels after longer (28 h) incubation times (data not shown). This indicates that the potentially high transglycosylation reaction of the β-xylosidase enzyme does not play a significant role in induction of the xylanolytic enzyme system of *A. niger*.

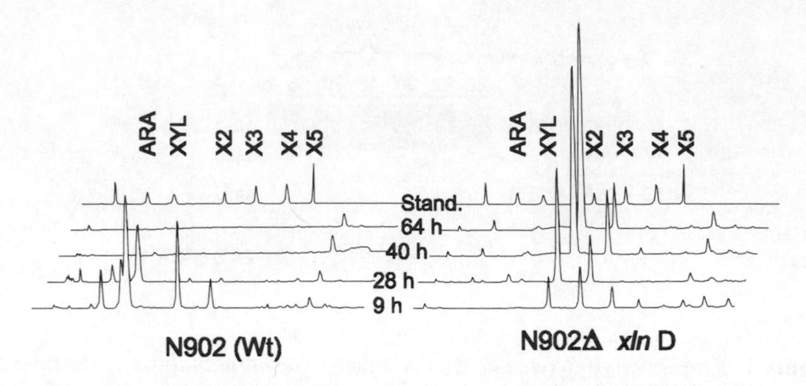

Figure 2 *Hydrolysis of xylan by* A. niger *wild type (N902) and a β-xylosidase disruption strain (N902:: Δ*xln*D). Culture filtrate was analysed by high-performance anion exchange chromatography using a Carbopac PA100 column and pulsed amperometrical detection. The strains were pregrown for 18 h on minimal medium containing 50 mM* D*-fructose, and then transferred to minimal medium containing 1% oat spelts xylan (Sigma). The standards used are: Ara, L-arabinose; Xyl,* D*-xylose; X2-X5, xylobiose-xylopentaose (Megazyme).*

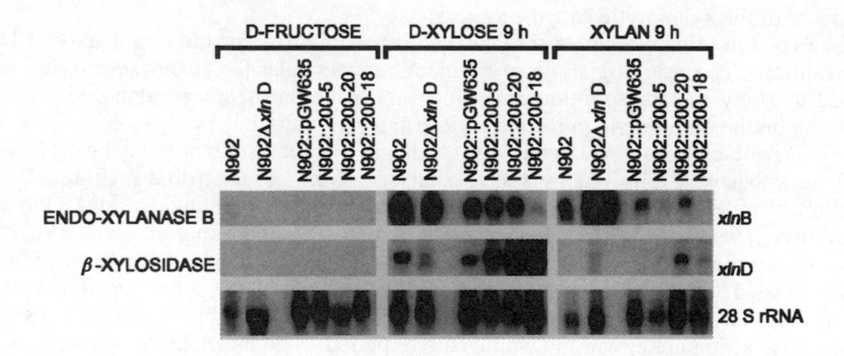

Figure 3 *Effect of β-xylosidase activity on the level of transcription of* xln*B encoding endoxylanase B. Mycelia were pregrown on 50 mM* D*-fructose for 18 h and then 3 g mycelium was transferred to 50 ml medium containing either 50 mM* D*-xylose or 1% xylan and incubated for 9 h. The strains were N902 (wild type), N902Δ*xln*D (β-xylosidase disruption mutant), N902::pGW635 (a control of N902 transformed with* pyrA *(pGW635)) and three β-xylosidase multiple copy transformants (N902::pIM200-#).*

4 THE TRANSCRIPTIONAL ACTIVATOR XlnR IS CONTROLLING XYLANOLYTIC ENZYME EXPRESSION

Upon formation of D-xylose as inducer, the transcription of the xylanolytic genes is activated, which results in expression of the correponding enzymes. The induction of transcription is mediated by a transcriptional activator, binding to an element in the promoter of the structral genes. By promoter deletion analysis of the *xln*A gene from *Aspergillus tubingensis* in *A. niger*, a fragment in this promoter could be identified, which contains a *cis*-acting element involved in induction of transcription of this gene. By cloning this fragment in front of the glucose oxidase encoding *gox*C gene, the presence of an upstream activating sequence (UAS_{XLN}) was shown, acting in *cis* on the regulation of transcription of the downstream gene. Although in the same analysis a different fragment containing elements involved in carbon catabolite repression could be identified, carbon catabolite repression acts also *via* the fragment containing the UAS_{XLN}.[6] Based on these results, a model for the regulation of xylanase expression was postulated (Figure 4), which resembles the double-lock mechanism as described for the expression of the alcohol dehydrogenase encoding *alc*A gene in *Aspergillus nidulans.*[7]

4.1 A Selection cassette for the isolation of regulatory mutants

Based on the model as shown in Figure 4, a selection cassette (pIM130) was constructed which allows the isolation of mutants in the regulatory circuit of the xylanolytic system. This cassette is based on the gene encoding orotidine-5'-phosphate-decarboxylase (*pyr*A in *A. niger*). In many organisms this gene is used as a selection marker in transformation, *e.g.* in *Aspergilli*, *Trichoderma* and *Saccharomyces*. In the constructed selection cassette, the transcription of the *pyr*A gene is under the control of a *cis*-acting element of a regulated gene, in our case the fragment containing the UAS_{XLN} of the endoxylanaseA encoding gene. Introduction of the selection cassette into a *pyr*A negative mutants strain makes the *pyr*A gene pathway-specifically expressed. Selection in this strain can be done in a bidirectional way, *i.e.* for a positive PYR^+ (uridine prototrophic) or a negative PYR^- (fluoro-orotic acid resistant) phenotype, under conditions with a normally opposite phenotype. In this way mutants can be selected in pathway-specific *trans*-acting factors. The corresponding genes, both positively and negatively acting transcription factors, can subsequently be isolated by mutant complementation.

4.2 The gene encoding the transcriptional regulator XlnR

Using the described selection cassette, 57 mutants were obtained after UV irradiation of spores, which were fluoro-orotic acid resistant under inducing conditions. After this primary selection, a second selection step was carried out on plates containing a medium with 1% oat spelts xylan as carbon source. Ten of these mutants showed poor growth on these plates and lacked halo formation around the colony, indicative for a strong reduction or absence of endoxylanase expression. Based on this phenotype, these mutants were given the prefix NXA, for Non-Xylanase producing. The expression of both endoxylanases and β-xylosidase in these mutants was analysed after growth in liquid cultures containing 50 mM D-xylose or 1% oat spelts xylan as carbon source (Table 1). This analysis showed that the isolated mutants have strongly decreased endoxylanase and β-xylosidase activities in comparison to the wild-type strain, while the expression of the α-L-arabinofuranosidase activity is not affected. This indicates a mutation in a *trans*-acting factor involved in the coordinate control of endoxylanase and β-xylosidase expression.

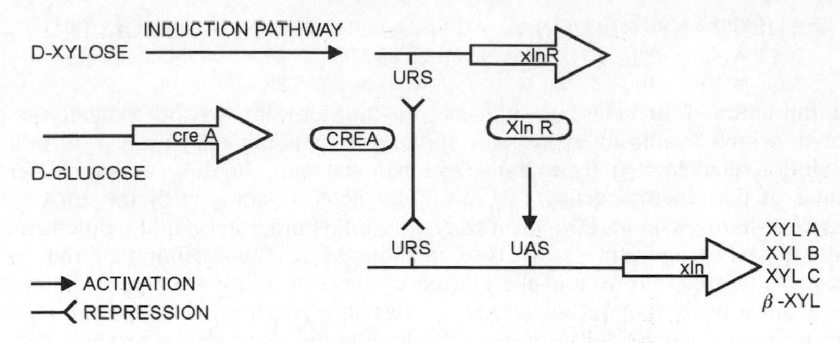

Figure 4 *Schematic model for the regulation of the xylanolytic enzyme system of* A. niger. *The model presumes a transcriptional regulator XlnR encoded by the* xln*R gene. Under conditions of carbon catabolite repression both the structural genes as well as the* xln*R gene are repressed by CREA.*

From the isolated mutants, NXA1-4 and NXA4-4 were used to clone the *xln*R gene by complementation, as is decribed in detail elsewere.[8] The cloned gene was able to complement all ten NXA mutants, suggesting that in all these mutants the same function had been mutated. Sequence analysis indicates that the *xln*R gene encodes a protein of the class of Zn binuclear cluster transcription factors, to which class also *e.g.* the *Saccharomyces cerevisiae* GAL4 transcription factor belongs. This was deduced from the amino acid sequence similarity at the putative DNA binding domain. Cloning by direct complementation may lead to the isolation of suppressor genes. Therefore, we have sequenced the *xln*R allele of three xylanase non-producing mutants, to relate the cloned *xln*R gene mutation to the NXA phenotype. DNA sequence analysis of the *xln*R gene of these three mutants revealed single base mutations. The mutations found were: a T→C substituting Leu650 for Pro in strain NXA1-4, a T→C substituting Leu823 for Ser in strain NXA1-15 and a T→G substituting Tyr864 for Asp in strain NXA2-5.

Table 1 *Xylanolytic activities in* A. niger *wild type and NXA mutants upon growth on arabinoxylan*

	Endoxylanase[a]		*β-Xylosidase*[a]		*α-Arabinofuranosidase*[a]
Carbon source[b]:	*Xylan*	*D-xylose*	*Xylan*	*D-xylose*	*Xylan*
Strain:					
NW205	$5*10^2$	$5*10^1$	0.35	0.40	0.33
NW205::130	$5*10^2$	$1*10^2$	0.36	0.51	0.40
NW205::130*nxa*1-4	2	1	0.01	0.01	0.36
NW205::130*nxa*4-4	2	0.3	0.01	0.01	0.24
NW205::130*nxa*1-6	2	0.3	0.01	0.01	0.31
NW205::130*nxa*3-14	1	0.4	0.01	0.01	0.29
NW205::130*nxa*4-8	2	0.4	0.01	0.01	0.38
NW205::130*nxa*2-10	1	0.1	0.01	0.01	0.19
NW205::130*nxa*2-8	1	0.3	0.01	0.01	0.29
NW205::130*nxa*2-5	1	0.2	0.01	0.01	0.28
NW205::130*nxa*1-15	2	0.2	0.01	0.01	0.23
NW205::130*nxa*4-14	2	0.2	0.01	0.01	0.28

[a] activities (nkatal ml^{-1}) determined in dialysed (1mM NaPi, pH 5,6) culture filtrate.
[b] one gram wet-weight mycelium was transferred for 5.5 hrs to 10 ml media containing 1% birchwood xylan or 10 mM D-xylose, after a preculture of 16 hrs in 50 mM D-fructose.

4.3 The effect of XylR on the expression of the xylanolytic enzyme system

Some of our studies, using transformants carrying multiple copies of one of the structural genes of the xylanolytic enzyme system, indicate that in these a decrease in total expression occurs, possibly due to titration of a factor involved in transcription of xylanolytic genes (see *e.g.* Figure 3). To investigate if this limitation is at the xylanolytic transactivator XlnR, we have constructed strains with increased copy numbers of *xln*R, and analysed these strains for xylanolytic expression during growth on arabinoxylan. Our results show the inability of the *xln*R$^-$ mutant to produce a number of proteins. The differences between the wild-type strain and this mutant, represent the components of the enzyme spectrum controlled by the transcriptional regulator XlnR (Figure 5). An increase in the *xln*R gene dosage leades to an increase of the overall expression of the xylanolytic proteins. From this it can be concluded that there is a limitation of the xylanolytic gene expression at the level of the transcriptional regulator XlnR. Moreover, the amounts of enzymes produced can also be influenced by an increase in inducer level, as can be concluded from the increase of the different proteins produced in the Δ*xln*D strain.

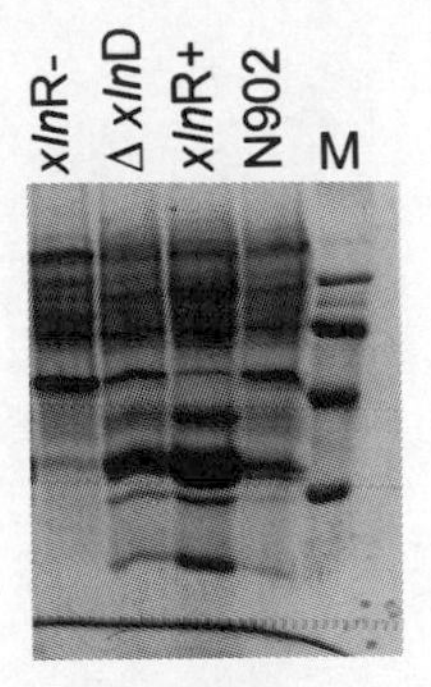

Figure 5 *Xylanase expressions in* A. niger *wild-type (N902), a* xlnR$^-$ *mutant, a strain having increased* xln*R gene copies and the* xln*D disruption mutant. The lane marked M contains molecular weight markers. The strains were grown for 4 days on arabinoxylan as a carbon source, after which the culture filtrate was concentrated 5 times. The concentrated culture filtrate was analysed by SDS PAGE followed by Coomassie Brilliant Blue staining of the proteins.*

References

1. P. Biely, *Trends in Biotechnol.*, 1985, **3**, 286.
2. M.P. Coughlan and G.P. Hazlewood, *Biotechnol. Appl. Biochem.*, 1993, **17**, 259.
3. N.A. Rodionova, I.M. Tavobilov and A.M. Bezborodov, *J. Appl. Biochem.*, 1983, **5**, 300.
4. M. Hrmova, E. Petrakova and P. Biely, *J. Gen. Microbiol.*, 1991, **137**, 541.
5. N.N.M.E. van Peij, J. Brinkmann, M. Vršanská, J. Visser and L.H. de Graaff, *Eur. J. Biochem.*, 1997, **245**, 164.
6. L.H. de Graaff, H.C. van den Broeck, A.J.J. van Ooijen and J. Visser, *Mol. Microbiol.*, 1994, **12**, 479.
7. B. Felenbok, *J. Biotechnol.*, 1991, **17**, 11.
8. N.N.M.E. van Peij, J. Visser and L.H. de Graaff, *Mol. Microbiol.*, in press.

Protein-linked Glycosyl Structures in Lower Eucaryotes

NMR TECHNIQUES IN THE STUDY OF PROTEIN-LINKED CARBOHYDRATES

André De Bruyn[*1,2], Jan Schraml[1,3], Marleen Maras[4], Roland Contreras[4] and P. Herdewijn[1]

[1]Lab. Medicinal Chemistry, Rega Institute, Katholieke Universiteit Leuven, Minderbroedersstraat, 10, B-3000 Leuven, België. [2]Dept. Organic Chemistry, Universiteit Gent, Krijgslaan 281 S4, B-9000 Gent, België. [3]Institute of Chemical Process Fundamentals, Czech Academy of Sciences, Prague 16502, Czech Republic. [4]Lab. of Molecular Biology, Universiteit Gent, K.L. Ledeganckstraat 35, B-9000 Gent, België.

1 INTRODUCTION

NMR techniques developed in our laboratory facilitate the assignment of the primary structure of complex carbohydrates. They are based on selective excitation of the individual units of the oligosaccharides and have following advantages:

1) A combination of several experiments (*e.g.* an experiment with heteronuclear irradiation) is sometimes easier to perform in a selective excited spectrum than in the 2D variant.
2) Pitfalls caused by overlap can be avoided by selective excitation of a certain band of resonances. For example, in a TOCSY experiment cross peaks are squares which may overlap making the assignment of the resonances belonging to one residue difficult. By selective excitation of a band of resonances (*e.g.* the glycosidic protons) and by decoupling in the $\omega 1$ direction, the squares can be reduced to lines, thus avoiding overlap.
3) Selective excitation can be used to enhance the sensitivity. Our experiments were performed with samples of 200 - 500 µg compound. With such small quantities, the necessary heteronuclear correlation experiments could be performed using sensitivity enhanced experiments. In this case an inverse technique was used.[1] Likewise sensitivity enhanced techniques based on a combination of the Hadamard algorithm[2] and selective excitation are described.

Furthermore, certain NMR parameters (coupling constants and NOE effects) may help in the elucidation of the tertiary structure.[3] From other parameters (T_1, T_2 and heteronuclear NOE), the $J(\omega)$ or spectral density function[4] can be derived and from this the dynamics of the glycoconjugate structure. With the Liparo and Szabo free-model interpretation of $J(\omega)$, the internal motion within a complex carbohydrate can be evaluated.[5] In this perspective the group of Goldman and Desvaux developed the so-called off-resonance ROESY technique in 1994.[6] Another interesting application is the use of transferred NOE to study the complexation between a carbohydrate and its receptor, a technique which was successfully applied in the work of Meyer and Peters[7] and Kroon-Batenburg.[8]

The structure of the following compounds (Figure 1), isolated after N-glycase treatment of cellobiohydrolase I from *Trichoderma reesei*, will be discussed.[9,10]

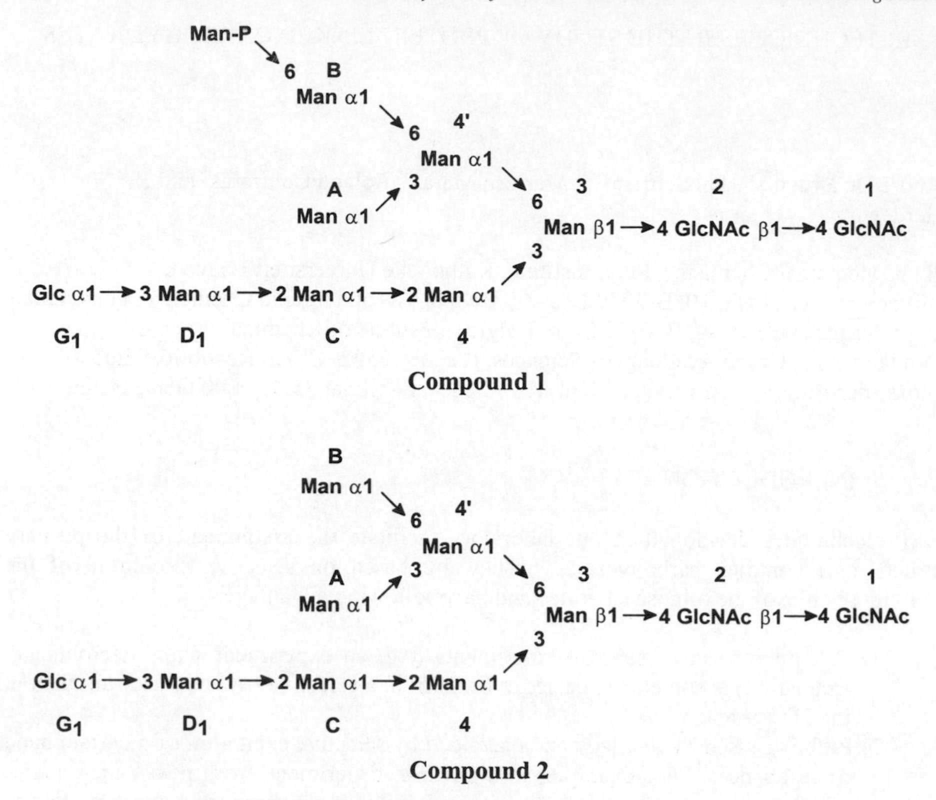

Figure 1 *The structure of the two glycans discussed in this article*

2 RESULTS AND DISCUSSION

Both compounds belong to the high mannose type glycans and the methods described here are specifically developed for this group of glycans. For glycans of other types similar approaches have been published. Of the common procedures used, two adequate methods are proposed to assign the primary structure of complex carbohydrates.

First, we try to find our way in the forest. A structure is proposed *via* a method introduced by Vliegenthart and coworkers,[11] that uses so-called structural reporter groups on one hand and chemical shift increments on the other. The other method allows verification of the proposal or to suggest another structure.

The ^{1}H NMR spectrum of compound 2 is given in Figure 2. The region δ 3.20 - 4.00 contains the bulk of the resonances. The structural reporter groups are the resonances outside this region, as *e.g.* the site of the resonances of the glycosidic protons, those of the H-2 protons, those of the mannose units and those of the methyl group in the acetamide groups of the GlcNAc residues. The resonance of the methyl group of the acetate impurity is used as reference.

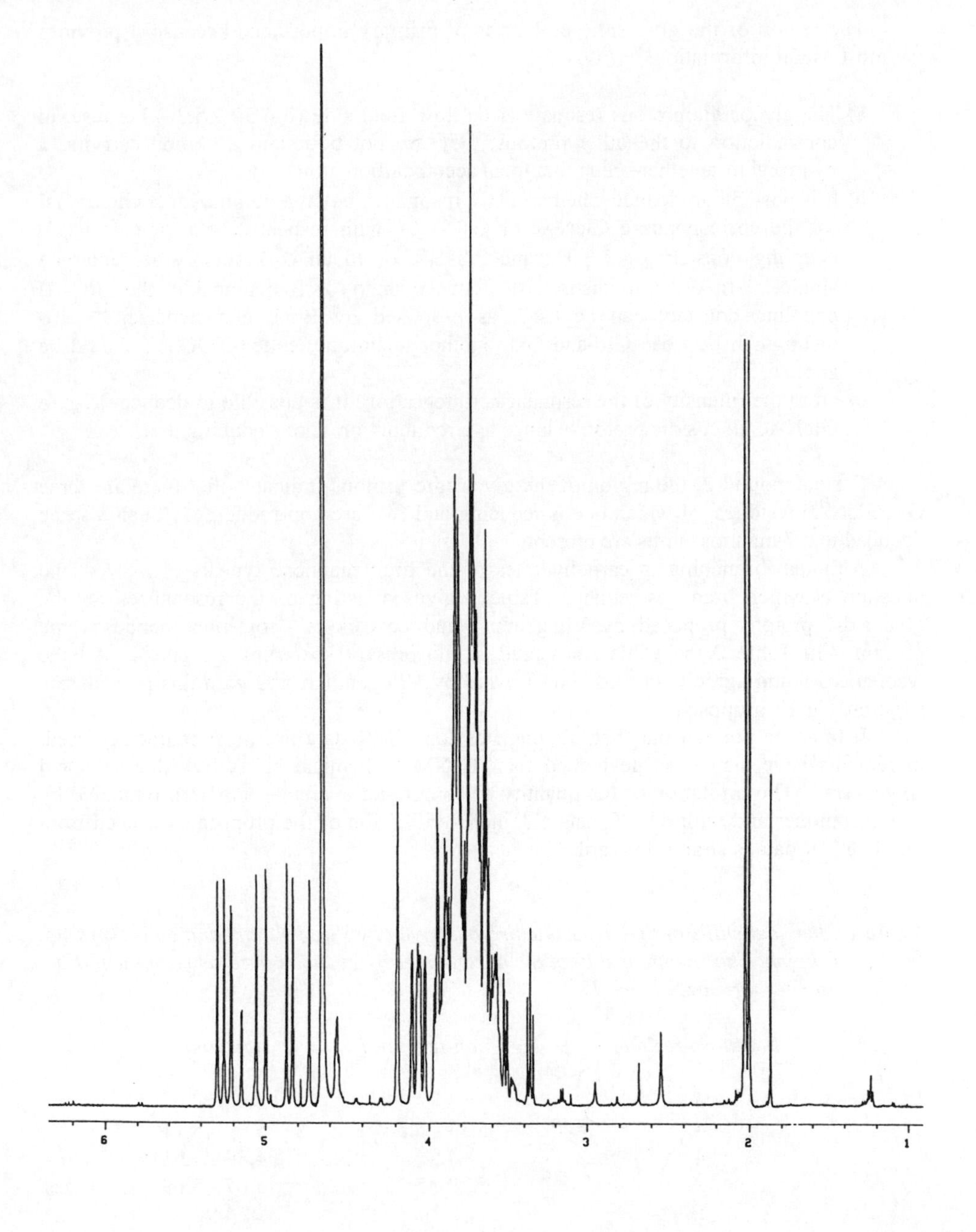

Figure 2 *^{1}H NMR spectrum of compound 2*

The region of the glycosidic protons is of primary importance because it provides the most useful information.

a) The glycosidic protons resonate at the low field side (δ 4.50 - 5.30) because, in contradiction to the other protons, they are not bound to a carbon carrying a hydroxyl or an ether group, but to an acetal carbon atom.
b) It is possible to deduce whether a H-1 resonance belongs to an α or β Glc or Gal (or the corresponding GlcNAc or GalNAc), with respectively a vicinal $^{3}J(1,2)$ coupling constant of 3.6 Hz and 7.8 Hz, or to an α/β mannoside (enlarged singlet). In order to distinguish between an α or β mannoside, the $^{1}J(C\text{-}H)$ coupling constant can be used, as proposed by Bock and Pedersen.[12] To distinguish between Glc and Gal, another technique, called TOCSY, should be applied.
c) From the intensity of the resonances (integration) it is possible to deduce if *e.g.* a GlcNAc glycosidic proton belongs to a reducing or a non-reducing unit.

For compound 2, the region of the glycosidic protons indicates that there are three Glc/GlcNAc residues, of which one is reducing and two are non-reducing. It can also be deduced that 7 mannose units are present.

Although branching in carbohydrates of the high mannose type is conserved, the question is which branches remain. Table 1 gives a listing of the resonances for the glycosidic protons proposed by Vliegenthart and coworkers[11] for high mannose type glycans. In Table 2 the values measured in the present spectrum are given. All the resonances found agree with those put forward by Vliegenthart and with this procedure a structure can be proposed.

It must be pointed out that, although proton NMR data are most frequently used, increments have also been developed for ^{13}C NMR chemical shifts by Allerhand and coworkers.[13] The limitation of the quantity of compound available has been remedied by new techniques and with the ^{13}C data obtained verification of the proposals deduced from the ^{1}H NMR data is straightforward.

Table 1 *Chemical shifts of the glycosidic protons given by Vliegenthart and coworkers*[11] *(in ppm. The chemical shift of the acetate methyl group is used as reference. For numbering: see Scheme 1).*

Mannose residue	*non-terminal*	*terminal*
3	4.77	
4	5.33 - 5.34	5.09 - 5.11
4'	4.87	4.89 - 4.91
A	5.39 - 5.40	5.07 - 5.09
B	5.14	4.90
C	5.29 - 5.30	5.05
D1, D2, D3	5.04 - 5.06	5.04 - 5.06

Table 2 *Chemical shifts of the glycosidic protons of compound 2, and the assignment of the mannose units by comparison with Table 1.*

Experimental value	residue	non-terminal	terminal
4.77	3	4.77	
4.87	4'	4.87	
4.91	B		4.90
5.04	D		
5.10	A		5.08
5.30	C	5.29	
5.34	4	5.33	

For further branching, *e.g.* on C-6 of unit 4, more specific data on the resonances for the glycosidic protons of the supplementary chain are provided by Trimble and Atkinson.[14]

From the data listed by Vliegenthart, only a proposal can be deduced that needs verification. When the compound is known, the data found can be compared with the published data of the compound. When the compound is unknown, further experiments are needed. In this case the so-called integrated approach can be used, which is a combination of COSY and TOCSY on one hand and HSQC and HMBC[1] on the other hand. COSY, TOCSY and HSQC are used for identification of the resonances.

The resonances of the successive protons in the pyranose ring can be assigned by following the intra-ring vicinal coupling constants. This information can be extracted step by step starting from the glycosidic protons in a phase-sensitive COSY experiment. However such a spectrum is usually overcrowded so that only three or four resonances of the seven can be assigned. Here a TOCSY experiment can help. With a TOCSY experiment we see on the trace of one resonance (*e.g.* that of the resonance of a glycosidic proton) all the resonances of the ring protons of one residue.

Instead of a 2D TOCSY experiment we propose a selective excited TOCSY experiment for each unit. This strategy has two advantages. First, we can array the mixing time, so that with an incrementing mixing time one resonance after the other appears. A second advantage is that sometimes the TOCSY experiment can be combined with another technique.

Since it is not possible to obtain a carbon-13 spectrum of 200 μg of a compound with the normal recording techniques using a 5 mm probe (*i.e.* a ^{1}H noise decoupled spectrum) even with NOE enhancement, a sensitivity enhanced experiment is needed. Techniques are now available allowing to transfer the magnetisation from a sensitive nucleus (a proton) to a non-sensitive nucleus (a carbon) to which it is bound, and then the FID of the non-sensitive nuclei is recorded. This technique is called DEPT.[15] For the identification of the resonances, a technique called HSQC[1] is used. In this case the correlation between carbon resonances and the resonances for the protons directly bound to it are observed. The proton resonances can be printed on the side of the spectrum.

The identification of the units is based on increments on the proton and the carbon resonances on the substitution site of the glycosidic bond. We can use ^{1}H NMR increments, where the proton on the substitution site, as well as the protons in β position of the substitution site show a downfield shift of 0.10-0.35 ppm in comparison with the non-substituted analogues. For the ^{13}C increments, the carbon resonance on the substitution site shows a downfield shift of 8 ppm.

The principle of the application of HMBC spectra is as follows (Figure 3). On the trace of each glycosidic proton, the resonances of C-5 and sometimes C-3 of the same glycosyl unit as H-1 are encountered as well as the resonance of the carbon by which the preceding unit is linked through the glycosidic bond. The latter shows a downfield shift of 8 ppm. When all resonances have been assigned, it is now easy to establish the structure.

For the linkages with the exocyclic CH_2OH group, an edited version of HMBC was used, where only the methylene proton resonances are observed.

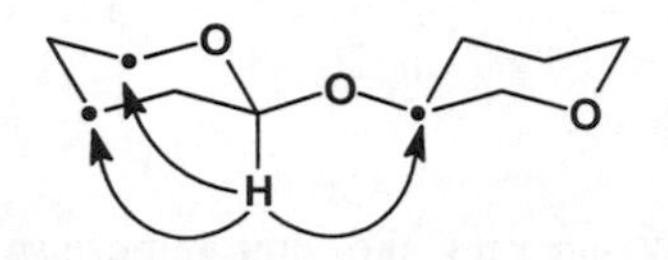

Figure 3 *Heteronuclear long-range effect detected from inspection of the cross peaks in a HMBC experiment on the trace of the resonance of a glycosidic proton*

Besides compound 2, another new compound (compound 1 in Figure 1) was isolated.[10] Its spectrum is reproduced in Figure 4. In addition to all the peaks found for the glycosidic protons of compound 2, now in the region of the glycosidic protons (at δ 5.43) also a doublet of doublets with coupling constants 8.1 Hz and 1.2 Hz was found. From its retention time on TLC, one could assume it to be a Man_6 derivative. This conclusion was contradicted by inspection of the ^1H NMR spectrum.

After inspection of this supplementary multiplet, it was concluded that the value of the large coupling constant points to a heteronuclear ^1H-^{31}P coupling constant, the small coupling constant being J(1,2) of a mannose unit. The presence of a phosphodiester link was verified by recording a ^{31}P spectrum and by irradiation of the phosphorus resonance while observing the ^1H NMR spectrum.

It must be noted that, according to the tables of Vliegenthart and coworkers, the resonance for H-1 of unit B, found at δ 4.90, points to a terminal residue. Since this unit is non-terminal, according to this table its value is expected at δ 5.14. This anomaly is due to the presence of the phosphodiester bond and illustrates the fact that the tables of Vliegenthart must be treated with caution.

In order to assign the unit and the site on this unit to which the supplementary mannosylphosphate is linked, two techniques are available. First, a HMBC spectrum was run. Since there is one phosphorus resonance, only the trace at the ^{31}P resonance of the HMBC spectrum must be considered.

A cross peak at the chemical shift values of H-1 and H-2 of the mannosylphosphate was observed, as well as a further resonance collapsing with the resonance of H-2, which could not be identified with this technique. Therefore a combination of selective excited TOCSY and ^{31}P decoupling was adopted.[16] When exciting the resonances for unit B, those for H-6R and H-6S are very complex. When irradiating the ^{31}P resonance, the resonances for these protons simplify to the expected pattern for an exocyclic CH_2OH group, as usually found in mannoses.

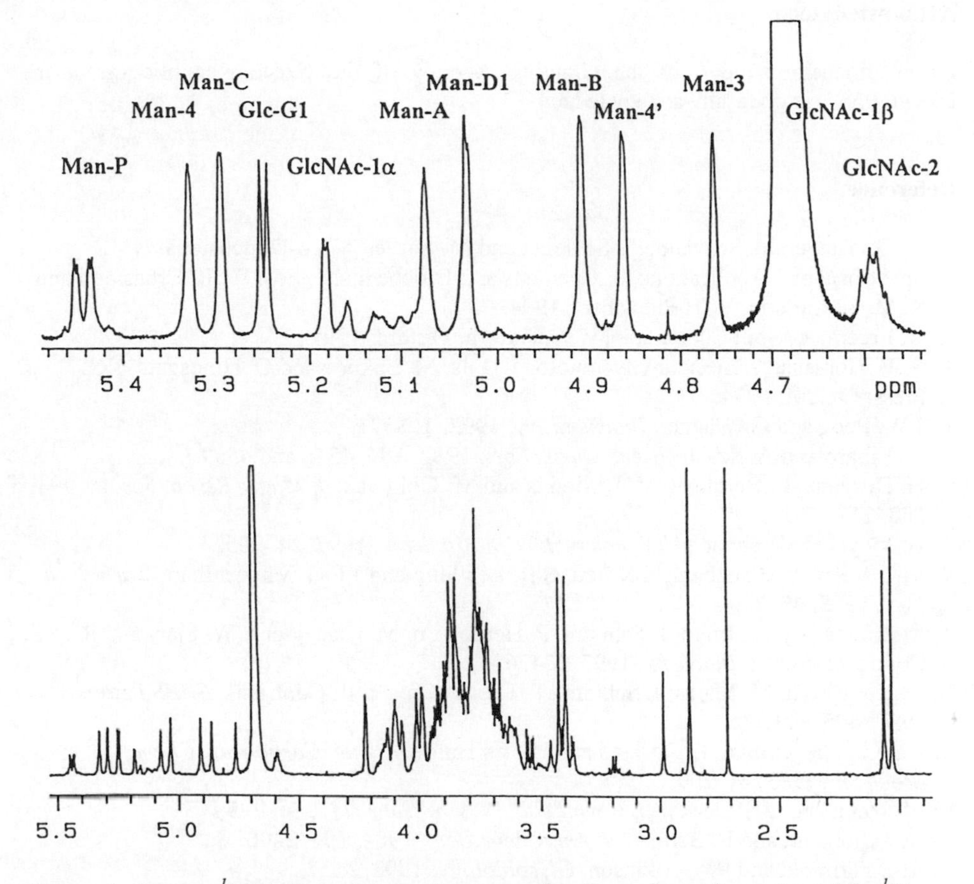

Figure 4 *Bottom: ^1H NMR spectrum of compound 1. Top: glycosidic region of the ^1H NMR spectrum of compound 1*

A report of a ^{31}P decoupled selective excited TOCSY spectrum was published in 1989 by the group of Vliegenthart.[17] They introduced a relayed modification of spin difference spectroscopy, called RESED. We have used selective excitation pulses introduced by Kupce and coworkers.[18] After the work of Vliegenthart and coworkers the pulse wave generator Pandora Box of Varian has been introduced, which allowed us to work with more complex molecules and smaller quantities (200µg). RESED was used for a phosphate monoester; our sequence was applied on the phosphodiester.

3 CONCLUSION

Two glycans with unknown structures were enzymatically released from cellobiohydrolase I (*Trichoderma reesei*) and identified by advanced NMR techniques using gradients. Techniques based on selective excitation were successfully applied.

Acknowledgment

Partial financial support of the Granting Agency of the Czech Republic (grant nr. 203/96/0567) is gratefully acknowledged.

References

1. C. Griesinger, H. Schwalbe, J. Schleuer and M. Sattler, 'Two-Dimensional NMR Spectroscopy. Applications for Chemists and Biochemists', Eds. W. R. Croasmun and R. M. K. Carlson, VCH Publishers, 1994.
2. A. Freeman, 'Spin Choreography', Spektrum, Oxford, 1997.
3. S.W. Homans, 'Molecular Glycobiology', Eds. M. Fukuda and O. Hindsgaul, IRL Press, Oxford, 1994.
4. J.W. Peng and G. Wagner, *Biochemistry*, 1992, **1**, 8571.
5. G. Liparo and A. Szabo, *J. Am. Chem. Soc.,* 1982, **104**, 4546 and 4559.
6. H. Desvaux, P. Berthault, M. Birlirakis and M. Goldman, *J. Magn. Reson. Ser A*, 1994, **108**, 219.
7. B. Meyer, T. Weimar and T. Peters, *Eur. J. Biochem*., 1997, **24**, 705.
8. L.M.J. Kroon-Batenburg, J. Kroon, B.R. Leeflang and J.F.G. Vliegenthart, *Carbohydr. Res*., 1993, **45**, 21.
9. M. Maras, A. De Bruyn, J. Schraml, P. Herdewijn, M. Claeyssens, W. Fiers and R. Contreras, *Eur. J. Biochem*., 1997, **254**, 617.
10. A. De Bruyn, M. Maras, J. Schraml, P. Herdewijn and R. Contreras, *FEBS Letters*, 1997, **405**, 111.
11. J.F.G. Vliegenthart, L. Dorland and H. van Halbeek, *Adv. Carbohydr. Chem. Biochem*., 1983, **41**, 209.
12. K. Bock and C. Pedersen, *J. Chem. Soc. Perkin Trans II*, 1974, 293.
13. A. Allerhand and E. Berman, *J. Am. Chem. Soc*., 1984, **106**, 2400.
14. R.B. Trimble and P.A. Atkinson, *Glycobiology*, 1992, **2**, 57.
15. D.M. Doddrell, D.T. Pegg and M.R. Bendall, *J. Magn. Reson.*,1984, **48**, 323.
16. J. Schraml, A. De Bruyn, R. Contreras and P. Herdewijn, *J. Carbohydr. Chem*., 1997, **16**, 65.
17. P. de Waard and J.F.G. Vliegenthart, *J. Magn. Reson*., 1989, **81**, 173.
18. E. Kupce, J. Boyd and I.D. Campbell, *J. Magn. Reson*., 1995, **106 Ser B**, 300.

ENGINEERING OF THE CARBOHYDRATE MOIETY OF FUNGAL PROTEINS TO A MAMMALIAN TYPE

M. Maras[1], A. De Bruyn[2], J. Schraml[2], P. Herdewijn[2], K. Piens[3], M. Claeyssens[3], J. Uusitalo[4], M. Penttilä[4], W. Fiers[1] and R. Contreras[1]

[1]Department of Molecular Biology and the Flemish Interuniversitary Institute of Biotechnology, K.L. Ledeganckstraat 35, B-9000 Gent, Belgium; [2]Laboratory of Medicinal Chemistry, Rega Institute, Catholic University of Leuven, Minderbroederstraat 10, B-3000 Leuven, Belgium; [3]Laboratory of Biochemistry, K.L. Ledeganckstraat 35, B-9000 Gent, Belgium; [4]VTT Laboratory, Rajamäki, Espoo, Finland

1 INTRODUCTION

Several research groups have shown that it is possible to produce human glycoproteins with therapeutical importance in filamentous fungi[1]. However, it is still not clear whether these recombinant proteins can really be used as effective therapeutic agents lacking serious negative side effects. One problem expected to occur is rapid clearing from the blood. Terminal mannoses on fungi-like high-mannose type oligosaccharides can be recognised by lectins that are present on liver cells, on macrophages and on different cells of the reticulo-endothelial system[2]. Another potential problem is antigenicity, provoked by unusual glycosidic substituents such as galactofuranose groups. Finally, the role or significance to the human body of fungal mannose linked phosphate groups is not clear. At the moment, procedures are designed to modify fungal glycosylation to mammalian types, *e.g.* complex glycans with terminal sialic acid residues. Before trying to manipulate the fungal glycosylation synthesis pathway, a good knowledge of glycan structures produced by the fungus under investigation is necessary. As an ideal starting situation for complex glycan synthesis, the choice of a suitable fungus which already has a large fraction of its glycans of the mammalian-like high-mannose type, is important. Some *Aspergillus* strains seem to synthesise large amounts of mentioned restrained sized high-mannose type glycans. For instance, on alfa-amylase from *Aspergillus niger*, the predominant oligosaccharide was characterised to be $Man_5GlcNAc_2$. Previous characterisation of $Man_5GlcNAc_2$ and $Man_9GlcNAc_2$ as the predominant N-glycans produced by *T. reesei*[3], led to the choice of the latter organism for analysing the convertibility of fungal glycans to the human type.

2 STRUCTRURES OF N-GLYCANS OF THE FUNGAL TYPES

2.1 Characterisation of N-glycans from different filamentous fungi

Only a limited number of glycoproteins from filamentous fungi have completely known oligosaccharide structures. A major conclusion from these studies is that glycosylation synthesis pathways in lower and higher eukaryotes must be very similar in their first steps. This leads to high-mannose type N-glycans commonly recognised in fungi (see above). However, some of the fungal oligosaccharides have never been found in human glycoproteins. In figure 1, a few examples of the latter, typical fungi-like N-glycans are shown. With different *Aspergillus* glycoproteins, the α-1,3 or α-1,6 mannose residues are found in positions where in cells from higher eukaryotes only α-1,2 mannose residues are present[4,5,6]. *Paracoccidoides* synthesises N-glycans with terminal galactose residues in their furanose form[7].

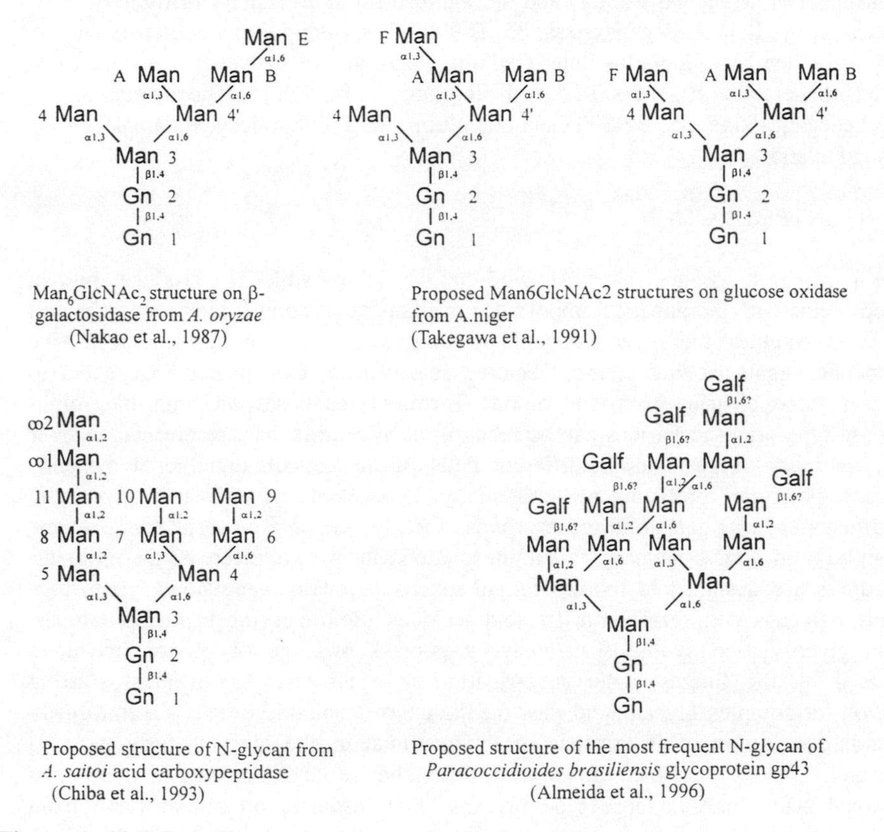

$Man_6GlcNAc_2$ structure on β-galactosidase from *A. oryzae* (Nakao et al., 1987)

Proposed Man6GlcNAc2 structures on glucose oxidase from A.niger (Takegawa et al., 1991)

Proposed structure of N-glycan from *A. saitoi* acid carboxypeptidase (Chiba et al., 1993)

Proposed structure of the most frequent N-glycan of *Paracoccidioides brasiliensis* glycoprotein gp43 (Almeida et al., 1996)

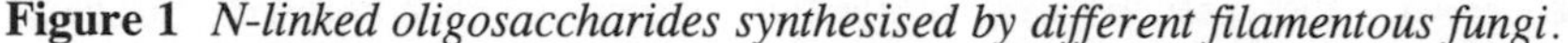

Figure 1 *N-linked oligosaccharides synthesised by different filamentous fungi.*

2.2 *Trichoderma reesei* N-glycans

With *Trichoderma reesei* RUTC30, one of the predominant oligosaccharide residues on cellobiohydrolase I was $GlcMan_7GlcNAc_2$, assumed to be the product of incomplete glucosidase processing of the precursor $Glc_3Man_9GlcNAc_2$. A second abundant compound was $ManPGlcMan_7GlcNAc_2$, with the phosphate in diester linkage between two mannoses[8]. With other strains of *Trichoderma reesei* different phosphorylated compounds are again likely to be present. Furthermore, a fraction of the N-glycans can not be trimmed by α-1,2 mannosidase, suggesting the presence of new linkage types and/or new sugar components. This implicates that with none of the

Trichoderma strains we have investigated, complete conversion to mammalian-type oligosaccharides can be achieved. The characterisation of the oligosaccharides of strain VTT D-80133 by NMR techniques is under progress.

2.3 In vitro conversion of fungal oligosaccharides to the mammalian type

One way to convert fungal N-glycans to a mammalian type is to mimick the mammalian N-glycosylation synthesis pathway. Fungal glycoproteins are treated with mammalian glycosyltransferases after preincubation with α-1,2 mannosidase. The mannosidase pretreatment is intended to create as much as possible acceptor substrate for the first glycosyltransferase: GlcNAc transferase I. With different *Trichoderma reesei* strains, conversion of fungal oligosaccharides to a mammalian hybrid type was clearly demonstrated[9]. However, conversion was relatively poor even after mannosidase pretreatment. In all cases investigated in our lab, the best conversions were obtained with strain VTT D-80133. Approximately 2.5 % of its N-glycans were converted to the hybrid type.

3 IN VIVO CONVERSION OF FUNGAL OLIGOSACCHARIDES TO THE MAMMALIAN TYPE

Expression of human GlcNAc transferase I in *T. reesei* could yield glycoproteins which carry terminal GlcNAc residues, since $Man_5GlcNAc_2$, the acceptor substrate for mentioned glycosyltransferase, is present. A similar experiment was done by Kalsner et al.[10] in *Aspergillus nidulans*, in which succesful expression of GlcNAc transferase I was achieved. However, no terminal GlcNAc was detected on secreted glycoproteins. Unsufficient amounts of UDP-GlcNAc can be a reason for lack of GlcNAc transfer, but also proper targeting of the glycosyltranferase might be a problem. Targeting of GlcNAc transferase I is now studied in more detail, since proper placement in one of the compartments of the glycosylation apparatus is important and can have profound effects on processing events in the latter compartments. One of the main questions here is whether succesful expression of properly targeted GlcNAc transferase I can prevent further fungi processing.

References

1. F.W. Hemming, *Biochem. Soc. Trans.*, 1995, **23**, 180-185.
2. K. Drickamer, *J. Biol. Chem.*, 1988, **263**, 9557-9560.
3. I. Salovuori, M. Makarow, H. Rauvala, J. Knowles and L. Kaariainen, *Bio/technology*, 1995, **5**, 152-156.
4. K. Takegawa, A. Kondo, H. Iwamoto, K. Fujiwara, Y. Hosokawa, I. Kato, K. Hiromi and S. Iwahara, *Biochem. Int.*, 1991, **25**, 181-190.
5. Y. Chiba, Y. Yamagata, S. Iijima, T. Nakajima and E. Ichishima, *Curr. Microbiol.*, 1993, **27**, 281-288.

6. Y. Nakao, Y. Kozutsumi, I. Funakoshi, T. Kawasaki, I. Yamashina, J.H.G.M. Mutsaers, H. Van Halbeek and J.F.G. Vliegenthart, *J. Biochem.*, 1987, **102**, 171-179.
7. I.C. Almeida, D.C.A. Neville, A. Mehlert, A. Treumann, M.A.J. Ferguson, J.O. Previato and L.R. Travassos, *Glycobiology*, 1996, **6**, 507-515.
8. M. Maras, A. De Bruyn, J. Schraml, ,P. Herdewijn, M. Claeyssens, W. Fiers and R. Contreras, *Eur. J. Biochem.*, 1997, **245**, 617-625.
9. M. Maras, X. Saelens, W. Laroy, K. Piens, M. Claeyssens, W. Fiers and R. Contreras, *Eur. J. Biochem*, 1997, in press.
10. I. Kalsner, W. Hintz, L.S. Reid and H. Schachter, *Glycoconjugate J.*, 1995, **12**, 360-370.

HYPEREXPRESSION AND GLYCOSYLATION OF *TRICHODERMA REESEI* EGIII

Benjamin Bower, Katherine Kodama, Barbara Swanson, Tim Fowler, Hendrik Meerman, Kathy Collier, Colin Mitchinson and Michael Ward

Genencor International, 925 Page Mill Rd., Palo Alto, CA 94304 USA

1 INTRODUCTION

One of the minor endoglucanases of the *Trichoderma reesei* cellulase complex is endoglucanase III (EGIII). *T. reesei* EGIII has a MW of 23.5 kDa, a pI of 7.4 and a pH optimum of 5.8. It does not contain a separate cellulose binding domain. The gene was previously cloned by degenerate primer PCR based on peptide fragments obtained through protein digestion.[1] The deduced EGIII amino acid sequence contains two putative N-linked glycosylation sites (**N**TT, **N**YS). Other groups have reported low molecular weight (20-25 kDa), high pI (~7.5) *T. reesei* endoglucanases although the relationship of these proteins to EGIII is not clear.[2-7] A number of these groups described the endoglucanase as being unglycosylated.

Industrial biotechnology applications often require the production of substantial quantities of enzymes. Typically, when an enzyme of interest has been identified, an expression system is used to increase levels of protein production. *T. reesei* and *Aspergillus niger* have been used for industrial enzyme production for many years and are well characterized production systems. Under the proper conditions both fungi are able to produce protein levels in the tens of grams per liter. Transformation systems exist for both fungi which has allowed genetic manipulation of individual genes. Both have been used as hosts for production of heterologous protein. We were interested in obtaining increased expression levels of EGIII.

T. reesei EGIII is used in textile processing. One problem encountered using cellulases in textile applications is fiber strength loss. We conceived of creating a larger EGIII molecule to determine if larger size could effect a reduction of strength loss. To increase the size of the EGIII molecule an EGIII dimer molecule was created.

In this paper we describe the results of engineering *T. reesei* to increase expression of EGIII. Also described is the expression of *T. reesei* EGIII in *A. niger*. And, lastly, expression of an EGIII dimer enzyme is described.

2 RESULTS

2.1 Expression of EGIII in *T. reesei*

An overexpressing cellulase mutant of *T. reesei,* strain RL-P37,[8] was used as the starting point for a series of gene deletions. The four major cellulase genes, *cbh1, cbh2, egl1* and *egl2*, were deleted sequentially by homologous recombination. Deletion vectors constructed for each gene typically consisted of the following: upstream sequence (of gene to be deleted), downstream sequence (of gene to be deleted) and the *T. reesei pyr4* gene as selectable marker; the *cbhI* deletion vector is shown (Figure 1).

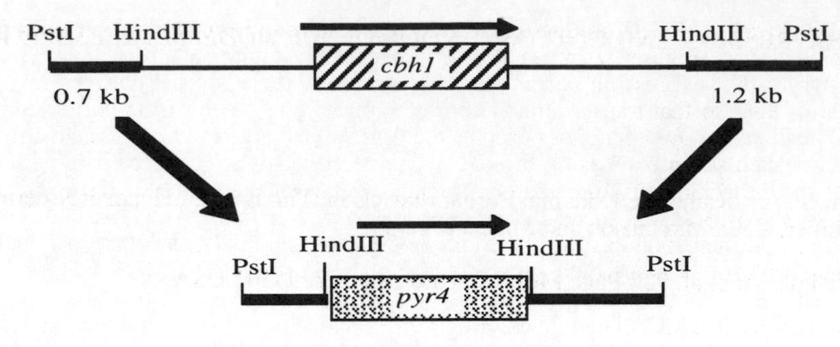

Figure 1 *Construction of cbhI deletion vector*

Each round of gene deletion required the selection of *pyr4⁻* inactivated strains which were conveniently isolated by selecting for growth on minimal media supplemented with fluoroorotic acid (FOA) and uridine. This was followed by transformation of the *pyr4⁻ strain* with a deletion vector. Transformants were selected for growth on minimal media lacking uridine followed by southern probing to confirm the gene deletion. Strains having various gene deletions were grown in 15 liter fermentors under cellulase inducing conditions, and the fermentation broths were examined by SDS-PAGE to determine the extracellular protein profiles. The deleted strains produced unique protein profiles distinct from the starting strain RL-P37 (Figure 2).

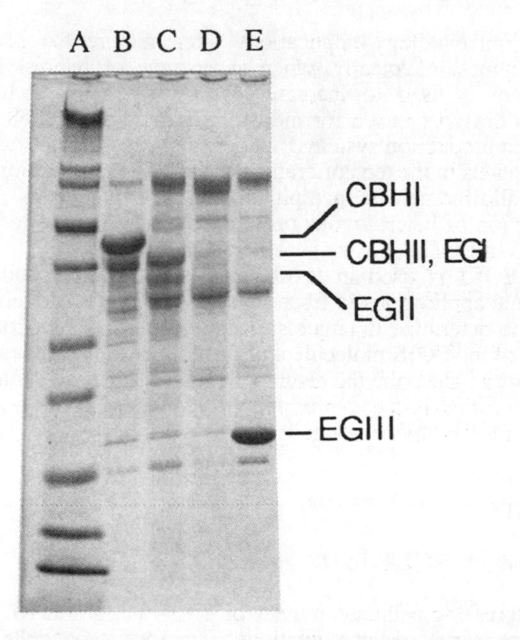

Figure 2 *SDS-PAGE (comassie stained) of fermentation broths of engineered* Trichoderma reesei *strains. Lane A, MW markers; B, parent strain; C, double deleted strain (*Δcbh1, Δcbh2*); D, quad deleted strain (*Δcbh1, Δcbh2, Δegl1, Δegl2*); E,* egl3 *overexpression in quad deleted strain, late fermentation sample.*

High levels of EGIII production were achieved through transformation with an *egl3* expression vector (Figure 3) into a RL-P37 quad deleted strain (Δ*cbh1*, Δ*cbh2*, Δ*egl1*, Δ*egl2, pyr4*⁻). The expression vector was constructed by replacing the *cbh1* coding region with that of *egl3* so that transcriptional control signals were those of the highly expressed *cbh1*. The *T. reesei pyr4* gene was used as the fungal selectable marker. Transformants of the quad deleted strain expressing the *egl3* expression vector were selected by growth on minimal media lacking uridine. Transformants were evaluated for EGIII production by growth in shake flasks in lactose containing media which induces cellulase expression. Shake flask fermentation broths were measured for endoglucanase activity using the substrate Remazol brilliant blue - carboxymethylcellulose (RBB-CMC). Quad deleted control strains without expression vector produce minimal endoglucanase activity as measured by RBB-CMC. The highest producing transformants were then grown in 15 liter fermentors for further evaluation.

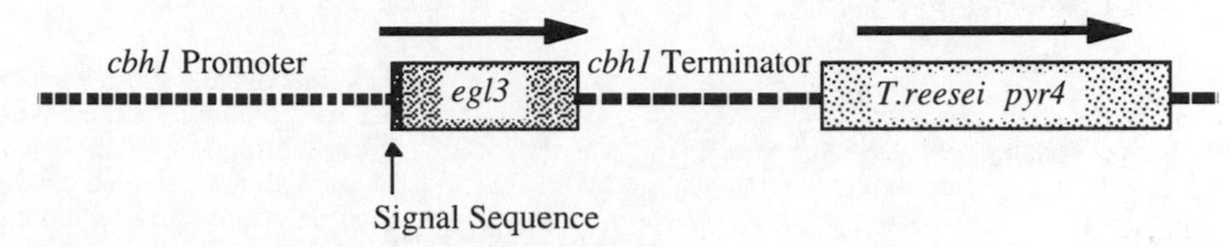

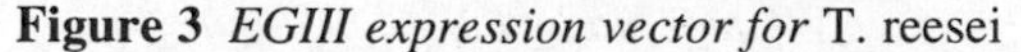

Figure 3 *EGIII expression vector for* T. reesei

Fermentation broths were examined using SDS-PAGE. At early timepoints, fermentation samples of *egl3* overexpression strains showed large increases in the quantity of the EGIII band which runs at approximately 23 kDa. Interestingly, these *egl3* overexpression strains also showed a major new band running at 28 kDa (Figure 4a, lanes 1,2). Western blotting analysis with purified EGIII antibody showed immunoreactivity of the new band at 28 kDa band in addition to the EGIII band at 23 kDa. This would support the identity of both proteins as being EGIII (Figure 4b). When fermentation samples displaying both the 23 kDa and 28 kDa protein bands were treated with Endoglycosidase H (EndoH), the 28 kDa band disappeared and there was a corresponding increase in the amount of the 23 kDa band. No additional or intermediate protein bands, besides the 28 and 23 kDa bands, were ever observed by SDS-PAGE analysis in any fermentation samples.

Concanavalin A Sepharose (ConA), which binds mannosylated glycoproteins, was used to separate the 23 and 28 kDa proteins. Fermentation broths were run through a ConA Sepharose column. The 28 kDa protein was retained by ConA Sepharose while the 23 kDa EGIII ran through. This result is consistent with the presence of mannose on the 28 kDa protein but not on the 23 kDa protein. Enzymatic stability comparisons were made between the two EGIII glycoforms separated on ConA. Both glycoforms electrophoresed to the same isoelectric point (7.4, pH 3-10 range gel). There was no apparent difference in thermostability (Figure 5), nor was there an apparent difference in the pH stability of the enzymes, nor a difference in specific activity (data not shown).

A timecourse of fermentation samples analyzed by SDS-PAGE shows deglycosylation of the 28 kDa band during fermentation (Figure 6). That is, the 28 kDa band of EGIII diminishes in intensity relative to the 23 kDa band during the fermentation run. Generally, by the end of fermentations, the 28 kDa band represented a small fraction of the total EGIII. As a result of storage of fermentation broths in either the presence or absence of cells at 5°C, glycosylated EGIII disappeared entirely leaving only a stable unglycosylated form of EGIII. This can be observed by comparing fermentation broth in Figure 2, lane E (run on the gel some weeks after fermentation), with the fermentation broths of Figure 4 (run recently after fermentation). This observation would indicate the presence of an extracellular deglycosylating activity in *T. reesei* fermentation broth.

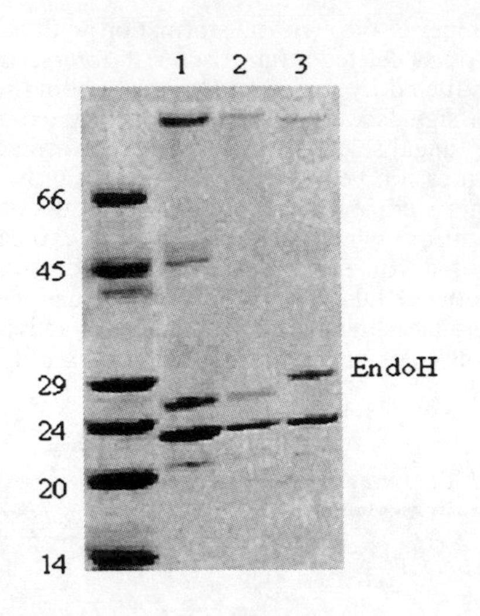

Figure 4a
SDS-PAGE (Coomassie blue stained) of T. reesei *fermentation sample. Lane 1, sample at 1:15 dilution; lane 2, sample at 1:50 dilution; lane 3, sample at 1:50 dilution and EndoH incubation*

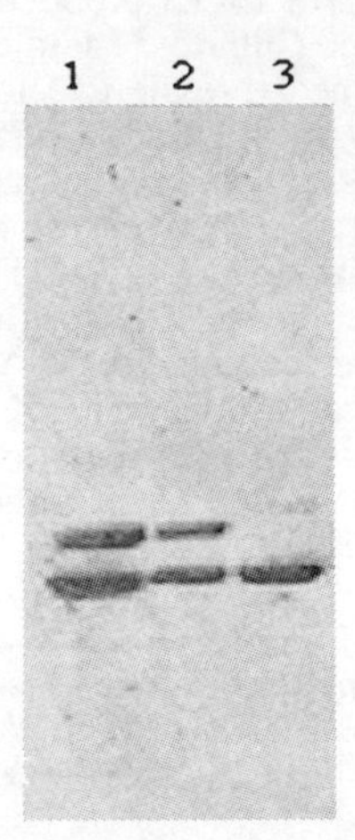

Figure 4b
Western blot of the gel in Figure 4a

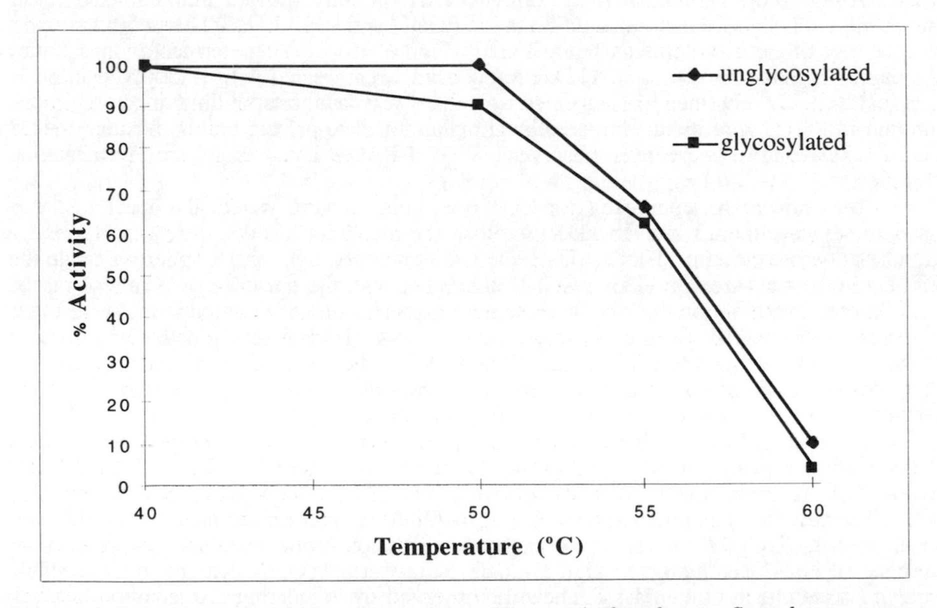

Figure 5 *Temperature stability comparison of* T. reesei *glycoforms. Samples were pre-incubated at indicated temperature for 30 minutes without substrate and then assayed at 40°C for 20 minutes, pH 5.5 with a soluble substrate.*

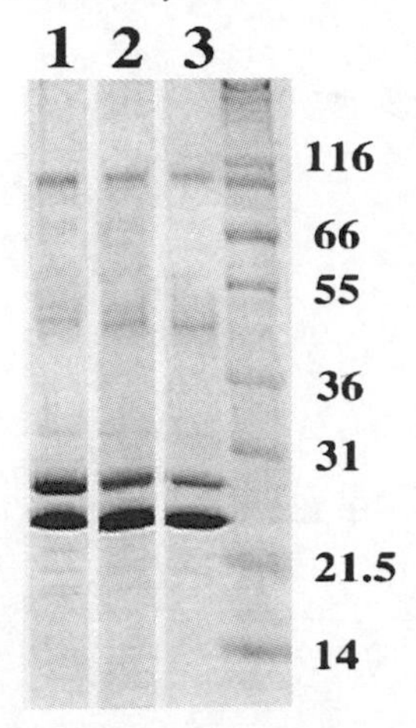

Figure 6 *SDS PAGE (Coomassie blue stain) of* T. reesei *EGIII transformant fermentation broth samples . Lane 1, 90 hour sample; lane 2, 138 hour sample; lane 3, 210 hour sample*

2.2 Expression of *T. reesei* EGIII in *A. niger*

A vector was constructed to express *T. reesei* EGIII in *A. niger*. The vector used the previously constructed plasmid *pGPTpyrG1* (Figure 7).[9] The *T. reesei* genomic clone of *egl3* was placed after the strong *A. niger* glucoamylase (*glaA*) promoter. The vector was transformed into *dgr246p2*, an *A. niger* strain previously used for foreign protein secretion.[10] Transformants were selected on media lacking uridine and then screened for high RBB-CMC activity in a shake flask culture. The best producers in shake flasks were grown in 15 liter fermentors. Fermentation broths were analyzed by SDS-PAGE to compare *A. niger* produced EGIII versus *T. reesei* produced EGIII (Figure 8). *A. niger* produced EGIII appears heterogeneously glycosylated based on the appearance of an indistinct smear running between 28 and 34 kDa with the absence of a band at 23 kDa. Treatment of *A. niger* broths with EndoH resolves some of the smear into a single 23 kDa band which runs at the same weight as deglycosylated *T. reesei*-produced EGIII. Control *A. niger* strains do not show the smear, nor do they show a band at 23 kDa upon Endo H treatment. No deglycosylation of EGIII was observed by SDS-PAGE during the fermentations of *A. niger* EGIII. No differences were seen when the specific activity was compared of EGIII produced by *A. niger* with or without EndoH treatment. This would indicate that the heterologous glycosylation does not effect its specific activity.

2.3 Expression of an EGIII Dimer in *T. reesei*

To produce an EGIII dimer enzyme an expression vector was designed. Two *egl3* coding regions were placed in series, joined 3' to 5' terminus by DNA encoding a *cbh1* linker. The *cbh1* promoter was used to express the dimer and *pyr4* was used as a selectable marker (Figure 9). The dimer expression vector was transformed into a quad deleted strain (Δ*cbh1*, Δ*cbh2*, Δ*egl1*, Δ*egl2, pyr4*$^-$). Transformants were selected on minimal media without uridine. Screening for high producing transformants as measured by RBB-CMC activity was done in shake flasks. The shake flask supernatans from one transformant was analyzed by SDS-PAGE (Figure 10a). A dimer enzyme can be seen as a band at about 50 kDa, consistent with the predicted molecular weight of 49 kDa. This dimer enzyme is immunoreactive with purified EGIII antibody on a Western blot (Figure 10b). The Western blot also shows a larger band running above the EGIII dimer band. Upon EndoH treatment, the upper band disappears, suggesting glycosylation of the dimer.

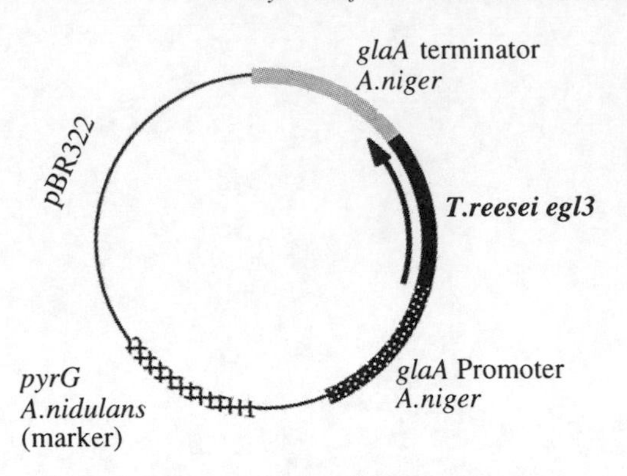

Figure 7 A. niger *expression vector for* T. reesei *EGIII*

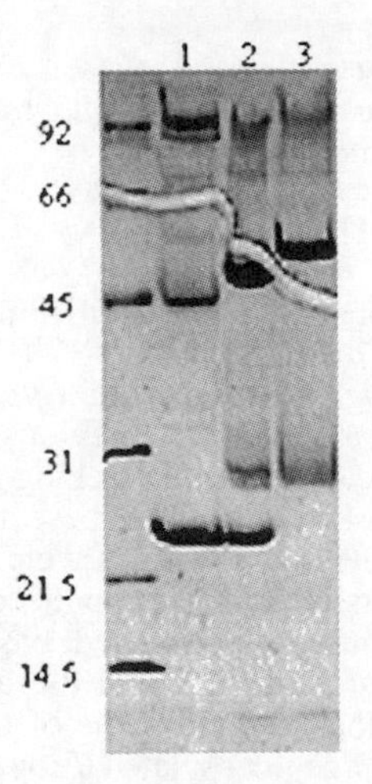

Figure 8 *SDS PAGE (Coomassie blue stain) of* A. niger *vs.* T. reesei *produced EGIII. Lane 1,* T. reesei *fermentation broth sample; lane 2,* A. niger *fermentation broth sample treated with Endo H; lane 3,* A. niger *fermentation broth sample (untreated)*

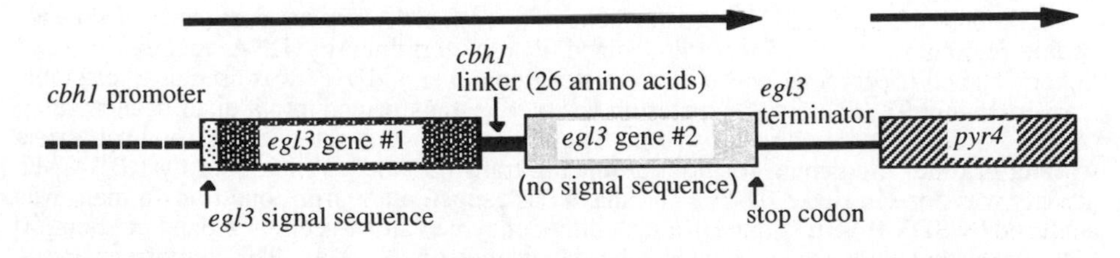

Figure 9 *EGIII dimer expression vector*

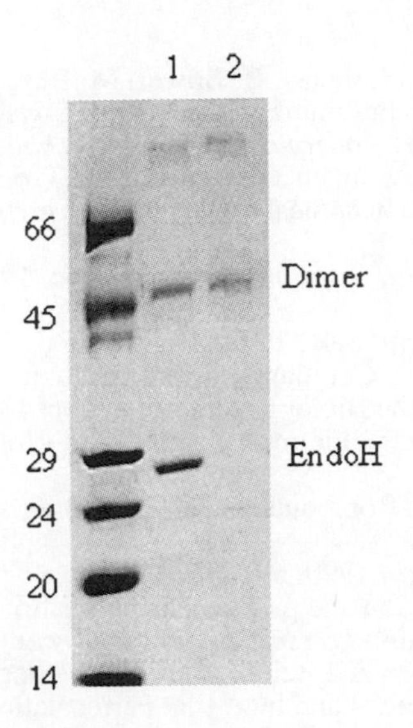

Figure 10a
SDS-PAGE (Coomassie blue stained) of EGIII dimer shake flask supernate. Lane 1, EndoH treated; lane 2, untreated

Figure 10b
Western blot of gel shown in Figure 10a

3 CONCLUSIONS

EGIII which is normally a minor component of the *T. reesei* cellulase complex can be expressed as the major secreted product of *T. reesei.*

A. niger is able to secrete the heterologous *T. reesei* EGIII efficiently. Using the strong *glaA* promoter to express *T. reesei egl3* results in secretion of EGIII as a major protein product. Production of EGIII by the best transformants was almost half that attained by the best *T. reesei* transformants (data not shown).

We found that EGIII is produced as a glycosylated enzyme in both *T. reesei* and *A. niger* expression systems. This finding is consistent with a coding sequence containing two putative N-linked glycosylation sites. We also found evidence for a deglycosylating activity present in *T. reesei* culture broth. Recent independent evidence implies a putative Endo H like activity in *T. reesei* as deduced from carbohydrate analysis of CBHI.[11] Similarly, the existence of two glycoforms of the extracellular enzyme fucosidase from *Fusarium oxysporum* has been reported.[12] No differences were observed between the two *T. reesei* glycoforms for specific activity, pH or temperature stability. *A. niger* produced EGIII appears to be heterogeneously glycosylated, but the specific activity is unaffected. Thus there is no evidence that glycosylation effects the biochemical properties of EGIII.

An EGIII Dimer enzyme was created by joining two *egl3* genes in series linked with a *cbh1* linker. The Dimer retains its specific activity (activity/weight) on the soluble substrate RBB-CMC. However, the best Dimer transformants yielded in fermentors only a low percentage of activity as compared to the best hyperexpressing transformants producing the native EGIII.

4 REFERENCES

1. M. Ward, S. Wu, J. Dauberman, G. Weiss, E. Larenas, B. Bower, M. Rey, K. Clarkson and R. Bott, Cloning, 'Sequence and Preliminary Structural Analysis of a Small, High pI Endoglucanase (EGIII) from *Trichoderma reesei*', *in* :*Trichoderma reesei* Cellulases and other Hydrolases, eds. Suominen and Reinikainen, Proceedings of the Tricel 93 Symposium, Foundation for Biotechnical and Industrial Fermentation Research, Helsinki, Finland, 1993, **8**, 153.
2. U. Hokansson, G. Petterson and L. Andersson, *Biochim. Biophys. Acta*, 1978, **524**, 385.
3. C. Gong, M. Ladisch and G. Tsao, *Adv. Chem. Ser.*, 1979, **181**, 261.
4. S.P. Shoemaker, J.C. Raymond and R. Bruner, 'Cellulases: diversity amongst improved *Trichoderma* strains', *in*: Trends in the Biology of Fermentations for Fuels and Chemicals, eds. Hollaender, Rabson, Rogers, Pietro, Valentine and Wolfe, Plenum, New York, USA, 1981, 89.
5. G. Beldman, M.F. Searle-van Leeuwen, F.M. Rombouts and F.G.J. Voragen, *Eur. J. Biochem*, 1985, **146**, 301.
6. A. Ulker and B. Sprey, *FEMS Microbiol. Lett.*, 1990, **69**, 215.
7. M. Hayn, R. Klinger and H. Esterbauer, 'Isolation and partial characterization of a low molecular weight endoglucanase from *Trichoderma reesei*', *in*:: *Trichoderma reesei* Cellulases and other Hydrolases, eds. Suominen and Reinikainen, Proceedings of the Tricel 93 Symposium, Foundation for Biotechnical and Industrial Fermentation Research, Helsinki, Finland, 1993, **8**, 147.
8. G. Sheir-Neiss and B.S. Montenecourt, *Appl. Microbiol. Biotechnol.*, 1984, **20**, 46.
9. R.M. Berka and C.C. Barnett, *Biotechnol. Adv.*, 1989, **7**, 127.
10. M. Ward, L.J. Wilson and K.H. Kodama, *Appl. Microbiol. Biotechnol.*, 1993, **39**, 738.
11. H. Nevalainen, 'Glycosylation of cellobiohydrolase I from *Trichoderma reesei*', *in*: Carbohydrases from *Trichoderma reesei* and other Microorganisms, Abstract book of the Tricel 97 symposium, University of Gent, Belgium, 1997, August 28-30, abstract L27. See also H. Nevalainen's proceeding in this volume.
12. Y. Yasunobu, K. Yamamoto and T. Tochikura, *Appl. and Envir. Microbiology*, 1990, **56**(4), 928.

GLYCOSYLATION OF CELLOBIOHYDROLASE I FROM *TRICHODERMA REESEI*

H. Nevalainen, M. Harrison, D. Jardine, N. Zachara, M. Paloheimo*, P. Suominen*, A. Gooley and N. Packer

Macquarie University Centre for Analytical Biotechnology (MUCAB), Macquarie Universit
Sydney, NSW 2109 Australia
*Roal Oy, Tykkimäentie 15, FIN-05200 Finland

1 INTRODUCTION

Trichoderma reesei strains presently used in the biotechnology industry are high cellulase-producing mutants, developed by random mutagenesis and screening[1,2], and their derivatives in which the enzyme profiles have been modified by genetic engineering[3,4].

The major cellulase secreted by *Trichoderma* is cellobiohydrolase I (CBHI) which comprises up to 60% of the total secreted protein[5]. It has a distinct N-terminal catalytic and C-terminal cellulose-binding domain (CBD) separated by a glycopeptide linker. Previous studies on CBHI have shown that the enzyme is modified by both *N*-and *O*-linked sugars[6,7,8]. However, detailed reports on the structure of these glycans have been rare and the technology to sequence through highly *O*-glycosylated regions of peptides has not been available until recently[9]. The widespread occurrence of highly *O*-glycosylated linkers[10,11] in hydrolytic enzymes implies that they fulfill an essential function[12,13].

2 CHARACTERISATION OF GLYCOSYLATION

2.1 General strategy for characterisation of protein glycosylation

An overall strategy for the characterisation of *N*- and *O*-linked glycosylation is presented in Figure 1. Proteins are first subjected to a suitable proteolytic cleavage, the resulting peptides are separated by liquid chromatography and analysed using mass spectrometry. Collision induced mass spectrometry (CID-MS) results in characteristic sugar oxonium ions, *e.g.* 162^+ m/z indicates the presence of a hexose sugar, which allow for the identification of glycopeptides. The nature of the monosaccharides modifying the protein can be determined using acid hydrolysis in combination with high performance anion exchange chromatography (HPAEC). Finally, the sites of both *N*- and *O*-linked glycosylation can be determined using Edman degradation[9].

Two proteases have been used for the detailed characterisation of the glycosylation on CBHI. Trypsin treatment produces suitable fragments from the catalytic core domain[14,15] and digestion with papain[16] separates the glycopeptide linker from the core allowing the isolation and subsequent analysis of the spacer region[15]. Comparison of the observed and calculated molecular masses of the peptides (from the protein database or predicted gene sequence) targets peptides that may be modified.

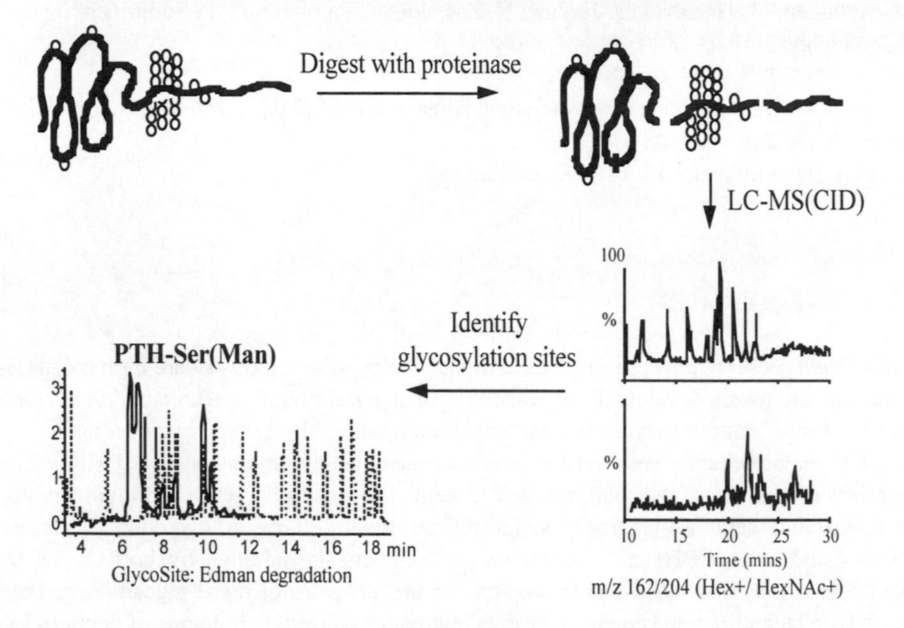

Figure 1 *General strategy for the characterisation of protein glycosylation*

2.2 *N*-linked sugars on CBHI

The *N*-linked sugars of CBHI have traditionally been studied after enzymatic release from the protein. Two high cellulase-producing mutant strains *T. reesei* VTT-D-80133[17] and RutC-30[18] have been characterised using this approach[19,20], both showing high mannose type *N*-glycans. In neither case, quantification of the sugar to protein ratio was made. However, the analysis of high mannose *N*-glycans from *T. reesei* RutC-30 by ^{1}H and ^{13}C NMR[20] necessitates considerable amounts of these structures. The report also revealed a new sugar modification, an outer branch α-1,6-phospodiester-mannose on the *N*-linked sugars of CBHI[20].

Our analysis of CBHI from a genetically-modified *T. reesei* strain ALKO2877[21] indicated very low amounts of these modified high mannose *N*-linked oligosaccharides[15]. We have shown that three of the four potential *N*-glycosylation motifs were occupied by a single *N*-acetyl glucosamine residue[15]. A similar finding[14] has been reported for CBHI purified from the strain *T. reesei* QM9414[22].

2.3 *O*-linked sugars on CBHI

In this paper we present the results of the characterisation of the *O*-linked glycans on CBHI from *T. reesei* QM9414 and compare them to those of ALKO2877. A sample of purified CBHI enzyme from QM9414 (obtained from J. Ståhlberg, BMC Uppsala) was digested with papain and the fragments separated by C8 reversed-phase chromatography as described previously[15]. This gave peaks representing the 46 kDa core and the 8-10 kDa linker peptide (Figure 2). The linker peptide was then analysed by electrospray ionisation mass spectrometry (Figure 3).

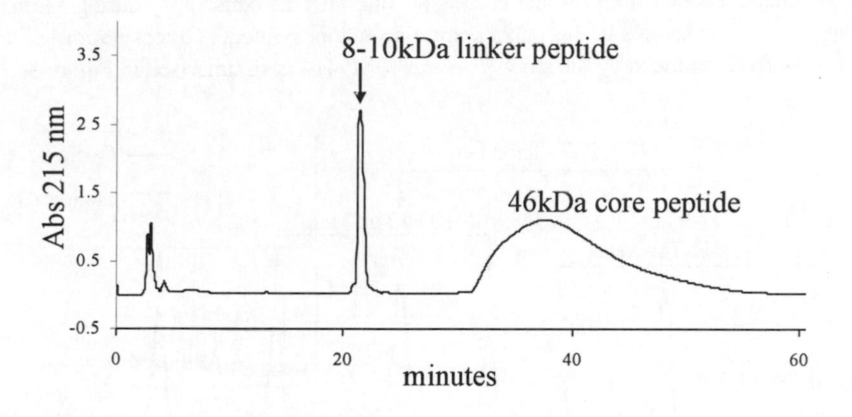

Figure 2 *Reversed phase HPLC separation of domains produced by papain hydrolysis of CBHI from* T. reesei *QM9414*

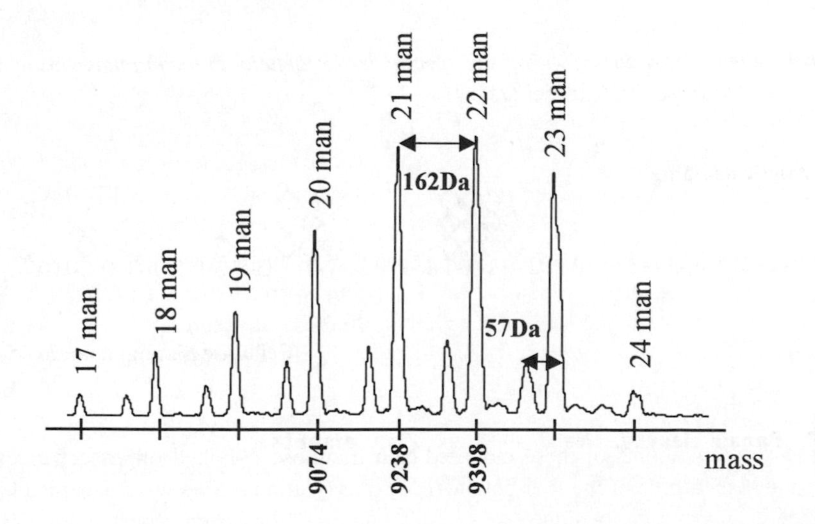

Figure 3 *Electrospray mass spectrometry of the 8-10 kDa linker peptide from* T. reesei *QM9414*

Mass spectrometry revealed heterogeneity in the *O*-glycosylation of the linker region. The number of hexoses (mass difference of 162 kDa) attached to the linker peptide varied from 17 to 24 with the average being 22 (Figure 3). A minor series of mass heterogeneity of 57 Da less than the major series corresponded to a glycopeptide which begins at Gly_{440} due to a cleavage site degeneracy by papain around Gly_{439} (see Figure 5). Subsequent monosaccharide analysis showed that the glycosylation of the 8-10 kDa peptide was predominantly mannose.

Edman degradation of the linker peptide was performed to assign the sites of *O*-glycosylation. All the threonines were glycosylated by one to three mannoses. An example of sequencing Thr_{443} is shown in Figure 4. O-glycosylated serines and threonines elute as characteristic diastereomeric peaks due to racemisation during Edman degradation[9,23]. The serines in the linker were also mannosylated. Glycosylation of the linker from CBHI produced by the strain *T. reesei* QM9414 is summarised in Figure 5.

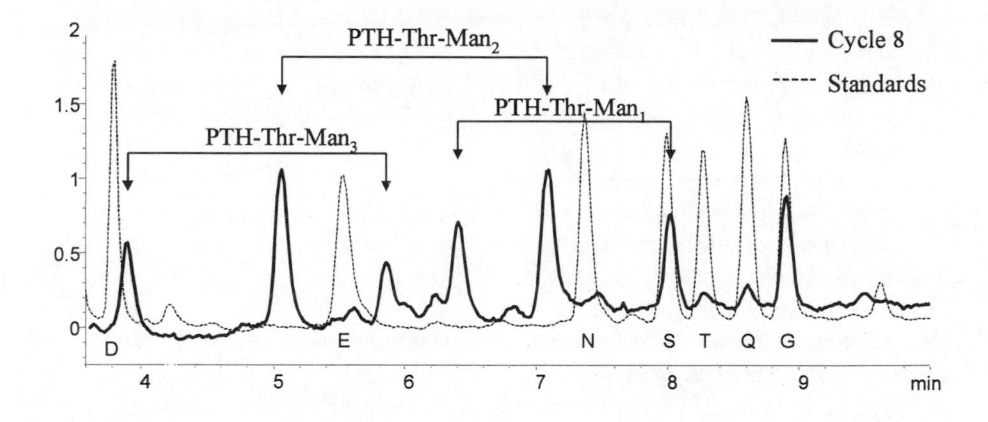

Figure 4 *Edman degradation shows site-specific heterogeneity in the mannosylation of the CBHI linker of* T. reesei *QM9414*

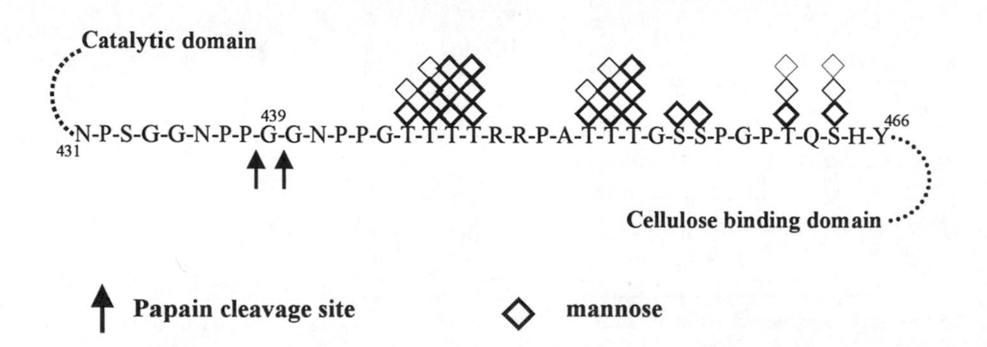

Figure 5 *Summary of glycosylation of the 8-10kDa linker peptide of CBHI from* T. reesei *QM9414*

3 COMPARISON OF GLYCOSYLATION OF CBHI IN DIFFERENT *T. REESEI* STRAINS

3.1 *N*-glycans

It seems evident that different strains of *T. reesei* *N*-glycosylate CBHI differently. However, detailed analysis of glycosylation has so far only been carried out on high cellulase producing mutants which have been developed by the application of several rounds of mutagenesis and screening (Figure 6). Thus it is possible that they carry as yet unknown mutation(s) in their protein glycosylation machinery. The occurrence of single *N*-acetyl glucosamines on the CBHI of *T. reesei* QM9414 and ALKO2877 may originate from a defect in the assembly of complete *N*-oligosaccharides in the dolichol pathway[24].

As the oligosaccharide transferase would still be functional, the truncated glycan would be attached to the protein resulting in sites occupied by single *N*-acetyl glucosamine residues. Another possibility is the presence of an endo-ß-*N*-acetyl-D-glucosaminidase (EndoH) type trimming activity in *Trichoderma* which would leave a single *N*-acetyl glucosamine sugar attached to the asparagine[25]. Interestingly, such enzyme activity has been found for example in *Aspergillus oryzae*, *Fusarium oxysporum* and *Mucor hiemalis*[26] and appears to be involved in deglycosylation of the endoglucanase III (EGIII) secreted from a *Trichoderma* host overproducing EGIII[27]. We are currently characterising this activity in *Trichoderma*.

Evidence for differential glycosylation of heterologous fungal proteins by *Trichoderma* hosts comes from studies on the expression and secretion of *Aspergillus* phytase and phosphatase enzymes[28,29].

When *A. niger* var. *awamori* phytase was produced in *T. reesei* ALKO2221[30], VTT-D-79125[17] and RutC-30 under the strong *cbh1* promoter on a lactose based medium, followed by the analysis by Western blotting, the protein seemed to be least glycosylated by VTT-D-79125[28]. *T. reesei* ALKO2221 and especially RutC-30 added considerably more sugar to the protein indicating that there are inherent glycosylation differences between the strains that are due to the nature of the culture medium. The excess sugar did not seem to affect the amount or activity of the enzyme secreted.

3.2 The linker region

3.2.1 Glycosylation of the linker region. Data for the detailed comparison of glycosylation and other modifications of the linker region of CBHI purified from *Trichoderma* is available only from the two strains, *T. reesei* ALKO2788 and QM9414, which are characterised in this work. The linkers from these strains differ in three aspects: (i) the number of mannose residues attached to the glycopeptide is 17-24 for QM9414 and 14-26 for ALKO2788, (ii) the digestion of QM9414 with papain produces only the 8-10 kDa linker peptide whereas from ALKO2788, two peptides of 4.4 kDa and 8-10 kDa can be formed, and (iii) unlike ALKO2788, the linker glycopeptide from QM9414 is not sulfated. Ion chromatography of an acid hydrolysate of QM9414 linker peptide revealed no peak corresponding to sulphate or phosphate (Figure 7), nor was the sulphate or phosphate ion observed by ESI/MS-MS of the linker region.

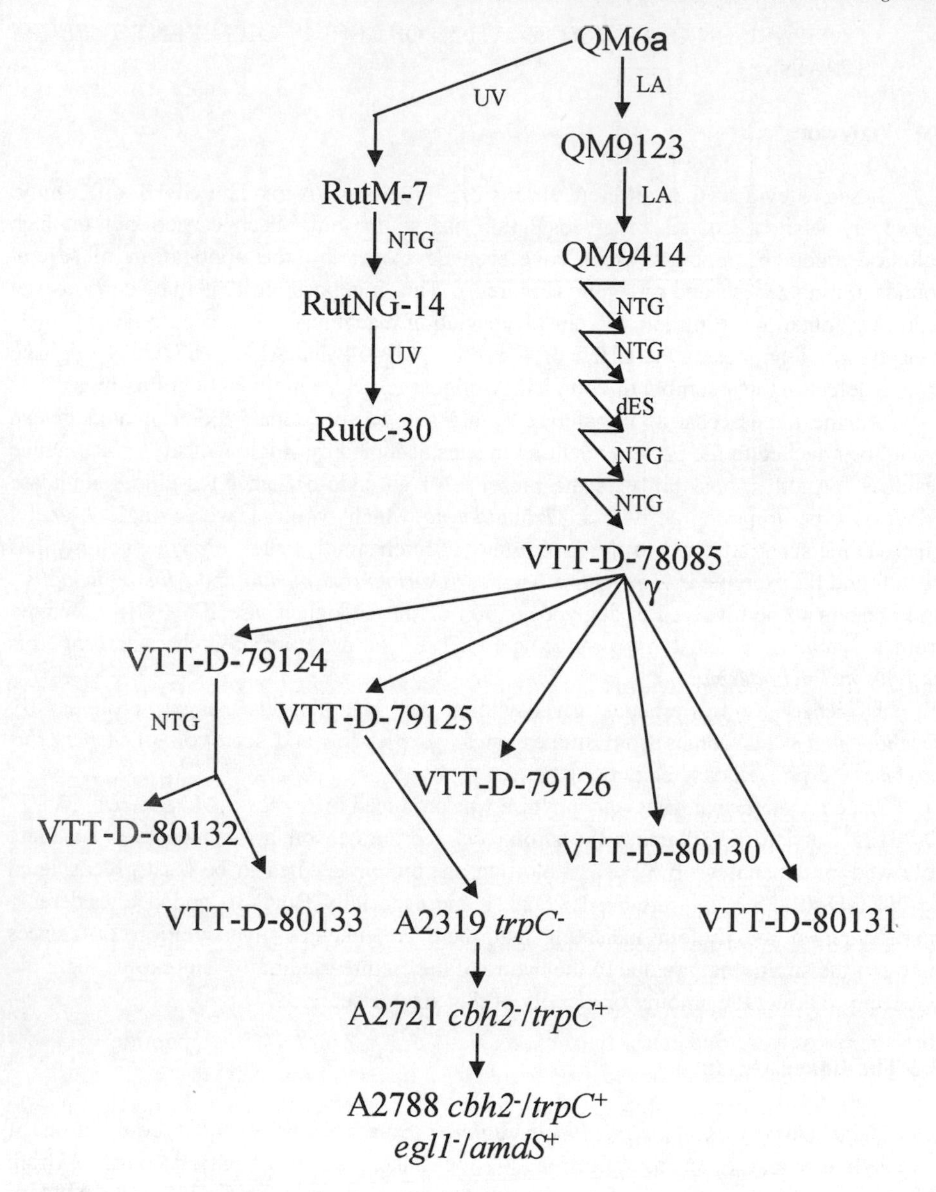

LA: Linear accelerator; dES: diethyl sulfate; NTG: N-methyl-N'-nitro-N-nitrosoguanidine; UV: ultraviolet light; γ: gamma irradiation.

Figure 6 *Partial family tree of high cellulase-producing* T. reesei *mutants*

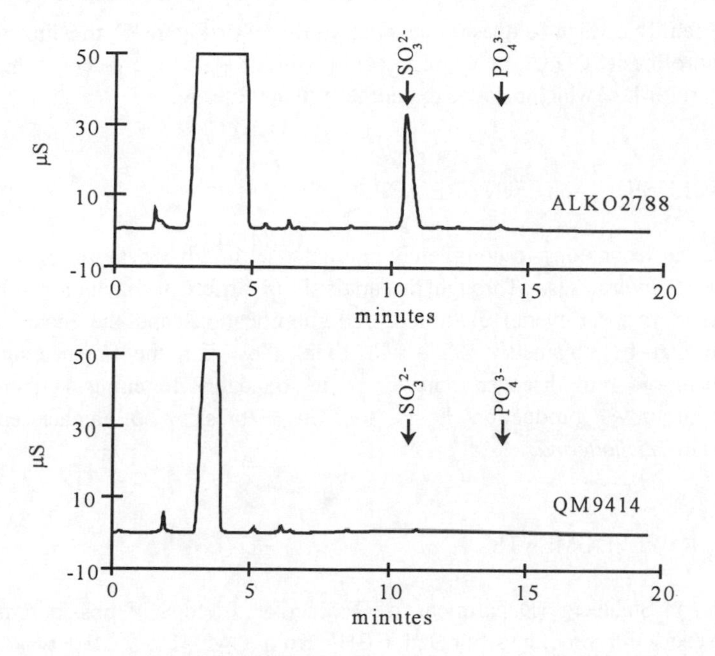

Figure 7 *Ion chromatography of the acid hydrolysate of the linker peptide from* T. reesei *QM9414 shows that it is not sulfated*

The structural role of the glycopeptide linker may be to provide sufficient spatial separation of the two domains. Studies on the glycopeptide linkers of glycosylhydrolases have indicated that separation of the CBD and catalytic domain is required for efficient digestion of crystalline substrates[11,12], and that the linker plays some role in thermostability and secretion[10]. Heavy *O*-glycosylation may contribute to function by providing stiffness to the spacer structure in order to prevent back-folding of the tail[31]. The fact that papain is able to cut the linker region of CBHI of ALKO2788 at Ser_{457}, where mannosylation is partial[15], whereas it does not cut the linker of QM9414, which is fully mannosylated, indicates that glycosylation may protect the peptide against proteolysis, as has been observed for several other glycopeptide linkers[32,33,34,35]. It has also been shown that the *O*-linked glycosylation (but not the *N*-linked) is necessary for the secretion of two endoglucanases, EGI and EGII, from *T. reesei*[36].

3.2.2 DNA sequence of the linker region. The linker region of the *cbh1* gene from nucleotides 1416 to 1595[37] was amplified and sequenced from seven *T. reesei* strains: the wild type QM6a [38], QM9414, VTT-D-79125, VTT-D-80133, RUTC-30, ALKO2721 (*cbh2*$^-$/*trpC*$^+$) and ALKO2877. Sequencing was carried out either from a plasmid containing the isolated *cbh1* gene (QM6a) or from a PCR fragment synthesised from the linker area by using standard procedures and fungal genomic DNA as a template. The resulting 198 bp PCR fragments were then purified and directly sequenced on an automated DNA sequencer (Applied Biosystems).

The DNA sequences obtained were identical to each other and confirmed the result from amino acid sequencing (Figure 5). Interestingly, the glycopeptide linker of the strains analysed above seems to differ from that of the hypercellulolytic *T. reesei* L27[37]. In the

ALKO-VTT family of high cellulase-producing strains (see Figure 6) the linker peptide contains two prolines (CCG CCT) at amino acid position 442 in the place of the arginine (CGT) reported for L27 which belongs to another mutant lineage.

4 CONCLUSION AND FUTURE ASPECTS

We now have the technology to completely characterise the glycosylation of industrially important fungal hydrolases. Through the analysis of different enzymes produced by different host strains, an overall picture of the modifications and the mode of fungal glycosylation can be obtained. This will then allow for the engineering of the glycosylation in order to discover more about its biological functions and enable the construction of novel production hosts for the expression of higher eukaryotic glycoproteins in *Trichoderma*.

5 ACKNOWLEDGMENTS

We thank Jerry Ståhlberg (Department of Molecular Biology, Uppsala University, Uppsala, Sweden) for providing purified CBHI from QM9414. MH was partially supported by a Macquarie University Research grant (MURG). Ms Outi Könönen (Roal Oy) and Ms Margaret Lawson (Macquarie University) are thanked for skillful technical assistance.

References

1. Nevalainen, M.E. Penttilä, A.M. Harkki, T.T. Teeri and J.K.C. Knowles, 'Molecular Industrial Mycology, Systems and Applications for Filamentous Fungi', R. Berka and S.A. Leong (eds.), Marcel Dekker Inc., New York, 1990, Chapter 6, p. 129.
2. Nevalainen, P. Suominen and K. Taimisto, *J. Biotechnol.*, 1994, **37**, 193.
3. Nevalainen and M. Penttilä, 'The Mycota II, Genetics and Biotechnology', Kück (ed.), Springer-Verlag Berlin Heidelberg, 1995, Chapter 18, p. 303.
4. Mäntylä, M. Paloheimo and P. Suominen, 'Trichoderma and Gliocladium', C. Kubicek and G. Harman (eds.), Taylor and Francis Ltd., London, 1997, in press.
5. Grizali and R.D. Brown, Jr., *Adv. Chem. Ser.*, 1979, **81**, 237.
6. Gum and R.D. Brown, *Biochim. Biophys. Acta*, 1979, **446**, 371.
7. Fägerstam, G. Pettersson and J. Engström, *FEBS Lett.*, 1984, **167**, 309.
8. Salovuori, M. Makarov, H. Rauvala, J. Knowles and L. Kääriäinen, *Bio/Technology*, 1987, **5**, 152.
9. Gooley and K.L. Williams, *Nature*, 1997, **385**, 557.
10. Baker-Libby, C.A.G Cornett, P.J. Reilly, and C. Ford, *Prot. Engin.,* 1994, **7**, 1109.
11. Shen, M. Schmuck, I. Pilz, N.R. Gilkes, D.G. Kilburn, R.C. Miller Jr., and R.A.J. Warren, *J. Biol. Chem.,*1991, **266**, 11335.

12. M. Sridosuk, T. Reinikainen, M. Penttilä, and T.T. Teeri, *J. Biol. Chem.,* 1993, **268**, 20756.
13. N. Jentoft, *TIBS*, 1990, **15**, 291.
14. K. Klarskov, K. Piens, J. Ståhlberg, P. B. Høj, J. Van Beeumen and M. Claeyssens, 1997, *Carbohydrate res.*, in press.
15. M. Harrison, A. Nouwens, D. Jardine, N. Zachara, A. Gooley, H. Nevalainen and N. Packer, 1997, submitted for publication.
16. P. Tomme, H. van Tilbeurgh, G. Pettersson, J. Van Damme, J. Vandekerckhove, J. Knowles, T.T. Teeri and M. Claeyssens, *Eur. J. Biochem.*, 1988, **170**, 575.
17. M. Bailey and H. Nevalainen, *Enzyme Microb. Technol.*, 1981, **3**, 153.
18. B.S. Montenecourt and D.E. Eveleigh, *Adv. Chem. Ser.*, 1979, **181**, 289.
19. I. Salovuori, M. Makarow, H. Rauvala, J. Knowles and L. Kääriäinen, *Bio/Technology*, 1987, **5**, 152.
20. M. Maras, A. De Bruyn, J. Schraml, P. Herdewijn, M. Claeyssens, W. Fiers and R. Contreras, *Eur. J. Biochem.*, 1997, **245**, 617.
21. A. Koivula, A. Lappalainen, S. Virtanen, A.L. Mäntylä, P. Suominen and T. Teeri, *Prot. Expr. Purif.*, 1996, **8**, 391.
22. M. Mandels, J. Weber and R. Parizek, *Appl. Microbiol.*, 1971, **2**, 59.
23. A. Pisano, N.H. Packer, J.R. Redmond, K.L. Williams and A.A. Gooley, 'Methods in Protein Structure Analysis', M.Z. Atassi and E. Appella (eds.), Plenum Press, New York, 1995, Chapter 7, p. 69.
24. R. Kornfeld and S. Kornfeld, *Annu. Rev. Biochem.*, 1985, **54**, 631.
25. J. Barreaud, S. Bourgerie, R. Julien, J.F. Guespin-Michel and Y. Karamanos, *J. Bacteriol.*, 1995, **177**, 916.
26. T. Suzuki, K. Kitajima, S. Inoue and Y. Inoue, *Glycoconjugate J.*, 1995, **12**, 183.
27. B. Bower, E. Larenas, B. Swanson and M. Ward, 'Carbohydrases from Trichoderma reesei and Other Microorganisms', TRICEL 97 Abstract Book, University of Ghent, Belgium, 1997, lecture abstract L17.
28. H. Nevalainen, M. Paloheimo, A. Miettinen-Oinonen, T. Torkkeli, M. Cantrell, C. Piddington and J. Rambosek, *PCT WO94/03612.*
29. A. Miettinen-Oinonen, T. Torkkeli, M. Paloheimo and H. Nevalainen, *J. Biotechnol.*, 1997, in press.
30. A. Mäntylä, R. Saarelainen, R. Fagerström, P. Suominen and H. Nevalainen, *2nd European Conference on Fungal Genetics, Lunteren, The Netherlands*, 1994, abstract B52.
31. J. Ståhlberg, PhD Thesis, University of Uppsala, 1991.
32. E. Gatti, L. Popolo, M. Vai, N. Rota and L. Alberghina, *J. Biol. Chem.*, 1994, **269**, 19695.
33. C.G. Davis, A. Elhammer, D.W Russel, W.J. Schneider, S. Kornfeld, M.S. Brown and J.L. Goldstein, *J. Biol. Chem.,* 1986, **261**, 2828.
34. K. Kozarsky, D. Kingsley and M. Krieger, *Proc. Natl. Acad. Sci. USA,* 1988, **85**, 4335.
35. P. Reddy, L. Caras, and M. Krieger, *J. Biol. Chem.,* 1989, **264**, 17329.
36. C.P. Kubicek, T. Panda, G. Schreferi-Kunar, F. Gruber and R. Messner, *Can. J. Bot.*, 1987, **33**, 698.

37. S. Shoemaker, V. Schweickart, M. Ladner, D. Gelfand, S. Kwok, K. Myambo and M. Innis, *Bio/Technology*, 1983, **1**, 691.
38. M. Mandels and E.T. Reese, *J. Bacteriol.*, 1957, **73**, 269.

Subject Index